THE ECOLOGY AND MANAGEMENT OF GRAZING SYSTEMS

The Ecology and Management of Grazing Systems

Edited by

J. Hodgson
Department of Plant Science
Massey University
New Zealand

and

A.W. Illius
Institute of Cell, Animal and Population Biology
University of Edinburgh
UK

CAB INTERNATIONAL

SF
85
.E36
1996

CAB INTERNATIONAL Tel: +44 (0)1491 832111
Wallingford Fax: +44 (0)1491 833508
Oxon OX10 8DE E-mail: cabi@cabi.org
UK Telex: 847964 (COMAGG G)

© CAB INTERNATIONAL 1996. All rights reserved. No part of this publication may be reproduced in any form or by any means, electronically, mechanically, by photocopying, recording or otherwise, without the prior permission of the copyright owners.

A catalogue record for this book is available from the British Library.

ISBN 0 85199 107 6

Typeset in Photina and Helvetica by AMA Graphics Ltd
Printed and bound in the UK by Biddles Ltd, Guildford

Contents

Contributors vii

Preface ix

Part I: Plants and Plant Populations

1 Tissue Flows in Grazed Plant Communities 3
   G. Lemaire and D. Chapman

2 Strategies of Plant Survival in Grazed Systems: A Functional Interpretation 37
   D.D. Briske

3 Plant Competition and Population Dynamics 69
   J.M. Bullock

4 Assessing and Interpreting Grass–Woody Plant Dynamics 101
   S. Archer

Part II: Animals and Animal Populations

5 Foraging Strategies of Grazing Animals 137
   E.A. Laca and M.W. Demment

6	Biochemical Aspects of Grazing Behaviour K.L. Launchbaugh	159
7	Ingestive Behaviour E.D. Ungar	185
8	The Ruminant, the Rumen and the Pasture Resource: Nutrient Interactions in the Grazing Animal H. Dove	219
9	Multispecies Grazing in the Serengeti M.G. Murray and A.W. Illius	247

Part III: Grazing Systems and Their Management

10	Complexity and Stability in Grazing Systems N.M. Tainton, C.D. Morris and M.B. Hardy	275
11	Management of Grazing Systems: Temperate Pastures G.W. Sheath and D.A. Clark	301
12	Management of Rangelands: Paradigms at Their Limits M. Stafford Smith	325
13	Management of Mediterranean Grasslands N.G. Seligman	359
14	Grasslands in the Well-watered Tropical Lowlands M.J. Fisher, I.M. Rao, R.J. Thomas and C.E. Lascano	393

Part IV: Conclusions

15	Progress in Understanding the Ecology and Management of Grazing Systems A.W. Illius and J. Hodgson	429

Index   459

Contributors

S. Archer, *Department of Rangeland Ecology and Management, Texas A&M University, College Station, TX 77843-2126, USA.*
D.D. Briske, *Department of Rangeland Ecology and Management, Texas A&M University, College Station, TX 77843-2126, USA.*
J.M. Bullock, *Furzebrook Research Station, NERC Institute of Terrestrial Ecology, Furzebrook Road, Wareham, Dorset BH20 5AS, UK.*
D. Chapman, *AgResearch Grasslands, Private Bag 11008, Palmerston North, New Zealand.*
D.A. Clark, *Dairying Research Corporation Ltd, Private Bag 3123, Hamilton, New Zealand.*
M.W. Demment, *Department of Agronomy and Range Science, University of California, Davis, CA 95616-8515, USA*
H. Dove, *CSIRO Division of Plant Industry, GPO Box 1600, Canberra, ACT 2601, Australia.*
M.J. Fisher, *Centro Internacional de Agricultura Tropical, Apartado Aéreo 6713, Cali, Colombia.*
M.B. Hardy, *Cedara Agricultural Development Institute, Private Bag X9059, Scottsville, Pietermaritzburg 3200, South Africa.*
J. Hodgson, *Department of Plant Science, Massey University, Private Bag 11222, Palmerston North, New Zealand.*
A.W. Illius, *Institute of Cell, Animal and Population Biology, Ashworth Laboratories, University of Edinburgh, West Mains Road, Edinburgh EH9 3JT, UK.*

E.A. Laca, *Department of Range and Wildlife Management, Texas Tech University, Lubbock, TX 79409-2125, USA.*

C.E. Lascano, *Centro Internacional de Agricultura Tropical, Apartado Aéreo 6713, Cali, Colombia.*

K.L. Launchbaugh, *Department of Range and Wildlife Management, Texas Tech University, Lubbock, TX 79409-2125, USA.*

G. Lemaire, *INRA, Station d'Ecophysiologie des Plantes Fourragères, 86600 Lusignan, France.*

C.D. Morris, *Range and Forage Institute, Private Bag X01, Scottsville, Pietermaritzburg 3209, South Africa.*

M.G. Murray, *Institute of Cell, Animal and Population Biology, Ashworth Laboratories, University of Edinburgh, West Mains Road, Edinburgh EH9 3JT, UK.*

I.M. Rao, *Centro Internacional de Agricultura Tropical, Apartado Aéreo 6713, Cali, Colombia.*

N.G. Seligman, *Agricultural Research Organization, Volcani Center, Bet Dagan, 50250, Israel.*

G.W. Sheath, *AgResearch, Whatawhata Research Centre, Private Bag 3089, Hamilton, New Zealand.*

M. Stafford Smith, *National Rangelands Program, CSIRO Division of Wildlife and Ecology, PO Box 2111, Alice Springs, NT 0871, Australia.*

N.M. Tainton, *Department of Grassland Science, University of Natal, Private Bag X01, Scottsville, Pietermaritzburg 3209, South Africa.*

R.J. Thomas, *Centro Internacional de Agricultura Tropical, Apartado Aéreo 6713, Cali, Colombia.*

E.D. Ungar, *Department of Agronomy and Natural Resources, Institute of Field and Garden Crops, Agricultural Research Organization, PO Box, Bet Dagan, Israel.*

Preface

Pastoral agriculture occupies around 20% of the land surface of the globe, and is directly or indirectly responsible for meeting the economic and material needs of a substantial proportion of its human population. It is also the predominant form of land use in many of the more fragile areas of the world, with a major contribution to make to the stability and sustainability of land resources. Yet it is the sector of land use which has received least research attention, primarily because of the complex nature of the interrelationships between soil, plant and animal resources involved and the relative intractability of the management issues that result.

The objective in this book is to draw together up-to-date information on the dynamic ecology of plant and herbivore communities and the relationships between them, and to place this information in the context of modern theories of ecosystem management. In addressing this objective we have been particularly concerned to emphasize the complementary nature of information from managed pastoral systems and natural ecosystems. International contributions provide authoritative evaluation of current evidence in a coordinated series of review chapters dealing with aspects of plant community structure and function, foraging strategy and nutritional ecology of animal populations, and ecosystem complexity and management. Examples are taken from the main grassland zones of the world, with emphasis on the extent to which climate and soil limitations dictate vegetation balance and management opportunity.

In commissioning the fourteen chapters in this book, we encouraged authors to be stimulating, rather than comprehensive in reviewing the available literature. On the whole, we believe that this objective has been admirably achieved, though at the inevitable cost of omission of some potentially important areas of research.

The book concludes with a chapter in which we have attempted a brief evaluation of the current status of research on grazing systems, with particular reference to the developing understanding of the dynamic ecology of such systems and the relevance of this understanding to systems management. In our view this is a vital area in which to maintain the recent momentum of research which links interests in natural ecology and agricultural ecology. It is also important to continue to develop the links between plant-related and animal-related interests.

We thank the publishers, CAB INTERNATIONAL, for the opportunity to indulge ourselves in the production of a book of this kind, and the authors who have made it possible. We trust that it will prove to be interesting and stimulating to students of, and practitioners in, pastoral systems, wherever they may be.

John Hodgson	**Andrew Illius**
Massey University	University of Edinburgh
Palmerston North	Edinburgh
New Zealand	UK

Plants and Plant Populations

Tissue Flows in Grazed Plant Communities

G. Lemaire[1] and D. Chapman[2]

[1]INRA, Station d'Ecophysiologie des Plantes Fourragères, 86600 Lusignan, France; [2]AgResearch Grasslands, Private Bag 11008, Palmerston North, New Zealand

INTRODUCTION

Individual plants in grazed communities are subject to sequential defoliations, the frequency and intensity of which mainly depend on the system of grazing management. Plant responses to the disturbance of defoliation can be viewed as having the goal of restoring and maintaining homeostatic growth patterns where all resources are used in a balanced way for optimal plant growth. Two types of response to defoliation can be distinguished at the individual plant level: (i) short-term physiological acclimatization to the restriction of carbohydrate supply for plant growth resulting from removal of photosynthetic tissues; and (ii) longer-term morphological adaptation, constituting an important part of the 'avoidance' mechanisms described by Briske (1986) for reducing the probability of defoliation and thereby conferring grazing resistance. Therefore, the effects of defoliation in grazed plant communities on tissue flow rates have to be analysed over the short term to understand the consequences of change in carbon (C) and nitrogen (N) fluxes for growth at the sward level and over the long term to understand the consequences of morphological adaptation of individual plants for sward structure and botanical composition.

Hence, our discussion focuses, firstly, on the C and N fluxes occurring in plant canopies and their regulation by sward state variables. Secondly, we show how the morphogenetic activity of grazed swards, determined both at the individual plant (appearance, elongation and senescence of leaves) and at the canopy (tiller population dynamics) level, provides explanations for changes in sward state variables in response to variation in the defoliation regime. Here, the concept of phenotypic plasticity is discussed in relation to plant genetic variation. Finally, we synthesize these components to analyse the 'global' relationships between grazing managements and their inherent defoliation patterns and between tissue flow and the efficiency of herbage utilization in grazed pasture communities.

CARBON AND NITROGEN FLUXES IN PLANT CANOPIES

Considering that C is the principal constituent of plant tissues, the rate of biomass accumulation in plant communities is determined by the rate at which C accumulates. The C accumulation rate is, in turn, influenced by the N content of plant tissue. Thus, the dynamics of the two plant growth resources are closely linked, and we choose to discuss them both. Given that C is acquired by leaves and N is acquired by roots, tissue fluxes in pastures can only be fully understood by analysing resource capture, allocation and use in both roots and shoots, even though only shoot growth is harvested for livestock production. Our discussion therefore also considers root : shoot interactions, working at the whole-plant level of system organization and above.

Carbon economy of the sward

Figure 1.1 represents a simplified carbon cycle within a sward. The key elements of the cycle, which are discussed in more detail in the following subsections, are: (i) the interception of light by photosynthetically active leaf material; (ii) the ensuing rate of canopy gross photosynthesis (CGP) and the rate of maintenance respiration; (iii) the allocation of net available C (gross photosynthesis minus maintenance respiration) to the synthesis of new shoot and root tissues, and the respiratory cost of new tissue synthesis; and (iv) the rates of senescence and decomposition of shoot and root material. In grazed swards, the living shoot mass exists in dynamic equilibrium through the action of three fluxes: growth, senescence and animal consumption. The sward leaf area index (LAI) depends directly on the equilibrium attained between these fluxes; the sward LAI in turn determines the carbon flow entering the system.

Light interception

The quantity of C fixed by a plant canopy per unit of time depends directly on the quantity of photosynthetically active radiation absorbed by green leaf material (PAR_a). This quantity is determined by the incident PAR (PAR_0) and by the absorption efficiency of the canopy (E_a):

$$PAR_a = E_a \, PAR_0 \tag{1.1}$$

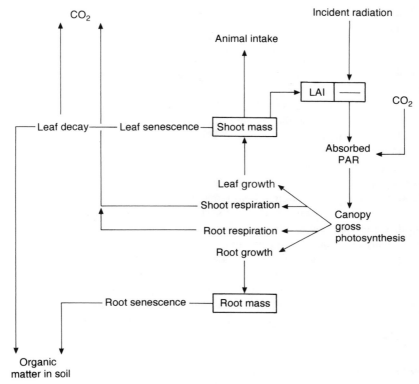

Fig. 1.1. Schematic representation of carbon fluxes in grazed sward. CO_2, carbon dioxide; LAI, leaf area index; PAR photosynthetically active radiation.

The absorption efficiency is determined by sward variables, such as the area of leaf available to intercept light per unit area of ground (LAI), the angle of leaf lamina displacement and light transmission properties of laminae, and by characteristics of solar radiation, such as diffuse/direct light ratio and the angle of incident radiation. A general model which relates E_a and the LAI of the sward, based on the Monsi and Saeki (1953) canopy transmittance studies, can be proposed:

$$E_a = K_1 (1 - e^{-K_2\, LAI}) \tag{1.2}$$

K_1 is a coefficient determined by the optical properties of leaves. A value of 0.95 is generally suitable for most crops (Varlet-Grancher *et al.*, 1989). K_2 is the light extinction coefficient, which depends on sward structural characteristics; its value can vary from 0.55 for swards with erect leaves like *Festuca arundinacea* (Bélanger *et al.*, 1992a) or *Dactylis glomerata* (Pearce *et al.*, 1965) to 0.8–0.9 for planophile species like *Medicago sativa* (Gosse *et al.*, 1982), *Trifolium repens* (Brougham, 1960) or *Digitaria decumbens* (Ludlow, 1985; Sinoquet and Cruz,

1993). For mixed swards the value for K_2 falls between values for erect and prostrate species (e.g. 0.66 for *Lolium perenne–T. repens* mixture (Brougham, 1958)).

For sward management purposes, an 'optimum' LAI has been defined as one at which 95% of incident radiation is intercepted by the canopy (Kasanga and Monsi, 1954; Donald, 1961). Equation (1.2) allows calculation of the optimum LAI and shows that the optimum will be higher for erect species (most grass species) than for planophile species (most legume species) (Brougham, 1960; Davidson and Birch, 1972). Extending this analysis, it is clear that a planophile leaf habit is better than an erect habit at low LAI because light interception will be more complete. The optimum LAI varies with season; it is higher when light intensity is high (e.g. spring and summer versus autumn and winter) and when the proportion of direct radiation in PAR_0 increases.

The C supply after infrequent and severe defoliation in intermittently grazed swards is therefore mainly dependent on the dynamics of leaf area expansion and on the delay in restoring the optimum LAI. In continuously grazed pastures, the C supply is mainly determined by the LAI at which the sward is maintained at equilibrium. Therefore, LAI appears to be an important sward state variable (Hodgson, 1985). For a given species and within a given range of defoliation managements, LAI of grazed swards can be estimated relatively simply by the mean of sward surface height measurements (Bircham and Hodgson, 1983), providing a useful tool for decision making in grazing management.

Photosynthesis and respiration

Many studies have been carried out on the rate of canopy photosynthesis of swards continuously or intermittently defoliated (Deinum *et al.*, 1981; Leafe and Parsons, 1981; Parsons *et al.*, 1983a; King *et al.*, 1984; Lantinga, 1985; Mazzanti, 1990; Gastal and Bélanger, 1993). These studies have quantified the effects of environmental variables on the photosynthesis of swards at a fixed LAI. The response curve of CGP can be expressed as the mean of two parameters: the maximum CGP (CGP_{max} in μmol (CO_2) m^{-2} s^{-1}) in saturated light conditions and the maximum light yield (LY_{max} in μmol CO_2 per mmol PAR), which represents the initial slope of the response curve. An important factor limiting CGP_{max} in vegetative grass swards is a reduction in the rate of light-saturated net photosynthesis of grass leaves which emerge and expand in low irradiance conditions compared with leaves emerging and expanding in high irradiance (Wilson and Cooper, 1969; Prioul, 1971; Woledge, 1971, 1973). Thus, as LAI increases in a regrowing canopy, successive grass leaves emerge into a shaded environment and their light-saturated photosynthetic capacity is impaired; CGP_{max} may therefore decline once complete light interception is reached (Robson *et al.*, 1988). Parsons and Robson (1981) showed that, at similar LAI, CGP_{max} was higher for swards in the reproductive stage than for vegetative swards. This change is due to modifications in sward structure when reproductive stem elongation pushes developing leaves into the well-illuminated horizon of the canopy, leading to sustained, high, light-saturated leaf photosynthetic capacity (Woledge, 1978).

Mazzanti (1990) and Gastal and Bélanger (1993) demonstrated that N deficiency affects only CGP_{max} and does not significatively change LY_{max} when compared with N-sufficient pastures at the same LAI. Nevertheless, this direct effect of N on canopy net photosynthesis is small compared with the effect of N on herbage growth, particularly on the dynamics of LAI expansion (Bélanger et al., 1992a) and changes in allocation patterns of assimilates between shoots and roots (Bélanger et al., 1992b). Differences in CGP_{max} between species of the same metabolic group are less important. Gastal and Bélanger (1993) reported similar values for F. arundinacea to those found by Parsons and Robson (1981) for L. perenne. Legumes, and particularly T. repens, show significantly higher CGP_{max} than grasses because petiole extension enables young developing clover leaves to escape the deleterious effect, observed for grasses, of shade in the base of a pasture on potential photosynthetic capacity in full light (Dennis and Woledge, 1982). Tropical grasses, with the C_4 photosynthetic pathway, have higher CGP_{max} (Ludlow, 1985), but small differences exist between species when they are compared at similar LAI.

Both CGP_{max} and LY_{max} increase with an increase in LAI (Gastal and Bélanger, 1993), therefore CGP increases rapidly until LAI reaches its optimum value. Leading up to the optimal LAI, an increasing amount of fixed C is diverted to support respiration in shaded tissue, and the optimum LAI coincides with the point where the C fixed by new leaf tissue just balances the extra respiratory demand created by the presence of the new leaf (extra shading). Thus, net assimilation (gross photosynthesis minus respiration) for the support of new tissue production reaches an asymptote, and typically equates to 55–60% of gross photosynthesis (Robson, 1973). About 25% of gross photosynthesis is used to support the synthesis of new tissues via biochemical pathways that operate at near maximum efficiency (Penning de Vries, 1972, 1975). In contrast, research on the pathways of maintenance respiration pointed to some opportunity for improving the efficiency of this process by exploiting the so-called 'alternative' oxidation pathway (Lambers, 1980; Lambers and Posthumus, 1980). Intraspecific variation for respiration rate in L. perenne was identified (Wilson, 1975), and led to the selection of ryegrass lines with 'slow' and 'fast' respiration rates (Wilson and Robson, 1981), which showed differences in growth characteristics and C economy (Robson, 1982a, b; Wilson and Jones, 1982). There is some doubt, however, that these lines did differ in their respiratory pathway activation (Pilbeam et al., 1986), and the existence of any firm physiological basis for manipulating the genetic control of maintenance respiration has yet to be established conclusively.

When compared at the same LAI, continuously grazed swards show lower CGP_{max} and LY_{max} than intermittently defoliated swards (King et al., 1984; Lantinga, 1985; Mazzanti, 1990). This effect has been explained by differences in leaf age structure between the two types of swards (Parsons et al., 1988), youngest leaves, which have higher photosynthetic capacity, being grazed more frequently than older leaves in continuous grazing situations (Barthram and Grant, 1984; Mazzanti and Lemaire, 1994).

Carbon allocation

The importance of knowledge of intraplant C allocation among competing meristematic sinks in quantifying the C economy of swards is well illustrated by the example of radiation use efficiency (RUE) analysis. Following the method of crop growth analysis developed by Monteith (1972), it is possible to establish a linear relationship between the gross production of new shoot mass and the quantity of PAR absorbed by the sward. The slope of this line represents the RUE (in g.MJ^{-1} PAR), which is a global estimation of the efficiency of the photosynthetic and respiratory processes. But, because only shoot growth is taken into account, comparisons of RUE between treatments can be misleading. For example, Bélanger *et al.* (1992a) recorded higher values of RUE in tall fescue swards in spring compared with summer, but this was explained by a higher proportion of C being allocated to shoot growth in reproductive swards (spring) than in vegetative swards (summer) (Bélanger *et al.* 1992b). Bélanger *et al.* (1992a) showed that the RUE of intermittently defoliated *F. arundinacea* swards is sharply reduced by N deficiency. This reduction is the consequence of both a decrease in canopy photosynthetic capacity (Gastal and Bélanger, 1993) and an increase in the proportion of assimilates allocated to roots (Bélanger *et al.*, 1992b). Estimates of RUE on continuously stocked swards of *F. arundinacea* have been made by Mazzanti (1990); values obtained in non-limiting N conditions were similar to those reported by Bélanger *et al.* (1992a) for intermittently defoliated swards. This indicates that, if comparisons of continuous versus intermittent stocking management were made at similar average LAI during a given period of time (i.e. at approximately similar amounts of PAR$_a$), then gross herbage production should be approximately equivalent in the two systems. The same general result was obtained by Parsons *et al.* (1988) using a model of sward growth (as discussed under Grazing Management and Tissue Dynamics in this chapter).

The absence of clear differences in RUE between continuously and intermittently defoliated swards is observed despite important differences in canopy photosynthetic capacities (see above). The explanation for this apparent contradiction could be found in differences in assimilate allocation patterns. Studies on numerous C$_3$ and C$_4$ forage grasses have demonstrated that root growth ceases after removal of 50% or more of the shoot system (Richards, 1993). Ryle and Powell (1975) observed an increase in ^{14}C allocation to growing shoot apices at the expense of roots after defoliation. Belanger *et al.* (1992b) showed that, during regrowth after complete defoliation of *F. arundinacea*, the proportion of assimilates allocated to roots increased from 10% to 20%. Richards (1993) indicated that priority allocation of C to shoots was the main adaptive response of plants to frequent defoliation, resulting in enhanced restoration of light interception and hence of C supply for new growth. The same author noted that, at the same time, defoliation reduced root respiration losses, but there are no data available to validate this for field conditions. The role of roots as a sink for assimilates and their interaction with whole-plant C supply emphasize the limitations of sward growth analysis by means of CO_2 exchange rates or measurements of shoot growth alone. C utilization by the different meristematic tissues of the plant has to be accounted for if the rate of new shoot tissue production in grazed communities is to be understood within the realistic context of variable defoliation and nutrient supply conditions.

Senescence and net herbage accumulation

The preceding sections show that, as swards begin regrowth from low LAI, the appearance and expansion of new leaves result in an increase in LAI, light interception and gross photosynthesis. In the early stages of regrowth, there is little or no death of leaf material, so that the net accumulation of herbage equates to the net assimilation rate of the canopy. The first leaves to die during regrowth are those produced at the beginning of the regrowth period; these leaves are smaller than those produced subsequently, so leaf death rate in mass flow terms initially lags behind the rate of new tissue production (Robson *et al.*, 1988). Net herbage accumulation rate is maximal during this lag period, but decreases until the rate of tissue death exactly balances the rate of new tissue production and a ceiling yield of live tissue is reached (Fig. 1.2). It is important to draw the distinction between respiration and senescence: although they both result in the loss of fixed C from the canopy, they are separated in time during the

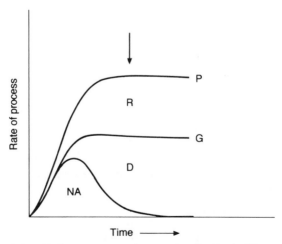

Fig. 1.2. The relationship between rates of gross photosynthesis (P), respiration (R), gross tissue production (G), net herbage accumulation (NA) and death (D) in a sward during regrowth. Arrow indicates time when ceiling yield is attained.

ontogeny of individual organs (biosynthesis respiration – maintenance respiration – senescence) and are regulated by different factors.

As discussed later, grass (and, to a lesser extent, legume) species have a genetically determined leaf life span. For *L. perenne*, this is about 330°C-days. *L. perenne* tillers in crowded swards support only three live leaves each; once three leaves have been produced, the oldest one begins to die. Thus, leaf senescence on a tiller starting regrowth with no leaves begins after 330°C-days, and the entire leaf complement of the tiller will have been replaced after about 660°C-days. In this time, six new leaves will have been produced and, if leaf size has reached a stable equilibrium within the first three-leaf appearance intervals, ceiling live leaf yield of a *L. perenne* canopy will also have been reached after about 660°C-days. In practice, the size of successive leaves tends to continue increasing beyond three-leaf appearance intervals if resource supply (especially light availability) is constant or increasing (Dale and Milthorpe, 1983; Van Loo, 1992), so that ceiling live leaf yield will not be reached until some time after 660°C days have elapsed.

Nitrogen uptake, nitrogen distribution and recycling

The C economy of plants is directly influenced by their N status. Photosynthetic activity of leaves is directly related to their N content (Robson and Parsons, 1978) and, as shown above, C supply of the sward depends on the level of N nutrition. C utilization in meristematic activities, such as leaf elongation rate (LER), also depends on plant N nutrition (Gastal *et al.*, 1992). Thus C fluxes within a sward are strongly influenced by N uptake and the partitioning and recycling of N within the sward.

When N is non-limiting the critical plant N content, i.e. minimum plant % N for maximum plant growth, appears to be a direct function of plant mass (W) (Lemaire and Salette, 1984):

$$\%N = a\,(W)^{-b} \tag{1.3}$$

where a and b are coefficients. Generalization of such a relationship to a wide range of species by Greenwood *et al.* (1990) showed that a unique value of 0.33 for b holds for C_3 and C_4 species, while different values for a (4.8 for C_3 and 3.6 for C_4) are necessary to account for the difference in photosynthetic and N use efficiency between the two metabolic pathways. From equation (1.3) it is possible to derive N uptake (N_{upt}):

$$N_{upt}\,(kg\ ha^{-1}) = 10a\,[W(t\ ha^{-1})]^{1-b} \tag{1.4}$$

Thus, in non-limiting N conditions, it is possible to relate the rate of N accumulation in a sward directly to the rate of herbage production. The decline in fractional N accumulation, i.e. the quantity of N necessary to produce a new unit of herbage production (g N per g dry matter (DM)), as herbage mass increases has been interpreted by Lemaire *et al.* (1991, 1992) as an intrinsic decrease in N requirements as the plant becomes larger and contains a greater proportion of structural materials, and as the N content of leaves which are progressively shaded during canopy development decreases. This second phenomenon corresponds to the remobilization of N from shaded to well illuminated leaves and leads to an optimization of N allocation for canopy photosynthesis (Field, 1983).

N remobilization from old leaves to young elongating leaves appears to be a general process which accompanies leaf senescence (Thomas and Stoddart, 1980). Lemaire and Culleton (1989) showed that between 75 and 80% of leaf N was recycled inside the plant and only 20–25% was lost and returned to soil through senescence of non-defoliated leaf. Hunt (1983) measured reabsorption of 30–50% of the N in senescing leaves of *L. perenne* and estimated that the reabsorbed N supplied 4–10% of the total amount of N required for annual net growth of the ryegrass population. Reabsorption efficiency was higher in swards receiving low rates of N fertilizer compared with high rates. It must be noted that, during senescence, the specific leaf weight decreases asymptotically as a consequence of continuous carbohydrate consumption by maintenance respiration, and reaches a value about half that of young mature leaves when the leaf is yellowing (Sheehy *et al.*, 1979). Thus, while the N content of leaves (expressed on % dry weight basis) may appear to halve moving from green to senescent states (e.g. Hunt, 1983), the quantity of N remobilized is actually nearer three-quarters of the amount contained in the green leaf. Working on this quantitative basis, J.R. Tallowin (Devon, UK, 1994, personal communication) measured mean remobilization of 70% of the N in green leaves of *L. perenne*, *Holcus lanatus*, *Agrostis stolonifera* and *Poa trivialis* in grazed pasture, with a tendency for proportionately more N to be remobilized in unfertilized swards than in swards receiving 400 kg N ha^{-1} $year^{-1}$. The remobilized N may equate to the metabolically active N contained in leaf tissue (e.g. in the photosynthetic enzyme rubisco and in the amino acids involved in cell division) while the 20–25% of N that is not remobilized may equate to leaf structural N; this hypothesis has not been explicitly tested.

On this basis it can be shown that, in undisturbed swards, the amount of N uptake required for production of new leaf material progressively declines as leaf senescence provides increasing quantities of recycling N. In continuously grazed conditions, because young well illuminated leaves with higher N content are more accessible to grazing livestock and have a higher probability of removal (Barthram and Grant, 1984; Clark *et al.*, 1984), a higher proportion of N required for new leaf production has to come from soil uptake and it could be predicted that the sward depends more on soil N availability when grazed more severely. This points to the possibility of interactions between defoliation management and N supply influencing the productivity of grazed pastures, a possibility that also has not been explicitly examined.

Several studies have highlighted the importance of N remobilization from roots after defoliation. In *M. sativa*, Lemaire *et al.* (1992) found that, 2 weeks after cutting, 40 kg N ha^{-1} was apparently removed from the tap root for shoot regrowth, corresponding to approximately one-third of the total N accumulated in shoots during 5 weeks of regrowth. Nitrogen reserves in the root and the crown, stubble or stolon have been shown to contribute 40–43% of the N in leaves after 14–24 days regrowth in *L. perenne* (Ourry *et al.*, 1990), *M. sativa* (Ourry *et al.*, 1994) and *T. repens* (Corre *et al.*, 1994). Using ^{15}N in controlled conditions, Kim *et al.* (1992) found similar patterns of N remobilization for *M. sativa* plants relying on either N_2 fixation or NO_3^- assimilation. Ourry *et al.* (1988) measured similar amounts of remobilization of N from roots and stubble of *F. arundinacea* after defoliation, but the recovery of N reserves during regrowth was more rapid than for *M. sativa*. Richards (1993) pointed out that the difference in defoliation recovery between species could be explained by whether shoot apices are removed by defoliation (such as *M. sativa*) or whether apices are not removed (such as grasses), the latter having a more rapid leaf area recovery rate. Rates of NO_3^- nitrate absorption and N_2 fixation are equally reduced after defoliation (Clement *et al.*, 1978; Kim *et al.*, 1992) and recover gradually as LAI and plant C supply increase.

Thus, severe defoliation leads to an important decline in plant N supply, and the leaf area recovery necessary for restoration of N supply involves remobilization of N reserves from roots and stubble. The implication of such a result for grazing management is very important. The regrowth of plants after defoliation is not directly determined by their level of carbohydrate reserves but by their N storage and remobilization capacity (Ourry *et al.*, 1994). As discussed by Richards (1993), the decline in soluble C in plants regrowing after defoliation is mostly due to respiration losses and only partly to direct remobilization of C for new tissue synthesis. Thus the quantity of reserve proteins and the rate at which they are recycled determine the rate of new leaf expansion.

N fluxes within grazed swards are greatly affected by the defoliation pattern. In intermittently and severely defoliated swards, N supply (N_2 fixation or NO_3^- absorption) alternates between high and low depending on the stage of regrowth. In continuously defoliated swards, N supply is more uniform and the rate of supply is determined by the average LAI. From an ecological point of view, severity of grazing determines the relative importance of the two N recycling pathways, i.e. internal recycling (N remobilization from senescent leaves) versus external recycling (via animal intake and urine and faeces return). Internal recycling is subject to less loss of N from the pasture ecosystem (leaching, gaseous loss) than external cycling. Thus, maximizing animal intake by increasing the utilization of herbage grown leads to lower efficiency of N utilization in the grazed system and may not be sustainable if N removal rates are not balanced by N input rates (Field and Ball, 1982; Parsons *et al.*, 1991).

MORPHOGENETIC ACTIVITY OF GRAZED SWARDS

The previous section has demonstrated that the pattern of C use by the plant and the processes of senescence are the main determinants of net herbage accumulation and harvestable herbage production. Plant morphogenesis can be defined as the dynamics of generation and expansion of the plant form in space (Chapman and Lemaire, 1993). It includes the rate of appearance of new organs, their rate of expansion and their rate of senescence and decomposition. Hence, for a given C supply, plant morphogenesis determines tissue fluxes and, in turn, internal C and N fluxes. Morphogenesis is defined at the level of the individual plant or autotrophic growth unit, such as the tiller for grasses or the rooted stolon for white clover. However, plants do not grow in pastures as isolated individuals but as members of a crowded community, where interplant and, in mixed-species pastures, interspecific competition has a major effect on plant growth and phenology. Hence, while morphogenetic activity may be analysed at the level of the individual growth unit, scaling up to the whole-sward level requires an understanding of competitive interactions and responses to leaf removal.

Morphogenesis of individual plants

Leaf tissue fluxes at the individual tiller level are best understood by considering the tiller axis as sequentially producing a chain of phytomers, each of which follows a preprogrammed series of developmental stages from primordium to mature organ and finally senescence (Silsbury, 1970). The sequential appearance of leaves on the tiller gives rise to the concept of 'site filling' to describe the development of daughter tillers in the leaf axils (Davies, 1974). The nodal structure of tiller development also allows root development to be related to leaf and tiller appearance (Matthew *et al.*, 1991), so that sward dynamics can be described and analysed in terms of clonal growth processes. For a vegetative grass sward in which only leaves are produced, plant morphogenesis can be described by three main characteristics: leaf appearance rate (LAR), leaf elongation rate (LER) and leaf lifespan. These characters are genetically determined, but are further influenced by environmental variables like temperature, nutrient supply and soil water status. Combinations of these elementary morphogenetic variables determine the three main structural characteristics of swards (Fig. 1.3):

1. *Leaf size*, which is determined by the ratio between LER and LAR because, for a given genotype, the duration of the elongation period for a leaf is a constant fraction of the leaf appearance interval (Robson, 1967; Dale, 1982).

2. *Tiller density*, which is partly related to LAR, which in turn determines the potential number of sites for tiller appearance (Davies, 1974). Thus, genotypes with high LAR (e.g. *L. perenne*) have high potential tillering rates and they produce swards with higher tiller densities than those with low LAR (e.g. *F. arundinacea*).

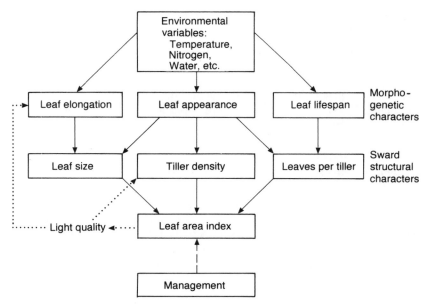

Fig. 1.3. Relationship between morphogenetic variables and sward structural characteristics.

3. *Number of green leaves per tiller*, which is the product of leaf lifespan and LAR.

Assuming a constant leaf area/leaf length ratio for a given genotype, the product of these three sward characteristics determines the LAI of the sward. Some other characteristics of swards, such as lamina/pseudostem ratio or bulk density of leaves, have only a small influence on LAI expansion, but are important considerations in plant–animal interactions because they influence the proportion of plant tissues which can be easily removed by grazing animals (Hodgson *et al.*, 1977). By altering light quality inside the canopy, increases in LAI can modify some morphogenetic variables at the individual plant level, such as LER and tillering rate, and consequently can change some structural characteristics of the sward, like tiller density and tiller size (Deregibus *et al.*, 1983).

LAR plays a central role in plant morphogenesis because of its direct influence on each of the three components of sward structure, as shown in Fig. 1.3. LAR is directly influenced by temperature and depends only loosely on N nutrition level for temperate grasses (Lemaire, 1988). For a given species, a more or less constant leaf appearance interval can be calculated in terms of degree-days: 110°C-days for *L. perenne* (Davies and Thomas, 1983), 230°C-days for *F. arundinacea* (Lemaire, 1985). Thus interspecific variation in LAR largely determines sward structure: high LAR corresponds to high density of small tillers (e.g. *L. perenne*) and low LAR to low density of larger tillers (e.g. *F. arundinacea*). LAR tends to decrease a little as the canopy develops during regrowth after complete defoliation. This phenomenon could be the consequence of the increasing length of sheath of successive leaves delaying the appearance of new laminae on the shoot axis. Thus the LAR of swards maintained at low LAI by frequent defoliation appears higher than that observed in intermittently defoliated swards.

LER responds immediately to any change in temperature at the shoot apex (Peacock, 1975a; Stoddart *et al.*, 1986). The response curve of LER to temperature changes rapidly during the transition from vegetative to reproductive growth of temperate grasses in spring (Peacock, 1975b; Parsons and Robson, 1980), resulting in much higher growth potential at a given temperature for reproductive swards in spring than for vegetative swards in summer or autumn (Gastal *et al.*, 1992). For most of the temperate grasses the response of LER to temperature appears to be exponential in the range 0–12°C of average daily temperature and linear above these values until an optimum is reached at about 20–25°C, depending on species. Because of the linear response of LAR to temperature, leaf size (determined by LER/LAR ratio) increases with increasing temperatures and, at similar temperature, is higher for reproductive swards in spring than for vegetative swards in summer and autumn. The effect of N nutrition level on LER is strong, resulting in three- to fourfold differences between extreme N nutrition levels in field conditions (Gastal *et al.*, 1992). LER is largely unaffected by a single defoliation removing only two of the three leaves per tiller, but is depressed by 15–20% when all leaves are removed (Davies, 1974). This result illustrates the increased strength of leaf meristems as a sink for assimilates after defoliation (see above).

Leaf senescence and leaf lifespan are affected by temperature in a similar manner to LAR. Thus, in steady-state conditions, an equilibrium between leaf appearance and leaf death is reached with a maximum number of living leaves per tiller. This maximum of leaves per tiller appears to be more or less genetically constant (e.g. three for *L. perenne*, 2.5 for *F. arundinacea*) and corresponds to the lifespan of leaves expressed in number of leaf appearance intervals. Thus, with a phyllochrone of 110°C-days and a maximum of three leaves, *L. perenne* has a leaf lifespan of about 330°C-days, in comparison with *F. arundinacea* which, with a phyllochrone of 230°C-days and 2.5 leaves per tiller, has a leaf lifespan of about 570°Cdays (Lemaire, 1988). This difference in leaf lifespan between species strongly influences their capacity to accumulate living shoot mass and reach high ceiling yields. It is important to note here that it is not the maximum number of living leaves which determines the ceiling yield of a sward, but only the leaf life duration, which is closely related to temperature. We must underline here the importance of knowledge of the leaf lifespan of different species for efficient grazing management of swards, because it strongly determines the proportion and amount of the gross herbage production which can be effectively harvested in a grazing management programme (Fig. 1.4, and see under Grazing Management and Tissue Dynamics in this chapter). Leaf lifespan is only slightly reduced by N deficiency (Gastal and Lemaire, 1988). Nevertheless, due to large effects of N nutrition on LER and leaf size, the senescence rate in absolute terms increases

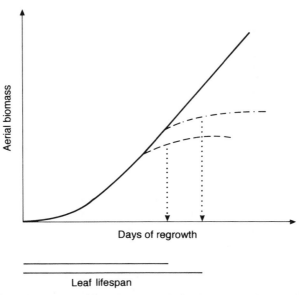

Fig. 1.4. Schematic representation of the evolution of primary production and harvestable production of two grass genotypes according to the length of life of their leaves. —, Primary production; -··-, harvestable production for a genotype with long leaf lifespan; - - -, harvestable production for a genotype with short leaf lifespan.

with the level of N fertilization (Mazzanti and Lemaire, 1994). Thus an increase in N fertilization without an appropriate adjustment to grazing management could lead to increased senescence and accumulation of dead matter in swards.

Tiller dynamics

Tiller density in grazed swards is a function of the equilibrium between tiller appearance rate (TAR) and tiller death rate. Potential TAR of isolated tillers has been studied in detail by different authors. The concept of 'site filling' (Davies, 1974; Neuteboom and Lantinga, 1989) allows derivation of the potential TAR from the LAR, as indicated in Fig. 1.3. In dense swards potential TAR may only be reached when the sward LAI is low, but actual TAR decreases as LAI develops and ceases at LAIs higher than 3–4 (Simon and Lemaire, 1987). N deficiency leads to low values of site filling (Lemaire, 1985), and suppresses TAR below its potential value, even in swards with low LAI.

Tiller death arises from different sources. A major cause of death is the removal of apices by grazing animals. This phenomenon is particularly important in reproductive swards when apices are elevated by stem internode elongation into the grazed horizon. But, even in vegetative swards of some species, such as *L.*

perenne, lax defoliation regimes can induce extension of basal internodes and increase the risk of decapitation of apices (Davies, 1988). Some tropical grasses are particularly vulnerable to removal of the stem apex (Chapman and Lemaire, 1993). Another important cause of tiller death in dense swards is C starvation resulting from competition for light. Davies *et al.* (1983) demonstrated that more dry matter is allocated to the growth of existing tillers and less to the development of new tillers in shaded plants. Young tillers are the first to die as a result of being overtopped and shaded by mature tillers (Ong, 1978) and some tiller buds probably abort before tiller emergence in shaded canopies.

The equilibrium between tiller appearance and tiller death is greatly dependent on the sward defoliation regime, which in turn determines the evolution of the LAI, which appears to be the key factor governing both appearance and death of tillers. For intermittently defoliated swards, tiller density increases after defoliation until an LAI of 3–4 is reached and then begins to decline. Despite the positive effect of N nutrition on TAR, N fertilizer use can lead to lower tiller density because of the rapid development of LAI and accelerated tiller death. In continuously grazed swards, the tiller density is mainly determined by the steady-state sward LAI, severely grazed swards having higher tiller density than laxly grazed swards (Grant *et al.*, 1983). In these situations, steady-state tiller density is influenced by N nutrition level if comparisons are made at similar LAI (Mazzanti *et al.*, 1994). Steady-state tiller density reflects the tillering capacity of the different species, which is linked to their LAR, as shown in Fig. 1.3. Thus, for a steady state LAI of 3 in an *F. arundinacea* sward, Mazzanti *et al.* (1994) obtained a tiller density of about 4000–6000 tillers m^{-2} which can be compared with 10,000–15,000 tillers m^{-2} typically observed in pure *L. perenne* swards maintained at similar LAI (Davies, 1988). This twice greater tiller density on perennial ryegrass corresponds to a twice higher LAR compared with tall fescue.

Phenotypic plasticity and adaptation to grazing

While inherent differences between species or cultivars in morphogenetic traits such as LAR may determine whole-sward characteristics, variation in the phenotype expressed by single genotypes in response to environmental and/or management variation also strongly influences sward structure. This phenomenon, termed phenotypic plasticity (Bradshaw, 1965), plays an important role in adaptation of swards to grazing management and is expressed as progressive and reversible modifications of morphological traits of individual plants.

For grass species, the plastic response to frequent and severe defoliation leads to a reduction in the size of individual tillers, accompanied by an increase in tiller density. The physiological mechanism governing this adaptation is not clearly understood. The maintenance of a high tillering rate under repeated defoliation may be due to the dampening of light quality variation within the canopy, and more specifically red/far red ratio diminution (Deregibus *et al.*, 1983). The concomitant reduction in tiller size is not so readily explained. Begg and Wright (1962) observed that leaf growth is strongly retarded as the leaf emerges in full light, and suggested that this response could be phytochrome-mediated. Davies (1974) and Davies et al. (1983) showed that the sheath tube, by protecting the elongating leaf from direct light, could greatly influence the LER, and consequently the final leaf length. Thus it is possible to explain variation in the length of new leaves by variation in the length of the sheath tube through which the leaves emerge (Grant et al., 1981). The physiological processes involved in this morphogenetic plasticity are not yet, to our knowledge, fully explained. This plastic response in plant morphology to defoliation appears to be an important process for providing a degree of grazing avoidance (see D.D. Briske, Chapter 2, this volume); positioning of the ligule below the level of defoliation allows the sward to maintain green leaf material below the grazing horizon and therefore to preserve actively photosynthesizing LAI. Thus the ability of species or cultivars to alter their sheath length in response to the defoliation regime appears to be of great importance in determining the adaptability of swards to different defoliation management programmes. Observations (G. Lemaire, L. Hazard and M. Ghesquiere, unpublished data) indicate that *F. arundinacea* is unable to reduce sheath length to a sufficiently low level to ensure persistence of leaf area when cut weekly to 3 cm, compared with *D. glomerata*, which is able to position its ligules down to ground level.

In the important forage legume *T. repens*, plasticity is expressed mainly through changes in lamina size, petiole length and stolon elongation and branching. These morphogenetic traits are influenced by the red/far red ratio of incident radiation in swards of increasing LAI (Robin et al., 1992; Varlet-Grancher et al., 1993). Thus, as competition for light within the canopy increases and the red/far red ratio lower in the canopy decreases, petiole length and lamina size increase, the emergence of axillary buds ceases and stolon internodes elongate more rapidly. The result of these changes is that leaf surfaces are positioned in the well illuminated layer of the canopy, and the stolon growing point can escape the shade and perhaps encounter a patch with better illumination in the pasture. In the short term, this adaptation allows the plant to optimize its C supply, but at the temporary expense of the production of new growing points, which are important for the continued persistence of clover (Chapman, 1983). These responses may be interpreted as 'foraging' for improved habitat quality (Harper, 1977; Sutherland and Stillman, 1988). Certainly, the stoloniferous growth habit of clover gives the species the capacity to exploit spatially patchy environments, whereas grasses remain largely confined to fixed, temporally patchy microhabitats. The stolon, as an organ, also plays a critical role in allowing plastic intraplant changes in resource allocation in response to environmental variation and defoliation (Chapman et al., 1991).

In continuously grazed swards at high LAI, the plastic response of increasing petiole length and leaf size as the red/far red ratio changes leads to a higher rate of removal of *T. repens* leaves, because sward surface horizons, where most of the clover leaf is positioned, are removed first during defoliation. In contrast, the capacity of plants to reduce petiole length in response to increased severity of grazing allows them to protect a large part of their leaf tissues from defoliation and to maintain a more stable C economy (Chapman, 1986). One important reason for the failure of large-leaved *T. repens* varieties to persist under intensive continuous sheep grazing is thought to be that their lower limit for petiole length reduction is insufficient to protect leaves from removal at high grazing pressure, leading to prolonged negative whole-plant C balance (Korte and Parsons, 1984). Conversely, large-leaved clover genotypes have a higher upper limit for maximum petiole length than small-leaved types, allowing them to compete more effectively for light with taller-growing grasses under intermittent defoliation. The interaction between phenotype and defoliation management can have a profound effect on clover performance in different grazing systems (Table 1.1).

At the whole-sward level, phenotypic plasticity is also expressed in the relationship between population density and the size of individuals in a grazed community. This inverse relationship has been described for many grazed swards (Grant *et al.*, 1983; Lambert *et al.*, 1986; Davies, 1988). Attempts have been made to relate the slope of this relationship obtained under grazing with that of the self-thinning law, i.e. $-3/2$ in log/log scale (Westoby, 1984). But, in reality, the $-3/2$ self-thinning relationship is best considered as a condition of canopy leaf area, occurring when canopy LAI has reached environmental potential (Sackville Hamilton *et al.*, 1995). This concept is not easily applied to intermittently grazed swards, except perhaps on a long-term coverage basis. In continuously stocked swards LAI is related to defoliation intensity, with the result that size–density compensation slopes are typically nearer $-5/2$ than $-3/2$ (Matthew *et al.*, 1995).

It is important to note that, for managed pasture communities in which density-dependent mortality is occurring, the population is reversible with respect

Table 1.1. Phenotypic plasticity in white clover (*Trifolium repens* L.): mean area of individual leaves of four clover cultivars, and mean clover content of swards, under continuous or rotational grazing (both at 22.5 ewes ha^{-1}), Palmerston North, New Zealand. (From Brock, 1988.)

		Grazing method			
		Individual leaf area (cm^2)		Clover content of pasture (% DM)	
Cultivar	Leaf size description	Rotational	Continuous	Rotational	Continuous
Grassland Tahora	Small	2.09	1.30	13.3	20.8
Grasslands Huia	Intermediate	2.75	1.15	11.0	13.1
Grasslands Pitau	Large–intermediate	4.08	1.30	15.1	7.0
Grasslands Kopu	Large	5.58	1.66	19.5	7.3
LSD$_{0.05}$†		0.35	2.28		

†Least significant difference of the mean at a probability of 0.05. DM, dry matter.

to the density–size relationship, in contrast to successional vegetation associations, which move up the self-thinning line only. Reversibility reflects the dynamic nature of growth and senescence in grass-dominated pastures, where there is continual turnover of leaves and growth units, and also the phenotypic plasticity of the component species. The morphology of replacement leaves or growth units will always reflect the environment in which they develop, as individual plants adapt to maintain an equilibrium between resource availability and resource allocation patterns. The equilibrium is clearly a dynamic one and the compensatory changes between tiller size and tiller density in response to defoliation regime result in little net difference in total grass leaf production per unit area of pasture (Chapman and Clark, 1984). Similar results have been obtained for continuously grazed pasture maintained at a range of sward heights (Grant *et al.*, 1981, 1983; Bircham and Hodgson, 1983) and similar outcomes in terms of leaf production per unit area of pasture can be observed when variations in tiller size/tiller density is genetic in origin, as shown by Mazzanti *et al.* (1994) in a comparison of two *F. arundinacea* cultivars.

Limits to the plastic response of forage plants

The above discussion on phenotypic plasticity of grass and legume species leads to the concept of limits to the plastic response of forage plants, and to consideration of how such limits may be defined and used to match forage species or genotypes to defoliation managements, in order to optimize long-term pasture growth within the constraints of temporal and spatial heterogeneity in pastures. These limits could determine the range of managements within which the compensation between, for example, size and density of individual growth units can operate completely to maintain optimal gross leaf tissue production. Similarly, knowledge of the cues and processes in white clover governing plastic trade-offs between directional stolon spread from less favourable microsites and consolidation of clonal presence in favourable microsites (through branching) could greatly enhance our understanding of spatial dynamics in grass/clover composition in grazed, mixed-species associations and could indicate how these dynamics may be manipulated to provide a more predictable pasture composition. Thus, the importance of phytochrome-mediated morphogenetic responses to variation in LAI and defoliation regime requires a focus in future investigation on local and global modifications to the light microenvironment in sward canopies, the reception and transmission of light quality signals by different plant organs (including integrated transmission among growth units within a plant) and the intensity and limits of morphogenetic responses for different genotypes, as proposed by Varlet-Grancher *et al.* (1993).

The manipulation of tiller dynamics can be an objective of pasture management, particularly in situations where tiller density tends toward the minimum needed to ensure perennation of the sward. Black and Chu (1989) showed that for prairie grass a hard/lax defoliation management encourages tiller appearance and increased herbage production in summer compared with either lax or hard defoliation treatments. Similarly, Lemaire and Culleton (1989) showed that autumn defoliation management associated with adequate N supply could greatly increase tiller appearance during winter and subsequent sward growth in spring for tall fescue. Demographic studies of tiller bud populations appear to be a good tool for predicting seasonal tiller density variation so that management strategy can be adapted to maintain adequate tiller density.

GRAZING MANAGEMENT AND TISSUE DYNAMICS

Frequency and intensity of defoliation in different grazing systems

The defoliation pattern of a pasture depends firstly on the defoliation management system: continuous versus intermittent. In intermittently grazed swards, where animals are allowed to graze accumulated herbage for fixed durations (usually in the range 12–72 h), the frequency of defoliation of leaves is closely related to the defoliation interval, which is determined by the whole farm management system. The intensity of defoliation of leaves in such a system can be expressed as the proportion of the initial length of leaf which has been removed by the end of the grazing period. This method holds in situations where the grazing period is sufficiently short for leaf elongation while animals are on the pasture to be ignored. Intensity of defoliation depends directly on the stocking density and on the duration of the grazing period, which are both features of the chosen management system.

Wade (1991) studied the frequency of defoliation events at the individual tiller level in a wide range of grazing systems (strip, rotational and continuous grazing) and stocking densities. He concluded that the relationship between the frequency of defoliation of individual tillers and the stocking density previously observed in continuously grazed swards (Wade and Baker, 1979) remained valid for intermittent grazing systems. The reciprocal of the average interval between two successive defoliations of the same tiller in a grazed sward equates to the proportion of tillers grazed each day, which in turn can be interpreted as the proportion of the sward area which has been grazed by animals each day (Wade *et al.*, 1989). Thus, it is possible to show that in strip grazing, with a stocking density of approximately 150×10^3 kg live weight (LW) ha^{-1}, the whole sward area available to the animals is grazed down on four successive occasions per day (400% of area grazed per day), whereas with continuous grazing (stocking density 1.3×10^3 to 8.1×10^3 kg LW ha^{-1}), only 6 to 20% of the plot area is grazed each day (Fig. 1.5). For continuous grazing, the area of sward grazed per day equates to defoliation intervals of 16 to 5 days, which are within the range reported by several authors for *L. perenne*-dominated swards (Hodgson, 1966; Hodgson and

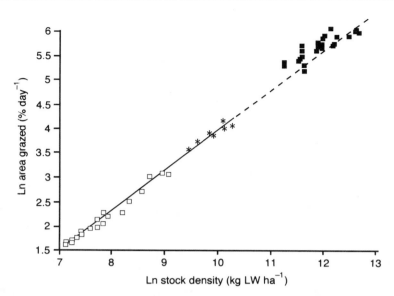

Fig. 1.5. Relationship between the area of pasture grazed per day and stocking density for continuous (□), paddock (∗) and strip (■) grazing systems. Ln, natural logarithm; LW, live weight. (From Wade, 1991.)

Ollerenshaw, 1969; Barthram and Grant, 1984) and for *F. arundinacea* (Mazzanti and Lemaire, 1994).

It is interesting to note here that for similar stocking densities the average defoliation interval of individual tillers is similar for *L. perenne* and *F. arundinacea* despite large differences in tiller size/tiller density between the two species, indicating that plastic responses of swards mediated by the trade-off between size and density of tillers do not necessarily affect the frequency at which individual tillers are defoliated; the effect may operate more through severity of tiller defoliation. The frequency of defoliation depends solely on stocking density. Mazzanti and Lemaire (1994) demonstrated that differences in the frequency of defoliation observed under contrasting N application rates in continuously grazed *F. arundinacea* swards were not a direct effect of N nutrition, but rather the consequence of a higher stocking density necessary to maintain an equivalent LAI on the higher N treatments. More detailed analysis of defoliation at the individual leaf level indicates that the youngest expanded leaf has a higher probability of defoliation than older ones (Barthram and Grant, 1984; Clark *et al.*, 1984). Mazzanti and Lemaire (1994) reported average defoliation intervals of 17, 15 and 27 days for leaves 1, 2 and 3 respectively (in increasing order of leaf age) in *F. arundinacea*, which simply reflects the relative position of different-age leaves within the canopy.

Intensity of defoliation can be studied at the level of individual defoliation events. Wade (1991) defined the intensity of defoliation as the reduction in the length of an extended tiller. This could also be described as the depth of defoliation because it identifies the layers of the sward which are removed by a single defoliation. Wade *et al.* (1989) demonstrated that with dairy cows grazing either continuously or intermittently over a wide range of sward surface heights, the mean depth of defoliation appears to be a relatively constant proportion (35%) of the mean extended tiller height, independent of the method of grazing and the stage of grazing down (Fig. 1.6). A similar result has been obtained by Edwards (1994). Thus, continuous and intermittent defoliation can be seen as simply representing different points on one continuum with respect to the relationship between tiller height and depth of grazing and not as discretely different processes as assumed in nearly all previous comparisons and analyses. The volume of sward grazed per day can be calculated as the product of depth of grazing × area grazed (the latter as a function of stocking density (Fig. 1.5)). By including information on the bulk density profile of the sward, it is then possible to estimate the quantity of herbage removed by grazing animals per day (Hoden *et al.*, 1991).

Continuous grazing creates a situation where the grazing down process is slow enough for the simultaneous reconstitution of the grazed layer of the sward in recently defoliated areas, whereas under intermittent defoliation the grazing down and regrowth processes are more clearly separated in time and therefore are more distinguishable. Wade (1991) demonstrated that the extended tiller height (closely related to sward surface height) and the bulk density profile of the canopy (which is related to tiller density) are the main sward characteristics which

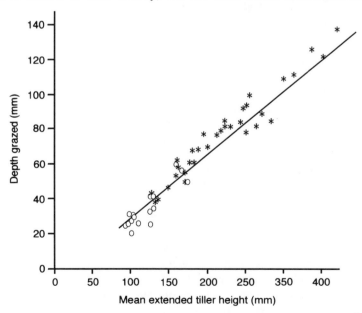

Fig. 1.6. Relationship between the depth of grazing within the sward and the extended height of grazed tillers in continuous (o) and paddock (*) grazing systems.

determine the maximum daily rate of removal of herbage by dairy cows. But he also pointed out that, as the grazing down process reached the sheath layer, the maximum daily herbage consumption rate decreased, as has been demonstrated previously (Chacon and Stobbs, 1976; Hodgson, 1982; Barthram and Grant, 1984). Thus sheath length is an important sward structural characteristic to take into account in grazing management for reasons of both its effect in restricting herbage intake rate and its effect on LER and LAR and the plastic responses of grass species to variation in defoliation management (see previous section – Phenotypic plasticity and adaptation to grazing).

Thus, the rate of leaf tissue removal by grazing animals can be quantified by reference to data on tiller density and the frequency and severity of defoliation of individual leaves. As demonstrated above, these three components are independent of each other, and only frequency of defoliation depends *directly* on stocking rate; tiller density is influenced by stocking rate, but indirectly and over longer time-scales.

Tissue flows and herbage utilization in different grazing systems

Herbage utilization can be analysed in terms of the balance between herbage growth and consumption and in terms of absolute quantities of herbage eaten by grazing animals.

Efficiency of utilization

The efficiency of herbage utilization in grazing systems can be defined as the proportion of gross herbage production which is removed by grazing animals before entering the senescent state, and it therefore depends on the proportion of leaf length which escapes defoliation and senesces. Optimizing the efficiency of utilization therefore requires an understanding of leaf lifespan in the pasture and of the factors influencing defoliation severity, as described above. As demonstrated by Mazzanti and Lemaire (1994), the proportion of leaf length that escapes defoliation and eventually senesces can be estimated by the ratio between leaf lifespan and defoliation interval, which determines the maximum number of times an individual leaf can be defoliated. Under continuous grazing, the proportion of leaf length removed at each defoliation is a relatively constant 50% (Mazzanti and Lemaire, 1994). Thus, with an average leaf lifespan of about 40 days for *F. arundinacea* and an average leaf defoliation interval of 20 days, a theoretical maximum herbage utilization efficiency of 75% is predicted; this is consistent with the maximum herbage use efficiency of 73% measured on continuously grazed tall fescue swards by Mazzanti and Lemaire (1994). These authors demonstrated that N deficiency leads to lower herbage use efficiency (57%

compared with 73% in optimum N supply conditions) only because of a longer defoliation interval (28 days compared with 20 days under optimum N supply conditions) as a direct consequence of a lower stocking density, as shown above. Thus it is possible to conclude that, in continuously grazed swards maintained at constant LAI, any reduction in leaf tissue production over and above that caused by N supply restrictions will lead to a further decrease in stocking density in order to maintain sward state, which in turn will lead to decreased herbage use efficiency of the system as shown in Fig. 1.7. The magnitude of such an effect could be dependent on the leaf lifespan of the grass species in the sward and this variable needs to be considered when designing systems to optimize utilization efficiency. Theoretically the reduction in grazing efficiency induced by a decrease in herbage growth and stocking density could be greater for species with a short leaf lifespan than for species with a longer leaf lifespan.

In intermittently grazed swards, the frequency of defoliation is determined by the frequency of moving animals from paddock to paddock, which is a function of paddock size, paddock number, net herbage accumulation rate and number of

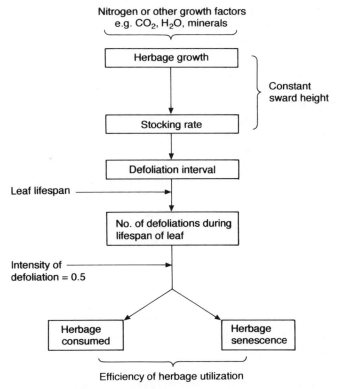

Fig. 1.7. Diagrammatic representation of the effect of rate of supply of growth resources on the efficiency of herbage utilization in a continuous-grazed sward.

animals. Thus, in such a system, the average length of the rest period can be adjusted in accordance with the leaf lifespan of the main species of the sward, in order to minimize leaf tissue losses by senescence, if the stocking density and the duration of the grazing period are sufficient to remove the maximum proportion of accumulated herbage. With such a system, it should be possible to maintain a high grazing efficiency despite decreases in herbage growth and stocking rate. Therefore, the reduction in stocking rate which could result from extensification of grassland systems should lead to the use of an intermittent grazing management system with an appropriate rest period (shorter than the average leaf lifespan) instead of a continuous grazing system. Under intermittent grazing, it should be possible to maintain a stable equilibrium between herbage consumption and herbage growth, and to avoid excess accumulation of senescent material and the development of patchiness in sward height, as animals reject areas with high dead-matter content.

Optimizing absolute amounts of utilized herbage

Optimizing the amount of herbage utilized requires consideration of all the variables influencing the efficiency of utilization, as described above, together with the added goal of maintaining maximum rates of green herbage accumulation. For some grazing systems, e.g. continuous grazing, this immediately highlights the conflict that can exist between managing pastures to maximize their growth (maintaining high LAI) versus managing pastures to maximize their harvested yield (high stocking rate and frequent defoliation). A compromise solution balancing these two objectives has to be struck; for UK conditions, a stable solution appears to be achieved within the range of sward LAI of approximately 2–4 (Bircham and Hodgson, 1983; Parsons *et al.*, 1983b).

To optimize harvested yield in intermittently defoliated systems over time periods greater than a single regrowth duration, theoretical analyses (Parsons *et al.*, 1988; Robson *et al.*, 1988) show that swards should be harvested when the phase of exponential total biomass accumulation during regrowth ceases, which coincides with maximum average growth rate (kg dry matter accumulated per day) of the sward for the entire regrowth period, but not maximum instantaneous growth rate (Fig. 1.8). While Fig. 1.8 portrays regrowth starting from a low LAI, optimal harvest timing still occurs when average growth rate is maximum for swards growing from a higher starting LAI. However, in these situations, the time elapsing from start of regrowth to optimum harvest will be less (Parsons *et al.*, 1988); in general, the higher the starting LAI, the shorter the interval to optimum harvest (Fig. 1.9). It is notable also that, as starting LAI increases (i.e. severity of defoliation decreases), analyses predict that maximum average growth rate will initially increase, and then decrease (Fig. 1.9), reflecting losses in potential production due to shading effects at high LAI on the photosynthetic capacity of new leaves. Thus, maintaining swards at high LAI will incur a loss in potential production which will limit actual herbage utilization irrespective of utilization

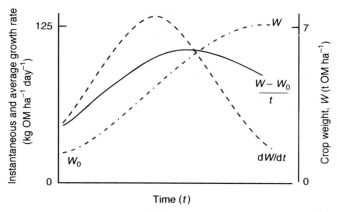

Fig. 1.8. Changes in instantaneous growth rate (dW/dt), standing crop weight (W) and average growth rate [($W - W_0$)/t] of a grass sward during regrowth from a low leaf area index. (After Parsons et al., 1988.) OM, organic matter.

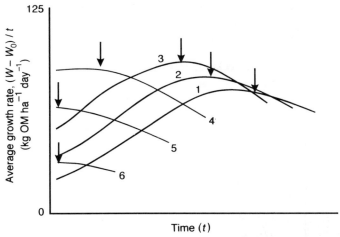

Fig. 1.9. Average growth rate of pastures regrowing from leaf area indices of 0.5, 0.8, 1.1, 3.4, 5.3 and 6.8 (numbered 1–6 respectively) as predicted by a model. Arrows indicate optimum time of harvest (see text). OM, organic matter. (After Parsons et al., 1988.)

efficiency. Indeed, herbage intake rate and hence herbage utilization efficiency will also tend to decrease in swards maintained at high LAI as a result of reduced grass tiller density and lower green leaf : stem ratios (Hodgson et al., 1977).

By plotting sward processes relative to average LAI sustained over a period of regrowth, Parsons et al. (1988) were able to compare swards regrowing from different initial LAIs on a common basis, and to demonstrate the changes in production that occur with change in average sward state (Fig. 1.10). On this basis, they were able to compare intermittent and continuous defoliation (the latter is the dashed line in Fig. 1.10) on an equivalent basis and show that they are essentially part of the same continuum of growth responses to average sward state. Thus, in the same way as shown for rate of herbage removal (see above), what superficially seem to be two discrete systems of defoliation management can be shown to represent simply different parts of the same general relationships governing resource capture and biomass utilization. While Fig. 1.10 seems to suggest some opportunity to capture the benefits of releasing a continuously grazed sward by imposing a rest from grazing (thereby exploiting the lag between new leaf production and leaf senescence), in practice the advantages accruing are small and difficult to translate into greater utilized production (Grant *et al.*, 1988; Parsons and Penning, 1988).

CONCLUSIONS

Theoretical principles for optimizing the amount of herbage grown and utilized and the efficiency of utilization in grazed pastures are well developed. These principles have been derived from detailed studies of tissue and mass flow dynamics in continuously and intermittently grazed swards, and their formalization has been enhanced by the use of mathematical models of mass flow. In practice, the choice of which system is adopted on-farm is rarely a simple one between either continuous or intermittent stocking, or between different severities or frequencies

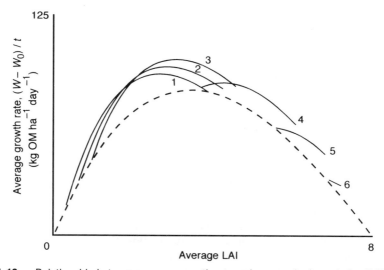

Fig. 1.10. Relationship between average growth rate and average leaf area index (LAI) of intermittently defoliated pastures regrowing from leaf area indices of 0.5, 0.8, 1.1, 3.4, 5.3 and 6.8 (solid lines, numbered 1–6 respectively) and of a continuously grazed pasture (broken line), as predicted by a model. OM, organic matter. (After Parsons *et al.*, 1988.)

of defoliation. This is because the objective of grazing management systems is not always to optimize average pasture or animal growth rate. For example, transferring standing herbage accumulated when there is a surplus of pasture growth over animal demand to a period when the reverse is true is a critical part of the management of all-grass, year-round feeding systems where seasonality in pasture growth and animal nutritional requirements is pronounced. Transferring feed in this way will inevitably result in a loss of potential production and reduce both the amount and efficiency of herbage utilization, as explained earlier. The management of such conflicting goals in temperate regions is discussed by G.W. Sheath and D.A. Clark (Chapter 11, this volume).

While theoretical knowledge may not always be directly applied in the formulation of grazing management systems, there is no doubt that several unifying relationships have emerged from research of the past two decades to provide a sounder understanding of the interactions between tissue flow dynamics and the defoliation management of pastures. Some of the important relationships are between: (i) leaf senescence rate and leaf lifespan, the latter determining the potential ceiling yield of the sward; (ii) N remobilization and the amount of N in new leaf material in early–mid-stages of regrowth; (iii) LAI and average sward growth rate; (iv) sward height and the depth of defoliation (the latter emerging as a constant equal to about 35% of the former); and (v) stocking density and the proportion of sward area grazed per day. In most instances, intermittent and continuous defoliation systems can be shown to depend on the same general quantitative relationship, and it is now clear that the growth response of pastures and the intake response of animals under either system of management can be explained and predicted from the same common basis. There is no great value in pursuing further comparison of tissue flows in intermittent versus continuous defoliation systems in temperate regions. Rather, future research effort should, among other possibilities, seek to: (i) extend the range of forage genetic resource for which quantitative information on leaf lifespan is available, and to examine the genetic basis of control of leaf lifespan; and (ii) explore the interaction between N supply and defoliation management of temperate pastures.

REFERENCES

Barthram, G.T. and Grant, S.A. (1984) Defoliation of ryegrass-dominated swards by sheep. *Grass and Forage Science* 39, 211–219.

Begg, J.E. and Wright, M.J. (1962) Growth and development of leaves from intercalary meristems in *Phalaris arundinacea* L. *Nature (London)* 194, 1097–1098.

Bélanger, G., Gastal, F. and Lemaire, G. (1992a) Growth analysis of a tall fescue sward fertilized with different rates of nitrogen. *Crop Science* 32, 1371–1376.

Bélanger, G., Gastal, F. and Warembourg, F. (1992b) The effects of nitrogen fertilization and the growing season on carbon partitioning in a sward of tall fescue. *Annals of Botany* 70, 239–244.

Bircham, J.S. and Hodgson, J. (1983) The influence of sward conditions on rates of herbage growth and senescence in mixed swards under continuous stocking management. *Grass and Forage Science* 38, 323–331.

Black, C.K. and Chu, A.C.P. (1989) Searching for an alternative way to manage prairie grass. *Proceedings of the New Zealand Grassland Association* 50, 219–223.

Bradshaw, A.D. (1965) Evolutionary significance of phenotypic plasticity in plants. *Advances in Genetics* 13, 115–155.

Briske, D.D. (1986) Plant responses to defoliation: morphological considerations and allocation priorities. In: Joss, J.W., Lynch, P.W. and Williams, O.B. (eds) *Rangelands: A Resource Under Siege*. Cambridge University Press, Cambridge, pp. 426–427.

Brock, J.L. (1988) Evaluation of New Zealand bred white clover cultivars under rotational grazing and set stocking with sheep. *Proceedings of the New Zealand Grassland Association* 49, 203–206.

Brougham, R.W. (1958) Interception of light by the foliage of pure and mixed swards of pasture plants. *Australian Journal of Agricultural Research* 9, 39–52.

Brougham, R.W. (1960) The relationship between the critical leaf area index, total chlorophyll content and maximum growth rate of some pasture and crop plants. *Annals of Botany* 24, 463–474.

Chacon, E. and Stobbs, T.H. (1976) Influence of progressive defoliation of a grass sward on the eating behaviour of cattle. *Australian Journal of Agricultural Research* 27, 709–727.

Chapman, D.F. (1983) Growth and demography of *Trifolium repens* stolons in grazed hill pastures. *Journal of Applied Ecology* 20, 597–608.

Chapman, D.F. (1986) Development, removal and death of white clover leaves under 3 grazing managements in hill country. *New Zealand Journal of Agricultural Research* 29, 39–47.

Chapman, D.F. and Clark, D.A. (1984) Pasture responses to grazing management in hill country. *Proceedings of the New Zealand Grassland Association* 45, 168–176.

Chapman, D.F. and Lemaire, G. (1993) Morphogenetic and structural determinants of plant regrowth after defoliation. In: Baker, M.J. (ed.) *Grasslands for Our World*. SIR Publishing, Wellington, pp. 55–64.

Chapman, D.F., Robson, M.J. and Snaydon, R.W. (1991) Influence of leaf position and defoliation on the assimilation and translocation of carbon in white clover (*Trifolium repens* L.). 1. Carbon distribution patterns. *Annals of Botany* 67, 295–302.

Clark, D.A., Chapman, D.F., Land, C. and Dymock, N. (1984) Defoliation of *Lolium perenne* and *Agrostis* spp. tillers and *Trifolium repens* stolons in set-stocked and rotationally grazed hill pastures. *New Zealand Journal of Agricultural Research* 27, 289–301.

Clement, C.R., Hopper, M.J., Jones, L.H.P. and Leafe, E.G. (1978) The uptake of nitrate by *Lolium perenne* from flowing nutrient solution. II. Effect of light, defoliation, and relationship to CO_2 flux. *Journal of Experimental Botany* 29, 1173–1183.

Corre, N., Ourry, A., Boucaud, J., Simon, J.C. and Salette, J. (1994) Mobilization of N reserve during regrowth of *Trifolium repens* L.: Quantification of N flows by ^{15}N labelling and purification of proteins involved in storage. *Proceedings of the 15th General Meeting of the European Grasslands Federation*. Wageningen, The Netherlands, June 6–10, pp. 66–73.

Dale, J.E. (1982) Some effects of temperature and irradiance on growth of the first four leaves of wheat *Triticum aestivum*. *Annals of Botany* 50, 851–858.
Dale, J.E. and Milthorpe, F.L. (eds) (1983) *The Growth and Functioning of Leaves*. Cambridge University Press, Cambridge.
Davidson, J.L. and Birch, J.W. (1972) Effects of defoliation on growth and carbon dioxide exchanges of subterranean clover swards. *Australian Journal of Agricultural Science* 9, 53–72.
Davies, A. (1974) Leaf tissue remaining after cutting and regrowth in perennial ryegrass. *Journal of Agricultural Science (Cambridge)* 82, 165–172.
Davies, A. (1988) The regrowth of grass swards. In: Jones, M.B. and Lazenby, A. (eds) *The Grass Crop: The Physiological Basis of Production*. Chapman and Hall, London, pp. 85–127.
Davies, A. and Thomas, H. (1983) Rates of leaf and tiller production in young spaced perennial ryegrass plants in relation to soil temperature and solar radiation. *Annals of Botany* 57, 591–597.
Davies, A., Evans, M.E. and Exley, J.K. (1983) Regrowth of perennial ryegrass as affected by simulated leaf sheaths. *Journal of Agricultural Science (Cambridge)* 101, 131–137.
Deinum B., 'T Hart, M.L. and Lantinga, E.A. (1981) Photosynthesis of grass swards under rotational and continuous grazing. *Proceedings of the 14th International Grasslands Congress*, 407–410.
Dennis, W.D. and Woledge, J. (1982) Photosynthesis by white clover leaves in mixed clover/ryegrass swards. *Annals of Botany* 49, 627–653.
Deregibus, V.A., Sanchez, R.A. and Casal, J.J. (1983) Effects of light quality on tiller production in *Lolium* spp. *Plant Physiology* 72, 900–912.
Donald, C.M. (1961) Competition for light in crops and pastures. In: Milthorpe, F.L. (ed.) *Mechanisms in Biological Competition*. Cambridge University Press, Cambridge, pp. 56–69.
Edwards, G.R. (1994) The creation and maintenance of spatial heterogeneity in plant communities: the role of plant–animal interactions. Unpublished PhD thesis, Oxford University.
Field, C. (1983) Allocating leaf nitrogen for the maximization of carbon gain: leaf age as a control on the allocation program. *Oecologia* 56, 341–347.
Field, T.R.O. and Ball, P.R. (1982) Nitrogen balance in an intensively utilised dairy farm system. *Proceedings of the New Zealand Grassland Association* 43, 64–69.
Gastal, F. and Bélanger, G. (1993) The effects of nitrogen and the growing season on photosynthesis of fieldgrown tall fescue canopies. *Annals of Botany* 72, 401–408.
Gastal, F. and Lemaire, G. (1988) Study of a tall fescue sward under nitrogen deficiency conditions. *Proceedings of the 12th General Meeting of the European Grassland Federation*, 323–327.
Gastal, F., Bélanger, G. and Lemaire, G. (1992) A model of the leaf extension rate of tall fescue in response to nitrogen and temperature. *Annals of Botany* 70, 437–442.
Gosse, G., Chartier, M., Varlet-Grancher, C. and Bonhomme, R. (1982) Interception du rayonnement utile à la photosynthèse chez la luzerne: variations et modélisation. *Agronomie* 2, 583–588.
Grant, S.A., Barthram, G.T. and Torvell, L. (1981) Components of regrowth in grazed and cut *Lolium perenne* swards. *Grass and Forage Science* 36, 155–168.

Grant, S.A., Barthram, G.T., Torvell, L., King, J. and Smith, H.K. (1983) Sward management, lamina turnover and tiller population density in continuously stocked *Lolium perenne*-dominated swards. *Grass and Forage Science* 38, 333–344.

Grant, S.A., Barthram, G.T., Torvell, L., King, J. and Elaton, D. (1988) Comparison of herbage production under continuous stocking and intermittent grazing. *Grass and Forage Science* 43, 29–39.

Greenwood, D.J., Lemaire, G., Gosse, G., Cruz, P., Draycott, A. and Neetson, J.J. (1990) Decline of % N in C_3 and C_4 crops with increasing plant mass. *Annals of Botany* 66, 425–436.

Harper, J.L. (1977) *Population Biology of Plants*. Academic Press, London.

Hoden, A., Peyraud, J.L., Muller, A., Delaby, L. and Faverdin, P. (1991) Simplified rotational grazing management of dairy cows: effects of rates of stocking and concentrate. *Journal of Agricultural Science (Cambridge)* 116, 417–428.

Hodgson, J. (1966) The frequency of defoliation of individual tillers in a set-stocked sward. *Journal of the British Grassland Society* 27, 258–263.

Hodgson, J. (1982) Influence of sward characteristics on diet selection and herbage intake by the grazing animal. In: Hacker, J.B. (ed.) *Nutritional Limits to Animal Production from Pasture*. CAB International, Wallingford, pp. 153–166.

Hodgson, J. (1985) The significance of sward characteristics in the management of temperate sown pastures. *Proceedings of the 16th International Grassland Congress*, 31–34.

Hodgson, J. and Ollerenshaw, J.H. (1969) The frequency and severity of defoliation of individual tillers in set-stocked swards. *Journal of the British Grassland Society* 24, 226–234.

Hodgson, J., Rodriguez Capriles, J.M. and Fenlon, J.S. (1977) The influence of sward characteristics on the herbage intake of grazing calves. *Journal of Agricultural Science (Cambridge)* 89, 743–750.

Hunt, W.F. (1983) Nitrogen cycling through senescent leaves and litter in swards of Ruanui and Nui ryegrass with high and low nitrogen inputs. *New Zealand Journal of Agricultural Research* 26, 461–471.

Kasanga, H. and Monsi, M. (1954) On the light transmission of leaves and its meaning for production of dry matter in plant communities. *Japanese Journal of Botany* 14, 304–324.

Kim, T.H., Ourry, A., Boucaud, J. and Lemaire, G. (1992) Changes in source/sink relationships for nitrogen during regrowth of alfalfa (*Medicago sativa* L.) following removal of shoots. *Australian Journal of Plant Physiology* 18, 593–602.

King, J., Sim, E.A. and Grant, S.A. (1984) Photosynthetic rate and carbon balance of grazed ryegrass pastures. *Grass and Forage Science* 39, 81–92.

Korte, C.J. and Parsons, A.J. (1984) Persistence of large-leaved white clover variety under sheep grazing. *Proceedings of the New Zealand Grassland Association* 45, 118–123.

Lambers, H. (1980) The physiological significance of cyanide-resistant respiration. *Plant, Cell and Environment* 3, 293–302.

Lambers, H. and Posthumus, F. (1980) The effect of light intensity and relative humidity on growth rate and root respiration of *Plantago lanceolata* and *Zea mays*. *Journal of Experimental Biology* 31, 1621–1630.

Lambert, M.G., Clark, D.A., Grant, D.A., Costall, D.A. and Gray, Y.S. (1986) Influence of fertiliser and grazing management on North Island moist hill country. 4. Pasture species abundance. *New Zealand Journal of Agricultural Research* 29, 23–31.

Lantinga, E.A. (1985) Productivity of grasslands under continuous and rotational grazing. Unpublished PhD thesis, University of Wageningen.
Leafe, E.L. and Parsons, A.J. (1981) Physiology of growth of a grazed sward. *Proceedings of the 14th International Grassland Congress,* 403–406.
Lemaire, G. (1985) Cinétique de croissance d'un peuplement de fétuque élevée (*Festuca arundinacea* Schreb.) pendant l'hiver et le printemps. Effets des facteurs climatiques. Thèse Doctorat és Sciences Naturelles, Université de Caen.
Lemaire, G. (1988) Sward dynamics under different management programmes. *Proceedings of the 12th General Meeting of the European Grassland Federation,* 7–22.
Lemaire, G. and Culleton, N. (1989) Effects of nitrogen applied after the last cut in autumn on a tall fescue sward. 2. Uptake and recycling of nitrogen in the sward during winter. *Agronomie* 9, 241–249.
Lemaire, G. and Salette, J. (1984) Relation entre dynamique de croissance et dynamique de prélèvement d'azote pour un peuplement de graminées fourragères. 1. Etude de l'effet du milieu. *Agronomie* 4, 423–430.
Lemaire, G., Onillon, B., Gosse, G., Chartier, M. and Allirand, J.M. (1991) Nitrogen distribution within a lucerne canopy during regrowth: relation with light distribution. *Annals of Botany* 68, 483–488.
Lemaire, G., Khaity, M., Onillon, B., Allirand, J.M., Chartier, M. and Gosse, G. (1992) Dynamics of accumulation and partitioning of N in leaves, stems and roots of lucerne (*Medicago sativa* L.) in dense canopy. *Annals of Botany* 70, 429–435.
Ludlow, M.M. (1985) Photosynthesis and dry matter production in C3 and C4 pasture plants, with special emphasis on tropical C3 legumes and C4 grasses. *Australian Journal of Plant Physiology* 12, 557–572.
Matthew, C., Xia, J.X., Chu, A.C.P., Mackay, A.D. and Hodgson, J. (1991) Relationship between root production and tiller appearance rates in perennial ryegrass (*Lolium perenne* L.). In: Atkinson, D. (ed.) *Plant Root Growth and Ecological Perspective.* British Ecological Society, special publication No. 10. Blackwell Scientific Publications, Oxford, pp. 281–290.
Matthew, C., Lemaire, G., Sackville Hamilton, N.R. and Hernandez-Garay, A. (1995) A modified self-thinning equation to describe size/density relationships for defoliated swards. *Annals of Botany,* 76, 579–587.
Mazzanti, A. (1990) Effet de l'azote sur la croissance de l'herbe d'une prairie de fétuque élevée et son utilisation par des moutons en pâturage continu. Thèse de Doctorat en Sciences, Université de Paris-Sud, Orsay.
Mazzanti, A. and Lemaire, G. (1994) Effect of nitrogen fertilization upon herbage production of tall fescue swards continuously grazed by sheep. 2. Consumption and efficiency of herbage utilisation. *Grass and Forage Science* 49, 352–359.
Mazzanti, A., Lemaire, G. and Gastal, F. (1994) The effect of nitrogen fertilisation on the herbage production of tall fescue swards grazed continuously with sheep. 1. Herbage growth dynamics. *Grass and Forage Science* 49, 111–120.
Monsi, M. and Saeki, T. (1953) Über den Lichtfaktor in den Planengesellshaften und seiner Bedeutung für die Stoffproduction. *Japanese Journal of Botany* 14, 22–52.
Monteith, J.L. (1972) Solar radiation and productivity in tropical ecosystems. *Journal of Applied Ecology* 9, 747–766.
Neuteboom, J.H. and Lantinga, E.A. (1989) Tillering potential and relationship between leaf and tiller production in perennial ryegrass. *Annals of Botany* 63, 265–270.

Ong, C.J. (1978) The physiology of tiller death in grasses. 1. The influence of tiller age, size and position. *Journal of the British Grassland Society* 33, 197–203.

Ourry, A., Boucaud, J. and Salette, J. (1988) Nitrogen mobilization from stubble and roots during regrowth of ryegrass. *Journal of Experimental Botany* 39, 803–809.

Ourry, A., Boucaud, J. and Salette, J. (1990) Partitioning and remobilization of nitrogen during regrowth of ryegrass subjected to nitrogen deficiency. *Crop Science* 30, 1251–1254.

Ourry, A., Kim, T.H. and Boucaud, J. (1994) Nitrogen reserve mobilization during regrowth of *Medicago sativa* L. Relationships between availability and regrowth yield. *Plant Physiology* 105, 831–837.

Parsons, A.J. and Penning, P.D. (1988) The effect of the duration of regrowth on photosynthesis, leaf death and the average rate of growth in a rotationally grazed sward. *Grass and Forage Science* 43, 15–27.

Parsons, A.J. and Robson, M.J. (1980) Seasonal changes in the physiology of S24 perennial ryegrass. 1. Response of leaf extension to temperature during the transition from vegetative to reproductive growth. *Annals of Botany* 46, 435–444.

Parsons, A.J. and Robson, M.J. (1981) Seasonal changes in the physiology of S24 perennial ryegrass. 2. Potential leaf and canopy photosynthesis during the transition from vegetative to reproductive growth. *Annals of Botany* 47, 249–258.

Parsons, A.J., Leafe, E.L., Collet, B., and Stiles, W. (1983a) The physiology of grass production under grazing. 1. Characteristics of leaf and canopy photosynthesis of continuously grazed swards. *Journal of Applied Ecology* 20, 117–126.

Parsons, A.J., Leafe, E.L., Collett, B., Penning, P.D. and Lewis, J. (1983b) The physiology of grass production under grazing. 2. Photosynthesis, crop growth and animal intake of continuously grazed swards. *Journal of Applied Ecology* 20, 127–139.

Parsons, A.J., Johnson, I.R. and Harvey, A. (1988) Use of a model to optimize the interaction between frequency and severity of intermittent defoliation and to provide a fundamental comparison of the continuous and intermittent defoliation of grass. *Grass and Forage Science* 43, 49–59.

Parsons, A.J., Orr, R.J., Penning, P.D. and Lockyer, D.R. (1991) Uptake, cycling and fate of nitrogen in grass–clover swards continuously grazed by sheep. *Journal of Agricultural Science (Cambridge)* 116, 47–61.

Peacock, J.M. (1975a) Temperature and leaf growth in *Lolium perenne*. 1. The thermal microclimate: its measurement and relation to plant growth. *Journal of Applied Ecology* 12, 115–123.

Peacock, J.M. (1975b) Temperature and leaf growth in *Lolium perenne*. 3. Factors affecting seasonal differences. *Journal of Applied Ecology* 12, 685–697.

Pearce, R.B., Brown, R.H. and Blaser, R.E. (1965) Relationships between leaf area index, light interception and net photosynthesis in orchardgrass. *Crop Science* 5, 553–556.

Penning de Vries, F.W.T. (1972) Respiration and growth. In: Rees, A.R., Cockshull, K.E., Hand, D.W. and Hurd, R.D. (eds) *Crop Processes in Controlled Environments.* Academic Press, London, pp. 327–347.

Penning de Vries, F.W.T. (1975) The cost of maintenance processes in plant cells. *Annals of Botany* 39, 77–92.

Pilbeam, C.J., Robson, M.J. and Lambers, H. (1986) Respiration in mature leaves of *Lolium perenne* as affected by nutrient supply and cutting. *Physiologia Plantarum* 66, 53–57.

Prioul, J.L. (1971) Réaction des feuilles de *Lolium multiflorum* à l'éclairement pendant la croissance et variation des résistances aux échanges gazeux photosynthétiques. *Photosynthetica* 5, 364–375.
Richards, J.H. (1993) Physiology of plants recovering from defoliation. In: Baker, M.J. (ed.) *Grasslands for Our World*. SIR Publishing, Wellington, pp. 46–54.
Robin, C., Varlet-Grancher, C., Gastal, F., Flenet, F. and Guckert, A. (1992) Photomorphogenesis of white clover (*Trifolium repens* L.): phytochrome mediated effects on ^{14}C-assimilate partitioning. *European Journal of Agronomy* 1, 235–240.
Robson, M.J. (1967) A comparison of British and North American varieties of tall fescue. 1. Leaf growth during winter and the effect on it of temperature and daylength. *Journal of Applied Ecology* 4, 475–484.
Robson, M.J. (1973) The growth and development of simulated swards of perennial ryegrass. II. Carbon assimilation and respiration in a seedling sward. *Annals of Botany* 37, 501–508.
Robson, M.J. (1982a) The growth and carbon economy of selection lines of *Lolium perenne* cv. S23 with 'fast' and 'slow' rates of dark respiration. I. Grown as simulated swards during a regrowth period. *Annals of Botany* 49, 321–329.
Robson, M.J. (1982b) The growth and carbon economy of selection lines of *Lolium perenne* cv. S23 with 'fast' and 'slow' rates of dark respiration. II. Grown as young plants from seed. *Annals of Botany* 49, 331–339.
Robson, M.J. and Parsons, A.J. (1978) Nitrogen deficiency in small closed communities of S24 ryegrass. 1. Photosynthesis, respiration, dry matter production and partition. *Annals of Botany* 42, 1185–1197.
Robson, M.J., Ryle, G.J.A. and Woledge, J. (1988) The grass plant – its form and function. In: Jones, M.B. and Lazenby, A. (eds) *The Grass Crop: The Physiological Basis of Production*. Chapman and Hall, London, pp. 25–83.
Ryle, G.J.A. and Powell, C.E. (1975) Defoliation and regrowth in the graminaceous plant: the role of current assimilate. *Annals of Botany* 39, 297–310.
Sackville Hamilton, N.R., Matthew, C. and Lemaire, G. (1995) Self-thinning: a re-evaluation of concepts and status. *Annals of Botany* 76, 569–577.
Sheehy, J.E., Cobby, J.M. and Ryle, G.J.A. (1979) The growth of perennial ryegrass: a model. *Annals of Botany* 43, 335–354.
Silsbury, J.H. (1970) Leaf growth in pasture grasses. *Tropical Grasslands* 4, 17–36.
Simon, J.C. and Lemaire, G. (1987) Tillering and leaf area index in grasses in the vegetative phase. *Grass and Forage Science* 42, 373–380.
Sinoquet, H. and Cruz, P. (1993) Analysis of light interception and use in pure and mixed stands of *Digitaria decumbens* and *Arachis pintoï*. *Acta Oecologica* 14, 327–339.
Stoddart, J.L., Thomas, H., Lloyd, E.J. and Pollock, C.J. (1986) The use of a temperature-profiled position transducer for the study of lowtemperature growth in Graminae. *Planta* 167, 359–363.
Sutherland, W.J. and Stillman, R.A. (1988) The foraging tactics of plants. *Oikos* 52, 239–244.
Thomas, H. and Stoddart, J.L. (1980) Leaf senescence. *Annual Review of Plant Physiology* 31, 83–111.
Van Loo, E.N. (1992) Tillering, leaf expansion and growth of plants of two cultivars of perennial ryegrass grown using hydroponics at two water potentials. *Annals of Botany* 70, 511–518.

Varlet-Grancher, C., Gosse, G., Chartier, M., Sinoquet, H., Bonhomme, R. and Allirand, J.M. (1989) Mise au point: rayonnement solaire absorbé ou intercepté par un couvert végétal. *Agronomie* 9, 419–439.

Varlet-Grancher, C., Moulia, B., Sinoquet, H. and Russell, G. (1993) Spectral modification of light within plant canopies: how to quantify its effects on the architecture of the plant stand. In: Varlet-Grancher, C., Bonhomme, A. and Sinoquet, H. (eds) *Crop Structure and Microclimate*. INRA, Paris, pp. 427–451.

Wade, M.H. (1991) Factors affecting the availability of vegetative *Lolium perenne* to grazing dairy cows with special reference to sward characteristics, stocking rate and grazing method. Thèse de Doctorat, Université de Rennes.

Wade, M.H. and Baker, R.D. (1979) Defoliation in set-stocked grazing systems. *Grass and Forage Science* 34, 73–74.

Wade, M.H., Peyraud, J.L., Lemaire, G. and Cameron, E.A. (1989) The dynamics of daily area and depth of grazing and herbage intake of cows in a five day paddock system. *Proceedings of the 16th International Grassland Congress*, 1111–1112.

Westoby, M. (1984) The self-thinning rule. *Advances in Ecological Research* 14, 167–225.

Wilson, D. (1975) Variation in leaf respiration in relation to growth and photosynthesis of *Lolium*. *Annals of Applied Biology* 80, 323–338.

Wilson, D. and Cooper, J.P. (1969) Effect of light intensity and CO_2 on apparent photosynthesis and its relationship with leaf anatomy in genotypes of *Lolium perenne* L. *New Phytologist* 68, 627–644.

Wilson, D. and Jones, J.G. (1982) Effects of selection for dark respiration of mature leaves on crop yields of *Lolium perenne* cv. S23. *Annals of Botany* 49, 313–320.

Wilson, D. and Robson, M.J. (1981) Varietal improvement by selection for dark respiration rate in perennial ryegrass. In: Wright, C.E. (ed.) *Plant Physiology and Herbage Production*. British Grassland Society, Hurley, pp. 209–211.

Woledge, J. (1971) The effect of light intensity during growth on the subsequent rate of photosynthesis in leaves of tall fescue (*Festuca arundinacea* Schreb.). *Annals of Botany* 35, 311–322.

Woledge, J. (1973) The photosynthesis of ryegrass leaves grown in a simulated sward. *Annals of Applied Biology* 73, 229–237.

Woledge, J. (1978) The effect of shading during vegetative and reproductive growth on the photosynthetic capacity of leaves in a grass sward. *Annals of Botany* 42, 1085–1089.

Strategies of Plant Survival in Grazed Systems: A Functional Interpretation

D.D. Briske

Department of Rangeland Ecology and Management, Texas A&M University, College Station, TX 77843-2126, USA

INTRODUCTION

The concept of grazing resistance, within the context of grassland and pasture management, describes the relative ability of plants to survive and grow in grazed systems (Briske, 1991). Although a reasonable understanding of the major plant attributes that confer grazing resistance has been attained, the concept is largely based on empirical evidence with a limited theoretical base (e.g. Scogings, 1995). Consequently, the potential contribution of this critical concept toward understanding plant–herbivore interactions has not been fully realized.

Development of the grazing resistance concept has been constrained by two major phenomena. The first involves the development of philosophical differences between applied and theoretical scientists investigating plant–herbivore interactions, which limited interaction and information exchange (Scogings, 1995). For example, the ecological and entomological literature contains a substantially greater theoretical basis for the occurrence and evolution of grazing resistance, although it is largely based on experimentation with invertebrate herbivores (e.g. Herms and Mattson, 1992; Pollard, 1992). The second involves the conceptual approach employed to investigate plant–herbivore interactions. Initially, research emphasis focused on herbivores and herbivore population dynamics, based on the assumption that grazing had a minimal effect on individual plants and populations (Fox, 1981). It is only more recently that emphasis has shifted to investigation of the chemical and structural properties of individual plants that affect their resistance to herbivores (Fox, 1981). Investigation of unique components of grazed systems with disparate research objectives

©1996 CAB INTERNATIONAL.
The Ecology and Management of Grazing Systems (eds J. Hodgson and A.W. Illius)

and approaches has slowed development of a comprehensive theory of plant–herbivore interactions.

A functional interpretation of vegetation responses to grazing based on plant resistance mechanisms has yet to be developed. For example, documented patterns of herbivore-induced compositional changes in grasslands and pastures often cannot be anticipated or explained solely on the basis of the grazing resistance of plants which comprise the community. Progress toward a functional interpretation of grazing resistance will be required to address critical questions and issues associated with plant–herbivore interactions. Are late-successional dominants more susceptible to grazing than mid- or early-successional subordinate species? Can plants be organized into functional groups based on the relative expression and associated costs of grazing resistance mechanisms? Must threshold levels of grazing intensity be surpassed to induce shifts in species composition? Answers to these questions are required to develop a more thorough understanding of plant–herbivore interactions and to implement effective conservation and management prescriptions on grasslands and pastures.

The objective of this chapter is to extend our current perspective and understanding of grazing resistance by relating it to the much larger body of information that exists within the ecological and entomological literature. Although this approach is inherently speculative, it is intended to expedite development of a functional interpretation of the herbivore-induced patterns of vegetation change observed in grasslands throughout the world. The chapter begins with an overview of grazing resistance, proceeds with a functional interpretation of species composition change in response to grazing, addresses the resistance strategies that may potentially influence plant responses to grazing, and concludes with the presentation of several examples indicating how management may modify grazing resistance to induce desired vegetation responses.

GRAZING RESISTANCE CONCEPT

Grazing resistance describes the relative ability of plants to survive and grow in grazed plant communities. Resistant plants or species are those which are inherently less damaged than others under comparable environmental conditions (Painter, 1958). The level of resistance is often determined by qualitative and/or quantitative expression of the attribute(s) conferring resistance to the plant (e.g. specific trait approach) (Simms, 1992). The alternative approach to describing grazing resistance is to define the extent of herbivore damage or biomass removal (e.g. bioassay approach). Our limited knowledge of the relationship between the various attributes used to estimate resistance and their correlation with plant fitness represents a major constraint to our understanding of the function and evolution of grazing resistance (Pollard, 1992).

Grazing resistance can be divided into avoidance and tolerance components, based on the general mechanisms conferring resistance (Briske, 1986, 1991;

Rosenthal and Kotanen, 1994; Briske and Richards, 1995). Grazing avoidance involves mechanisms that reduce the probability and severity of grazing, while grazing tolerance consists of mechanisms that promote growth following defoliation (Fig. 2.1). Avoidance mechanisms are composed of architectural attributes, mechanical deterrents and biochemical compounds which reduce tissue accessibility and palatability. Tolerance mechanisms are composed of the availability and source of residual meristems and physiological processes capable of promoting growth following defoliation (Fig. 2.1). An evaluation of the specific mechanisms contributing to grazing avoidance and tolerance improves concept organization and clarity, but does little to enhance the functional interpretation of grazing resistance.

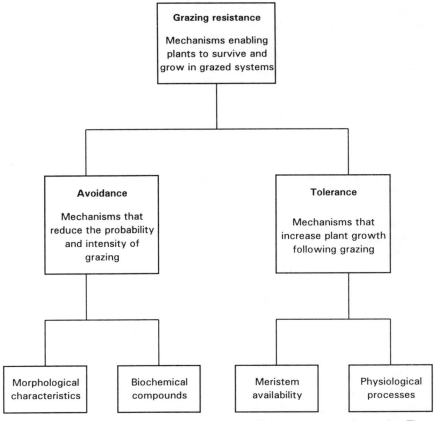

Fig. 2.1. Organization of grazing resistance into avoidance and tolerance strategies. The avoidance strategy decreases the probability and intensity of grazing while the tolerance strategy increases growth following grazing (from Briske, 1986, 1991).

The occurrence of varying degrees of grazing resistance among plants is well established by the predictable patterns of species selection by herbivores (e.g. Heady, 1964; Jarman and Sinclair, 1979; Hanley and Hanley, 1982; Heitschmidt *et al.*, 1990) and species replacement in response to long-term grazing (e.g. Dyksterhuis, 1946; Ellison, 1960; Noy-Meir *et al.*, 1989). However, in most cases, it is uncertain whether grazing-induced species replacement occurs in response to the selective utilization of plant species (i.e. avoidance mechanisms) or the occurrence of unequal growth capabilities among species following grazing (i.e. tolerance mechanisms). Although both mechanisms are known to occur, the predominant mechanism or relative combination of mechanisms remains unknown for most species combinations and plant communities (e.g. Lawrey, 1983; van der Meijden *et al.*, 1988). Selective grazing of the late-successional dominant *Schizachyrium scoparium* has been recognized as the predominant mechanism contributing to species replacement in the southern true prairie of Texas (Brown and Stuth, 1993; Anderson and Briske, 1995). Frequent and intensive grazing appears to negate the greater competitive ability and equivalent grazing tolerance of the late-successional dominant compared with associated mid-successional species. Alternatively, the greater grazing resistance displayed by *Agropyron desertorum*, in comparison with *Pseudoroegneria spicata* (syn. *Agropyron spicatum*), in the intermountain west of North America, can be attributed to different degrees of grazing tolerance (Caldwell *et al.*, 1981; Richards and Caldwell, 1985; Caldwell *et al.*, 1987). In this comparison, the potential contribution of avoidance mechanisms is minimized by comparable canopy architectures, phenology and leaf nitrogen content between the two species and the imposition of equivalent defoliation intensities.

Species grazed less severely (i.e. avoidance mechanisms), capable of growing more rapidly following defoliation (i.e. tolerance mechanisms) or possessing a combination of these two mechanisms realize a competitive advantage within a community (Caldwell *et al.*, 1981; Richards, 1984; Briske, 1991). These species, through the possession of a greater canopy area, are able to intercept greater amounts of solar energy, assimilate greater amounts of carbon and further enhance their competitive ability. Consequently, grazing-resistant species are frequently able to pre-empt resources that may have been utilized by associated grazing-sensitive species prior to grazing (Mueggler, 1972; Caldwell *et al.*, 1987; Briske and Richards, 1995).

Recognition of grazing avoidance or tolerance as the predominant resistance strategy inducing species composition shifts has important implications for grassland and pasture management. It is especially important in cases where the reduction and/or local extinction of late-successional dominants in severely grazed grasslands may have been misinterpreted as a consequence of lesser grazing tolerance, as opposed to lesser grazing avoidance, in relation to the associated mid-successional species. If the late-successional dominants are perceived as being relatively intolerant of grazing, the only viable management strategy to maintain dominance of the late-successional species is to reduce the intensity and/or frequency of grazing within the community. Alternatively, if a lesser expression of grazing avoidance by the late-successional dominants is the predominant mechanism inducing species replacement, then managerial decisions to regulate the uniformity of grazing among species may be implemented to maintain dominance of the late-successional species. Intensive grazing when the relative expression of avoidance mechanisms is lowest for subordinate, mid-successional species and highest for dominant, late-successional species can minimize, although not eliminate, selective grazing among grasses (e.g. Briske and Heitschmidt, 1991; Brown and Stuth, 1993).

FUNCTIONAL INTERPRETATION OF GRAZING RESISTANCE

Inferences drawn from herbivore-induced patterns of species replacement

Herbivore-induced shifts in species composition have been documented from native grasslands and savannas throughout the world (Ellison, 1960; Archer, 1989; Bosch, 1989; Noy-Meir *et al.*, 1989; Westoby *et al.*, 1989; Milton *et al.*, 1994). Compositional changes most frequently involve the replacement of late-successional dominants by early- or mid-successional species (Dyksterhuis, 1946; Canfield, 1957), while structural changes frequently involve the replacement of tallgrasses by mid- or shortgrasses (Arnold, 1955; Mitchley, 1988; Belsky, 1992). If species replacement continues, grassland communities may become vulnerable to an ingress of ruderals and herbaceous and woody perennials, which support fewer domestic herbivores and may eventually decrease the production potential of the site (Ellison, 1960; Archer and Smeins, 1991; Milton *et al.*, 1994).

The compositional changes induced by grazing suggest that late-successional grassland dominants possess the least developed avoidance strategy and are therefore the most palatable or preferred species within the community (e.g. Sal *et al.*, 1986). This supposition is also inherent in the most current theory of plant defence, indicating that relatively rapidly growing, competitive plants express avoidance mechanisms to a lesser extent than slower-growing, less competitive plants (Coley *et al.*, 1985). However, little conclusive experimental evidence exists to support this interpretation and a majority of the information originates from research with invertebrate herbivores. Generalist herbivores have been documented to prefer early-successional species (Cates and Orians, 1975), mid-successional species (Davidson, 1993) or late-successional species (Otte, 1975), or to possess no significant preference for plants of various successional stages (Rathcke, 1985). However, the survey of Davidson (1993) is the most comprehensive, including a broad range of plants and both vertebrate and invertebrate herbivores. The inconsistencies associated with herbivore preferences for plants within various successional stages may reside in the fact that typical generalist herbivores do not exist (Crawley, 1983, pp. 207, 344). Generalist herbivores themselves are often characteristic of particular successional communities (Maiorana, 1978; Rathcke, 1985).

Are late-successional dominants more susceptible to grazing than subordinate species?

The concept of grazing resistance is at least partially based on the assumption that the possession of resistance mechanisms represents an associated cost to the plant (Simms and Rausher, 1987, 1989; Simms, 1992; Rosenthal and Kotanen, 1994). This assumption is most easily quantified in plants which produce secondary compounds to deter herbivores. For example, seedlings of the neotropical tree *Cecropia peltata*, possessing high tannin concentrations, experienced less grazing but displayed lower growth rates in the absence of grazing than did seedlings which possessed lower tannin concentrations (Coley, 1986). Slower growth rates in seedlings possessing high tannin concentrations were attributed to the cost of tannin production and can be interpreted as the cost of grazing avoidance. Similar conclusions have been drawn from other investigations in which secondary compounds were produced in various quantities by plants to deter herbivores (e.g. Windle and Franz, 1979; Dirzo and Harper, 1982).

Mechanical deterrents may also convey large costs to plants, but the evidence is much less extensive than it is for secondary compounds (van der Meijden *et al.*, 1988; Björkman and Anderson, 1990; but see Ågren and Schemske, 1993). Although tolerance mechanisms require resources to replace biomass removed by herbivores (e.g. Wareing *et al.*, 1968; Ourry *et al.*, 1988), resource investment directly contributes to growth, rather than diverting resources from growth as in the case of avoidance mechansims (e.g. Davidson, 1993). Therefore, avoidance mechanisms can be assumed to represent a greater production cost and a greater trade-off with competitive ability than tolerance mechanisms (van der Meijden *et al.*, 1988) (Fig. 2.2). However, the nature of this relationship is largely unexplored and is far from being quantified. Trade-offs between grazing resistance and competitive ability are assumed to be greatest in resource-rich environments (Coley *et al.*, 1985; Herms and Mattson, 1992). Alternatively, grazing resistance is assumed to be more compatible with competitive ability and therefore less costly in resource-limited environments. The potential trade-off between the costs and benefits of grazing resistance occurs because competition is the predominant selective agent constraining the evolution of grazing resistance (Herms and Mattson, 1992). In other words, plants do not become completely resistant to herbivores because the cost of grazing resistance must, at some point, exceed the benefit conveyed by resistance (Pimentel, 1988).

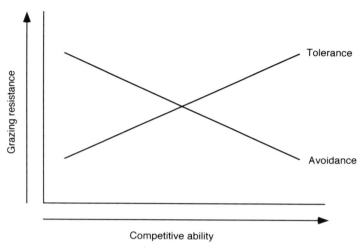

Fig. 2.2. Hypothesized relative contributions of tolerance and avoidance strategies to grazing resistance in plants possessing a range of competitive abilities. Avoidance mechanisms are assumed to represent a greater trade-off with competitive ability than tolerance mechanisms because they divert resources from plant growth. Therefore, late-successional dominants are proposed to rely on the tolerance strategy to a greater extent than early- or mid-successional species (see Fig. 2.4).

The resource availability hypothesis suggests that plants have adjusted their growth rates to match the level of resource availability in their habitats (Coley *et al.*, 1985; Coley, 1988; Yamamura and Tsuji, 1995) (Fig. 2.3). This hypothesis is based on the assumption that the realized growth rate represents the net effect of the growth reduction associated with the cost of grazing resistance and a growth increase associated with protection from herbivores. An optimal resource investment in grazing resistance is assumed to increase as the potential maximum growth rate decreases because: (i) it becomes more costly to replace resources removed by herbivores; (ii) a fixed proportion of biomass removal represents a larger fraction of net production; and (iii) a comparable reduction in growth rate resulting from the investment in plant defence represents a greater absolute growth reduction for fast-growing plants (Coley *et al.*, 1985). It can be inferred from this hypothesis that tolerance mechanisms comprise a greater proportion of herbivory resistance than avoidance mechanisms in late-successional dominants based on their relatively high maximal growth rates.

Mesic grasslands occupy relatively productive environments characterized by intermediate temperature and precipitation regimes (Walter, 1979). Consequently, many grassland dominants can be categorized as competitive strategists based on their relatively large stature, rapid growth rates and high rates of tissue turnover (Grime, 1979, p. 9). Effective competitive abilities are required

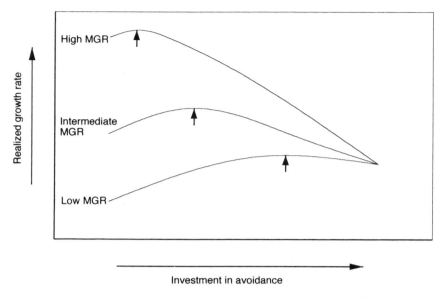

Fig. 2.3. Schematic representation of the resource availability hypothesis indicating that resource investment in avoidance mechanisms (i.e. plant defence) increases as the maximal growth rate (MGR) of plants decreases. Lines represent maximal growth rates for a broad range of plant species and arrows represent the point at which investment in avoidance mechanisms maximizes the realized growth rate (from Coley *et al.*, 1985).

for late-successional species to maintain their dominance within the community (Wilson and Keddy, 1986; Tilman and Wedin, 1991). However, a similar suite of attributes are also associated with the effective expression of herbivory tolerance (Herms and Mattson, 1992; Rosenthal and Kotanen, 1994). Competitive strategists generally possess nitrogen contents and palatability rankings intermediate to those of stress tolerators and ruderals (Cates and Orians, 1975; Southwood *et al.*, 1986; Van Arendonk and Poorter, 1994).

It is hypothesized that late-successional dominants rely on tolerance mechanisms for grazing resistance to a greater extent than early- or mid-successional species because they are closely correlated with the life history attributes of the competitor strategy (e.g. Davidson, 1993). In contrast, large investments in avoidance mechanisms are considered to divert resources from growth and potentially decrease both competitive ability and the expression of tolerance mechanisms. However, tolerance mechanisms can be suppressed prior to avoidance mechanisms because a grazing intensity can be imposed which removes biomass more rapidly than tolerance mechanisms can replace it. The suppression of tolerance mechanisms will eventually reduce the grazing resistance of dominants to a level less than that of the associated mid- or early-successional species within the community. The grazing-induced suppression of tolerance mechanisms within populations of late-successional dominants is proposed to be the predominant process contributing to species replacement within mesic grasslands. The physiological processes contributing to the decline in grazing tolerance with increasing grazing severity are reasonably well understood (Briske and Richards, 1994, 1995).

Does a threshold grazing intensity exist at which late-successional dominants are replaced by subordinate species?

The contribution of tolerance mechanisms to grazing resistance in late-successional dominants may potentially increase with increasing grazing intensities to some undefined point (Fig. 2.4). For example, compensatory photosynthesis, resource allocation, nutrient absorption and growth have been documented to increase at various defoliation intensities for numerous plant species (Briske and Richards, 1994, 1995). Although an increase in grazing tolerance may occur for subordinate species, it is assumed to occur to a lesser extent than for dominant species. However, the actual contribution of these compensatory processes to grazing tolerance has yet to be definitively established (e.g. Caldwell *et al.*, 1981; Nowak and Caldwell, 1984). Eventually, however, frequent, intensive grazing will suppress grazing tolerance mechanisms by chronic leaf removal. Although increases in the expression of avoidance mechanisms with increasing grazing intensities have also been demonstrated, they are much less frequent than for tolerance mechanisms (Abrahamson, 1975; Milewski *et al.*, 1991). Few examples of grazing-induced increases in avoidance mechanisms have been documented in grasses (see under Inducible Defences below). A notable exception is the development of decumbent canopies with frequent, intensive grazing (see under Architectural Plasticity below). However, this architectural response does not directly convey a cost associated with resource diversion from plant growth, but rather a potential loss in production (Briske and Richards, 1995). Therefore, the contribution of avoidance mechanisms is assumed to remain constant with increasing grazing intensity for both dominant and subordinate species (Fig. 2.4).

The grazing intensity at which the contribution of tolerance mechanisms is reduced to the point where grazing resistance of dominant plants equals grazing resistance of the subordinate species can be considered the resistance threshold (Fig. 2.4). If the resistance threshold is exceeded, the grazing resistance of the dominants is reduced to less than that of the subordinate species and the dominants will lose their competitive advantage and become locally extinct (O'Connor, 1991) or persist as small, scattered plants (Butler and Briske, 1988). The grazing intensity at which the contribution of tolerance mechanisms is eliminated represents the tolerance threshold (Fig. 2.4). Plant growth is entirely suppressed by

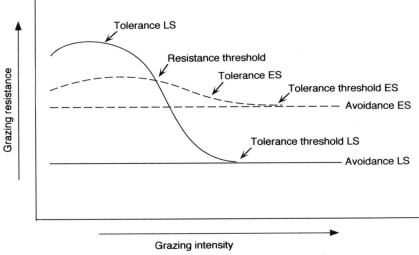

Fig. 2.4. Hypothesized functional interpretation of grazing induced species composition changes in mesic grasslands. Grazers initially suppress the expression of grazing tolerance in late-successional dominants (LS) because they can potentially remove biomass more rapidly than tolerance mechanisms can replace it. The grazing intensity at which the contribution of the tolerance strategy is reduced to the point where grazing resistance of dominant plants equals grazing resistance of the early- and mid-succession species (ES) represents the resistance threshold. Grazing resistance of the dominants is reduced to less than that of the subordinate species when the resistance threshold is exceeded and the dominants will lose their competitive advantage and decrease in abundance. The grazing intensity at which the contribution of the tolerance strategy is eliminated represents the tolerance threshold. At this grazing intensity, plant growth is suppressed by photosynthetic and/or meristematic limitations and plant death will result if this grazing intensity is maintained.

photosynthetic and/or meristematic limitations at this grazing intensity. Plant death will result if this grazing intensity is maintained.

Herbivores will begin to successively graze species with a progressively greater expression of avoidance mechanisms after the species relying on tolerance mechanisms have been reduced in abundance. Species with the most well developed avoidance strategy will be grazed to a lesser extent and their relative abundance will remain constant or increase in response to continued intensive grazing. Both the resistance and tolerance thresholds will vary among species in response to their relative expression of avoidance and tolerance strategies (see under Grazing Resistance Strategies below). In the context of this functional interpretation, communities can be envisioned as assemblages of plant species possessing a series of resistance and tolerance thresholds that vary in approximate proportion to their relative competitive abilities (Fig. 2.4).

GRAZING RESISTANCE STRATEGIES

Grazing resistance strategies are often organized into four general categories: (i) escape in time and space; (ii) confrontation (mechanical and biochemical defences); (iii) associative defence; and (iv) tolerance (Kennedy and Barbour, 1992). In this treatment, the first three categories will be presented as constitutive, spatial and temporal mechanisms of an avoidance strategy and the tolerance strategy will be divided into morphological and physiological mechanisms (Table 2.1). These strategies are capable of variously affecting the resistance and tolerance thresholds previously proposed. Resistance strategies also emphasize that frequently more than one plant attribute is involved in determining the relative grazing resistance within and among plant species (Pollard, 1992; Simms, 1992). The relative contribution of individual plant attributes to grazing resistance has been reviewed previously (Theron and Booysen, 1966; Pollard, 1992; Kennedy and Barbour, 1992; Briske and Richards, 1995). For purposes of brevity and continuity with previous sections, grazing resistance of grasses will be emphasized.

It is important to recognize that the grazing resistance strategies expressed by plants vary greatly between native and introduced and/or intensively managed systems (Kennedy and Barbour, 1992). The most obvious distinction is that resistance strategies in native systems are not influenced as greatly by anthropogenic considerations. For example, the mechanism of escape in time and space in intensively managed grazing systems is constrained by limited herbivore movement and managerially imposed grazing periods. Therefore, caution is

Table 2.1. Categories of mechanisms within the avoidance and tolerance strategies of grazing resistance. Resistance strategies imply that more than one plant attribute is frequently involved in determining the relative grazing resistance within and among plant species. These resistance strategies are capable of variously affecting the resistance and tolerance thresholds proposed in the functional interpretation (see Fig. 2.4).

Avoidance strategy	Tolerance strategy

Constitutive mechanisms
  Mechanical deterrents
  Biochemical compounds
  Defensive symbiosis
	Morphological mechanisms
	  Meristem source/number
	  Seed availability

Spatial mechanisms
  Growth form
  Architectural plasticity
  Species associations
	Physiological mechanisms
	  Compensatory process
	  Compensatory growth

Temporal mechanisms
  Inducible defences
  Asynchronous growth
  Developmental resistance	

required when attempting to draw inferences or extrapolate data concerning grazing resistance between these two diverse systems.

Avoidance strategy

Constitutive mechanisms

Constitutive mechanisms remain relatively constant in space and time (Feeny, 1976; Rhoades, 1983). These mechanisms can be considered to form the 'background' resistance level of plants (Ernest, 1994).

Mechanical

Although mechanical attributes of plants are widely assumed to function as deterrents to grazing by vertebrate herbivores, their influence is not well documented. For example, experimental evidence to verify that awns and sharpened calluses of caryopses influence livestock preferences is non-existent. Trichomes and epidermal waxes are known to be important deterrents to insect herbivores (Becerra and Ezcurra, 1986; Woodman and Fernandes, 1991), but they appear to have only limited influence on mammalian herbivores (Theron and Booysen, 1966). Similarly, the presence of vascular bundles within the Kranz leaf anatomy of C_4 grasses has been demonstrated to reduced palatability to insect herbivores (Caswell *et al.*, 1973; Heidorn and Joern, 1984), but this anatomical attribute is only known to reduce tissue digestibility in livestock (Wilson *et al.*, 1983; Akin, 1989). Leaf toughness is regarded as among the most important mechanical attributes influencing grazing by both vertebrate and invertebrate herbivores (Theron and Booysen, 1966; Coley, 1983).

Cattle have been demonstrated to graze caespitose grasses with intermediate basal areas to a greater extent than plants with larger or smaller basal areas (Norton and Johnson, 1983). Animal preference and/or accessibility to live biomass within large plants is restricted by the accumulation of culms and senescent leaf material (Norton and Johnson, 1983; Ganskopp *et al.*, 1992). When these deterrents are removed, grazing intensity becomes proportional to the canopy volume of plants. Small plants may not be grazed proportionately because herbivores are less likely to locate them within the community (Feeny, 1976; Norton and Johnson, 1983).

Biochemical compounds

Grasses are generally considered to be relatively depauperated with regard to biochemical compounds capable of deterring herbivores (McNaughton, 1983; Briske, 1991). Although this may be an accurate generalization relative to all other angiosperms, grasses are known to possess a broad array of secondary compounds

(Redak, 1987; Vicari and Bazely, 1993). Although relatively little is known about the ecological function of secondary compounds in grasses, their occurrence has been demonstrated to deter grazers. For example, alkaloids in *Phalaris arundinacea* have been shown to reduce sheep (Simons and Marten, 1971) and cattle preference (Marten *et al.*, 1976) and the occurrence of phenolics deterred grazing by voles (*Microtus* spp.) (Kendall and Sherwood, 1975; Lindroth and Batzli, 1984) and geese (Buchsbaum *et al.*, 1984). Grasses are also known to contain cyanide and cyanogenic glycosides (Georgiadis and McNaughton, 1988) as well as condensed tannins (Redak, 1987; du Toit *et al.*, 1991; Chesselet *et al.*, 1992), but the relative contribution of these compounds to grazing resistance has yet to be determined.

Defensive symbiosis

Ergot alkaloids produced by systemic fungi within infected grasses and sedges may protect plants against grazers and seed predators (Cheplick and Clay, 1988). Most endophytic fungi are distributed intercellularly throughout above-ground plant tissues including leaves, stems, flowers and seeds. Alkaloids produced in this symbiotic relationship are known to deter grazing by both insect and mammalian grazers and are potentially toxic to livestock and humans (Clay, 1990; Towers and Siegel, 1993; Latch, 1994). This symbiotic association has been interpreted as a defence mechanism in which fungi defend host plants from grazers, thereby defending their own resources (Clay, 1988; Vicari and Bazely, 1993). For example, infected *Festuca arundinacea* plants had higher survival, tiller production, biomass production and flowering than non- infected plants in a 3-year field study (Clay, 1990). Endophytes induced similar responses in *Sporobolus poiretii*, but flowering was suppressed in comparison with non-infected plants. Although the known distribution of this symbiotic interaction is currently limited (Cheplick and Clay, 1988), it has been suggested that endophytic fungi are as common among plants as are mycorrhizae (Carroll, 1988).

Spatial mechanisms

Spatial avoidance mechanisms influence the vertical and horizontal distribution of plant canopies to limit herbivore accessibility.

Growth form

Species replacement in grasslands frequently involves the replacement of tall-grasses by midgrasses and of midgrasses by shortgrasses (Arnold, 1955; Mitchley, 1988; Belsky, 1992). This pattern of grazing induced replacement is partially based upon the relative accessibility and relative amount of biomass or canopy volume removed from these various growth forms. The accessibility of plant tissues within close proximity of the soil surface varies with the anatomical structure and prehensile abilities of specific herbivores (Stobbs, 1973; Hofmann, 1988). Therefore, large-statured grasses may potentially encounter greater biomass removal than small-statured grasses when defoliated to a comparable height.

Architectural plasticity

Many species of tall- and midgrasses are capable of developing a prostrate or decumbent architecture in response to frequent, intensive defoliation. Examples include *Hordeum bulbosum* (Noy-Meir *et al.*, 1989), *Cynodon plectostachyus* (Georgiadis and McNaughton, 1988) and *Cenchrus ciliaris* (Hodgkinson *et al.*, 1989). Decumbent canopies are better able to resist grazing because less biomass is accessible to herbivores (Stobbs, 1973; Hofmann, 1988) and a greater amount of photosynthetic and meristematic tissues remain to facilitate growth following grazing (Detling and Painter, 1983; Carman and Briske, 1985).

Long-term herbivory of several perennial grass populations is known to have selected against morphotypes possessing an erect canopy architecture (Painter *et al.*, 1989; Briske and Richards, 1995). The plant morphotypes remaining in the population are characterized by a large number of small tillers with reduced leaf numbers and blade areas (Alexander and Thompson, 1982; Detling and Painter, 1983). A majority of the evidence suggests that grazing-induced selection of morphotypes is genetically based, but developmental plasticity also appears to be involved (Briske and Richards, 1994). Although the concept of biotic selection by herbivores was recognized early in this century (Gregor and Sansome, 1926; Kemp, 1937), its potential ecological significance to plant–herbivore interactions has only recently been appreciated (Detling and Painter, 1983).

It is generally assumed that grazing-induced selection primarily affects architectural attributes of plants, rather than physiological processes (Detling and Painter, 1983; Jaramillo and Detling, 1988). Consequently, the inference drawn from morphotypic selection in grazed populations is that the avoidance strategy, rather than the tolerance strategy, makes the greatest contribution to grazing resistance (Detling and Painter, 1983; Jaramillo and Detling, 1988). However, root systems of grazing morphotypes in *Pascophylum smithii* populations had a three-fold greater rate of nitrogen accumulation (Polley and Detling, 1988) and tillers of grazing morphotypes of *P. smithii* were more productive following defoliation than were populations with no history of grazing (Painter *et al.*, 1989). Therefore, it must be concluded that both the avoidance and tolerance strategies of grazing resistance can be affected by grazing-induced morphotypic selection.

Species associations

The association of less palatable species with more palatable species may also influence the relative frequency and intensity of grazing (McNaughton, 1978; Hay, 1986; Tuomi and Augner, 1993). The protection afforded to grasses growing within the canopy of low-growing shrubs is a widely observed example (Davis and Bonham, 1979; Jaksi and Fuentes, 1980). Although this phenomenon is not well documented in grasslands or pastures, there is no reason to suspect that it does not occur among herbaceous plant assemblages (McNaughton, 1978). For example, association of a palatable grass, *Bouteloua gracilis*, with an unpalatable grass, *Arisitida* spp., reduced the probability of grasshoppers grazing the more palatable species in one of two years, but increased the grazing intensity of the more palatable species when they were located by grasshoppers (Holmes and Jepson-Innes, 1989). The benefits derived from associative defences is based on the assumption that the growth reduction resulting from competition with unpalatable plants is less than the growth reduction resulting from severe grazing (Hay, 1986).

Temporal mechanisms

Temporal mechanisms influence the extent to which the expression of grazing avoidance varies throughout the growing season and with time following plant defoliation.

Inducible defences

Avoidance mechanisms which increase with increasing grazing intensity are termed inducible defences (Rhoades, 1985). The vast majority of cases involve qualitative biochemical compounds, but a few examples of increasing mechanical deterrents have also been identified (Karban and Myers, 1989). For example, browsed *Acacia* shrubs have been documented to produce a greater density of longer thorns than unbrowsed branches (Young, 1987; Milewski *et al.*, 1991). Inducible defences may have evolved to minimize the cost of avoidance when herbivores are not present (Karban and Myers, 1989). However, minimal evidence exists to demonstrate that grazing-induced responses actually increase grazing resistance and plant fitness.

Silica (SiO_2), within the epidermal cells of grasses, has been proposed as an inducible avoidance mechanism. Grasses subjected to long-term grazing often possess higher silica concentrations than do grasses with a limited history of grazing (McNaughton and Tarrants, 1983; McNaughton *et al.*, 1985; Brizuela *et al.*, 1986). However, the short-term responsiveness of silica to defoliation is less definitive (Brizuela *et al.*, 1986) and the ability of silica to deter a broad range of mammalian herbivores has not been well established (Theron and Booysen, 1966; Shewmaker *et al.*, 1989; but see Gali-Muhtasib *et al.*, 1992). Current evidence indicates that silica in grasses is a more effective deterrent to invertebrate than vertebrate herbivores (Vicari and Bazely, 1993).

Cyanide has also been documented to increase within 18 hours of defoliation in potted plants of *C. plectostachyus* relative to undefoliated plants (Georgiadis and McNaughton, 1988). Although cyanide levels in plant shoots varied within and among collection sites, it was not significantly correlated with previous grazing history, soil nitrogen content or plant water potential. The authors concluded that cyanogenesis functioned as a deterrent to grazers and suggested that cyanide levels may become sufficiently high to become lethal to cattle.

Grasshopper grazing has been demonstrated to increase phenolic concentrations by 47% in foliage of *Pascopyrum smithii* (Redak and Capinera, 1994). The increase in phenolics and associated reductions in foliage quality contributed to a decrease in consumption rate and weight gains of grasshoppers.

Asynchronous growth and development

The concept of phenological escape (*sensu* Kinsman and Platt, 1984) has been studied extensively by comparing seed production and predation among years (Evans *et al.*, 1989; Wright, 1994). Periodicity of various phenological events, including bud break and flowering, has also been interpreted as temporal avoidance in relation to herbivorous insects (Evans *et al.*, 1989; Tuomi *et al.*, 1989). Phenological development is known to influence the relative degree of grazing avoidance expressed by plant species to mammalian herbivores. For example, reproductive culm development and the accumulation of dead leaves are known to reduce animal preference for grasses (Willms *et al.*, 1980; Norton and Johnson, 1983; Ganskopp *et al.*, 1992). However, 12 years of livestock grazing did not significantly modify the periodicity of flowering or fruiting in a montane grassland in Argentina (Diaz *et al.*, 1994).

Asynchronous growth periods among plant species can also influence the frequency and intensity of grazing. Plants which are intensively grazed throughout their entire growing season are placed at a competitive disadvantage with plants that are dormant or quiescent throughout a portion of the grazing period. For example, the maintenance of green biomass by perennial grasses following senescence of annual grasses increases their susceptibility to livestock grazing in Mediterranean grasslands (I. Noy-Meir, personal communication). Alternatively, ephemeral annuals may encounter minimal grazing based on their short life history and inconspicuous nature relative to perennial plants (Feeny, 1976).

Developmental resistance

Developmental resistance refers to changes in plant tissues with increasing age that can alter the behaviour, growth and survival of herbivores that consume them (Kearsley and Whitham, 1989). Greater palatability and digestibility of younger tissues, in comparison with older mature tissues, is widely recognized (Georgiadis and McNaughton, 1990; Fryxell, 1991). Structural carbohydrates often increase and nitrogen often decreases with ageing in grass tissues (Mattson, 1980; Van Soest, 1982). Consequently, plants at various developmental stages may possess various degrees of avoidance resulting from developmental resistance. Developmental resistance may be an important mechanism contributing to the development of patch grazing (Bakker *et al.*, 1983; Kellner and Bosch, 1992) and the occurrence of 'grazing lawns' (McNaughton, 1984; Georgiadis and McNaughton, 1990).

Tolerance strategy

Morphological mechanisms

Number and source of meristems

Grass growth is dependent upon the availability and activity of intercalary meristems, apical meristems and axillary buds, but their relative contribution differs in magnitude and chronology (Briske and Richards, 1995). Leaf extension proceeds most rapidly from intercalary meristems because cell division has previously occurred within leaf primordia and growth occurs rapidly from cell expansion (Cook and Stoddart, 1953; Hyder, 1972; Briske, 1991). Leaf growth from apical meristems occurs at a slower rate because of the time required for cell differentiation prior to cell expansion (Skinner and Nelson, 1994). Leaf growth is slowest from axillary buds because of the time required for bud activation and leaf primordium differentiation. The relative contribution of these meristematic sources to plant growth varies among species and is influenced by environmental variables and stage of phenological development (Coughenour *et al.*, 1985; Olson and Richards, 1988). For example, species producing a large proportion of reproductive or culmed vegetative tillers are best suited to intermittent defoliation rather than continuous grazing (Branson, 1953; Hyder, 1972). Intermittent defoliation provides a sufficient period of time for tillers to express maximum vegetative production prior to the termination of growth following floral induction or apical meristem removal by grazers.

Differences in grazing tolerance between rhizomatous and caespitose grasses and among species within each group is largely a function of meristem availability at the time of defoliation. Rhizomatous species frequently possess large numbers of active meristems throughout the growing season and are most sensitive to defoliation when tiller densities are at a seasonal low (Hull, 1987). Synchronous tiller development increases the susceptibility of caespitose grasses to a greater loss of active shoot meristems when grazed after internode elongation (Branson, 1953; Westoby, 1980; Olson and Richards, 1988). Synchronous tiller development also contributes to wide fluctuations in grazing tolerance with the progression of phenological plant development. For example, the grazing-sensitive *P. spicata* is quite tolerant of defoliation in the early spring when culmless, because active intercalary and apical meristems are located at or near ground level. However, defoliation tolerance decreases rapidly following internode elongation (Richards and Caldwell, 1985; Busso *et al.*, 1990). Height of apical meristems among vegetative tillers of temperate perennial grasses varies appreciably (Branson, 1953; Wilman *et al.*, 1994). Seasonal variation in defoliation tolerance is much less pronounced in species with asynchronous tiller development. The most grazing tolerant caespitose grasses frequently possess asynchronous tiller development, producing a situation similar to that described for rhizomatous grasses (e.g. Hodgkinson *et al.*, 1989; Mott *et al.*, 1992).

Number and viability of seed

Seed production and development of a seed bank can be interpreted as either a tolerance or an avoidance strategy to grazing. Although the short vegetative life cycle of annuals may function as a temporal avoidance mechanism in some cases (e.g. Feeny, 1976), the production of seed and a seed bank will be treated as a mechanism to promote plant establishment and growth following severe grazing (e.g. Noble and Slatyer, 1980).

Vegetative growth (i.e. tillering) is often assumed to be a more prevalent form of reproduction than plant establishment from seed in both semiarid and mesic grasslands (e.g. Belsky, 1992). Investigations conducted in the tallgrass (Rabinowitz, 1981; Johnson and Anderson, 1986), midgrass (Kinucan and Smeins 1992) and shortgrass (Coffin and Lauenroth, 1989) prairies of North America consistently demonstrate a lack of correspondence between the existing vegetation of late-successional grassland communities and the species composition of the seed bank. Although caryopses are frequently produced, few appear to retain their viability within the soil for greater than 1 year (Thompson and Grime, 1979; Pyke, 1990). In addition, the number of seedlings recruited within established grasslands is frequently low and this occurs only sporadically during years of favourable moisture and temperature conditions (Salihi and Norton, 1987; Pyke, 1990; Jonsdottir, 1991). However, infrequent sexual reproduction may be sufficient to maintain genetic diversity and contribute to population regeneration following plant mortality associated with large-scale disturbances.

However, grassland and pasture persistence is dependent upon plant recruitment from seed in numerous systems, including both annual (Andrew and Mott, 1983; Noy-Meir *et al.*, 1989) and perennial species populations (O'Connor, 1991). The greater commitment to reproduction in annual than in perennial grasses (Adams and Wallace, 1985) is also reflected in the larger contribution of annuals to the seed bank relative to perennial grasses even though perennials may be more abundant (Major and Pyott, 1966; McIvor and Gardener, 1991). Nevertheless, the seed bank of annual grasses is relatively short lived, similar to that of perennial species (Russi *et al.*, 1992). Grazing can reduce seed production in both annual and perennial grasses by affecting resource availability for reproduction, alteration of the microenvironment for seed germination and seedling establishment, and the direct removal of flowers and seeds (O'Connor, 1991; Noy-Meir and Briske, 1996).

Physiological mechanisms

Compensatory processes

Several physiological and morphological mechanisms potentially capable of increasing plant growth following defoliation have been identified. Compensatory photosynthesis, resource allocation, nutrient absorption and shoot growth have been documented in a variety of species by numerous investigators (Briske and Richards, 1994, 1995). However, the expression of a single compensatory process does not explain the differential responses of grazing-tolerant and grazing-sensitive grasses (e.g. Caldwell *et al.*, 1981; Hodgkinson *et al.*, 1989). For example, both the grazing-tolerant and the grazing-sensitive caespitose grasses, *A. desertorum* and *P. spicata* respectively, display comparable rates of compensatory photosynthesis (Nowak and Caldwell, 1984).

Compensatory growth

The frequency of occurrence and the magnitude and significance of these compensatory processes to individual plant growth and community productivity following grazing are less clearly understood (Briske and Richards, 1995). These compensatory mechanisms may frequently prevent plant growth from being suppressed in direct proportion to the frequency and intensity of defoliation (McNaughton, 1983, 1985), but only infrequently increase total growth beyond that of undefoliated plants (Belsky, 1986; Milchunas and Lauenroth, 1993). Although the available data substantiate the occurrence of compensatory growth, they also demonstrate the specific and ill-defined conditions necessary to induce this response (Briske and Richards, 1995). For example, compensatory growth in *Ipomopsis arizonica* decreased in response to increasing competition, decreasing resource availability and grazing during the latter portions of the growing season (Maschinski and Whitham, 1989).

MANAGEMENT IMPLICATIONS

Development and implementation of successful management prescriptions require at least a partial understanding of the tolerance and avoidance strategies utilized by the major species and/or functional plant groups within grasslands and pastures. The existence of various resistance and tolerance thresholds among species presents management opportunities to influence plant utilization and growth responses among species populations within plant communities. Several examples of how management strategies may influence the resistance and tolerance thresholds of plants are presented below.

Stocking rate

Stocking rate directly influences plant utilization and the relative expression of grazing resistance among species. Late-successional dominants are frequently grazed more severely than subordinate species because they have the most rapid growth rate and a lesser expression of avoidance mechanisms (see Functional Interpretation of Grazing Resistance above). Although increasing grazing intensity can minimize selective grazing by reducing the opportunity for expression of herbivore preference, it is difficult to eliminate selective grazing entirely. For example, stocking rates required to achieve comparable utilization among two perennial grasses in a Texas grassland exceeded stocking rates required to maintain adequate forage intake by cattle (Brown and Stuth, 1993).

Season of grazing

Recognition that both avoidance and tolerance strategies vary throughout the growing season presents opportunities to differentially impact species populations in grazed systems. Grazing at a time when dominant plants express tolerance and/or avoidance strategies to the greatest extent would increase the grazing intensity at which the resistance and tolerance thresholds would be attained. Intensively grazing caespitose grasses following culm elongation represents a good example of how grazing at an inappropriate time can detrimentally influence plant growth and population persistence (e.g. Olson and Richards, 1988; Mott *et al.*, 1992).

In communities where species growth patterns are asynchronous, season of grazing will determine which species are grazed most severely during the growth period. Species grazed less intensively during the growth period or grazed when dormant will attain their resistance and tolerance thresholds at higher grazing intensities than those species grazed throughout the growing season (e.g. West *et al.*, 1979). Season of grazing is also an important consideration in species that are dependent on seed production for persistence (O'Connor, 1991; Noy-Meir and Briske, 1996).

Species of herbivore

Herbivores possess unique preferences for various plant species or plant groups within communities (Heady, 1964; Theron and Booysen, 1966). Therefore, specific herbivores will have a different impact on species populations and plant communities by utilizing various proportions of the available species. Incorporation of two or more species of herbivores will potentially utilize a greater portion of plant species within the community. A more uniform pattern of herbivore utilization between late-successional dominants and subordinate species will increase the grazing intensity at which the resistance and tolerance thresholds of the late-successional dominants will be attained.

Prescribed burning

Burning can modify the relative expression of grazing resistance among species in two distinct ways. First, fire can temporarily eliminate the expression of various avoidance mechanisms (e.g. culms and senescent leaves) and increase herbivore accessibility to younger tissues. This can minimize selective grazing among species until plants avoidance mechanisms are re-established. For example, bison were observed to more intensively graze caespitose, as opposed to rhizomatous grasses in recently burned grasslands (Vinton *et al.*, 1993; Pfeiffer and Steuter, 1994). Second, fire may modify relative species abundance and thereby alter the relative expression of avoidance and tolerance strategies among species within a community. For example, frequent burning is known to increase abundance of the late-successional dominants, *S. scoparium* and *Andropogon gerardii*, in the mixed and tallgrass prairies of North America respectively (Collins, 1987, 1992). A greater abundance of late-successional dominants would imply a greater dependence on tolerance, rather than on avoidance strategies, for population maintenance in response to grazing.

SUMMARY AND CONCLUSIONS

The concept of grazing resistance, within the context of grassland and pasture management, describes the relative ability of plants to survive and grow in grazed systems. Although a reasonable understanding of the major plant attributes that confer grazing resistance has been achieved, the concept is largely based on empirical evidence with a limited theoretical base. Consequently, the potential contribution of this critical concept towards understanding plant–herbivore interactions has not been fully realized. Progress toward a greater functional interpretation of grazing resistance will be required to address critical questions and issues associated with plant–herbivore interactions.

A functional interpretation of widely documented herbivore-induced shifts in species composition is proposed. The concept is based on the relative expression of resistance mechanisms and their associated costs in various ecological plant groups. It hypothesizes that late-successional dominants rely on the tolerance strategy to a greater extent than early- or mid-successional species because they are closely correlated with life history attributes of the competitor strategy. Although tolerance mechanisms require resources to replace biomass removed by herbivores, they do not represent a diversion of resources from growth, but rather directly contribute to plant growth. Therefore, avoidance mechanisms can be assumed to represent a greater production cost and a greater trade-off with competitive ability than can tolerance mechanisms.

However, the contribution of tolerance mechanisms to grazing resistance can be suppressed prior to those of avoidance mechanisms because grazers can potentially remove biomass more rapidly than tolerance mechanisms can replace it. The grazing intensity at which the contribution of tolerance mechanisms is reduced to the point where the grazing resistance of dominant plants equals the grazing resistance of the subordinate species represents the resistance threshold. If the resistance threshold is exceeded, the grazing resistance of the dominants is reduced to less than that of the subordinate species and the dominants will lose their competitive advantage and become locally extinct or persist only as small, scattered plants. The grazing intensity at which the expression of tolerance mechanisms is eliminated represents the tolerance threshold. Plant growth is entirely suppressed by photosynthetic and/or meristematic limitations at this grazing intensity, and plant death will result if grazing intensity is maintained. Herbivores will begin to successively graze species with progressively greater expressions of avoidance mechanisms after species relying on tolerance mechanisms have decreased in abundance.

Grazing resistance is categorized into constitutive, spatial and temporal mechanisms of avoidance strategy, while tolerance strategy is divided into morphological and physiological mechanisms. The concept of resistance strategies emphasizes that frequently more than one plant attribute is involved in determining the relative grazing resistance within and among plant species. These strategies are capable of variously affecting the resistance and tolerance thresholds proposed in the functional interpretation of grazing resistance. The existence of various resistance and tolerance thresholds among species presents management opportunities to influence plant utilization and growth responses among species populations within plant communities. Development and implementation of successful management prescriptions requires at least a partial understanding of the tolerance and avoidance strategies utilized by the major species and/or functional plant groups within grasslands and pastures.

REFERENCES

Abrahamson, W.G. (1975) Reproductive strategies in dewberries. *Ecology* 56, 721–726.

Adams, D.E. and Wallace, L.L. (1985) Nutrient and biomass allocation in five grass species in an Oklahoma tallgrass prairie. *American Midland Naturalist* 113, 170–181.

Ågren, J. and Schemske, D.W. (1993) The cost of defense against herbivores: an experimental study of trichome production in *Brassica rapa*. *American Naturalist* 141, 338–350.

Akin, D.E. (1989) Histological and physical factors affecting digestibility of forages. *Agronomy Journal* 81, 17–25.

Alexander, K.I. and Thompson, K. (1982) The effect of clipping frequency on the competitive interaction between two perennial grass species. *Oecologia* 53, 251–254.

Anderson, V.J. and Briske, D.D. (1995) Herbivore-induced species replacement in grasslands: is it driven by herbivory tolerance or avoidance? *Ecological Applications* 5(4), 1014–1024

Andrew, M.H. and Mott, J.J. (1983) Annuals with transient seedbanks: the population biology of indigenous sorghum species of tropical north-west Australia. *Australian Journal of Ecology* 8, 265–276.

Archer, S. (1989) Have southern Texas savannas been converted to woodlands in recent history? *American Naturalist* 134, 545–561.

Archer, S. and Smeins, F.E. (1991) Ecosystem-level processes. In: Heitschmidt, R.K. and Stuth, J.W. (eds) *Grazing Management: An Ecological Perspective*. Timber Press, Portland, Oregon, pp. 109–139.

Arnold, J.F. (1955) Plant life-form classification and its use in evaluating range conditions and trend. *Journal of Range Management* 8, 176–181.

Bakker, J.P., de Leeuw, J. and van Wieren, S.E. (1983) Micro-patterns in grassland vegetation created and sustained by sheep-grazing. *Vegetatio* 55, 153–161.

Becerra, J. and Ezcurra, E. (1986) Glandular hairs in the *Arbutus xalapensis* complex in relation to herbivory. *American Journal of Botany* 73, 1427–1430.

Belsky, A.J. (1986) Does herbivory benefit plants? A review of the evidence. *American Naturalist* 127, 870–892.

Belsky, A.J. (1992) Effects of grazing, competition, disturbance and fire on species composition and diversity in grassland communities. *Journal of Vegetation Science* 3, 187–200.

Björkman, C. and Anderson, D.B. (1990) Trade-off among antiherbivore defenses in a South American blackberry (*Rubus bogotensis*). *Oecologia* 85, 247–249.

Bosch, O.J.H. (1989) Degradation of the semi-arid grasslands of southern Africa. *Journal of Arid Environments* 16, 165–175.

Branson, F.A. (1953) Two new factors affecting resistance of grasses to grazing. *Journal of Range Management* 6, 165–171.

Briske, D.D. (1986) Plant response to defoliation: morphological considerations and allocation priorities. In: Joss, P.J., Lynch, P.W. and Williams, O.B. (eds) *Rangelands: A Resource Under Siege*. Cambridge University Press, Sydney, pp. 425–427.

Briske, D.D. (1991) Developmental morphology and physiology of grasses. In: Heitschmidt, R.K. and Stuth, J.W. (eds) *Grazing Management: An Ecological Perspective*. Timber Press, Portland, Oregon. pp. 85–108.

Briske, D.D. and Heitschmidt, R.K. (1991) An ecological perspective. In: Heitschmidt, R.K. and Stuth, J.W. (eds) *Grazing Management: An Ecological Perspective*. Timber Press, Portland, Oregon, pp. 11–26.

Briske, D.D. and Richards, J.H. (1994) Physiological responses of individual plants to grazing: current status and ecological significance. In: Vavra, M., Laycock, W. and Pieper, R. (eds) *Ecological Implications of Livestock Herbivory in the West*. Society for Range Management, Denver, Colorado, pp. 147–176.

Briske, D.D. and Richards, J.H. (1995) Plant responses to defoliation: a physiological, morphological, and demographic evaluation. In: Bedunah, D.J. and Sosebee, R.E. (eds) *Wildland Plants: Physiological Geology and Developmental Morphology*. Society for Range Management, Denver, Colorado, pp. 635–710.

Brizuela, M.A., Detling, J.K. and Cid, M.S. (1986) Silicon concentration of grasses growing in sites with different grazing histories. *Ecology* 67, 1098–1101.

Brown, J.R. and Stuth, J.W. (1993) How herbivory affects grazing tolerant and sensitive grasses in a central Texas grassland: integrating plant response across hierarchical levels. *Oikos* 67, 291–298.

Buchsbaum, R., Valiela, I. and Swain, T. (1984) The role of phenolic compounds and other plant constituents in feeding by Canada geese in a coastal marsh. *Oecologia* 63, 343–349.

Busso, C.A., Richards, J.H. and Chatterton, N.J. (1990) Nonstructural carbohydrates and spring regrowth of two cool-season grasses: interaction of drought and clipping. *Journal of Range Management* 43, 336–343.

Butler, J.L. and Briske, D.D. (1988) Population structure and tiller demography of the bunchgrass *Schizachyrium scoparium* in response to herbivory. *Oikos* 51, 306–312.

Caldwell, M.M., Richards, J.H., Johnson, D.A., Nowak, R.S. and Dzurec, R.S. (1981) Coping with herbivory: photosynthetic capacity and resource allocation in two semi-arid *Agropyron* bunchgrasses. *Oecologia* 50, 14–24.

Caldwell, M.M., Richards, J.H., Manwaring, J.H. and D.M. Eissenstat. (1987) Rapid shifts in phosphate acquisition show direct competition between neighbouring plants. *Nature* 327, 615–616.

Canfield, R.H. (1957) Reproduction and life span of some perennial grasses of southern Arizona. *Journal of Range Management* 10, 199–203.

Carman, J.G. and Briske, D.D. (1985) Morphologic and allozymic variation between long-term grazed and non-grazed populations of the bunchgrass *Schizachyrium scoparium* var. *frequens*. *Oecologia* 66, 332–337.

Carroll, G. (1988) Fungal endophytes in stems and leaves: from latent pathogen to mutualistic symbiont. *Ecology* 69, 2–9.

Caswell, H., Reed, F., Stephenson, S.N. and Werner, P.A. (1973) Photosynthetic pathways and selective herbivory: a hypothesis. *American Naturalist* 107, 465–480.

Cates, R.G. and Orians, G.H. (1975) Successional status and the palatability of plants to generalized herbivores. *Ecology* 56, 410–418.

Cheplick, G.P. and Clay, K. (1988) Acquired chemical defenses in grasses: the role of fungal endophytes. *Oikos* 52, 309–318.

Chesselet, P., Wolfson, M.M. and Ellis, R.P. (1992) A comparative histochemical study of plant polyphenols in southern African grasses. *Journal of the Grassland Society of South Africa* 9, 119–125.

Clay, K. (1988) Fungal endophytes of grasses: a defense mutualism between plants and fungi. *Ecology* 69, 10–16.

Clay, K. (1990) Comparative demography of three graminoids infected by systemic, clavicipitaceous fungi. *Ecology* 71, 558–570.

Coffin, D.P. and Lauenroth, W.K. (1989) Spatial and temporal variation in the seed bank of a semiarid grassland. *American Journal of Botany* 76, 53–58.

Coley, P.D. (1983) Herbivory and defensive characteristics of tree species in a lowland tropical forest. *Ecological Monographs* 53, 209–233.

Coley, P.D. (1986) Cost and benefits of defense by tannins in a neotropical tree. *Oecologia* 70, 238–241.

Coley, P.D. (1988) Effects of plant growth rate and leaf lifetime on the amount and type of anti-herbivore defense. *Oecologia* 74, 531–536.

Coley, P.D., Bryant, J.P. and Chapin III, F.S. (1985) Resource availability and plant antiherbivore defense. *Science* 230, 895–899.
Collins, S.L. (1987) Interaction of disturbances in tallgrass prairie: a field experiment. *Ecology* 68, 1243–1250.
Collins, S.L. (1992) Fire frequency and community heterogeneity in tallgrass prairie vegetation. *Ecology* 73, 2001–2006.
Cook, C.W. and Stoddart, L.A. (1953) Some growth responses of crested wheatgrass following herbage removal. *Journal of Range Management* 6, 267–270.
Coughenour, M.B., McNaughton, S.J. and Wallace, L.L. (1985) Responses of an African tall-grass (*Hyparrhenia filipendula* stapf.) to defoliation and limitations of water and nitrogen. *Oecologia* 68, 80–86.
Crawley, M.J. (1983) *Herbivory: The Dynamics of Animal–Plant Interactions.* University of California Press, Los Angeles, California.
Davidson, D.W. (1993) The effects of herbivory and granivory on terrestrial plant succession. *Oikos* 68, 23–35.
Davis, J.H. and Bonham, C.D. (1979) Interference of sand sagebrush canopy with needle and thread. *Journal of Range Management* 32, 384–386.
Detling, J.K. and Painter, E.L. (1983) Defoliation responses of western wheatgrass populations with diverse histories of prairie dog grazing. *Oecologia* 57, 65–71.
Diaz, S., Acosta, A. and Cabido, M. (1994) Grazing and the phenology of flowering and fruiting in a montane grassland in Argentina: a niche approach. *Oikos* 70, 287–295.
Dirzo, R. and Harper, J.L. (1982) Experimental studies on slug–plant interaction. IV. The performance of cyanogenic and acyanogenic morphs of *Trifolium repens* in the field. *Journal of Ecology* 70, 119–138.
du Toit, E.W., Wolfson, M.M. and Ellis, R.P. (1991) The presence of condensed tannin in the leaves of *Eulalia villosa. Journal of the Grassland Society of South Africa* 8, 74–76.
Dyksterhuis, E.J. (1946) The vegetation of the Fort Worth Prairie. *Ecological Monographs* 16, 1–28.
Ellison, L. (1960) Influence of grazing on plant succession of rangelands. *Botanical Review* 26, 1–78.
Ernest, K.A. (1994) Resistance of creosotebush to mammalian herbivory: temporal consistency and browsing-induced changes. *Ecology* 75, 1684–1692.
Evans, E.W., Smith, C.C. and Gendron, R.P. (1989) Timing of reproduction in a prairie legume: seasonal impacts of insects consuming flowers and seeds. *Oecologia* 78, 220–230.
Feeny, P. (1976) Plant apparency and chemical defense. *Recent Advances in Phytochemistry* 10, 1–40.
Fox, L.R. (1981) Defense and dynamics in plant–herbivore systems. *American Zoologist* 21, 853–864.
Fryxell, J.M. (1991) Forage quality and aggregation by large herbivores. *American Naturalist* 138, 478–498.
Gali-Muhtasib, H.U., Smith, C.C. and Higgins, J.J. (1992) The effect of silica in grasses on the feeding behavior of the prairie vole, *Microtus ochrogaster. Ecology* 73, 1724–1729.
Ganskopp, D., Angell, R. and Rose, J. (1992) Response of cattle to cured reproductive stems in caespitose grass. *Journal of Range Management* 45, 401–404.

Georgiadis, N.J. and McNaughton, S.J. (1988) Interactions between grazers and a cyanogenic grass, *Cynodon plectostachyus*. *Oikos* 51, 343–350.

Georgiadis, N.J. and McNaughton, S.J. (1990) Elemental and fibre contents of savanna grasses: variation with grazing, soil type, season and species. *Journal of Applied Ecology* 27, 623–634.

Gregor, J.W. and Sansome, F.W. (1926) Experiments on the genetics of wild populations. Part I. Grasses. *Journal of Genetics* 17, 349–364.

Grime, J.P. (1979) *Plant Strategies and Vegetation Processes*. John Wiley & Sons, New York.

Hanley, T.A. and Hanley, K.A. (1982) Food resource partitioning by sympatric ungulates on Great Basin rangeland. *Journal of Range Management* 35, 152–158.

Hay, M.E. (1986) Associational plant defenses and the maintenance of species diversity: turning competitors into accomplices. *American Naturalist* 128, 617–641.

Heady, H.F. (1964) Palatability of herbage and animal preference. *Journal of Range Management* 17, 76–82.

Heidorn, T. and Joern, A. (1984) Differential herbivory on C_3 versus C_4 grasses by the grasshopper *Ageneotettix deorum* (*Orthoptera: Acrididae*). *Oecologia* 65, 19–25.

Heitschmidt, R.K., Briske, D.D. and Price, D.L. (1990) Pattern of interspecific tiller defoliation in a mixed-grass prairie grazed by cattle. *Grass and Forage Science* 45, 215–222.

Herms, D.A. and Mattson, W.J. (1992) The dilemma of plants: to grow or defend. *Quarterly Review of Biology* 67, 283–335.

Hodgkinson, K.C., Ludlow, M.M, Mott, J.J. and Baruch, Z. (1989) Comparative responses of the savanna grasses *Cenchrus ciliaris* and *Themeda triandra* to defoliation. *Oecologia* 79, 45–52.

Hofmann, R.R. (1988) Anatomy of the gastro-intestinal tract. In: Church, D.C. (ed.) *The Ruminant Animal: Digestive Physiology and Nutrition*. Prentice-Hall Publications, Englewood Cliffs, New Jersey.

Holmes, R.D. and Jepson-Innes, K. (1989) A neighborhood analysis of herbivory in *Bouteloua gracilis*. *Ecology* 70, 971–976.

Hull, R.J. (1987) Kentucky bluegrass photosynthate partitioning following scheduled mowing. *Journal of the American Society of Horticultural Science* 112, 829–834.

Hyder, D.N. (1972) Defoliation in relation to vegetative growth. In: Youngner, B.V. and McKell, C.M. (eds) *The Biology and Utilization of Grasses*. Academic Press, New York, pp. 302–317.

Jaksi , F.M. and Fuentes, E.R. (1980) Why are native herbs in the Chilean Matorral more abundant beneath bushes: microclimate or grazing? *Journal of Ecology* 68, 665–669.

Jaramillo, V.J. and Detling, J.K. (1988) Grazing history, defoliation, and competition: Effects on shortgrass production and nitrogen accumulation. *Ecology* 69, 1599–1608.

Jarman, P.J. and Sinclair, A.R.E. (1979) Feeding strategy and the pattern of resource-partitioning in ungulates. In: Sinclair, A.R.E. and Norton-Griffiths, M. (eds) *Serengeti: Dynamics of an Ecosystem*. University of Chicago Press, Chicago, pp. 130–163.

Johnson, R.G. and Anderson, R.C. (1986) The seed bank of a tallgrass prairie in Illinois. *American Midland Naturalist* 115, 123–130.

Jonsdottir, G.A. (1991) Tiller demography in seashore populations of *Agrostis stolonifera*, *Festuca rubra* and *Poa irrigata*. *Journal of Vegetation Science* 2, 89–94.

Karban, R. and Myers, J.H. (1989) Induced plant responses to herbivory. *Annual Review of Ecology and Systematics* 20, 331–348.

Kearsley, M.J.C. and Whitham, T.G. (1989) Developmental changes in resistance to herbivory: implications for individuals and populations. *Ecology* 70, 422–434.

Kellner, K. and Bosch, O.J.H. (1992) Influence of patch formation in determining the stocking rate of southern African grasslands. *Journal of Arid Environments* 22, 99–105.

Kemp, W.B. (1937) Natural selection within plant species as exemplified in a permanent pasture. *Journal of Heredity* 28, 329–333.

Kendall, W.A. and Sherwood, R.T. (1975) Palatability of leaves of tall fescue and reed canarygrass and some of their alkaloids to meadow voles. *Agronomy Journal* 67, 667–671.

Kennedy, G.G. and Barbour, J.D. (1992) Resistance variation in natural and managed systems. In: Fritz, R.S. and Simms, E.L. (eds) *Plant Resistance to Herbivores and Pathogens*. University of Chicago Press, Chicago, pp. 13–41.

Kinsman, S. and Platt, W.J. (1984) The impact of a herbivore upon *Mirabilis hirsuta*, a fugitive prairie plant. *Oecologia* 65, 2–6.

Kinucan, R.J. and Smeins, F.E. (1992) Soil seed bank of a semiarid Texas grassland under three long-term (36 years) grazing regimes. *American Midland Naturalist* 128, 11–21.

Latch, G.C.M. (1994) Influence of *Acremonium* endophytes on perennial grass improvement. *New Zealand Journal of Agricultural Research* 37, 311–318.

Lawrey, J.D. (1983) Lichen herbivore preference: a test of two hypotheses. *American Journal of Botany* 70, 1188–1194.

Lindroth, R.L. and Batzli, G.O. (1984) Plant phenolics as chemical defenses: effects of natural phenolics on survival and growth of prairie voles (*Microtus orchrogaster*). *Journal of Chemical Ecology* 10, 229–244.

McIvor, J.G. and Gardener, C.J. (1991) Soil seed densities and emergence patterns in pastures in the seasonally dry tropics of northeastern Australia. *Australian Journal of Ecology* 16, 159–169.

McNaughton, S.J. (1978) Serengeti ungulates: feeding selectivity influences the effectiveness of plant defense guilds. *Science* 199, 806–807.

McNaughton, S.J. (1983) Compensatory plant growth as a response to herbivory. *Oikos* 40, 329–336.

McNaughton, S.J. (1984) Grazing lawns: animals in herds, plant form, and coevolution. *American Naturalist* 124, 863–886.

McNaughton, S.J. (1985) Ecology of a grazing ecosystem: the Serengeti. *Ecological Monographs* 55, 259–294.

McNaughton, S.J. and Tarrants, J.L. (1983) Grass leaf silicification: natural selection for an inducible defense against herbivores. *Proceedings of the National Academy of Sciences, USA* 80, 790–791.

McNaughton, S.J., Tarrants, J.L., McNaughton, M.M. and Davis, R.H. (1985) Silica as a defense against herbivory and a growth promotor in African grasses. *Ecology* 66, 528–535.

Maiorana, V.C. (1978) What kinds of plants do herbivores really prefer? *American Naturalist* 112, 631–635.

Major, J. and Pyott, W.T. (1966) Buried viable seeds in two California bunchgrass sites and their bearing on the definition of a flora. *Vegetatio* 13, 253–282.

Marten, G.C., Jordan, R.M. and Hovin, A.W. (1976) Biological significance of reed canarygrass alkaloids and associated palatability variation to grazing sheep and cattle. *Agronomy Journal* 68, 909–914.

Maschinski, J. and Whitham, T.G. (1989) The continuum of plant responses to herbivory – the influence of plant association, nutrient availability, and timing. *American Naturalist* 134, 1–19.

Mattson, W.J., Jr (1980) Herbivory in relation to plant nitrogen content. *Annual Review of Ecology and Systematics* 11, 119–161.

Milchunas, D.G. and Lauenroth, W.K. (1993) Quantitative effects of grazing on vegetation and soils over a global range of environments. *Ecological Monographs* 63, 327–366.

Milewski, A.V., Young, T.P. and Madden, D. (1991) Thorns as induced defenses: experimental evidence. *Oecologia* 86, 70–75.

Milton, S.J., Dean, W.R.J., du Plessis, M.A. and Siegfried, W.R. (1994) A conceptual model of arid rangeland degradation. *Bioscience* 44, 70–76.

Mitchley, J. (1988) Control of relative abundance of perennials in chalk grassland in southern England: II. Vertical canopy structure. *Journal of Ecology* 76, 341–350.

Mott, J.J., Ludlow, M.M., Richards, J.H. and Parsons, A.D. (1992) Effects of moisture supply in the dry season and subsequent defoliation on persistence of the savanna grasses *Themeda triandra*, *Heteropogon contortus* and *Panicum maximum*. *Australian Journal of Agricultural Research* 43, 241–260.

Mueggler, W.F. (1972) Influence of competition on the response of bluebunch wheatgrass to clipping. *Journal of Range Management* 25, 88–92.

Noble, I.R. and Slatyer, R.O. (1980) The use of vital attributes to predict successional changes in plant communities subject to recurrent disturbances. *Vegetatio* 43, 5–21.

Norton, B.E. and Johnson, P.S. (1983) Pattern of defoliation by cattle grazing crested wheatgrass pastures. In: Smith, J.A. and Hayes, V.W. (eds) *Proceedings of the 14th International Grasslands Congress*. Westview Press, Boulder, Colorado, pp. 462–464.

Nowak, R.S. and Caldwell, M.M. (1984) A test of compensatory photosynthesis in the field: implications for herbivory tolerance. *Oecologia* 61, 311–318.

Noy-Meir, I. and Briske, D.D. (1996) Fitness components of grazing-induced population reduction in a dominant annual, *Triticum dicoccoides* (wild wheat). *Journal of Ecology* (in press)

Noy-Meir, I., Gutman, M. and Kaplan, Y. (1989) Responses of Mediterranean grassland plants to grazing and protection. *Journal of Ecology* 77, 290–310.

O'Connor, T.G. (1991) Local extinction in perennial grasslands: a life history approach. *American Naturalist* 137, 753–773.

Olson, B.E. and Richards, J.H. (1988) Tussock regrowth after grazing: intercalary meristem and axillary bud activity of *Agropyron desertorum*. *Oikos* 51, 374–382.

Otte, D. (1975) Plant preference and plant succession: a consideration of evolution of plant preference in *Schistocerca*. *Oecologia* 51, 271–275.

Ourry, A., Boucaud, J. and Salette, J. (1988) Nitrogen mobilization from stubble and roots during re-growth of defoliated perennial ryegrass. *Journal of Experimental Botany* 39, 803–809.

Painter, E.L., Detling, J.K. and Steingraeber, D.A. (1989) Grazing history, defoliation, and frequency-dependent competition: effects on two North American grasses. *American Journal of Botany* 76, 1368–1379.

Painter, R.H. (1958) Resistance of plants to insects. *Annual Review of Entomology* 3, 267–290.

Pfeiffer, K.E. and Steuter, A.A. (1994) Preliminary response of Sandhills prairie to fire and bison grazing. *Journal of Range Management* 47, 395–397.

Pimentel, D. (1988) Herbivore population feeding pressure on plant hosts: feedback evolution and host conservation. *Oikos* 53, 289–302.

Pollard, A.J. (1992) The importance of deterrence: responses of grazing animals to plant variation. In: Fritz, R.S. and Simms, E.L. (eds) *Plant Resistance to Herbivores and Pathogens*. The University of Chicago Press, Chicago, pp. 216–239.

Polley, H.W. and Detling, J.K. (1988) Herbivory tolerance of *Agropyron smithii* populations with different grazing histories. *Oecologia* 77, 261–267.

Pyke, D.A. (1990) Comparative demography of co-occurring introduced and native tussock grasses: persistence and potential expansion. *Oecologia* 82, 537–543.

Rabinowitz, D. (1981) Buried viable seeds in a North American prairie: the resemblance of their abundance and composition to dispersing seeds. *Oikos* 36, 191–195.

Rathcke, B. (1985) Slugs as generalist herbivores: tests of three hypotheses on plant choices. *Ecology* 66, 828–836.

Redak, R.A. (1987) Forage quality: secondary chemistry of grasses. In: Capinera, J.L. (ed.) *Integrated Pest Management on Rangeland: A Short Grass Prairie Perspective*. Westview Press, Boulder, Colorado, pp. 38–55.

Redak, R.A. and Capinera, J.L. (1994) Changes in western wheatgrass foliage quality following defoliation: consequences for a graminivorous grasshopper. *Oecologia* 100, 80–88.

Rhoades, D.F. (1983) Herbivore population dynamics and plant chemistry. In: Denno, R.F. and McClure, M.S. (eds) *Variable Plants and Herbivores in Natural and Managed Systems*. Academic Press, New York, pp. 155–219.

Rhoades, D.F. (1985) Offensive–defensive interactions between herbivores and plants: their relevance in herbivore population dynamics and ecological theory. *American Naturalist* 125, 205–238.

Richards, J.H. (1984) Root growth response to defoliation in two *Agropyron* bunchgrasses: field observations with an improved root periscope. *Oecologia* 64, 21–25.

Richards, J.H. and Caldwell, M.M. (1985) Soluble carbohydrates, concurrent photosynthesis and efficiency in regrowth following defoliation: a field study with *Agropyron* species. *Journal of Applied Ecology* 22, 907–920.

Rosenthal, J.P. and Kotanen, P.M. (1994) Terrestrial plant tolerance to herbivory. *Trends in Ecology and Evolution* 9, 145–148.

Russi, L., Cocks, P.S. and Roberts, E.H. (1992) Seed bank dynamics in a Mediterranean grassland. *Journal of Applied Ecology* 29, 763–771.

Sal, A.G., De Miguel, J.M., Casado, M.A. and Pineda, F.D. (1986) Successional changes in the morphology and ecological responses of a grazed pasture ecosystem in central Spain. *Vegetatio* 67, 33–44.

Salihi, D.O. and Norton, B.E. (1987) Survival of perennial grass seedlings under intensive management of semiarid rangelands. *Journal of Applied Ecology* 24, 145–153.

Scogings, P.F. (1995) Synthesizing plant tolerance and avoidance of herbivory: do ecologists risk re-inventing the wheel? *Trends in Ecology and Evolution* 10, 81.

Shewmaker, G.E., Mayland, H.F., Rosenau, R.C. and Asay, K.H. (1989) Silicon in C-3 grasses: effects on forage quality and sheep preference. *Journal of Range Management* 42, 122–127.

Simms, E.L. (1992) Costs of plant resistance to herbivory. In: Fritz, R.S. and Simms, E.L. (eds) *Plant Resistance to Herbivory and Pathogens.* University of Chicago Press, Chicago, pp. 392–425.

Simms, E.L. and Rausher, M.D. (1987) Costs and benefits of plant resistance to herbivory. *American Naturalist* 130, 570–581.

Simms, E.L. and Rausher, M.D. (1989) The evolution of resistance to herbivory in *Ipomoea purpurea*. II. Natural selection by insects and costs of resistance. *Evolution* 43, 573–585.

Simons, A.B. and Marten, G.C. (1971) Relationship of indole alkaloids to palatability of *Phalaris arundinacea* L. *Agronomy Journal* 63, 915–919.

Skinner, R.H. and Nelson, C.J. (1994) Epidermal cell division and the coordination of leaf and tiller development. *Annals of Botany* 74, 9–15.

Southwood, T.R.E., Brown, V.K. and Reader, P.M. (1986) Leaf palatability, life expectancy and herbivore damage. *Oecologia* 70, 544–548.

Stobbs, T.H. (1973) The effect of plant structure on the intake of tropical pastures. I. Variation in the bite size of grazing cattle. *Australian Journal of Agricultural Research* 24, 809–819.

Theron, E.P. and Booysen, P. de V. (1966) Palatability in grasses. *Proceedings of the Grassland Society of South Africa* 1, 111–120.

Thompson, K. and Grime, J.P. (1979) Seasonal variation in the seed banks of herbaceous species in ten contrasting habitats. *Journal of Ecology* 67, 893–921.

Tilman, D. and Wedin, D. (1991) Plant traits and resource reduction for five grasses growing on a nitrogen gradient. *Ecology* 72, 685–700.

Towers, N.R. and Siegel, M.R. (1993) Coping with mycotoxins that constrain animal production. *Proceedings of 17th International Grassland Congress,* Palmerston North, New Zealand, pp. 1369–1375.

Tuomi, J. and Augner, M. (1993) Synergistic selection of unpalatability in plants. *Evolution* 47, 668–672.

Tuomi, J., Niemela, P., Jussila, I., Vuorisalo, T. and Jormalainen, V. (1989) Delayed budbreak: a defense response of mountain birch to early-season defoliation? *Oikos* 54, 87–91.

Van Arendonk, J.J.C.M. and Poorter, H. (1994) The chemical composition and anatomical structure of leaves of grass species differing in relative growth rate. *Plant, Cell and Environment* 17, 963–970.

Van der Meijden, E., Wijn, M. and Verkaar, H.J. (1988) Defence and regrowth, alternative plant strategies in the struggle against herbivores. *Oikos* 51, 355–363.

Van Soest, P.J. (1982) *Nutritional Ecology of the Ruminant.* O&B Books, Corvallis, Oregon.

Vicari, M. and Bazely, D.R. (1993) Do grasses fight back? The case for antiherbivore defences. *Trends in Ecology and Evolution* 8, 137–141.

Vinton, M.A., Hartnett, D.C., Finck, E.J. and Briggs, J.M. (1993) Interactive effects of fire, bison (*Bison bison*) grazing and plant community composition in tallgrass prairie. *American Midland Naturalist* 129, 10–18.

Walter, H. (1979) *Vegetation of the Earth and the Ecological Systems of the Geobiosphere,* 2nd edn. Springer-Verlag, New York.

Wareing, P.F., Khalifa, M.M. and Treharne, J.K. (1968) Rate limiting processes in photosynthesis at saturating light intensities. *Nature* 220, 453–457.

West, N.E., Rea, K.H. and Harniss, R.O. (1979) Plant demographic studies in sagebrush-grass communities of southeastern Idaho. *Ecology* 60, 376–388.

Westoby, M. (1980) Relations between genet and tiller population dynamics: survival of *Phalaris tuberosa* tillers after clipping. *Journal of Ecology* 68, 863–869.

Westoby, M., Walker, B. and Noy-Meir, I. (1989) Opportunistic management for rangelands not at equilibrium. *Journal of Range Management* 42, 266–274.

Willms, W., Bailey, A.W. and Mclean, A. (1980) Effect of burning or clipping *Agropyron spicatum* in the autumn on the spring foraging behaviour of mule deer and cattle. *Journal of Applied Ecology* 17, 69–84.

Wilman, D., Gao, Y. and Michaud, P.J. (1994) Morphology and position of the shoot apex in some temperate grasses. *Journal of Agricultural Science (Cambridge)* 122, 375–383.

Wilson, J.R., Brown, R.H. and Windham, W.R. (1983) Influence of leaf anatomy on the dry matter digestibility of C_3, C_4 and C_3/C_4 intermediate types of *Panicum* species. *Crop Science* 23, 141–146.

Wilson, S.D. and Keddy, P.A. (1986) Species competitive ability and position along a natural stress/disturbance gradient. *Ecology* 67, 1236–1242.

Windle, P.N. and Franz, E.H. (1979) The effects of insect parasitism on plant competition: greenbugs and barley. *Ecology* 60, 521–529.

Woodman, R.L. and Fernandes W. (1991) Differential mechanical defense: herbivory, evapotranspiration, and leaf hairs. *Oikos* 60, 11–19.

Wright, M.G. (1994) Unpredictable seed-set: a defence mechanism against seed-eating insects in *Protea* species (Proteaceae). *Oecologia* 99, 397–400.

Yamamura, N., and Tsuji, N. (1995) Optimal strategy of plant antiherbivore defense: implications for apparency and resource-availability theories. *Ecological Research* 10, 19–30.

Young, T.P. (1987) Increased thorn length in *Acacia drepanolobium* – an induced response to browsing. *Oecologia* 71, 436–438.

Plant Competition and Population Dynamics

J.M. Bullock

Furzebrook Research Station, NERC Institute of Terrestrial Ecology, Furzebrook Road, Wareham, Dorset BH20 5AS, UK

INTRODUCTION

The aim of this chapter is to review our knowledge of the processes influencing the dynamics and persistence of plant populations in grazed plant communities. This includes the effects on a plant population of abiotic and biotic environmental variables, including grazing, and the process of density-dependent population regulation. The majority of this chapter discusses a third process: interspecific plant competition. The evidence is reviewed for competition in the field and the modern theories concerning the determinants of species competitive ability, including the effects of grazing on competition. This will lead to a discussion of the competitive exclusion principle, the processes promoting species coexistence and the effects of grazing on diversity. Most of the examples are from temperate grasslands, but the principles discussed apply to all plant communities and all grazed systems.

PLANT POPULATION ECOLOGY

Population dynamics

The number of individual plants in a population varies over time due to changes in the number of new individuals born into the population and in the number of deaths. The dynamics of a plant population can be described quite simply by the equation

$$N_{t+1} = N_t + B - D \qquad (3.1)$$

Population size (N) changes between time t and a later time $t + 1$ as a result of births (B) and deaths (D).[1] The change over a generation is the 'net reproductive rate' of the population. This is illustrated in Fig. 3.1, which shows the dynamics of the tiller populations of two grass species over 2 years in a pasture at Little Wittenham Nature Reserve (LWNR) in southern England (Bullock et al., 1994c). The balance of births and deaths every month determined the extent of population increase or decrease and after the 2 years the cumulative effects of these births and deaths had decreased the population size of *Lolium perenne* and increased the population size of *Agrostis stolonifera*.

Another way to express population change is

$$N_{t+1} = \lambda N_t \qquad (3.2)$$

where λ is the so-called 'finite rate of population change', which is unity when births = deaths and ranges between zero and infinity. When the population is increasing $\lambda > 1$ and a diminishing population is indicated by $\lambda < 1$. Thus, the *Lolium* population in Fig. 3.1 is decreasing, $\lambda = 0.87$, and the *Agrostis* population is increasing, $\lambda = 1.38$.

There were large month-to-month changes in the population sizes of the two grasses, mostly due to seasonal climatic variation. The birth rates were positively correlated with mean daytime temperatures and death rates were higher in the winter, leading to population increases in the summer and decreases over the winter. These seasonal patterns may vary between years due to climatic changes. The July–August peaks in population sizes seen in the second year did not occur in the first year, due to a drought which decreased birth rates in both species. There were also year-to-year changes in population sizes. For instance, the *Lolium* population increased slightly in the first year ($\lambda = 1.07$) but decreased in the second year ($\lambda = 0.81$).

Another factor influencing the net reproductive rate of pasture plants may be the grazing regime. Figure 3.2 compares the population dynamics of two *Agrostis stolonifera* populations at LWNR, one grazed by sheep during the winter, the other ungrazed over winter. Grazing decreased the population size over 2 years ($\lambda = 0.42$), whereas the ungrazed population increased ($\lambda = 1.38$). This was caused by lower birth rates in the grazed population, due to the loss of biomass to the grazers (Bullock et al., 1994c).

Temporal and spatial differences in λ can be caused by a myriad of environmental factors (Harper, 1977; Silvertown and Lovett Doust, 1993), including climatic variables, such as rainfall, daylength, temperature and intensity of sunlight, and other abiotic factors, such as soil type, drainage, soil nutrient concentrations, soil pH, exposure and aspect. Biotic factors include grazers and other

---

[1] New individuals may also arrive by immigration, and emigration may also cause the loss of individuals, but, for the moment, I shall ignore these two processes.

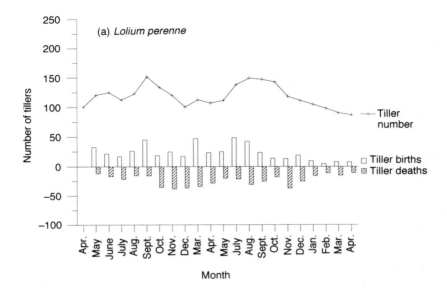

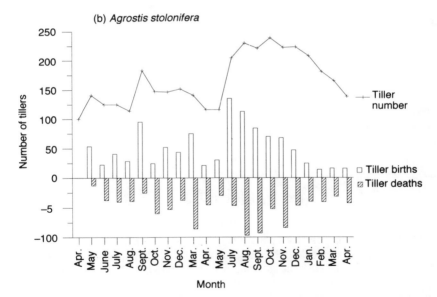

Fig. 3.1. The tiller dynamics of two grasses over 2 years in a permanent pasture at Little Wittenham Nature Reserve (LWNR) in southern England (Bullock *et al.*, 1994c). All values scaled to initial tiller population = 100.

herbivores, fungi and other microorganisms and, of course, humans. For example, the population growth of *Trifolium repens* in a Welsh pasture was found to be

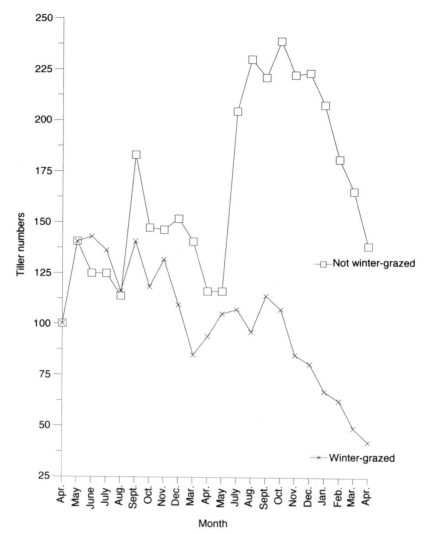

Fig. 3.2. Tiller dynamics of two populations of *Agrostis stolonifera* over 2 years at Little Wittenham Nature Reserve (LWNR) (Bullock *et al.*, 1994c). One population was grazed by sheep during the winter while the other was ungrazed. The initial tiller numbers of both populations are standardized to 100.

affected by air and soil temperature, daylight hours, rainfall and stocking rate (Sackville Hamilton and Harper, 1989). *Senecio jacobaea* in a Dutch grazed dune system showed changes in population size related to the amount of shade and summer rainfall and the intensity of attack by the lepidopteran herbivore *Tyria jacobaea* (van der Meijden *et al.*, 1985).

Grazing effects on plant population dynamics

Vertebrate grazing can affect plant population dynamics through a number of direct and indirect mechanisms. Winter grazing decreased the tiller population size of *A. stolonifera* at LWNR by decreasing tiller birth rates (Fig. 3.2). In the same experiment, winter grazing had a number of effects on the herb *Cirsium vulgare* (Fig. 3.3) (Bullock *et al.*, 1994a). Winter grazing decreased seed survival through soil disturbance, increased germination by opening gaps in the sward, increased seedling and rosette survival by decreasing shading from other plants and decreased flowering rate and seed production by decreasing average plant size. These positive and negative grazing effects combined to produce a positive effect of winter grazing on the population growth rate (Fig. 3.3).

More generally, there are a number of direct effects of grazing on plants which may subsequently affect their population dynamics. Crawley (1983) should be consulted for a more detailed discussion of these.

1. *Growth.* Removal of biomass and photosynthetic tissue may reduce plant growth. Allocations of resources to growth of certain plant parts may also change. For example, diversion of resources to induced chemical defences and the repair of damaged tissue may also reduce growth. Alternatively, grazing may stimulate plant growth by removing senescent tissue (Belsky, 1986). These factors may also change the plant's phenology and the seasonal pattern of growth.

2. *Adult mortality.* Mortality may be increased by removal of tissue and by opening wounds susceptible to attack by pathogens. Longevity may be increased in semelparous species if grazers remove flowers.

3. *Fecundity.* Flowering probability, seed number, seed size and the number and size of vegetative offspring may be decreased by reduced growth under grazing or the removal of flowers and seed heads. The onset of reproduction may also be delayed.

These effects on individual plants are generally negative, as might be expected by the removal of biomass. The effects on the net reproductive rate of the population are always negative. Even if grazing increases individual longevity by removal of flowering stems, the reduction in reproductive output will decrease the population growth rate. There has been some discussion of the idea that stimulation of individual growth by grazing may increase the population net reproductive rate (Belsky, 1986). However, only one study has provided any evidence for this hypothesis (Paige and Whitham, 1987), and this evidence is in dispute (Bergelson and Crawley, 1992). Grazing also has a number of indirect effects on population dynamics, but these tend to occur through effects on competitive interactions and are discussed later.

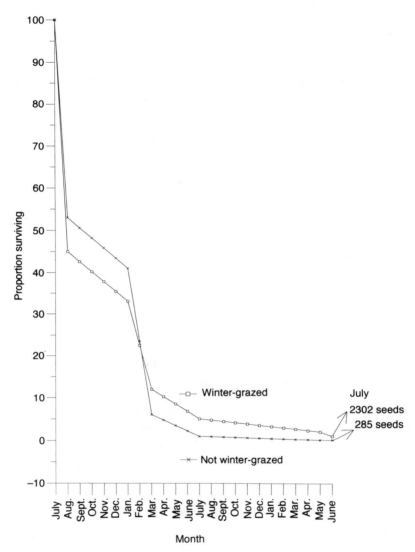

Fig. 3.3. The effects of winter grazing on the semelparous short-lived perennial herb spear thistle, *Cirsium vulgare*, at Little Wittenham Nature Reserve (LWNR) (Bullock *et al.*, 1994a). For two populations, one winter-grazed and the other not winter-grazed, the figure shows the survival of 100 seeds dispersed in July through the winter to germination in January/February, the survival of these seedlings to form rosettes in June, the survival of these rosettes to the next June and the total seed production by the surviving flowering rosettes in July.

Density dependence

The effects of the environmental factors described above are density-independent. Therefore, they cause a proportionate change in population size independent of the initial population size. Thus, grazers may kill a certain proportion of the tiller population or increased temperature may cause a certain increase in the rate of production of new tillers. However, the densities of the populations themselves affected the grass population dynamics at LWNR. Increased tiller densities decreased the per capita tiller production and increased the percentage tiller mortality. Thus, tiller numbers decreased at high densities but increased at low densities.

Although difficult to illustrate with field data, pot experiments, where other factors are held constant, can demonstrate density dependence in action. Figure 3.4 shows the results of two experiments where tillers of the pasture grass *Holcus lanatus* were planted in a range of densities, from 20 tillers m^{-2} to 40,000 tillers m^{-2} (J.M. Bullock, unpublished data). At low densities both populations had a high rate of increase over 10 weeks, but this decreased at higher densities, until the populations showed hardly any change at the highest planting densities. This effect of density on the population growth can be described by a modification of equation (3.2):

$$N_{t+1} = \lambda N_t (1 + aN_t)^{-1} \tag{3.3}$$

where λ is now the measured rate of population increase at very low densities and $(1 + aN)^{-1}$ has been shown to describe well the decline in population growth at higher densities (Watkinson, 1980, 1984). The crowding coefficient, a, is a measure of the depressive effect of each additional individual on population growth; if $a = 0$ population change is density-independent.

Equation (3.3) indicates that the final plant density (see Fig. 3.4) approaches an asymptote at high densities. This asymptote is the carrying capacity, K, of the population – the maximum density the environmental conditions allow the population to attain. In equation (3.3), $K = (\lambda - 1)/a$. As shown above, the environmental conditions determine population change over time and therefore the influence of the environmental conditions on the population can be expressed through their effects on K or, more directly, on λ and a. Thus the clipping of tillers in the pot experiments on *H. lanatus* led to decreases in λ and a, reducing K from 47,619 tillers m^{-2} to 46,364 tillers m^{-2}. However, there may not be an exact compensation for density. Thus, a constant final plant density (K) may never be reached (undercompensation), or the final density may begin to decrease at very high densities (overcompensation). Equation (3.3) is often modified to account for this:

$$N_{t+1} = \lambda N_t (1 + aN_t)^{-b} \tag{3.4}$$

where $b > 1$ indicates overcompensation and $b < 1$ indicates undercompensation for density (Watkinson, 1984; Bullock *et al.*, 1995b).

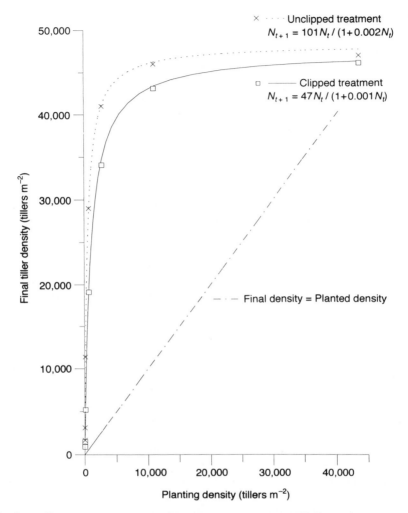

Fig. 3.4. The tiller production of pot populations of the grass *Holcus lanatus* planted at different densities. Two treatments were applied: one involved regular clipping and the other treatment was unclipped. The symbols indicate experimental results and the curved lines show the fitted relationships which are described in the equations (see equation 3.4).

Density-dependent population change occurs because the plants are competing for limiting resources. To survive and grow, plants require light, carbon dioxide (CO_2), oxygen (O_2), water, the major soil nutrients nitrogen, phosphorus, potassium, calcium, magnesium, sulphur and iron and soil trace elements such as manganese, copper and zinc. Each unit (photon, molecule or atom) of these resources can be taken up by only one plant and each unit can be used for growth or maintenance of the plant or reproduction. Each of these resources has a supply rate determined by, for instance, light intensity, rainfall, soil mineralization rates or litter decomposition rates, and the supply rate of one of the resources is such that it limits the growth and survival of plants while the other resources are still in adequate supply. If the supply rate of this limiting resource is increased, the growth and survival of plants increases to a new level where this or another resource again limits the plants. Thus, at very low densities there are plenty of resources to go round and growth and survival are not restricted by density. As densities increase, plants begin to compete for the same units of limiting resource, until at very high densities all the limiting resources are taken up by the plants and an increase in density leads to an increase in the death rate or a decrease in the average reproductive output of plants.

Density dependence can act on a variety of stages in the life cycle. In the *Lolium* and *Agrostis* populations, density affected the death rates of adult tillers and the vegetative production of tillers by adult tillers (Bullock *et al.*, 1994c). Seedling establishment was density-dependent in populations of *Cirsium vulgare* at LWNR, but later seedling survival was not (Gillman *et al.*, 1993). In a Baltic seashore meadow, density affected vegetative tiller production and flowering frequency in the grass *Festuca rubra*, but tiller death rate was unaffected (Jonsdottir, 1991). Putwain *et al.* (1968) found that vegetative reproduction was density-dependent in the perennial herb *Rumex acetosella* in a Welsh pasture. Density can affect germination, seedling growth and survival, adult growth and survival, the probability of flowering, flower numbers, seed numbers, seed viability and seed size. These effects can be classified into three types: effects on individual growth, on mortality and on reproduction. These effects are interconnected. The self-thinning rule is derived from experiments on a wide variety of plants which have shown that a plot of mean plant weight against planting density has a negative slope (often with a gradient around $-3/2$), indicating decreased growth with increased density (Weller, 1987). Plants can only increase in size if other plants are lost through mortality. Smaller plants are less likely to flower and, if they do so, will produce fewer flowers, fewer seeds and/or smaller seeds (Samson and Werk, 1986). Smaller tillers also produce fewer vegetative offspring (Bullock *et al.*, 1994d). Thus, density can affect B and D (equation 3.1) of a population both directly and indirectly.

INTERSPECIFIC COMPETITION

A definition of competition

Another factor affecting population dynamics is interspecific interactions. The population of a species co-occurs with populations of other species and it may interact with these other populations in a number of ways. The effect of each species on the net reproductive rate of the other can be positive (+), negative (−) or neutral (0). There may be no interaction between species (0,0) or there are four kinds of interaction (Begon et al., 1990; Silvertown and Lovett Doust, 1993); mutualism (+,+), parasitism (+,−), commensalism (+,0) and competition (−,−). A fifth type, amensalism (−,0), is really an extreme type of asymmetric competition (see below).

There are a number of definitions of competition (Grime, 1979; Tilman, 1988; Grace, 1990), but fundamentally it involves negative effects on the net reproductive rates of the two species or more involved. The definition of Begon et al. (1990) is a more elaborate version of this definition: 'competition is an interaction between individuals, brought about by a shared requirement for a resource in limited supply, and leading to a reduction in the survivorship, growth and/or reproduction of the competing individuals concerned'. This definition resembles the discussion above of density dependence in plant populations, and interspecific competition can be seen as a form of density dependence (also known as intraspecific competition), where an individual of the competing species has a greater or lesser effect on the target species than does an individual of the target species itself. The equation representing two-species competition is therefore a straightforward modification of equation (3.4).

$$N_{i,t+1} = \lambda_i N_{i,t}(1 + a_i(N_{i,t} + \alpha_{ij}N_{j,t}))^{-b_i} \tag{3.5}$$

where the subscript i denotes the species under study and the effect of species j on species i is represented by the equivalence coefficient α_{ij} (Firbank and Watkinson, 1990; Bullock et al., 1994b). If $\alpha_{ij} = 1$, then each individual of species j has the same effect on species i as does an individual of species i (obviously α_{ii} and $\alpha_{jj} = 1$). If $\alpha_{ij} > 1$, the effect of species j on species i is greater than that of species i on itself and, if $\alpha_{ij} < 1$, the effect of species j on species i is less than that of species i. Figure 3.5 illustrates how different competitor densities can depress the population growth of the target population.

The effect of species i on the population growth of species j is described by a complementary equation:

$$N_{j,t+1} = \lambda_j N_{j,t}(1 + a_j(N_{j,t} + \alpha_{ji}N_{i,t}))^{-b_j} \tag{3.6}$$

and the equation can be extended to describe the effects of each of n species on species i in a multispecies system, i.e. a plant community.

$$N_{i,t+1} = \lambda_i N_{i,t}(1 + a_i(N_{i,t} + \alpha_{i1}N_{1,t} + \alpha_{i2}N_{2,t} + \ldots + \alpha_{in}N_{n,t}))^{-b_i} \tag{3.7}$$

The occurrence of competition in the field

Interspecific competition has been demonstrated many times in pot and glasshouse experiments (Firbank and Watkinson, 1990; Shipley, 1994), and the controlled conditions of these experiments have allowed models such as equation

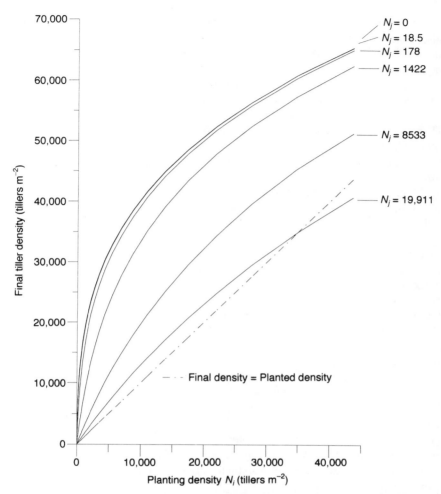

Fig. 3.5. The effects of different planting densities of a competitor genotype (indicated by N_j) on the tiller production of pot populations of a *Holcus lanatus* genotype planted at different densities (N_i) (Bullock *et al.*, 1994b). The lines are fitted from the empirical data, using equation 3.6. Final density of genotype $i = 144.6 N_{i,t}(1 + 0.02(N_i + 2.22 N_j))^{-0.69}$.

(3.5) to be fitted to the data. Field experiments, involving the manipulation of species densities, are necessary in order to determine the importance of competition in real communities, but these experiments are much less precise and model-fitting is a rarity. Hundreds of field competition experiments have been carried out in a wide variety of plant communities and they have demonstrated that increased densities of other species generally decrease the performance of individuals of the target species (see reviews by Connell, 1983; Schoener, 1983; Goldberg and Barton, 1993; Gurevitch *et al.*, 1992).

A large number of competition experiments carried out in grasslands have also generally demonstrated competition. Fowler (1981) removed species in a mown North Carolina (USA) grassland and found positive effects on the growth of a number of species. Growth of plants of *T. repens* and of *L. perenne* in two Canadian pastures was increased by clearing other vegetation (Turkington and Mehrhoff, 1990). Reader *et al.* (1994) carried out a standardized field competition experiment in 12 herb and grassland communities over three continents, Australia, USA and Europe. Seedlings of *Poa pratensis* were transplanted into plots where other plants were removed and other plots where the sward was left intact. Competition affected seedling growth in 68% of the experiments but survival was affected at only one site.

A proportion of competition experiments do not demonstrate competition and it has been suggested that the intensity of competition (i.e. the degree of the reduction in performance due to competition) will be lower in more nutrient-poor habitats due to lower productivity (Grime, 1979). Tilman (1988) suggests, in contrast, that there is no straightforward relationship between fertility and the intensity of competition. Rather, the form of competition changes as the nature of the limiting resource varies with fertility. At low productivity, competition is for a limiting soil nutrient and, at higher productivities, light becomes the limiting resource. A number of experiments have tested whether the intensity of competition varies with productivity (Wilson and Shay, 1990; Wilson and Tilman, 1991; Campbell and Grime, 1992; Turkington *et al.*, 1993) and have produced conflicting results. However, Grace (1993) pointed out that the results from these experiments have been interpreted differently. Competition effects on the *absolute* biomass of target individuals are greater in more fertile habitats, because of the higher individual biomass of plants under a higher productivity. However, when *relative* reductions in biomass are measured, to discount those effects of fertility which are not related to interspecific competition, there is no relationship with productivity in any experiment. Reader *et al.* (1994) found no relationship between productivity and intensity of competition in the *P. pratensis* experiment.

Determinants of competitive ability

If two species or more are in competition, what are the physiological, morphological and ecological characteristics which determine that one species is a better competitor than the other? This simple question is the basis of one of the most fraught discussions in modern plant ecology. A better competitor can be defined as the species which exhibits the lesser reduction in net reproductive rate in the presence of a population of the other competing species, i.e. it is able to suppress the other species more than it is itself suppressed by the other species. Goldberg (1990) pointed out that this may occur through two mechanisms, which correspond to the two processes involved in competition: (i) plants have an effect (depletion) on the abundance of a resource; and (ii) the same plants show a response (decreased performance) to these changes in the abundance of the resource. Plants can therefore suppress competitors by having a greater competitive effect by depleting the liming resource more rapidly, or by having a lesser competitive response by being able to continue growth at depleted resource levels. The two modern theories of traits determining competitive ability are related to these two competitive processes.

Grime's theory

Grime's theory (Grime, 1979; 1987; Grime *et al.*, 1988) is based on the effects of two types of factor on plant growth: (i) stress involves constraints on plant growth through low supply rates of soil nutrients, light or water, or sub- or supraoptimal temperatures; and (ii) disturbance involves the loss of plant tissue due to herbivory, pathogens, wind or frost damage, fire or other factors. The intensities of stress and disturbance combine to give three basic types of habitat (Grime suggests that no plant can survive in conditions of high stress and high disturbance). High stress and low disturbance result in a stressed environment, causing decreased growth rates. Low stress and high disturbance lead to a disturbed environment, causing increased death rates and decreased plant size. Low stress and low disturbance result in a competitive environment because plants can grow rapidly and to a large size, resulting in an intermingling of above-ground parts and of roots, leading to competition for light and soil nutrients. These are three extremes in a continuum of habitat types (represented by Grime as a triangle), and the amount of disturbance and stress in a habitat determines its position in this continuum.

Grime developed these ideas into plant strategy theory, the idea that these three habitat types favour plants with different life-history traits. In disturbed habitats ruderal species are favoured. These are plants with rapid growth and early semelparous reproduction, which allows them to take advantage of nutrients released by disturbance and to produce the maximum number of offspring before the next disturbance. The stress-tolerator plants favoured in stressed environments are slow-growing with slow rates of tissue turnover – for example, they may be evergreen. They are iteroparous, first reproduction comes late in the life cycle and they have a low reproductive output. This strategy allows them to cope with low rates of resource supply and capture, long periods of low growth and poor seedling establishment. 'Competitors' are robust and large perennial plants with high potential growth rates and thus high rates of tissue turnover. Reproduction begins early in life and the plants are iteroparous with a high investment in reproduction – by seed or vegetatively. This, Grime suggests, results in a high ability to compete for all types of resources. They have a robust and rapidly ascending and expanding canopy, which makes them effective competitors for light, and rapidly spreading root and shoot systems allow pre-emptive capture of soil resources.

This summary reveals trade-offs among the traits adapted to the different environmental types. For instance, the rapid uptake of resources for rapid growth and the concomitant need for high resource levels by competitors means that they are unable to tolerate low resource levels, while their perenniality and tall growth form gives them a low tolerance of disturbance. The slow growth rate of stress tolerators and the short lifespan of ruderals makes them both poor competitors.

Tilman's theory

Tilman (1988, 1990a) states that as a plant population increases through plant growth and births the amount of the limiting resource (R) declines. This is because the difference between the rates of resource uptake by plants and resource supply decreases. This continues until the amount of the limiting resource is such that population biomass can increase no further and the population stabilizes. R^* is the resource level at this equilibrium point. Thus, if $R = R^*$, the population is stable, if $R > R^*$, the population increases and, if $R < R^*$, the populations declines. This is obviously a restatement of the process of density-dependent population change described above (Fig. 3.4).

The different species have different values of R^* for the limiting resource. All populations increase and thus R is decreased. The species with the lowest value of R^* is able to decrease R to this level and therefore R becomes $< R^*$ for all the other species. These other species decline to extinction and the species with the lowest R^* is the superior competitor.

In different habitats with the same limiting resource, the value of R^* for a species may change due to differences in the resource supply rates and in the loss rates of resources. Higher loss rates cause an increase in R^*. This leads to a classification of habitats along two gradients: resource supply rates (e.g. fertility levels) and resource loss rates (e.g. rates of mortality or loss of plant parts due to disturbance). These are the same gradients used by Grime (1979), but Tilman (1988) says that the intensity of competition does not necessarily vary among habitats. Another possible difference among habitats is in the identity of the limiting resource (e.g. light, nitrogen, phosphorus, etc.). The traits conferring a low R^*, and thus high competitive ability, therefore vary among habitats.

Whereas Grime's (1979) theory of plant traits is based on an intuitive and correlative approach, Tilman (1990a) used a modelling approach to determine the effects of a number of physiological traits on R^* in an environment where a soil nutrient is limiting. Competitive ability is increased by a lower maximum rate of nutrient uptake, a lower rate of loss of the nutrient to senescence (low tissue turnover or minimizing herbivory), a lower tissue nutrient concentration, a greater ability to recycle nutrients from senescing tissues, and high conversion and transfer efficiencies. Growth form affects R^* as well. Another model by Tilman (1990b) showed that fertile habitats with low disturbance benefit a high allocation to stem (competition for light), high fertility with high disturbance favours high allocation to leaves, and a high allocation to roots is favoured in infertile habitats (nutrient competition). Therefore, in contrast to Grime (1979), Tilman (1988) suggests that abilities to compete for different resources are negatively correlated as they require different patterns of resource allocation and different physiological traits.

Comparison of the theories

Goldberg (1990) pointed out that the competitive ability discussed by Grime is in terms of a large competitive effect – the ability to capture units of resource pre-emptively. This reveals that Grime's definition of competitive ability is based on a rather restricted definition of competition. In Grime (1979) he defines competition as 'the activity whereby certain types of plant achieve high rates of resource capture in productive, relatively undisturbed environments'. It was shown above that there is little evidence that the intensity of competition is affected by productivity (i.e. different degrees of stress). In habitats where soil nutrients or other resources are relatively low, stress tolerators perform better because some other species cannot grow in these conditions (Grime's interpretation), but also because the stress tolerators have the competitive advantage through their ability to conserve and utilize nutrients efficiently. The competitive ability of stress tolerators therefore arises through a low competitive response (Goldberg, 1990). Disturbance can decrease the intensity of competition if plants are killed (see under Gap dynamics below), but there is no evidence that it does so if plants are simply damaged (e.g. by grazers). In fact, the need for soil nutrients for regrowth could increase the intensity of competition (as in Tilman's theory).

Therefore Grime's strategy theory is really defining those traits that confer competitive advantage in particular types of habitat, and the nomenclature of 'competitor' species is misleading. In productive, undisturbed habitats where light is the limiting resource, 'competitor' types are the better competitors. In other habitats, different traits increase competitive ability. Seen in this way, Grime provides a very useful list of traits which he suggests confer advantage (including competitive ability) in different habitat types.

Tilman describes competitive ability in terms of competitive response – the ability to tolerate low resource levels (Goldberg, 1990). The traits Tilman describes as conferring a competitive advantage when the limiting resource is a soil nutrient are the same as those of Grime's (1979) stress tolerators. For this reason the differences between the theories of Grime and Tilman are partly semantic (see also Grace, 1990). However, the differences are less easy to resolve if competition is for light. Superior competitors for light according to Tilman's theory are those species able to tolerate the lowest light levels, whereas according to Grime they are those species able to pre-emptively capture light, i.e. the taller species. In a study of competition for light among seven species, Goldberg and Landa (1991) found that competitive response and effect were determined by different traits. The ability to capture light pre-emptively was greater for species with a high growth rate and a more erect growth form, whereas the ability to tolerate shading was correlated with a low maximum potential growth rate. However, it is likely that in competition for light, competitive ability is related to a large competitive effect. If two plants are competing, the taller is able to capture almost all the light and this disproportionate capture of resources is termed asymmetric competition (Weiner, 1990). However, when competition is for a limiting soil nutrient, even if one plant with a more developed root system or a greater rate of uptake per unit root area captures more of the nutrient, the partition of resources is proportionately more equal (symmetric competition). Therefore toleration of low resource levels may be good strategy for nutrient competition, but pre-emptive capture of resources may be more effective for light competition (see also Goldberg, 1990). Thus, Mitchley and Grubb (1986) in a study of 14 chalk grassland species, Gaudet and Keddy (1988) in a study of 44 wetland species and Keddy and Shipley (1989) in a study of seven wetland species all correlated competitive ability with the height attained by a species in monoculture. Tilman (1988, 1990b) seems to recognize this in his models of plant competition where competitive ability for light is determined by the ability to overtop other plants rather than by an R^* for light. Tilman (1990a) also admits that it would be difficult to measure the light R^* of a species.

Therefore, there are similarities between Grime's and Tilman's theories which are obscured by semantic differences, but if we accept that competition occurs in most environments, including disturbed and nutrient-poor conditions, then particular traits will increase competitive ability in certain environments. There is room for improvement in both theories and, perhaps, for a synthesis. Tilman's theory seems to lose its generality if competition for light is considered, but Grime's theory would be strengthened by a more rigorous definition of competition and a framework, e.g. involving modelling, which would allow the production of testable hypotheses. R^* for a species can be measured very easily. In the USA, Tilman and Wedin (1991) and Wedin and Tilman (1993) studied four species from an old field where soil nitrogen was the limiting factor. The species were grown in monocultures in field plots, and soil nitrogen concentrations were measured when the monocultures had reached an equilibrium biomass. These concentrations were used as a measure of R^* for nitrogen and were used to successfully predict the outcomes of two-species competition experiments according to Tilman's theory. Grime's theory has yet to be tested in such a precise manner.

Competitive ability under grazing

It is widely thought that grazing may change the outcome of competition between two or more species (Crawley, 1983; Louda *et al.*, 1990a). This change may occur through two mechanisms, one indirect and the other direct. The indirect mechanism involves grazing causing a decline in the population of one of the species (i.e. grazing causes λ_i to become < 1 (equation 3.5)). Thus, although each individual of species *i* has the same competitive effect on individuals of other species as in the absence of grazing, there are fewer individuals of species *i* and the other species experience competitive release from species *i*. This is often seen where grazers kill plants, allowing seedlings of other species to establish in the resulting gaps (e.g. Louda *et al.*, 1990b; Silvertown *et al.*, 1992; Bullock *et al.*, 1994a). Alternatively, grazing has a direct effect on the competitive interaction and the relative competitive abilities of species are actually changed, i.e. the values of α (equation 3.5) are different. These changes reflect the differences among species in the traits that confer competitive advantage under particular conditions.

The indirect and direct effects are often conflated and manipulative experiments are needed to separate the two effects of grazing (e.g. Watt and Hagger, 1980; Alexander and Thompson, 1982; Berendse, 1985; Silvertown *et al.*, 1994). Bullock *et al.* (1994d) carried out competition experiments between pairs of genotypes of the grass *H. lanatus* and found clipping effects on the values λ, *a*, *b* and α (equation 3.5) of both genotypes, i.e. both indirect and direct effects on competition.

Changes in outcome of competition under grazing therefore occur through differential effects of grazing on the competing species. D.D. Briske (Chapter 2, this volume) discusses the causes of these differences and I shall not repeat his detail. There may be selective grazing through active preference or avoidance, or through passive selection, e.g. of taller-growing species. Once grazed, species may show different tolerances to grazing (e.g. the basal meristems of grasses give them an advantage over many herbs with aerial meristems) or different compensatory abilities (e.g. different rates of regrowth). There are many examples of the consequences of such differences (see also D.D. Briske, Chapter 2, this volume), although, as stated above, these do not separate the direct and indirect effects of grazing on the competitive outcome. Three grasses in a southern English acid grassland had different palatabilities to rabbits, *Anthoxanthum odoratum* > *Agrostis capillaris* > *Festuca rubra*, which explained the different responses of the species to exclosure of the rabbits. The cover of *Anthoxanthum* increased while *Agrostis* and *Festuca* decreased (Crawley, 1990). In a Texan grassland, grazing decreased the cover of the dominant grass *Schizachyrium scoparium* and increased the cover of the other dominant *Paspalum plicatulum*, because, although *Schizachyrium* was more tolerant of grazing, it was preferentially grazed (Brown and Stuth, 1994). Noy-Meir *et al.* (1989) found that cattle exclosures in an arid grassland in Israel resulted in decreases in the cover of prostrate and rosette species while the erect species increased. Sheep exclosures on some Welsh hill grasslands resulted in declines in the abundances of the low-growing species (Hill *et al.*, 1992). Light grazing of a southern English grass/clover ley allowed *T. repens* to increase at the expense of *L. perenne* because the more erect *Lolium* plants lost more biomass to grazers (Parsons *et al.*, 1991).

Considering direct effects of grazing on competition, if a weaker competitor is favoured by grazing, the differences in competitive abilities will be decreased or even reversed. However, if the stronger competitor is favoured by grazing, the difference in competitive abilities will increase. For example, Welch (1986) found that the grass *Nardus stricta* increased in abundance over time in a number of ungrazed British upland grasslands, but it is avoided by livestock and therefore showed greater increases in abundance in grazed sites. However, switches in competitive ability under grazing seem more common and this is because the ability to avoid or tolerate grazing may result in trade-offs in competitive ability under ungrazed conditions (see D.D. Briske, Chapter 2, this volume). Allocation to antiherbivore defences may decrease the allocation to growth, and Grime (1979) suggests that his 'competitor' species maximize their growth rates, and thus their competitive abilities, at the cost of low investment in antiherbivore chemicals. The diversion of nutrients and the development of biochemical pathways to produce defensive chemicals may also reduce physiological efficiencies and increase loss rates, which will increase R^* for nutrient competition under Tilman's (1988) theory. An erect growth form, increasing ability to compete for light, also increases the probability of being grazed. Finally, a robust growth form with a high investment in stem involves a high investment of resources and may not be the most efficient growth form to cope with repeated defoliation. For these reasons, a model by Tilman (1990b) showed that allocation to leaves rather than stem improved competitive ability in a disturbed (e.g. grazed) environment.

As well as differential effects of grazing on species performance, changes in relative competitive ability under grazing may occur through effects on the nature of the competition. If the canopy is reduced in height and the architecture is simplified by grazing, light may become less limiting, whereas the loss and damage of plant parts may increase the demand for water and soil nutrients. Therefore grazing may change the limiting resource from light to, say, soil nitrogen. The relative competitive abilities of different species may change depending on the resource competed for (Tilman, 1988), and thus the outcome of the competition may be changed. Milchunas *et al.* (1992) have also suggested that grazing increases competition for water and soil nutrients. However, there have been no tests of this possibility.

COMPETITION AND COMMUNITY STRUCTURE

Competitive exclusion

The previous sections have shown that competition occurs in plant communities and that species differ in their competitive abilities. Several studies have related species competitive abilities to their abundances in the field. Mitchley and Grubb (1986) found that the rank abundances of 14 English chalk grassland species in the field were positively correlated with their performances in competition against a target species in pot experiments. At LWNR, Silvertown *et al.* (1994) found that relative competitive abilities of four grass species showed some relationship with their relative abundances in the field. Tilman's experiments to test his theory involved two-species competition experiments for all possible combinations of four species, and their abundances in the field were correlated with their competitive abilities and with their R* values for nitrogen (Tilman and Wedin, 1991; Wedin and Tilman, 1993). In contrast, Duralia and Reader (1993) found that the competitive abilities of three USA tallgrass prairie species were not related to their abundances in the field.

These results seem to indicate that competition structured these communities, and that we can predict community structure and species abundances by studying the competitive interactions of the species (although Duralia and Reader's (1993) results indicate that one cannot generalize). However, Tilman's competition experiments described above all predicted 'competitive exclusion' – that one species in the pair would always drive the other to extinction. Despite this, all four species coexisted in the field site (Tilman and Wedin, 1991; Wedin and Tilman, 1993).

In fact, competitive exclusion is the usual prediction of two-species competition experiments (e.g. Law and Watkinson, 1987). This can be explained by the 'competitive exclusion principle'. Also known as Gause's principle, this states that species with the same niche cannot coexist. Niche can be defined as the ranges of all environmental factors that influence growth, survival and reproduction within which a species can maintain a population. These generalities are rather vague but they are better defined for the special case of plants. The plant species in a grassland community generally have the same niche (see also Silvertown and Law, 1987). They all require light, water, nitrogen, phosphorus, potassium, calcium, magnesium, etc. (see above) and they all usually grow at the soil–air interface. Thus they are competing for the same units of resource from the same spatial dimension. Contrast this to animal species (e.g. sheep vs. cows) which, even when they are feeding on the same pool of species at the same place and time, will often have different niches because they feed on a different selection of species (e.g. Connolly, 1987). Because plant species often have the same niche and are competing for the same units of resource, and often one species is the better competitor (see above), that species will exclude the other.

Competitive exclusion for two-species interactions can be expressed mathematically in terms of the balance of interspecific and intraspecific effects, described by α_{ij}, α_{ji}, K_i and K_j (see Silvertown and Lovett Doust, 1993, for a detailed description) – that is, the effects of the density of each species on the density of the other species, and the maximum density attainable by each species in the absence of the other. Competitive exclusion of species j by species i is predicted if two criteria are satisfied: the inhibitory effects of species i on j (interspecific effects) are greater than the inhibitory effects of j on itself (intraspecific effects) (expressed in shorthand as 'inter > intra'); but the reverse is true for species i (intra > inter). If inter > intra for both species, there will be competitive exclusion, but the identity of the winner depends on initial densities. If intraspecific inhibition is greater than interspecific effects for both species, then the species will coexist.

Shipley (1994) reviewed 299 two-species competition experiments. Competitive exclusion was predicted in 280 of these. In 227 inter > intra for one species, but the second species showed the opposite effect. In 53 inter > intra for both species. Only in 19 was coexistence predicted with intra > inter for both species. Gurevitch *et al.* (1992) and Goldberg and Barton (1992) found no consistent patterns as to the relative strengths of intraspecific and interspecific effects in reviews of field competition experiments. This is consistent with the hypothesis that, for a majority of two-species interactions, inter > intra for one species and intra > inter for the other, and that competitive exclusion will occur.

Coexistence

If exclusion is the usual outcome of competition, how do species coexist in the field? In Tilman's field site, not only did his four study species coexist (Wedin and Tilman, 1993), but so did over 100 other plant species (Tilman, 1994). Despite large differences in competitive abilities, tens of species coexist in temperature pastures such as chalk grasslands (Mitchley and Grubb, 1986) or improved pastures (Silvertown *et al.*, 1994). Obviously, some species are excluded by competition. For example. *C. vulgare* was excluded through competition with the dominant grasses in the least intensively grazed paddocks at Little Wittenham Nature Reserve (Bullock *et al.*, 1994a).

There is a large literature concerning the processes allowing species coexistence, and this is summarized below. Many or all of these processes may be occurring in a community, and their occurrence may depend on which species are interacting and may also vary spatially and temporally for each species interaction. These processes fall into two types. Equilibrium processes are those where the interspecific interaction does not lead to competitive exclusion, i.e. the interaction is such that there is species coexistence at equilibrium. Non-equilibrium processes are those where the system tends to competitive exclusion, but these processes perturb the system and keep it from equilibrium. There is another process which does not fall into either category: competitive equivalence.

Competitive equivalence

Van der Maarel and Sykes (1993) studies a grassland on the island of Öland in Sweden and suggested that the species were competitively equivalent (i.e. inter = intra for both species). In their 'carousel' model they state that such equivalence allows the coexistence of species. It is difficult to accept this model as having general applicability. Species generally differ in their competitive abilities (see above). More importantly, equivalence would lead to random fluctuations in the populations of each species and, as for alleles in the analogous process of genetic drift, species would go extinct and the equilibrium state of this system is a monoculture.

Equilibrium processes

Mutualism

If mutualism (+,+) or commensalism (+,0) is occurring, rather than competition, the presence of other species enhances the persistence of a species in the system. Therefore, the equilibrium state is for coexistence of these species. Goldberg and Barton (1992) found indications of some mutualistic interactions in 33% of the field experiments they reviewed on plant interactions. Plant–plant mutualisms have been little studied (Bronstein, 1994), but one example may be nurse plants, which protect other plants (especially seedlings), e.g. from temperature extremes or wind damage or by maintaining a high humidity.

Niche separation

Niche separation will occur when species use different resources or show spatial or temporal separation in resource use, even at the highest densities of both species. This is expressed mathematically when α_{ij}, $\alpha_{ji} < 1$, and may allow coexistence for certain values of K_i and K_j (e.g. if they are similar). Classically, it has been thought that species coexist through many subtle niche differences (MacArthur and Levins, 1967). Tilman (1988) showed theoretically that, if different species are limited by different resources, this is a form of niche separation and as many species may coexist as there are limiting resources. He showed this occurring for diatom species (Tilman, 1982), but it has not been tested explicitly for higher plants. In addition, there are very few potential limiting resources and this theory could not explain the many coexisting species found in plant communities.

Although niche separation has been suggested by some authors (e.g. Turkington and Harper, 1979; Fowler and Antonovics, 1981), there is little evidence for it in grassland communities (see Silvertown and Law, 1987; Mahdi et al., 1989). However, Berendse (1981, 1982) found that intraspecific competitive effects were greater than interspecific effects for both species in a competition experiment between *Plantago lanceolata* and *A. odoratum*. This was because *Anthoxanthum* rooted at a shallower depth than *Plantago*.

Non-equilibrium processes

Spatial dynamics

The sessile nature of higher plants and their tendency for spatially limited seed dispersal and clonal spread means that most plant populations exhibit spatial patterning in their dynamics. This influences competitive interactions and coexistence through two related processes: aggregation and gap dynamics. As discussed here, these processes are generated purely by the intrinsic properties of the plant populations. Spatial patterning by extrinsic environmental factors occurs as well and is discussed later.

Aggregation   Patchy species distributions are often seen in grasslands (Turkington and Harper, 1979; Mahdi and Law, 1987; Thorhallsdottir, 1990). They can be caused by clonal growth or restricted seed dispersal, both of which mean that new plants are likely to arise closer to conspecifics than expected by chance. Hassell *et al.* (1994) and Molofsky (1994) both demonstrated this process in models of spatial dynamics. Aggregation will mean that most, or all, of the neighbours of a plant will be conspecifics and there are few interspecific contacts. Therefore, fewer plants in the population experience interspecific competition. For instance, Bergelson (1990) found higher seedling densities of *Capsella bursa-pastoris* or *Senecio vulgaris* when grown with *Poa annua* planted in a clumped pattern than with *Poa* planted randomly. Shmida and Ellner (1984) modelled the population dynamics of species with different competitive abilities and showed that patchiness in species distributions due to limited seed dispersal reduced the effects of competition. This process will slow competitive exclusion, but because one species is always the superior competitor it will eventually exclude the other species, no matter how clumped they are (Silvertown *et al.*, 1994).

Gap dynamics   Dying plants leave gaps, which provide an area in the community which, for a short period, is free from the competition experienced by plants in the closed sward. This period of freedom from competition may be especially important for the poorer competitors in the community and will allow species coexistence if the ability of a species to colonize a gap (e.g. by clonal growth or seed dispersal) forms a trade-off against competitive ability. For example, in a Welsh pasture Schmid and Harper (1985) found that *Prunella vulgaris* had long stolons and was able to invade gaps rapidly. *Bellis perennis* had a clumped growth form, which allowed it to outcompete *Prunella* at high densities, but it invaded gaps only slowly.

Tilman (1994) found a negative correlation between colonization ability (years to colonize an old field) and competitive ability (R^* for nitrogen) among species at his study site, and developed a model of a multispecies system incorporating this trade-off. This predicted multiple species coexistence through gap dynamics. The better colonizers were excluded in competition with the superior competitors but were able to escape competition for a period of time by invading the gaps left by dying plants. The superior competitors then gradually encroached and caused local exclusion of the colonizers.

Spatial heterogeneity

The outcome of competition can be changed by a number of environmental variables, such as grazing (Bullock *et al.*, 1994b), watering and fertilizer treatments (Fowler, 1982) or the presence of mycorrhizae (Allen and Allen, 1990; Hetrick *et al.*, 1994) and other fungi (Clay, 1990). Patchiness can occur in a number of these variables, including grazing, soil nutrients, moisture levels, microorganism abundances, etc. (Gibson, 1988; Mahdi *et al.*, 1989; Putnam *et al.*, 1991; Jastow and Miller, 1993). Therefore, the competitive abilities of species may vary between patches such that no one species is competitively dominant over the whole community. The Shmida and Ellner (1984) model (see above) also demonstrated coexistence if there was spatial variation in environmental conditions and thus in species competitive abilities.

Gaps may be caused by disturbance as well as the intrinsic dynamics of the populations (see above). Large herbivores create gaps by grazing and trampling (Silvertown and Smith, 1988; Bullock *et al.*, 1995a) and by dung and urine deposition (Day and Detling, 1990; Putnam *et al.*, 1991). Gaps can also be caused by digging by fossorial mammals (Platt, 1975; Hobbs and Mooney, 1991), anthills (King, 1977) and a myriad of other factors (Pickett and White, 1985). The 'intermediate disturbance hypothesis' develops the theory stated under Gap dynamics above, that disturbances and trade-offs between competitive ability and colonization ability allow coexistence (see Petraitis *et al.*, 1989, for a review), and this is the basis of many studies of coexistence in grasslands (Martinsen *et al.*, 1990; O'Connor, 1991; Aguilera and Lauenroth, 1993).

Temporal variation

In the same way that spatial variation in competitive ability can promote coexistence, temporal variation in environmental factors may have the same effect. Fowler (1982) found that the outcome of competition between pairs of grassland species was affected by the time of year the experiment was carried out. Fowler (1986) showed that the relative abundances of the dominant species in a Texas grassland varied between years in relation to the rainfall. Seasonal or yearly variations in stocking rate of grazers may cause temporal variation in competition (e.g. Silvertown *et al.*, 1994). Grazers also show temporal changes in their patterns of species preference, which is influenced by their recent diet, their physiological state and the time of day (Thornley *et al.*, 1994).

Spatiotemporal variation

Many of the extrinsic factors discussed above vary both spatially and temporally and as a result many plant populations are seen to show temporal variation in their spatial distribution patterns (e.g. van der Meijden *et al.*, 1985), in what Grubb (1986) calls 'shifting clouds of abundance'. These two processes therefore act in concert to promote coexistence, as the competitive ability of species varies in a spatiotemporal fashion. For example, grazing is spatially heterogeneous and grasslands exist as mosaics of recently grazed and ungrazed patches. As grazers pass over the sward the pattern of the mosaic changes. Mitchley (1988) suggested that prostrate species are favoured at the time of grazing but in the recently ungrazed patches the erect species compete better for light.

Lottery models simulate the population dynamics of, and interactions among, species with different competitive abilities (Shmida and Ellner, 1984; Chesson, 1991, 1994; Pacala and Tilman, 1994). These differences cause a tendency towards competitive exclusion. However, the species have different life histories and responses to environmental variables, which cause their relative birth and death rates to vary differently over time in response to temporal (seasonal or yearly) environmental fluctuations, over space in response to spatial environmental heterogeneity, or over space and time in response to spatiotemporal heterogeneity. All forms of heterogeneity have been modelled and all predict coexistence as long as the species are sufficiently dissimilar to allow adequate partitioning of the environmental heterogeneity.

Grubb (1977, 1986) developed an intuitive precursor to these lottery models with his concept of the 'regeneration niche'. Although not a niche in the sense developed in this chapter, Grubb (1977) states that species have different responses to the local environmental conditions and different abilities to establish new seedlings, and therefore temporal and spatial heterogeneity allows species coexistence.

Frequency dependence

Coexistence is promoted if forces act to selectively decrease the abundances of commoner species and increase the abundances of rarer species. This form of frequency dependence, known as apostatic selection (i.e. selection against the commonest species), must involve a switching of negative influences against any species that tends to dominate the system. Apostatic herbivory is seen in some invertebrates (Dirzo, 1984; Cottam, 1985) but has not been detected in vertebrate grazers (Lundberg *et al.*, 1990). However, the tendency for vertebrates to graze taller plants (Illius *et al.*, 1992) may lead to apostatic grazing against the successful species which are growing well.

Mass effects

At the start of this chapter immigration and emigration were relegated to a footnote. However, the dispersal of seeds among communities may have a very important role in maintaining populations. According to the 'mass effect', species which cannot maintain a population in a community without immigration (for instance, due to competitive exclusion) may be supplemented by inward dispersal of seeds. Shmida and Ellner (1984) demonstrated that such mass effects maintained the populations of a number of rare species in a semidesert community.

CONCLUSION – GRAZING AND DIVERSITY

Much agricultural research on the responses of grassland communities to grazing is concerned with the changes in species' relative abundances caused by the grazing regime. These changes can be caused by direct or indirect responses to grazing. Direct effects on demography (survival, fecundity, etc.) are caused by biomass removal, trampling, etc. by grazers. Indirect effects can arise through alterations in plant–plant interactions. Intraspecific density responses may be changed and the outcome of interspecific interactions may be altered. Competition is probably the major plant–plant interaction in most communities, and has been detected consistently in grasslands. The outcome of competition is determined by species characteristics and the environment. The traits (sometimes coadapted, resulting in 'plant strategies') of the different species and the particular environment interact to affect the demographic performances and competitive abilities of the species. An understanding of these processes can be used to predict which species may theoretically 'win' competition and exclude other species and how grazing may alter the competitive interactions. However, although competitive exclusion is a common prediction of theoretical and empirical studies, many species coexist in nature.

The study of grasslands exposes a further problem. Not only do species coexist, but grazing has often been shown to increase the number of species coexisting: in Welsh hill pastures (Hill *et al.*, 1992), in English chalk grasslands (Tansley and Adamson, 1925; Gibson *et al.*, 1987), in English acid grassland (Putnam *et al.*, 1991), in English improved pasture (Bullock *et al.*, 1994e) and in many other communities throughout the world (Bakker, 1989). This cannot be caused by straightforward grazing effects on species' competitive abilities. Such effects will simply lead to changes in the outcome of competition and thus in which species dominate the system, and not in the number of coexisting species. Indeed, where grazing enhances the competitive ability of a dominant and diversifying influences have little effect, the diversity of the system will decrease (see Crawley, 1988), as in the case of *Nardus stricta* in upland grasslands (Welch, 1986; also see above). In a number of cases, therefore, grazing does not promote coexistence (e.g. Belsky, 1992).

For grazing to increase diversity it must enhance one of the processes described above that promote coexistence. As Pacala and Crawley (1992) showed in a lottery model of community dynamics under grazing, coexistence is promoted by frequency-dependent grazing and/or creation of spatial and temporal heterogeneity by grazing. Most studies of the community effects of grazers invoke, implicitly or explicitly, the creation of heterogeneity by the animals, particularly by gap creation. Another diversifying mechanism is possible if more than one species of grazer are present. Grazing animal species often differ in their patterns of plant species preference (Grant *et al.*, 1985; McNaughton and Georgiadis, 1986) and therefore they will have different effects on the competitive interactions and this will promote coexistence.

At the moment, the relative importance of the mechanisms which can allow species coexistence in grassland is unknown. An understanding of these aspects of population and community ecology requires detailed knowledge of the demography of species, of species interactions at a range of densities and in a variety of environments, and of spatiotemporal environmental and demographic changes.

REFERENCES

Aguilera, M.O. and Lauenroth, W.K. (1993). Neighborhood interactions in a natural population of the perennial bunchgrass *Bouteloua gracilis*. *Oecologia* 94, 595–602.

Alexander, K.I. and Thompson, K. (1982) The effect of clipping on the competitive interaction between two perennial grasses. *Oecologia* 53, 251–254.

Allen, E.B. and Allen, M.F. (1990) The mediation of competition by mycorrhizae in successional and patchy environments. In: Grace, J.B. and Tilman, D. (eds) *Perspectives on Plant Competition*. Academic Press, San Diego, pp. 367–389.

Bakker, J.P. (1989) *Nature Management by Grazing and Cutting*. Kluwer Academic Publishers, Dordrecht.

Begon, M., Harper, J.L. and Townsend, C.R. (1990) *Ecology: Individuals, Populations and Communities*, 2nd edn. Blackwell Scientific Publications, Oxford.

Belsky, A.J. (1986) Does herbivory benefit plants? A review of the evidence. *American Naturalist* 127, 870–892.

Belsky, A.J. (1992) Effects of grazing, disturbance and fire on species composition and diversity in grassland communities. *Journal of Vegetation Science* 3, 187–200.

Berendse, F. (1981) Competition between plant populations with different rooting depths. II. Pot experiments. *Oecologia* 48, 334–341.

Berendse, F. (1982) Competition between plant populations with different rooting depths. III. Field experiments. *Oecologia* 53, 50–55.

Berendse, F. (1985) The effect of grazing on the outcome of competition between plant species with different nutrient requirements. *Oikos* 44, 35–39.

Bergelson, J. (1990) Life after death: site pre-emption by the remains of *Poa annua*. *Ecology* 71, 2157–2165.

Bergelson, J. and Crawley, M.J. (1992) Herbivory and *Ipomopsis aggregata*: the disadvantages of being eaten. *American Naturalist* 139, 870–882.

Bronstein, J.L. (1994) Our current understanding of mutualism. *Quarterly Review of Biology* 69, 31–51.

Brown, J.R. and Stuth, J.W. (1994) How herbivory affects grazing tolerant and sensitive grasses in a central Texas grassland – integrating plant response across hierarchical levels. *Oikos* 67, 291–298.

Bullock, J.M., Clear Hill, B. and Silvertown, J. (1994a) Demography of *Cirsium vulgare* in a grazing experiment. *Journal of Ecology* 82, 101–111.

Bullock, J.M., Mortimer, A.M. and Begon, M. (1994b) The effects of clipping on intergenotypic competition in *Holcus lanatus* – a response surface analysis. *Journal of Ecology* 82, 259–270.
Bullock, J.M., Clear Hill, B. and Silvertown, J. (1994c). Tiller dynamics of two grasses – responses to grazing, density and weather. *Journal of Ecology* 82, 331–340.
Bullock, J.M., Mortimer, A.M. and Begon, M. (1994d) Physiological integration among tillers of *Holcus lanatus*: age-dependence and responses to clipping and competition. *New Phytologist* 128, 737–747.
Bullock, J.M., Clear Hill, B., Dale, M.P. and Silvertown, J. (1994e) An experimental study of vegetation change due to grazing in a species-poor grassland and the role of the seed bank. *Journal of Applied Ecology* 31, 493–507.
Bullock, J.M., Clear Hill, B., Silvertown, J. and Sutton, M. (1995a) Gap colonization as a source of grassland community change: effects of gap size and grazing on the rate and mode of colonization by different species. *Oikos* 72, 273–282.
Bullock, J.M., Mortimer, A.M. and Begon, M. (1995b) Carryover effects of interclonal competition in the grass *Holcus lanatus*: a response surface analysis. *Oikos* 72, 411–421.
Campbell, B.D. and Grime, J.P. (1992) An experimental test of plant strategy theory. *Ecology* 73, 15–29.
Chesson, P.L. (1991) A need for niches? *Trends in Ecology and Evolution* 6, 26–28.
Chesson, P.L. (1994) Multispecies competition in variable environments. *Theoretical Population Biology* 45, 227–276.
Clay, K. (1990) The impact of parasitic and mutualistic fungi on competitive interactions among plants. In: Grace, J.B. and Tilman, D. (eds) *Perspectives on Plant Competition*. Academic Press, San Diego, pp. 391–412.
Connell, J.H. (1983) On the prevalence and relative importance of interspecific competition: evidence from field experiments. *American Naturalist* 122, 661–696.
Connolly, J. (1987) On the use of response models in mixture experiments. *Oecologia* 72, 95–103.
Cottam, D.A. (1985) Frequency-dependent grazing by slugs and grasshoppers. *Journal of Ecology* 73, 925–933.
Crawley, M.J. (1983) *Herbivory: The Dynamics of Animal–Plant Interactions*. Blackwell Scientific Publications. Oxford.
Crawley, M.J. (1988) Herbivores and plant population dynamics. In: Davy, A.J., Hutchings, M.J. and Watkinson, A.R. (eds) *Plant Population Ecology*. Blackwell Scientific Publications, Oxford, pp. 367–392.
Crawley, M.J. (1990) Rabbit grazing, plant competition and seedling recruitment in acid grassland. *Journal of Applied Ecology* 27, 803–820.
Day, T.A. and Detling, J.K. (1990) Grassland patch dynamics and herbivore grazing preference following urine deposition. *Ecology* 71, 180–188.
Dirzo, R. (1984) Herbivory: a phytocentric overview. In: Dirzo, R. and Sarukhan, J. (eds) *Perspectives on Plant Population Ecology*. Sinauer Associates, Sunderland, Massachusetts, pp. 141–165.
Duralia, T.E. and Reader, R.J. (1993) Does abundance reflect competitive ability? A field test with three prairie grasses. *Oikos* 68, 82–90.
Firbank, L.G. and Watkinson, A.R. (1990) On the effects of competition: from monocultures to mixtures. In: Grace, J.B. and Tilman, D. (eds) *Perspectives on Plant Competition*. Academic Press, San Diego, pp. 165–191.

Fowler, N.L. (1981) Competition and coexistence in a North Carolina grassland. II. The effects of experimental removal of species. *Journal of Ecology* 69, 843–854.

Fowler, N.L. (1982) Competition and coexistence in a North Carolina grassland. III. Mixtures of component species. *Journal of Ecology* 70, 77–92.

Fowler, N.L. (1986) Density-dependent population regulation in a Texas grassland. *Ecology* 67, 545–554.

Fowler, N.L. and Antonovics, J. (1981) Competition and coexistence in a North Carolina grassland. I. Patterns in undisturbed vegetation. *Journal of Ecology* 69, 825–841.

Gaudet, C.L. and Keddy, P.A. (1988) A comparative approach to predicting competitive ability from plant traits. *Nature* 334, 242–243.

Gibson, C.W.D., Dawkins, H.C., Brown, V.K. and Jepsen, M. (1987). Spring grazing by sheep: effects on seasonal changes during early old field succession. *Vegetatio* 70, 33–43.

Gibson, D.J. (1988) The maintenance of plant and soil heterogeneity in dune grassland. *Journal of Ecology* 76, 497–508.

Gillman, M., Bullock, J.M., Silvertown, J. and Clear Hill, B. (1993) Density-dependent models of *Cirsium vulgare* population dynamics using field-estimated parameter values. *Oecologia* 96, 282–289.

Goldberg, D.E. (1990) Components of resource competition in plant communities. In: Grace, J.B. and Tilman, D. (eds) *Perspectives on Plant Competition*. Academic Press, San Diego, pp. 27–49.

Goldberg, D.E. and Barton, A.M. (1992) Patterns and consequences of interspecific competition in natural communities: a review of field experiments with plants. *American Naturalist* 139, 771–801.

Goldberg, D.E. and Landa, K. (1991) Competitive effect and response – hierarchies and correlated traits in the early stages of competition. *Journal of Ecology* 79, 1013–1030.

Grace, J.B. (1990) On the relationship between plant traits and competitive ability. In: Grace, J.B. and Tilman, D. (eds) *Perspectives on Plant Competition*. Academic Press, San Diego, pp. 51–65.

Grace, J.B. (1993) The effects of habitat productivity on competition intensity. *Trends in Ecology and Evolution* 8, 229–230.

Grant, S.A., Suckling, D.E., Smith, H.K., Torvell, L., Forbes, T.D.A. and Hodgson, J. (1985) Comparative studies of diet selection by sheep and cattle: the hill grasslands. *Journal of Ecology* 73, 987–1004.

Grime, J.P. (1979) *Plant Strategies and Vegetation Processes*. John Wiley and Sons, Chichester.

Grime, J.P. (1987) The C-S-R model of primary plant strategies: origins, implications and tests. In: Gottlieb, L.D. and Jain, S.K. (eds) *Evolutionary Plant Biology*. Chapman and Hall, London, pp. 371–394.

Grime, J.P., Hodgson, J.G. and Hunt, R. (1988) *Comparative Plant Ecology*. Unwin Hyman, London.

Grubb, P.J. (1977) The maintenance of species-richness in plant communities: the importance of the regeneration niche. *Biological Reviews* 52, 107–145.

Grubb, P.J. (1986) Problems posed by sparse and patchily distributed species in species-rich plant communities. In: Diamond, J. and Case, T.J. (eds) *Community Ecology*. Harper and Row, New York, pp. 207–225.

Gurevitch, J., Morrow, L.L., Wallace, A. and Walsh, J.S. (1992). A meta-analysis of competition in field experiments. *American Naturalist* 140, 539–572.

Harper, J.L. (1977) *Population Biology of Plants*. Academic Press, London.
Hassell, M.P., Comins, H.N. and May, R.M. (1994) Species coexistence and self-organising spatial dynamics. *Nature* 370, 290–292.
Hetrick, B.A.D., Hartnett, D.C., Wilson, G.W.T. and Gibson, D.J. (1994) Effects of mycorrhizae, phosphorus availability, and plant density on yield relationships among competing tallgrass prairie grasses. *Canadian Journal of Botany* 72, 168–176.
Hill, M.O., Evans, D.F. and Bell, S.A. (1992) Long-term effects of excluding sheep from hill pastures in North Wales. *Journal of Ecology* 80, 1–13.
Hobbs, R.J. and Mooney, H.A. (1991) Effects of rainfall and gopher disturbance on serpentine annual grassland dynamics. *Ecology* 72, 59–68.
Illius, A.W., Clark, D.A. and Hodgson, J. (1992). Discrimination and patch choice by sheep grazing grass–clover swards. *Journal of Animal Ecology* 61, 183–194.
Jastow, J.D. and Miller, R.M. (1993) Neighbor influences on root morphology and mycorrhizal fungus colonization in tallgrass prairie plants. *Ecology* 74, 561–569.
Jonsdottir, G.A. (1991) Effects of density and weather on tiller dynamics in *Agrostis stolonifera, Festuca rubra* and *Poa irrigata. Acta Botanica Neerlandica* 40, 311–318.
Keddy, P.A. and Shipley, B. (1989) Competitive hierarchies in plant communities. *Oikos* 49, 234–241.
King, T.J. (1977) The plant ecology of anti-hills in calcareous grasslands. *Journal of Ecology* 65, 235–256.
Law, R. and Watkinson, A.R. (1987) Response-surface analysis of two-species competition: an experiment on *Phleum arenaria* and *Vulpia fasiculata. Journal of Ecology* 75, 871–886.
Louda, S.V., Keeler, K.H. and Holt, R.D. (1990a) Herbivore influences on plant performance and competitive interactions. In: Grace, J.B. and Tilman, D. (eds) *Perspectives on Plant Competition*. Academic Press, San Diego, pp. 414–443.
Louda, S.V., Potvin, M.A. and Collinge, S.K. (1990b) Predispersal seed predation, postdispersal seed predation and competition in the recruitment of a native thistle in sandhills prairie. *American Midland Naturalist* 124, 105–121.
Lundberg, P., Astrom, M. and Danell, K. (1990) An experimental test of frequency-dependent food selection – winter browsing by moose. *Holarctic Ecology* 13, 177–182.
MacArthur, R.H. and Levins, R. (1967) The limiting similarity, convergence and divergence of coexisting species. *American Naturalist* 101, 377–385.
McNaughton, S.J. and Georgiadis, N.J. (1986) Ecology of African grazing and browsing mammals. *Annual Review of Ecology and Systematics* 17, 39–65.
Mahdi, A. and Law, R. (1987) On the spatial organization of plant species in a limestone grassland community. *Journal of Ecology* 75, 459–476.
Mahdi, A., Law, R. and Willis, A.J. (1989) Large niche overlaps among coexisting plant species in a limestone grassland community. *Journal of Ecology* 77, 386–400.
Martinsen, G.D., Cushman, J.H. and Whitham, T.G. (1990) Impact of pocket gopher disturbance on plant species diversity in a shortgrass prairie. *Oecologia* 83, 132–138.
Milchunas, D.G., Lauenroth, W.K. and Chapman, P.L. (1992) Plant competition, abiotic, and long- and short-term effects of large herbivores on demography of opportunistic species in a semiarid grassland. *Oecologia* 92, 520–531.
Mitchley, J. (1988) Control of relative abundance of perennials in chalk grassland in southern England. II. Vertical canopy structure. *Journal of Ecology* 76, 341–350.
Mitchley, J. and Grubb, P.J. (1986) Control of relative abundance of perennials in chalk grassland in southern England. *Journal of Ecology* 74, 1139–1166.

Molofsky, J. (1994) Population dynamics and pattern formation in theoretical populations. *Ecology* 75, 30–39.

Noy-Meir, I., Gutman, M. and Kaplan, Y. (1989) Responses of mediterranean grassland plant to grazing and protection. *Journal of Ecology* 77, 290–310.

O'Connor, T.G. (1991) Patch colonisation in a savanna grassland. *Journal of Vegetation Science* 2, 245–254.

Pacala, S.W. and Crawley, M.J. (1992) Herbivores and plant diversity. *American Naturalist* 140, 243–260.

Pacala, S.W. and Tilman, D. (1994) Limiting similarity in mechanistic and spatial models of plant competition in heterogeneous environments. *American Naturalist* 143, 222–257.

Paige, K.N. and Whitham, T.G. (1987) Overcompensation in response to mammalian herbivory: the advantage of being eaten. *American Naturalist* 129, 407–416.

Parsons, A.J., Harvey, A. and Johnson, I.R. (1991) Plant–animal interactions in a continuously grazed mixture. 2. The role of differences in the physiology of plant growth and of selective grazing on the performance and stability of species in a mixture. *Journal of Applied Ecology* 28, 635–658.

Petraitis, P.S., Latham, R.E. and Niesenbaum, R.A. (1989) The maintenance of species diversity by disturbance. *Quarterly Review of Biology* 64, 393–418.

Pickett, S.T.A. and White, P.S. (1985) *The Ecology of Natural Disturbance and Patch Dynamics*. Academic Press, Orlando.

Platt, W.J. (1975) The colonization and formation of equilibrium plant species associations on badger disturbances in a tall grass prairie. *Ecological Monographs* 45, 285–305.

Putnam, R.J., Fowler, A.D. and Tout, S. (1991) Patterns of use of ancient grassland by cattle and horses and effects on vegetational structure and composition. *Biological Conservation* 56, 329–347.

Putwain, P.D., Machin, D. and Harper, J.L. (1968) Studies in the dynamics of plant populations. II. Components and regulation of a natural population of *Rumex acetosella* in grassland. *Journal of Ecology* 56, 421–431.

Reader, R.J., Wilson, S.D., Belcher, J.W., Wisheu, I., Keddy, P.A., Tilman, D., Morris, E.C., Grace, J.B., McGraw, J.B., Olff, H., Turkington, R., Klein, E., Leung, Y., Shipley, B., van Hulst, R., Johansson, M.E., Nilsson, C., Gurevitch, J., Grigulis, K. and Beisner, B.E. (1994) Plant competition in relation to neighbor biomass: an intercontinental study with *Poa pratensis*. *Ecology* 75, 1753–1760.

Sackville Hamilton, N.R. and Harper, J.L. (1989) The dynamics of *Trifolium repens* in a permanent pasture. I. The population dynamics of leaves and nodes per shoot axis. *Proceedings of the Royal Society of London, Series B* 237, 133–173.

Samson, D.A. and Werk, K.S. (1986) Size-dependent effects in the analysis of reproductive effort in plants. *American Naturalist* 127, 667–680.

Schmid, B. and Harper, J.L. (1985) Clonal growth in grassland perennials. *Journal of Ecology* 73, 793–808.

Schoener, T.W. (1983) Field experiments on interspecific competition. *American Naturalist* 122, 240–285.

Shipley, B. (1994) Evaluating the evidence for competitive hierarchies in plant communities. *Oikos* 69, 340–345.

Shmida, A. and Ellner, S. (1984) Coexistence of plant species with similar niches. *Vegetatio* 58, 29–55.

Silvertown, J. and Law, R. (1987) Do plants need niches? *Trends in Ecology and Evolution* 2, 24–26.
Silvertown, J. and Lovett Doust, J. (1993) *Introduction to Plant Population Biology.* Blackwell Scientific Publications, Oxford.
Silvertown, J. and Smith, B.A. (1988) Gaps in the canopy: the missing dimension in vegetation dynamics. *Vegetatio* 77, 57–60.
Silvertown, J., Watt, T.A., Smith, B. and Treweek, J.R. (1992) Complex effects of grazing treatment on the establishment, growth, survival and reproduction of an annual in a species-poor grassland community. *Journal of Vegetation Science* 3, 35–40.
Silvertown, J., Lines, C.E.M. and Dale, M.P. (1994) Spatial competition between grasses – rates of mutual invasion between four species and the interaction with grazing. *Journal of Ecology* 82, 31–38.
Tansley, A.G. and Adamson, R.S. (1925) Studies of the vegetation of the English Chalk. III. The chalk grasslands of the Hampshire–Sussex border. *Journal of Ecology* 13, 177–223.
Thorhallsdottir, T.E. (1990) The dynamics of a grassland community: a simultaneous investigation of spatial and temporal heterogeneity at various scales. *Journal of Ecology* 78, 884–908.
Thornley, J.H.M., Parsons, A.J., Newman, J. and Penning, P.D. (1994) A cost–benefit model of grazing intake and diet in a two-species temperate grassland sward. *Functional Ecology* 8, 5–16.
Tilman, D. (1982) *Resource Competition and Community Structure.* Princeton University Press, Princeton.
Tilman, D. (1988) *Plant Strategies and the Dynamics and Structure of Plant Communities.* Princeton University Press, Princeton.
Tilman, D. (1990a) Mechanisms of plant competition for nutrients: the elements of a predictive theory for competition. In: Grace, J.B. and Tilman, D. (eds) *Perspectives on Plant Competition.* Academic Press, San Diego, pp. 117–141.
Tilman, D. (1990b) Constraints and trade offs: towards a predictive theory of competition and succession. *Oikos* 58, 3–15.
Tilman, D. (1994) Competition and biodiversity in spatially structured habitats. *Ecology* 75, 2–16.
Tilman, D. and Wedin, D. (1991) Dynamics of nutrient competition between successional grasses. *Ecology* 72, 1038–1049.
Turkington, R. and Harper, J.L. (1979) The growth, distribution and neighbour relationships of *Trifolium repens* in a permanent pasture. *Journal of Ecology* 67, 201–218.
Turkington, R. and Mehrhoff, L.A. (1990) The role of competition in structuring pasture communities In: Grace, J.B. and Tilman, D. (eds) *Perspectives on Plant Competition.* Academic Press, San Diego, pp. 308–349.
Turkington, R., Klein, E. and Chanway, C.P. (1993) Interactive effects of nutrients and disturbance: an experimental test of plant strategy theory. *Ecology* 74, 863–878.
van der Maarel, E. and Sykes, M.T. (1993) Small-scale plant species turnover in a limestone grassland: the carousel model and some comments on the niche concept. *Journal of Vegetation Science* 4, 179–188.
van der Meijden, E., Jong, T.J., Klinkhamer, G. and Kooi, R.E. (1985) Temporal and spatial dynamics in populations of biennial plants. In: Haeck, J. and Woldendorp, J.W. (eds) *Structure and Functioning of Plant Populations 2.* North-Holland, Amsterdam, pp. 391–403.

Watkinson, A.R. (1980) Density-dependence in single-species populations of plants. *Journal of Theoretical Biology* 83, 345–357.

Watkinson, A.R. (1984) Yield–density relationships: the influence of resource availability on growth and self-thinning in populations of *Vulpia fasciculata*. *Annals of Botany* 53, 469–482.

Watt, T.A. and Hagger, R.J. (1980) The effect of defoliation upon yield, flowing and vegetative spread of *Holcus lanatus* growing with and without *Lolium perenne*. *Grass and Forage Science* 35, 227–234.

Wedin, D. and Tilman, D. (1993) Competition among grasses along a nitrogen gradient: initial conditions and mechanisms of competition. *Ecological Monographs* 63, 199–229.

Weiner, J. (1990) Asymmetric competition in plant populations. *Trends in Ecology and Evolution* 5, 360–364.

Welch, D. (1986) Studies in the grazing of heather moorland in north-east Scotland. V. Trends in *Nardus stricta* and other unpalatable graminoids. *Journal of Applied Ecology* 23, 1047–1058.

Weller, D.E. (1987) A reevaluation of the $-3/2$ power rule of plant self-thinning. *Ecological Monographs* 57, 23–43.

Wilson, S.D. and Shay, J.M. (1990) Competition, fire and nutrients in a mixed-grass prairie. *Ecology* 71, 1959–1967.

Wilson, S.D. and Tilman, D. (1991) Components of plant competition along an experimental gradient of nitrogen availability. *Ecology* 72, 1050–1065.

Assessing and Interpreting Grass–Woody Plant Dynamics

S. Archer

Department of Rangeland Ecology and Management, Texas A&M University, College Station, TX 77843-2126, USA

INTRODUCTION

In many arid and semiarid systems, grazing by domestic herbivores is a primary land use for commercial enterprises, pastoral societies and subsistence cultures. Ecosystem sustainability for livestock production requires management which maintains the soil resource and ensures a favourable balance between palatable and unpalatable vegetation. In many arid and semiarid systems, this means: (i) regulating grazing to maintain cover and production of palatable, perennial grasses that are the forage base for livestock or wildlife; and (ii) limiting invasion or encroachment by unpalatable woody vegetation. Improper management may contribute to detrimental changes that ultimately reduce both plant and animal productivity and diversity, increase the need for expensive supplemental feeding, increase the potential for soil erosion and increase the probability that expensive rehabilitation practices will be required to stabilize or restore sites. On a global scale, grazing-induced alterations of plant cover and soil processes may constitute important feedback to climate and atmospheric chemistry (Schlesinger *et al.*, 1990). General effects of grazing on energy flow and nutrient cycling (Detling, 1988), biodiversity (McNaughton, 1993, 1994; West, 1993), vegetation composition and soil properties (Archer and Smeins, 1991; Thurow, 1991; Skarpe, 1991; Milchunas and Lauenroth, 1993; Milton *et al.*, 1994, Pieper, 1994) and successional dynamics (Westoby *et al.*, 1989; Friedel, 1991; Stafford Smith and Pickup, 1993) have been recently reviewed. Other chapters in this book focus on responses of herbaceous vegetation to grazing. This chapter examines woody plant–grass dynamics and reviews approaches for enhancing our understanding of the rates,

©1996 CAB INTERNATIONAL.
The Ecology and Management of Grazing Systems (eds J. Hodgson and A.W. Illius)

dynamics and causes of increased abundance of unpalatable woody vegetation on grazed landscapes.

Displacement of grasses by woody plants over the past century has been widely reported for arid and semiarid rangelands (see Table 1 in Archer, 1995a). Even so, our knowledge of the rates, dynamics, patterns and extent of this phenomenon is limited. Available data indicate these directional shifts in life-form abundance have been: (i) rapid, with substantial changes occurring over 50- to 100-year time spans; (ii) non-linear and accentuated by episodic climatic events (drought or above-normal rainfall); (iii) locally influenced by topoedaphic factors; and (iv) non-reversible over time-frames relevant to management.

Explanations for historical increases in abundance of woody plants in dryland ecosystems centre around changes in climatic, grazing and fire regimes and atmospheric carbon dioxide (CO_2) enrichment (see Archer, 1994; Miller and Wigand, 1994; Archer et al., 1995). Influences of domestic and native herbivores on the balance between grasses and woody plants have been specifically addressed, with emphasis on the critical seedling establishment phase of the woody plant life cycle (Archer, 1995b). However, broad-scale understanding of grass–woody plant dynamics is limited by a paucity of spatially explicit information at the landscape level of resolution. Spatial heterogeneity and temporal variability in rangelands impose significant constraints on our ability to inventory, monitor, predict and manage vegetation and soils at scales of time (decades) and space (hundreds of hectares) relevant to management (Stafford Smith and Pickup, 1993; M. Stafford Smith, Chapter 12, this volume). Here, I briefly review approaches for interpreting vegetation dynamics of grazed systems across hierarchical scales of time and space with an emphasis on woody plants. I then focus on the application of an array of underutilized tools that can be used alone or in concert to: (i) further our quantitative and conceptual understanding of spatial and temporal heterogeneity; (ii) develop comprehensive monitoring schemes; (iii) evaluate land management impacts on vegetation; and (iv) temper expectations with regard to range improvement practices and rehabilitation efforts.

LIVESTOCK AND WILDLIFE

Although livestock are typically the focus of attention in managed systems, activities of native or feral herbivores, both above and below ground, should not be overlooked. In many rangeland settings, managers have little information about or control over the population dynamics of these animals. When proportions of browsers or grazers shift in response to environmental change or management, the balance between grasses and woody plants shifts accordingly (Sinclair, 1979). In some cases, management for livestock enhances the abundance of native grazers, thus putting additional pressure on vegetation and soils. Activities of inconspicuous nocturnal granivores (Brown and Heske, 1990) or consumption of plant roots by nematodes (Coleman et al., 1976) or grubs (Lura and Nyren, 1992) may have a comparable to greater effect on vegetation than livestock. The influence of insects, arthropods, rodents and soil invertebrates on vegetation dynamics relative to that of the more conspicuous large herbivores is seldom known. Activities of these organisms not accounted for in field studies may obscure our understanding of livestock grazing effects on plant community dynamics, thus making it difficult to compare studies meaningfully.

DEFOLIATION OF GRASSES VERSUS GRAZING ON LANDSCAPES

Plant species composition and productivity within a region are largely a function of the prevailing climate. However, substantial variation occurs across landscapes, and broad-scale climatic factors cannot account for the spatial patterns which shape vegetation form and function at a local scale. Soils and topography exert a strong influence on patterns of plant distribution, growth and abundance through regulation of water and nutrient availability. Grazing influences are superimposed on this background of topoedaphic heterogeneity and climatic variability to further influence vegetation structure and ecosystem processes. Species adapted to the prevailing climate and soils might be the competitive dominants of the community under conditions of minimal grazing, but may assume subordinate roles or even face local extinction as grazing intensity increases.

At community and landscape levels of resolution, grazing influences on ecosystem processes and plant community dynamics are both direct and indirect, and vary across the landscape depending on the type of grazing animal, seasonal patterns of animal distribution, soils, topography and distance from resources such as water or shade. Direct effects of herbivores are those associated with the consumption or trampling of plant tissues and subsequent changes in growth, biomass allocation and vegetative and sexual reproduction. The role of herbivores as agents of seed dispersal (Janzen, 1984; Brown and Archer, 1987) and predation (Brown and Heske, 1990) is also potentially important in regulating plant population dynamics. Indirect effects of grazing include alteration of microenvironment, changes in soil physical and chemical properties, hydrology and erosion, disruption of algal or lichen crusts and the redistribution and transformation of nutrients across the landscape (Thurow, 1991; Williams and Chartres, 1991; Ludwig and Tongway, 1993). These, in turn, may feed back to affect plant growth, reproduction and seedling establishment and intensify defoliation impacts. Alterations in plant density and cover by grazers can intensify run-off/run-on patterns across the landscape and accentuate natural heterogeneity. Systems where soil resources are plant-controlled rather than terrain-controlled may be particularly sensitive to grazing (Ludwig *et al.*, 1994; Tongway and Ludwig, 1994). Alterations of microclimate, local hydrology and nutrients by grazing may favour unpalatable, nitrogen (N_2)-fixing woody plants (e.g. *Prosopis, Acacia* spp.) and evergreen growth forms tolerant of low nutrient conditions and water stress.

Preferential utilization of plants which vary in their palatability or sensitivity to defoliation can directionally alter the nature and intensity of plant competitive interactions and influence population dynamics and hence species composition (see Chapters 2 and 3 by D.D. Briske and J.J. Bullock respectively, this volume). Alterations in species composition and productivity combine to influence soil physical properties, nutrient cycling and microclimate. Species effects on nutrient cycling can be as important as or more important than abiotic factors in controlling ecosystem fertility (Hobbie, 1992). Changes in species composition associated with selective grazing typically result in replacement of palatable plants by unpalatable plants and reductions in litter quality. This reflects the fact that plants which produce easily decomposable litter are also those which will be heavily grazed, because the same chemical properties that determine litter decay also determine palatability and digestibility (Pastor and Naiman, 1992). Changes in amount and quality of litter associated with changes in species composition can lead to reductions in microbial biomass, mineralization and respiration. Plants which remain or increase with grazing may further reduce rates of nutrient cycling, thus accentuating defoliation stress and influencing species composition, plant cover and production. Changes in soil nutrient distribution subsequent to establishment of woody plants may feed back to increase the likelihood of additional woody plant encroachment, increase the spatial heterogeneity of nutrient distribution and accelerate water and wind erosion (Schlesinger *et al.*, 1990). The rate and direction of succession following relaxation of grazing may largely depend on the degree to which soil properties and processes have been affected. Unfortunately, there are no clear generalizations which emerge with respect to grazing impacts on soils (Milchunas and Lauenroth, 1993). This may reflect the fact that most studies have been initiated only after sites have already experienced varying and unknown degrees of historical grazing.

Plant distributions and patch structure vary along environmental gradients within pastures and management units. The likelihood of being grazed by a specific class of herbivore and the plants' response to grazing vary along these gradients, depending on water and nutrient availability and neighbourhood effects. Fluctuation in rainfall may accentuate or mitigate grazing impacts on vegetation. Grass consumption also reduces the amount and continuity of fine fuels and hence the frequency, pattern and areal extent of fire (Wright and Bailey, 1982; Savage and Swetnam, 1990). As a result, successful management of the balance between grasses and woody vegetation requires a spatially explicit understanding of the interactive roles of climate, soils, topography, fire and herbivory over time (Fig. 4.1). Given the complexity of interactions among soil properties, resource availability and climatic stresses on plant growth and survival, it can be difficult to ascribe adaptive significance to traits that enable plants to tolerate or evade herbivory. For example, some graminoid traits which may have originally evolved in response to selection pressures imposed by water stress, fire or

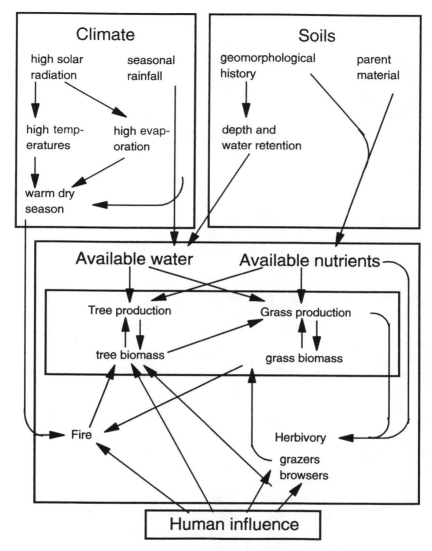

Fig. 4.1. Understanding and interpreting changes in grass–woody plant abundance requires a spatially explicit knowledge of interacting biotic and abiotic factors (from Scholes and Walker, 1993).

competition also confer benefits to grazed plants (Coughenour, 1985). As a result, screening of plant genotypes or the placement of species into 'functional groups' with respect to grazing requires evaluation of responses to multiple stresses and broad sets of interacting parameters.

To understand fully the influence of grazing impacts on ecosystem dynamics, it is essential to identify levels in a hierarchy of landscape organization, the key processes that occur at each level, interrelationships among levels, and interactive influences of other disturbance or environmental factors (Archer and Tieszen, 1986). Modification of microclimate, plant competitive interactions, soil processes and fire frequency associated with the defoliation and preferential utilization of grasses can benefit unpalatable woody plants in several ways. These include increased probabilities of successful seedling establishment, greater growth rates, decreased time to reproductive maturity, increased frequency and magnitude of seed production and extended longevity (Archer, 1995b, and references therein).

STATES AND TRANSITIONS

Management and conservation of grazed rangelands depends on knowledge of likely vegetation states and the transitions affecting those states (Fig. 4.2). In many arid and semiarid regions, the 'state and transition' model (Westoby et al., 1989) provides a more suitable conceptual framework for interpreting vegetation dynamics than the traditional equilibrium-based successional models. Vegetation 'states' are recognizable and relatively stable assemblages of species occupying a site; transitions between states are triggered either by natural events or by management actions. This approach is flexible, incorporates cyclic and successional processes and stochastic responses of vegetation to climate or biotic disturbance.

Applications of this concept relative to equilibrium-based models have been widely reviewed (Archer, 1989; Friedel, 1991; Laycock, 1991; Ellis, 1992; Dankwerts et al., 1993; Joyce, 1993; Walker, 1993; Borman and Pyke, 1994; Whalley, 1994; M. Stafford Smith, Chapter 12, this volume). Specific examples from different rangeland systems are accumulating (George et al., 1992; Huntsinger and Bartolome, 1992; Jones, 1992; Milton et al., 1994; special issue of *Tropical Grasslands* 28(4), 1994). While the state and transition approach is valuable for management and classification, the definition of states is largely heuristic and proposed mechanisms for transitions between states are often hypothetical rather than empirical. Studies quantifying rates and probabilities of transition among states are uncommon. Similarly, the role of 'triggering' events which might initiate or drive transitions are not well understood. As a result, it is difficult to predict the longevity of a given vegetation state and the level (frequency, intensity, duration) of stress, disturbance or environmental change required to shift vegetation from one state to another. Generalizations regarding the rate or extent to which 'recovery' to a previous state occurs after disturbance or following relaxation of grazing are equally elusive. Transitions will depend on complex interactions between species life-history attributes, availability of propagule sources, the extent to which soils have changed, and climatic conditions.

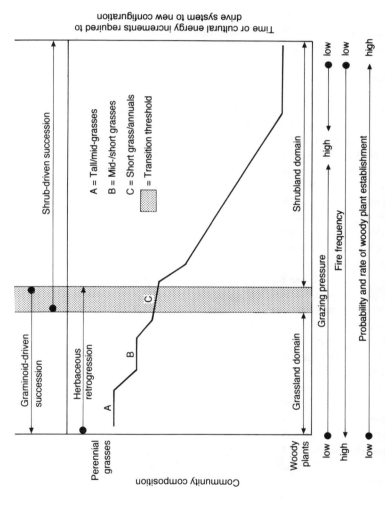

Fig. 4.2. Conceptual model of grass and woody plant abundance in grazed ecosystems, postulating the existence of a threshold of disturbance (grazing-driven) in triggering a successional transition between vegetation states dominated by grasses and woody plants (from Archer, 1989). The threshold of herbaceous utilization required to enable some woody species to successfully establish from seed can be readily exceeded, even at low levels of grazing in some systems (Archer, 1995b).

Thresholds

The conceptual model in Fig. 4.2 illustrates how grazing animals can direct plant succession to effect changes in vegetation structure. This model is predicated on the existence of transition thresholds, the evidence of which is based on widespread observations of abrupt, non-linear changes in vegetation composition in arid and semiarid systems (Buffington and Herbel, 1965; Herbel *et al.*, 1972; Archer *et al.*, 1988; Friedel, 1991). These thresholds may exist for various herbaceous transitions (perennial ↔ annual; tallgrass ↔ shortgrass), for transitions among woody elements (palatable ↔ unpalatable; suffruticose ↔ fruticose ↔ arborescent; deciduous ↔ evergreen) and for herbaceous ↔ woody transitions.

Vegetation within a particular state or 'domain' may be relatively stable and resistant to change. Intensification of grazing will alter herbaceous composition as animals preferentially utilize certain species, some of which may be relatively intolerant of defoliation (Briske, Chapter 2, this volume). These species are replaced by less preferred or more grazing-tolerant species. Gaps formed by mortality of grazed plants or a decline in their canopy or basal area represent opportunities for establishment of other species. As a result, diversity may increase (Milchunas *et al.*, 1988). If grazing pressure is relaxed prior to some critical threshold, succession may return the site to its earlier composition, the rate of recovery being influenced by climatic conditions and the extent to which soil structure and fertility may have changed. If grazing intensity is maintained, herbaceous biomass (above and below ground) may decline and the capacity of grazed plants to competitively exclude other plants will diminish. Increases in size and density of unpalatable plants, woody or herbaceous, will further intensify grazing pressure on remaining palatable herbaceous plants. As woody plants establish, new successional processes and positive (self-reinforcing) feedbacks drive the system to a new state (e.g. Archer *et al.*, 1988; Schlesinger *et al.*, 1990). The woody plant seed bank may increase and established trees and shrubs often have high vegetative regenerative potentials and extended longevities (decades to centuries). Reductions in palatable grass seed production and a deterioration of their seed bank make a return to grass domination unlikely, even when livestock are removed. Stochastic climatic events (drought, frost) may hasten grazing-mediated transitions (e.g. O'Connor, 1993, 1994). Transitions may be accelerated if keystone plant species establish and initiate strong positive feedbacks which drive successional processes or change disturbance regimes. For example, *Prosopis glandulosa*, an invasive shrub of grasslands and savannas of southwestern North America, alters soils and microclimate subsequent to its establishment in grass stands and facilitates the dispersal and establishment of additional woody species (Archer, 1995a). Introduced annual *Bromus* spp. readily invade grazed shrub steppe in North American cold deserts and have increased the frequency and areal extent of fire in a self-reinforcing fashion to trigger conversion to annual grasslands over extensive areas (Billings, 1994).

Little is known of the nature of transition thresholds or how to anticipate when we might be approaching one so that management can be adjusted to avert undesirable change. Multivariate analyses have been used to reduce complex species composition data into a few functional groups (Friedel *et al.*, 1988; Bosch and Booysen, 1992). The potential therefore exists to characterize the configuration of communities that exist on either side of a threshold in terms of the relative proportions of a few groups, under specified seasonal conditions (Friedel, 1991). Identification of demographic variables (size class distribution, critical minimum basal area, tiller and plant density, seed production) that portend thresholds between states might yield a mechanistic basis for monitoring to anticipate change (see chapters 2 and 3).

Figure 4.2 proposes a critical grazing threshold beyond which the probability of woody plant recruitment increases markedly. In many cases, the threshold of herbaceous utilization required to enable woody plants to establish from seed appears to be readily exceeded, even at low levels of grazing (Archer, 1995b). Increases in grass biomass, achieved experimentally or by relaxing grazing, can slow rates of woody plant seedling establishment and growth, but may not prevent it. For example, savannas of the Edwards Plateau of central Texas have been heavily and continuously grazed since the mid-1800s. In 1948, several grazing systems were implemented on an experimental station in this region. Cover of unpalatable evergreen shrubs has increased two- to fourfold since 1948, despite the relaxation or exclusion of livestock grazing (Table 4.1). Ironically, the greatest increases were on pastures protected from livestock grazing. Such data suggest that, by the time progressive grazing management practices were implemented in 1948, these systems were already in the woody plant 'domain of attraction' (Fig. 4.2); changes in soils, microclimate, seed bank and vegetative regeneration potential were such that succession toward woodland was under way and perhaps inevitable. In cases such as these, grazing management schemes may have to aggressively incorporate the use of fire. Proper grass utilization and maintenance of herbaceous composition alone may not be sufficient to successfully curtail woody plant encroachment.

Table 4.1. Woody species composition (% of total canopy cover) and total canopy cover for three pastures in 1949 and 1983. All areas had been continuously and heavily grazed since the mid-1800s, until establishment of the pastures in 1948. Pastures were grazed by cattle, sheep and goats (60–20–20 ratios); exclosure was protected from livestock grazing but not wildlife. (From Smeins and Merrill, 1988.)

	Continuous		Rotation		Exclosure	
Species	1949	1983	1949	1983	1949	1983
Quercus spp.	89	41	90	50	93	41
Juniperus spp.	7	40	4	39	3	32
Other species	4	19	6	11	4	27
Total canopy cover	14	10	10	30	8	35

Woody plant encroachment is a subtle process that operates at decadal timescales. Forces setting the process of invasion in motion may occur long before results are readily apparent. By the time results are manifested, cost-effective management options may have been precluded. Communities and landscapes may have a gross, outward appearance of stability for many years and then change radically over a short period of time. In some systems this reflects the importance of rare or infrequent events which trigger episodes of seed production, seed dispersal or seedling establishment. It can also reflect patterns of plant growth and development, whereby 'seedlings' persist, inconspicuously distributed throughout the herbaceous vegetation (Archer, 1995b). After several years, there is a dramatic shift in allocation to shoot growth. Such plants may not be apparent to the casual observer until many years after their establishment, by which time they are highly persistent members of the plant community. Given these patterns of growth, it is important to closely monitor rangelands where bush encroachment is a potential problem.

Vegetation change following relaxation of grazing

Changes in rangeland vegetation tend to be slow in dry environments and observational time scales that exceed a human lifespan are required to separate directional trends from fluctuations associated with weather-driven variability (Collins *et al.*, 1987). Long-term data sets from Utah indicate that recovery of palatable shrubs and perennial grasses (three- to tenfold increase in canopy cover) has occurred since 1933, following implementation of federal legislation which led to reductions in livestock grazing (Yorks *et al.*, 1992, 1994). Similarly, differences in albedo, soil temperature, soil moisture retention and vegetation 'greenness' along the USA–Mexico border appear related to relaxation of grazing pressure in the USA since passage of the Taylor Grazing Act in 1934 (Bryant *et al.*, 1990). However, widespread observations indicate that, once critical thresholds are crossed, grazing-induced changes in composition of arid and semi-arid rangelands will not be reversed simply by removal of livestock, especially where palatable plants are rare and unpalatable perennials predominate (see Archer, 1989; Westoby *et al.*, 1989; Archer and Smeins, 1991; Walker, 1993). This may reflect differences in seed production among grazed, palatable plants (low, infrequent) and less grazed, unpalatable species (high, frequent), competitive suppression of seedlings of palatable species by high densities of established, long-lived unpalatable species, lack of suitable microsites for establishment of palatable species (variable microclimate, soil compaction, reduced infiltration, loss of microsymbionts), and loss of species involved in keystone processes related to soil nutrient availability, pollination, dispersal, or mediation of competitive interactions (e.g. Bond, 1993).

Reversal of transitions may require active intervention by land managers and may involve clearing of bush (mechanical, chemical), fertilizing and seeding. All are costly, are risky in terms of the probability of achieving goals and have the potential to exacerbate existing problems. In many cases, unpalatable plants are not a problem *per se*. Rather, they may be symptomatic of past management transgressions (McKell, 1989). While vegetation dominated by unpalatable perennials may not be desirable, such plants reflect the prevailing environmental conditions (e.g. long-term heavy grazing) and may be important for energy flow, nutrient cycling, wildlife habitat and soil stabilization. Their removal should not be contemplated without due consideration of what will replace them. Where soils, seed bank and vegetative regeneration potentials favour post-intervention re-establishment of unpalatable woody vegetation (Fig. 4.2), a long-term, carefully planned, strategically timed sequence of vegetation manipulation technologies may be required to drive the system back to some previous state (Scifres *et al.*, 1985; Noble *et al.*, 1991). However, chemical and mechanical manipulation may not be ecologically sound, socially acceptable, biologically effective or economically feasible on a large scale. Given the effort and expense required to reduce cover or biomass of unpalatable woody plants, it would be desirable to manage grazing lands to minimize their establishment. Experience to date suggests that the adage 'an ounce of prevention is worth a pound of cure' is certainly applicable. However, climatic variability and the unpredictable occurrence of extreme climatic events may effect rapid shifts in plant recruitment and mortality. These may unexpectedly promote grass die-off or enhance woody plant seed production and seedling establishment, leaving managers little opportunity to adjust animal numbers/composition or implement a prescribed burn. Socioeconomic externalities may further interact to impede or constrain deployment of desired management practices.

KNOWING THE PAST, UNDERSTANDING THE PRESENT, PLANNING THE FUTURE

Causes for change cannot be addressed until we have an adequate understanding of the extent, pattern and rate of change that has occurred. Presumed composition and geographic distribution of presettlement vegetation is often used, either explicitly or implicitly, as a control or baseline to assess impacts of land use. Unfortunately, our knowledge of presettlement vegetation is sketchy; hence our foundation for determining the extent of the impact which livestock grazing may have had on soils and vegetation is often weak. Lack of historical perspective can place short-term studies in the 'invisible present', where a lack of temporal perspective can produce misleading conclusions (Magnuson, 1990). Assessments of stability and equilibrium are typically artefacts of the spatial and temporal scale at which we observe (DeAngelis and Waterhouse, 1987). Equilibrium states can occur at certain scales and contain disequilibrium at smaller scales. A historical perspective on vegetation dynamics is required to distinguish between short-term (seasonal, annual) fluctuation and long-term (decadal) directional change.

Original site factors/conditions significantly affect the structure and function of present-day vegetation and control the ways humans use sites and how natural processes affect them (Foster, 1992). For example, prehistoric faunal extinctions (Janzen, 1986; Owen-Smith, 1987), activities of early humans (Blackmore et al., 1990) and historical fluctuations in native browsers and grazers (Sinclair, 1979) have significantly influenced the pattern and abundance of woody plants on modern landscapes. In southwestern North America, desert grasslands, which established under 300 years of cooler, moister Little Ice Age conditions, may be ill-suited for the warmer, drier climates of the last century and destined for replacement by xerophytic shrubs (Neilson, 1986). However, given the substantial 'biological inertia' of perennial plants, changes in vegetation may have lagged behind changes in climate and were not yet been manifested in the early 1800s. Grazing by livestock may have accelerated a vegetation change in progress at the time of settlement.

An accurate understanding of the extent and cause of changes which have occurred in systems grazed by livestock are necessary if we are to: (i) mitigate future undesirable impacts of grazing; and (ii) realistically assess restoration potentials. Conclusive studies linking human activity and ecosystem change require a combination of field experimentation coupled with comprehensive analyses of land-use history and long-term vegetation records. The subsequent sections review and evaluate techniques with the potential to reconstruct spatial and temporal patterns of vegetation and to relate these to environmental factors, land use and cultural conditions. Techniques in stable isotope chemistry, biogenic opals, dendroecology and historical aerial photography offer opportunities to generate spatially explicit reconstructions of vegetation history and to determine rates and dynamics of changes. As the number of such studies increases, our understanding of vegetation dynamics at landscape and regional scales will grow.

Traditional assessments of historical change in vegetation

Comparisons with relict stands

Relict stands on isolated mesas, on road or railway rights of way, in cemeteries or in long-term enclosures are often used as indicators of 'pristine' conditions. However, these: (i) are not necessarily representative of past communities or optimal conditions; (ii) may have been established after anthropogenic disturbances had influenced vegetation or soils; (iii) are typically small in size; or (iv) are confined to select topoedaphic conditions. This potentially produces artificialities in plant or animal production and population dynamics, disturbance regimes and microclimate, which can influence species composition or abundance. Extrapolation to other landscape units or sites within the region is therefore risky.

Historical records

Descriptions of vegetation from diaries of early explorers and settlers can be used to assess the historical impacts of livestock grazing (Malin, 1953). These are subject to many sources of error and bias (Forman and Russell, 1983). Discrepancies between present-day composition of relict stands and descriptions by early travellers cast doubt on the reliability of one or both as indicators of the extent or pattern of vegetation change. In addition, rates of change required to produce shifts in vegetation from the time of historical observation to the present may not agree with measured or ecologically realistic rates of change (Hoffman and Cowling, 1990; Palmer *et al.*, 1990).

Historical ground photographs

Matched or repeat ground photographs from early to recent times provide another means of visually comparing past and present vegetation. However, oblique, ground-level shots with narrow fields of view cover only small, select portions of a landscape, making it difficult to generalize about the areal extent or pattern of change (Bahre, 1991, p. 14). Shifting mosaics, resulting from cyclical replacements of species, may give the appearance of directional change, depending on the time-scale of observation (Remmert, 1991). Some portions of landscapes may be dynamic and responsive to changes in disturbance or environment, whereas other portions remain static, perhaps controlled by topoedaphic constraints. Serial photographs might therefore tell different stories, depending on when and where they were taken.

Stable carbon isotopes

Naturally occurring stable isotopes of carbon (^{13}C and ^{12}C) in CO_2 are differentially incorporated into vegetation in the process of photosynthesis. Plants with the C_3 photosynthetic pathway discriminate against ^{13}C to a greater extent than plants with the C_4 pathway (Bender, 1968). Tissues of plants with the C_3 pathway have a characteristic $^{13}C/^{12}C$ ratio (expressed as $\delta^{13}C$) of c. $-27‰$, whereas organic matter of C_4 plants is c. $-12‰$ (Smith and Epstein, 1971). Tropical and subtropical systems are dominated by C_4 grasses, their proportionate contribution to the flora decreasing with increasing latitude (Teeri and Stowe, 1976) and elevation (Tieszen *et al.*, 1979; Boutton *et al.*, 1980). Woody plants and herbaceous dicotyledons, with few exceptions, have the C_3 pathway. The $\delta^{13}C$ values of plant tissues are only modified slightly during decomposition. Accordingly, the proportionate contribution of C_3 and C_4 plants as carbon sources contributing to the tissues of heterotrophs and the organic matter of soils can be quantified by measuring the $^{13}C/^{12}C$ ratio in samples (see Tieszen and Boutton, 1989).

The isotopic 'memory' of soils can be queried by analysing the $\delta^{13}C$ of bulk soil organic carbon (SOC) or the carbon associated with various soil particle size-class fractions (sand, silt, clay), which differ in their turnover rates. The resultant signature is a direct reflection of the proportionate input of C_3 and C_4 vegetation to SOC integrated over long periods. If current vegetation has been a long-term occupant of the site, the SOC of soils should be comparable to that being put in by foliage, stems and roots. If a shift in the proportion of C_3 and C_4 plants has occurred, changes in SOC will lag behind changes in vegetation composition and reflect the input from previous plants long after they are gone. The extent to which $\delta^{13}C$ of vegetation and SOC are in equilibrium with each other is thus a quantitative indicator of vegetation history (Tieszen and Archer, 1990).

Changes in $\delta^{13}C$ with soil depth are an indirect measure of time, which can be corroborated by ^{14}C analyses. Quantification of $\delta^{13}C$ with depth can therefore provide a continuous record of vegetation composition from the past through the present. Palynological, archaeological or pack-rat midden techniques of vegetation reconstruction are site-specific and contain artificialities resulting from long-distance dispersal, differential preservation of materials and human or animal selection biases, which limit quantitative interpretation. In contrast, the $\delta^{13}C$ technique can be applied in a spatially explicit fashion and will quantitatively represent the proportionate biomass contribution of C_3 and C_4 plants to a given location over time. The following sections highlight some applications in grazed ecosystems.

Have historical increases in atmospheric CO_2 favoured C_3 shrubs over C_4 grasses?

It is difficult to assess the relative contribution of the various factors that may lead to directional shifts in vegetation. One novel explanation offered to account for historical vegetation change on rangelands centres around the hypothesis that increases in atmospheric CO_2 since the industrial revolution (*c.* 30%) have favoured C_3 plants over C_4 plants (Idso, 1992; Polley *et al.*, 1992; Johnson *et al.*, 1993). The historical displacement of C_4 grasses in tropical and subtropical regions by C_3 woody plants may not reflect changes in climate, fire or grazing regimes, but a differential response of their photosynthetic physiologies to increases in CO_2. This hypothesis is difficult to test, because assessments of vegetation change attributable to environmental effects on photosynthetic pathways are often confounded by life-form (e.g. grass vs. shrub vs. tree) or growth-form (e.g. evergreen vs. deciduous) differences in growth rate, phenology, canopy and root architecture and stress tolerance unrelated to C_3 or C_4 physiology. A more rigorous evaluation of the historical CO_2 enrichment hypothesis could be achieved if life-form or growth-form differences could be minimized to isolate the effect of photosynthetic pathway on historical changes in plant distribution and abundance.

Have there been historical shifts in C_3–C_4 distribution and abundance in accordance with the CO_2 enrichment hypothesis where plant life-form or growth-form differences are minimal? In the southwestern USA, the suffrutescent shrubs *Atriplex confertifolia* (shadscale) and *Ceratoides lanata* (winterfat) are widespread and achieve local dominance. *A. confertifolia* is C_4, whereas *C. lanata* is C_3. In other respects, these plants are quite similar. Both are members of the *Chenopodiaceae* and comprehensive studies have revealed few differences in productivity, water use efficiency and soil moisture utilization (Caldwell *et al.*, 1977). Dzurec *et al.* (1985) quantified $\delta^{13}C$ of soil organic matter along transects spanning contiguous, monospecific stands of each species. ^{13}C values of roots and soil organic matter under *Ceratoides* were in equilibrium with the current plant community. In contrast, $\delta^{13}C$ values of roots and soils under *Atriplex* portions of the transects were more negative than would be expected for a C_4-dominated community. Results indicate that the C_4 shrub, *A. confertifolia*, has increased in importance. This is contrary to predictions of the historical CO_2 enrichment hypothesis and suggests that other factors (Boutton *et al.*, 1994; Archer *et al.*, 1995) may have been more important in producing vegetation change on grazed rangeland.

Has livestock grazing contributed to regional desertification?

The origin and geographic extent of some biomes and their regional associations is the subject of frequent debate. In some cases, climatic and edaphic factors may determine the composition and extent of grasslands and savannas (Walker, 1987). In other instances, grasslands and savannas may be the result of forest and woodland conversion by indigenous people and settlers (Gadgil and Meher-Homji, 1985; Stott, 1991). The extensive grasslands of southern Africa may have existed for millennia, the result of climatic, edaphic or pyric determinants. Alternatively, these grasslands may be the result of extensive removal of trees by indigenous peoples and European settlers (Ellery and Mentis, 1992). Grazing in the drier regions of the country may have contributed to the expansion of arid Karoo shrublands into grasslands since settlement (Acocks, 1953). Thus, the extent of geographical change in shrubland and grassland boundaries, if any, is not clear (Hoffman and Cowling, 1990).

Bond *et al.* (1994) examined the geographical extent of South African grasslands using $\delta^{13}C$ techniques. The sites inspected were typically dominated (> 50% cover) by shrubs. Stable shrublands with little C_4 grass biomass were characteristic of the southwestern regions ($\delta^{13}C$ values strongly C_3 throughout the profile). Isotopic signatures indicated that the proportion of shrub biomass has increased in the central Karoo. Soils in the northeast were characterized by C_4 carbon at depths below 10 cm, indicating long-term past domination by C_4 grasses. Summer rainfall for the 11 sites across the region was strongly correlated with $\delta^{13}C$ values at each soil depth. Shrubs dominated where summer rainfall was below 150 mm; areas receiving above 280 mm were dominated by C_4 grasses.

The slope of the relationship between summer rainfall and $\delta^{13}C$ decreased with increasing soil depth, suggesting that the importance of summer rainfall in determining shrub–grass biomass has decreased over time. The results indicate that changes in land use (livestock grazing) have reduced grass abundance relative to the climatic potential.

This study supports the view that grass cover has declined under grazing pressure, but not that grasslands covered most of the central Karoo before settlement. This knowledge will facilitate evaluation of land management impacts on vegetation and temper expectations for range improvement and rehabilitation efforts. Establishment of grassland in portions of the Karoo which have historically been shrublands is probably unrealistic; efforts should be concentrated in the higher summer rainfall zones where grasses have historically flourished.

Resolution of conflicting assessments of historical woodland boundaries

Boundaries between grasslands and shrubland, woodland or forest systems can be dynamic or static, depending on soils and geomorphology, disturbance regimes, climatic stability and the spatial or temporal scales of observation (Longman and Jeník, 1992). In southwestern North America, post-Pleistocene, mid-latitude woody plant communities appear to have retreated upslope and to have been replaced by grasslands, a response to warmer, drier climatic conditions (Betancourt *et al.*, 1990). Historical–modern ground photographs suggest that *Quercus* woodlands in Arizona, USA have continued to recede upslope over the past century (Hastings and Turner, 1965). A shift toward a more xeric climate during this period is presumed to have caused this change. However, climatic records do not indicate significant changes in rainfall and other repeat photography suggests that *Quercus* woodland boundaries have been stable (Bahre, 1991). No clear generalizations emerge which enable us to infer how vegetation might have changed on landscapes for which there are no historical photographs.

$\delta^{13}C$ analysis of SOC provides the capability of assessing site-specific patterns of changes in grass–woody plant abundance. When applied to soils from stands in southeastern Arizona, it was determined that C_3 *Quercus* and *Prosopis* trees occupied soils whose isotopic signature reflected prior domination by C_4 grasses (McPherson *et al.*, 1993). Discrepancies in isotopic composition between the current vegetation and SOC indicate that patches dominated by woody plants are recent. The isotopic data provide direct evidence that woodland margins have been advancing at this site. This is contrary to historical photography, which suggests that woodland boundaries have been either static or retreating.

Historical accounts and archaeological records often indicate that woody vegetation was present in grasslands, but restricted to riparian corridors and intermittent drainages and, as gallery forest stands, associated with escarpments and steep topography. The former sites could have favoured woody vegetation by affording deeper soils and better water relations, the latter by conferring a degree of protection from fire. It is inferred that woody plants have subsequently spread from these historical enclaves and increased in abundance in other portions of the landscape. $\delta^{13}C$ analysis of SOC supports this contention in some areas, but not in others. For example, along the Niobrara River and its short tributary streams, past woodlands appear to have been more narrowly restricted to lower canyon slopes than the current woodlands (Steuter et al., 1990). Isolated islands of woodland vegetation were also identified within grasslands on the upper canyon slopes prior to European settlement. These historical patches have since been engulfed by woodlands expanding from the lower slopes. In this situation, $\delta^{13}C$ analyses confirm historical observations. In other instances, assumptions of historical occupancy of certain landscape elements by woodlands do not appear valid. $\delta^{13}C$ reconstructions in southern Texas savannas indicate that closed-canopy woodlands of present-day intermittent drainages were dominated by C_4 grasses (Boutton et al., 1993). This represents a case where vegetation history would have been incorrectly inferred from generalizations based on historical reports.

Grassland-to-woodland succession: corroboration of mechanisms

An understanding of successional processes and identification of states and transitions is of interest in cases where vegetation changes are thought to have occurred. A chronosequence of bush clump development in the succession from grasslands to woodland has been proposed for savanna parklands of southern Texas, USA (Archer et al., 1988). Their scenario is based upon inferences derived from 'space-for-time substitution' studies of vegetation structure (Van der Maarel and Werger, 1978). It is desirable to independently corroborate the proposed chronosequence, because inferences from this static approach can be misleading (Austin, 1980; Shugart et al., 1981). If shrub clusters have been a long-term constituent of the landscape, the $\delta^{13}C$ of SOC beneath them should fall in the -27 to $-32‰$ range. However, if C_3 shrubs have displaced C_4 grasses: (i) SOC $\delta^{13}C$ values would be less negative than -27 to $-32‰$; (ii) the degree of departure from the expected $\delta^{13}C$ would decrease as time of site occupancy by shrubs increases; and (iii) SOC $\delta^{13}C$ values would become less negative with depth along the chronosequence.

An analysis of SOC $\delta^{13}C$ confirmed these predictions (Archer, 1990). The SOC beneath herbaceous zones was strongly C_4 and reflected the composition of the current vegetation throughout the profile ($\delta^{13}C = -14$ to $-18‰$). In contrast, mean $\delta^{13}C$ values in the upper horizon of soils beneath clusters at early and late stages of development was -21 and $-23‰$, respectively, reflecting the passage of time and development of *Prosopis* plants and clusters. Among soils supporting woody vegetation, the contribution of C_3-derived carbon decreased with depth to 60 cm, converging on the values observed for the herbaceous zones. The observed SOC $\delta^{13}C$ values provide direct evidence that woody plants have displaced grasses on these landscapes and that the chronosequence proposed by Archer et al. (1988) is reasonable. Further, they lend credibility to models (Archer, 1989;

Scanlan and Archer, 1991) which indicate that succession from grassland to woodland began about 100–150 years ago.

Biogenic opals

Opaline phytoliths (SiO_2) are formed in plants when silicon passively enters the transpiration stream and precipitates in foliage. These microscopic particles, also known as biogenic opals, plant opals, silica bodies or bioliths, are added to the soil via litter fall. Grasslands commonly contribute five to 20 times more opal than woody plant communities, and the shapes and sizes of opals derived from grasses differ from those derived from woody plants. Accordingly, soil opals have been used to detect changes between grass and woody plant domination (Kalisz and Boettcher, 1990) and to document the stability of boundaries between adjacent grassland and woodland communities (Kalisz and Stone, 1984; Fisher et al., 1986). The morphology of opals produced by grasses in the Chloridoid, Panicoid and Festucoid tribes differ, and their relative abundance in the soil can be quantified and compared with the extant vegetation. Opal phytoliths are highly resistant to weathering, and problems of long-distance transport, which occur with pollen-based reconstructions, are minimized. Some care is required in interpreting phytolith assemblages, as their composition can be affected by decay, fire, herbivory and fluvial/colluvial deposition (Piperno, 1988; Fredlund and Tieszen, 1994). As with the $\delta^{13}C$ approach, analysis of soil opals can be conducted in a spatially explicit fashion, both by soil depth and across landscapes or regions. In contrast to $\delta^{13}C$ analysis of SOC which provides only a low resolution of compositional change (C_3 vs. C_4 plants), biogenic opals can be used to determine shifts between tallgrass and shortgrass composition (Twiss et al., 1969), replacement of perennial grasslands by annual grasslands (Bartolome et al., 1986) and grassland vs. woodland composition shifts in temperate zones where there are few C_4 plants.

Dendroecology

Spatially explicit changes in grass–woody plant abundance can be documented by $\delta^{13}C$ or opal phytolith analysis, but they do not provide detailed time lines for the rates and dynamics of change. ^{14}C dating can be employed, but this is of limited use because of the expense, low precision and poor resolution, particularly for recent (the past 200 years) carbon. Analysis of annual rings of woody plants is a powerful tool in reconstructing the rates and dynamics of plant growth and stand history. Traditionally, tree ring research has focused on long-lived forest species with an emphasis on climate reconstructions. Ecological applications quantifying fire history and differences in species growth and population structure in relation to soils, disturbance, succession and annual rainfall are accumulating (Henry and Swan, 1974; Stewart, 1986; Swetnam and Lynch, 1989; Johnson and Young,

1992; Villalba *et al.*, 1994). Although the potential exists to use techniques in dendroecology in shrub- and woodland systems (Ferguson, 1964; Roughton, 1972; Wyant and Reid, 1992; Keeley, 1993), surprisingly few studies have been undertaken.

No rings, false rings, double rings

Utilization of woody plants for 'dendroecology' requires species that produce distinguishable rings, which can be dated with dendrochronology, and attain sufficient age to provide the time control required for a particular investigation (Fritts and Swetnam, 1989). Species vary widely in the extent to which growth rings are discernible, and growth anomalies such as 'missing rings' and 'false rings' occur. It is necessary to verify the annual nature of ring deposition if techniques in dendrochronology are to be reliably applied. This is particularly important for woody plants, whose ranges extend into tropical and subtropical environments, where temperatures are mild year-round and patterns of ring deposition may be in response to wet and dry periods (Jacoby, 1989). Many dryland woody plants are also capable of vegetative regeneration from lignotubers, below-ground stems or roots following disturbance. As a result, ages of current stems or trunks (ramets) may not necessarily reflect the antiquity of the plant (genet), which may produce many generations of stems (Table 4.2) (Wellington *et al.*, 1979; Vasek, 1980; Grimm, 1983). Present-day stems of such plants may contain information on site history acquired during their lifetime, but do not necessarily represent the potentially longer history of plant occupancy.

In many dryland systems, the suitability of woody plants for tree ring analysis is not known and must be ascertained. Flinn *et al.* (1994) determined that the dryland tree legume *P. glandulosa* produced annual rings across a broad north–south temperature gradient and east–west annual rainfall gradient by examining rings on

Table 4.2. Comparison of *Prosopis glandulosa* (mesquite) stem (ramet) ages determined from annual ring counts with estimated plant (genet) age determined from ^{14}C ageing of underground burls or lignotubers which gave rise to those stems (S. Archer, unpublished data). Lignotubers of plants 1–4 dated 'modern' (< 200 years) in accordance with trunk ages. However, the burl of plant 5 was of significantly greater antiquity than the trunk it produced. Ages determined from annual ring counts of above-ground stems may not reflect the true age of woody plants, which vegetatively regenerate after disturbance.

Plant	Estimated age years	
	Lignotuber	Trunk
1	190 ± 75	73 ± 1
2	210 ± 80	81 ± 1
3	50 ± 53	73 ± 6
4	185 ± 75	79 ± 6
5	510 ± 75	67 ± 3

plants from stands with known management histories. They also found that special sanding and staining techniques helped highlight annual rings of *P. glandulosa*.

Stands of known age may be impossible to locate. How, then, can annual ring production be validated? McAuliffe (1988) utilized a novel approach whereby he scarred basal stems of *Larrea tridentata* by removing a 1-mm strip of outer bark and cambium. Secondary xylem deposition subsequently occurred around the entire perimeter of the trunk, except where the cambium was removed. The scar thus provided a permanent marker distinguishing the wood deposited before and after the cambium was removed. Cross-sections were subsequently harvested at various dates after scarring, the amount of xylem deposition since the date of cambium scarring was determined, and annual ring deposition was verified.

Ecological applications in rangelands

Dendroecology can be used, where annual ring production has been confirmed, to quantify regional synchronization between climatic conditions and wildfire (Swetnam and Lynch, 1989), to reconstruct local fire histories (Arno and Wilson, 1986) and to demonstrate how declines in fire frequency associated with the advent of livestock grazing have been accompanied by an increase in woody plant establishment in grasslands and savannas (Madany and West, 1983; Arno and Gruell, 1986; Savage and Swetnam, 1990). Timing of woody plant establishment and rates and patterns of stand development in grazed systems have also been quantified using dendrochronology. McPherson and Wright (1990) used dendrochronology and climatic records to demonstrate that consecutive years of above-average rainfall were required to trigger *Juniperus pinchotii* establishment on western Texas grasslands; one wet year is required for seed production and a second wet year for seedling establishment. Steinauer and Bragg (1987) documented accelerated rates of tree establishment in prairies since settlement and showed the rates of establishment were greatest on north-facing slopes. Sequential invasion of *Artemisia nova* shrublands, first by *Juniperus osteosperma* and later by *Pinus monophylla*, was quantified by Blackburn and Tueller (1970), using tree ring analysis. Such studies indicate the utility of dendroecology for assessing site-specific changes in grass and woody plant abundance. Given their potential for providing unique ecological information, annual ring production in arid and semiarid shrub and arborescent species should be thoroughly investigated.

Repeat aerial photography

The areal extent of spatial reconstructions of vegetation change based on $\delta^{13}C$, opal phytoliths and dendroecology are limited by the time, labour and financial costs associated with collecting and processing of samples. When available, historical aerial photographs are a means of obtaining more extensive, landscape-level assessments of rates, dynamics and patterns of change in herbaceous and woody plant distribution. Constraints on the use of sequential aerial photography include: (i) historical resolution (photographs seldom predate the 1930s or 1940s); (ii) frequency (time elapsed between successive photo dates is variable and not necessarily related to ecologically significant events); (iii) spatial resolution (photography may not be sufficiently detailed to detect change, or the scale may vary between dates, making it difficult to obtain accurate comparisons); (iv) image quality (quality of imagery may be insufficient to enable accurate determinations and the quality may vary between dates); and (v) parallax distortion, tilt and terrain effects (Bolstad, 1992). However, when available, sequential aerial photography can be used to quantify and integrate the outcome of interactions among short-term, small-scale processes and climatic fluctuation on large-scale, long-term vegetation patterns. As such, it provides a useful tool for quantifying vegetation dynamics on landscapes at spatial and temporal scales relevant to perennial plant life histories, secondary succession and land management.

States, transitions and boundary dynamics

As discussed previously, management and conservation of grazed rangelands depends on knowledge of likely vegetation states and transitions affecting those states (Fig. 4.2). Application of state and transition models in research and management are constrained by a lack of quantitative details regarding the identification of states and probabilities of change between states on a particular landscape unit. One approach involves the computation of transition probabilities. Matrix projection models have been traditionally utilized to analyse, interpret and project plant demography and life-cycle attributes (Caswell, 1989). Applications on rangelands include analysis of grass demography with respect to fire (Silva *et al.*, 1991), grazing and drought (O'Connor, 1993) and fertilization and seeding (Scott *et al.*, 1990). The approach has also seen limited use in predicting species interactions and succession with repeated, ground-based measurements (Redetzke and Van Dyne, 1976; Austin, 1980; Burrows *et al.*, 1985) or size/age/fecundity relationships (McAuliffe, 1988; Yeaton and Bond, 1991). As the following case-studies indicate, this approach can be used in conjunction with aerial photography in lieu of long-term permanent plot data to identify, quantify and predict patch and community 'states and transitions' at spatial and temporal scales pertinent to succession and landscape management (see Hall *et al.*, 1991, for applications in satellite imagery).

Changes in savanna tree cover

To quantify thornbush encroachment in a heavily grazed Botswana shrub savanna, Van Vegten (1983) distinguished and mapped eight woody plant canopy cover classes (ranging from < 1% to > 75%) on aerial photographs from 1950, 1963 and 1975. The woody plant biomass represented by each cover class was estimated from field sampling. Average woody plant biomass nearly tripled on this site over the 25-year period. A similar increase in woody plant cover was noted in heavily grazed subtropical savannas in southern Texas, USA (Fig. 4.3a). Drought and grazing interacted to influence the dynamics of woody plant encroachment (Archer, 1994), which was spatially variable across the landscape (Archer, 1995a). The outcome of these interactions was quantified by aerial photography at spatial and temporal scales not possible using ground-level clipping experiments and monitoring regimes.

Reconstructing the past, predicting the future

The successional development of woodland communities depicted in Fig. 4.3a has been elucidated (Archer *et al.*, 1988) and $\delta^{13}C$ analyses confirm that woody plant complexes occupy sites once dominated by C_4 grasses (see under Stable Carbon Isotopes above). Is it possible that successional processes producing a shift from grassland to woodland could have occurred since settlement of this area in the mid to late 1800s? To address this question, rates of canopy expansion of woody clusters representing successional age states were determined from historical aerial photography, related to annual rainfall and modelled to determine woody plant size–age relationships over an array of annual precipitation regimes (Archer, 1989). The model predicted that sizes of present-day woody plant assemblages could have been achieved within the past 100 years under the annual rainfall regimes characteristic of the region. Estimates of woody plant growth rates derived from canopy expansion measurements on aerial photos (0.8–1.9 mm year^{-1} radial trunk growth) were consistent with measurements from dendrometer bands over 6 years (0.2 to 2.0 mm year^{-1}) (S. Archer, unpublished).

Use of aerial photography to parameterize woody plant growth models and determine age-class distributions provides a population biology perspective on woody plant encroachment and stand development, but does not represent the net result of spatial variation in recruitment, growth and mortality of plants across the landscape over time. For this perspective, Scanlan and Archer (1991) used aerial photography to quantify probabilities of transition between seven vegetation age states corresponding to previously defined seral stages in succession from grassland to woodland. These transitions were used in a matrix projection model to assess how landscape composition might vary over time. Vegetation states were assessed in grids of 20 m × 20 m cells superimposed on aerial photos. Transitions were calculated for a drought period (1941–1960) and a normal annual rainfall period (1960–1983) and were found to differ significantly between these two periods. Subsequent theoretical analyses indicated the analytic solutions of this non-homogeneous matrix projection approach provided a reasonable explanation of observed vegetation dynamics (Li, 1995). Past and future landscape structure was modelled by randomly selecting 'normal' and 'dry' transitions at 20-year time steps. The model was run for a series of rainfall scenarios, ranging from 'normal' chosen at each time step to 'dry' chosen at each time step. Forward and reverse simulations were used to project and reconstruct vegetation change. Linkage of

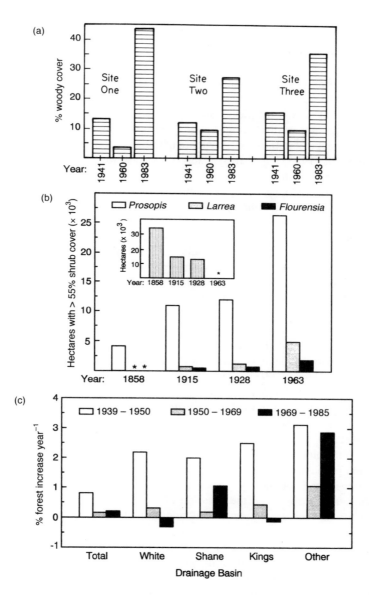

Fig. 4.3. Changes in woody plant abundance in grasslands and savannas representing three climatic zones in North America have been non-linear and spatially variable. (a) *Prosopis–Acacia* savanna in southern Texas; a severe drought occurred in this region in the 1950s (from Archer *et al.*, 1988); (b) Chihuahuan Desert, New Mexico. Insert depicts acreage with no woody plants. Asterisks denote zero acres (from Buffington and Herbel, 1965); and (c) gallery forest expansion rates in a Kansas tallgrass prairie. Primary woody plants were *Quercus, Celtis* and *Ulmus* (from Knight *et al.*, 1994).

this successional model to an ecosystem biogeochemistry model has enabled evaluation of the landscape-level impacts of grazing on carbon and nitrogen dynamics (Hibbard, 1995).

Shifting mosaics and boundary dynamics

Historical aerial photographs substantiate the contention that riparian corridors, intermittent drainages and escarpments are an important source of propagules for the lateral spread of woody plants into uplands with the advent of livestock grazing and fire suppression. Knight *et al.* (1994) document an increase in the number and acreage of gallery forest patches in tallgrass prairie in North America between 1939 and 1985 (Fig. 4.3(c)). Rates of increase were not linear, nor were they uniform from one drainage basin to another.

Biological interactions acting in concert with disturbance and variation in topography and substrate may produce complex transitional changes among community mosaics. These changes can be difficult to evaluate because they occur at spatial scales and over time-frames not amenable to traditional experimentation or monitoring. Patterns of cyclic succession that occur over decadal time-frames and at landscape spatial scales may be erroneously interpreted as directional succession if observations are made over shorter time intervals at the patch level. However, by comparing changes in mosaic patterns on landscapes with different management histories, some insights regarding land-use influences on rates of transition among vegetation states can emerge.

A recent example involves quantification of shifts among grassland, coastal sage scrub, chaparral and oak woodland communities in central coastal California (Callaway and Davis, 1993). Shifts were determined by comparing aerial photographs from 1947 and 1989. Geomorphic substrate, soil type, aspect and topography were recorded in randomly located 'plots' on the 1947 photographs. Vegetation cover was classified in the plots distributed across areas which differed in their fire and grazing history and compared with that recorded for the same plots relocated on 1989 photographs. A matrix projection model based on transition probabilities was then developed. Results (Fig. 4.4) indicated that: (i) transitions among community types were high, even in the absence of grazing and burning; (ii) fire reduced the invasion of grassland by coastal sage scrub, converted coastal sage scrub to grassland and limited oak woodland expansion; (iii) effects of fire on grass–woody plant ratios varied with soil type; (iv) grassland to coastal sage to oak woodland to grassland transitions occurred, suggesting cyclic succession; (v) chaparral shrubland and oak woodland rarely replaced grassland directly, but both rapidly replaced the coastal sage scrub, which directly replaced grassland; and (vi) livestock grazing had little influence on transitions from grassland to coastal sage scrub to oak woodland.

As the above studies indicate, aerial photography can help elucidate and quantify the outcome of interactions between biology, disturbance and the physical environment (see also Richardson and Brown, 1986; Williams *et al.*, 1987).

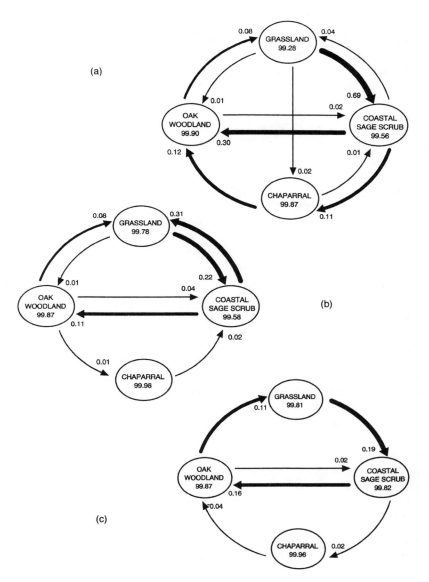

Fig. 4.4. Annual transition rates among vegetation states with different grazing and burning histories in central California determined from aerial photography (1947 vs. 1989). Numbers in ovals represent the probability, as a percentage, that a given community will remain the same; numbers on the arrows estimate the probability that a community will change in the indicated direction. (a) No fire, no livestock grazing; (b) fire, no livestock grazing; and (c) livestock grazing, no fire (from Callaway and Davis, 1993).

SUMMARY

An understanding of factors affecting the composition and productivity of communities through time is of fundamental interest to plant ecologists. However, the world's plant cover is complex and variable and much is not readily accessible for scientific study. We are typically forced to extrapolate our knowledge of plant and community response to grazing obtained from short-term, small-scale studies with little understanding of their historical context or with little knowledge of how to apply them across a landscape, to other landscapes or over longer time periods. Expansion and proliferation of unpalatable woody plants is often associated with livestock grazing in arid and semiarid regions. However, there has been little quantification of the rate, dynamics, pattern and extent of these vegetation changes. As a result, we are often left to speculate whether this sort of vegetation change has occurred and what the proximate causes might have been.

In many instances confusion, contradictions or inconsistencies regarding impacts of grazing on vegetation composition and dynamics can be resolved if: (i) site or land-use history is known; and (ii) processes are expressed in a spatially explicit manner which considers soils, topography and geomorphology at time intervals appropriate for evaluation of species interactions, plant life histories and climatic variation. Stable carbon isotope chemistry, biogenic opal inventories, dendroecology and repeat aerial photography are underutilized tools capable of providing information needed to reconstruct, understand and interpret vegetation dynamics in rangeland ecosystems. Used alone or in concert, these tools enable quantification of past changes in plant distribution in a spatially explicit framework over time-frames and spatial scales relevant to management of grazed landscapes. As databases generated from studies using these approaches accumulate, we can explicitly refine and better evaluate hypotheses and conceptual models of vegetation dynamics in grazed ecosystems. When used in conjunction with mechanistic investigations of factors influencing patterns and processes at plant and population levels of resolution, demographic data, biogeochemical simulation modelling and climatic records, a more complete, accurate and spatially explicit representation of grazing impacts on landscape structure can emerge. Armed with a more accurate historical perspective, we can design better monitoring schemes, more objectively evaluate land management impacts on vegetation, improve our predictive capabilities and temper our expectations with regard to range improvement practices and rehabilitation efforts.

ACKNOWLEDGEMENTS

My thanks to J. Hodgson, G.R. McPherson, B. Northup, T. O'Connor and R.I. Yeaton for reviewing earlier drafts and making helpful suggestions.

REFERENCES

Acocks, J.P.H. (1953) *Veld Types of South Africa*. Memoirs, Botanical Survey of South Africa 28, Botanical Research Institute, Pretoria.
Archer, S. (1989) Have southern Texas savannas been converted to woodlands in recent history? *American Naturalist* 134, 545–561.
Archer, S. (1990) Development and stability of grass/woody mosaics in a subtropical savanna parkland, Texas, USA. *Journal of Biogeography* 17, 453–462.
Archer, S. (1994) Woody plant encroachment into southwestern grasslands and savannas: rates, patterns and proximate causes. In: Vavra, M., Laycock, W. and Pieper, R. (eds) *Ecological Implications of Livestock Herbivory in the West*. Society for Range Management, Denver, pp. 13–68.
Archer, S. (1995a) Tree–grass interactions in a *Prosopis*–thornscrub savanna parkland: reconstructing the past and predicting the future. *Ecoscience* 2, 83–99.
Archer, S. (1995b) Herbivore mediation of grass–woody plant interactions. *Tropical Grasslands* 29, 218–235.
Archer, S. and Smeins, F.E. (1991) Ecosystemlevel processes. In: Heitschmidt, R.K. and Stuth, J.W. (eds) *Grazing Management: An Ecological Perspective*. Timberline Press, Portland, Oregon, pp. 109–139.
Archer, S.R. and Tieszen, L.L. (1986) Plant response to defoliation: hierarchical considerations. In: Gudmundsson, O. (ed.) *Grazing Research at Northern Latitudes*. Plenum Press, New York, pp. 45–59.
Archer, S., Scifres, C.J., Bassham, C.R. and Maggio, R. (1988) Autogenic succession in a subtropical savanna: conversion of grassland to thorn woodland. *Ecological Monographs* 80, 272–276.
Archer, S., Schimel, D.S. and Holland, E.A. (1995) Mechanisms of shrubland expansion: land use, climate or CO_2? *Climatic Change* 29, 91–99.
Arno, S.F. and Gruell, G.E. (1986) Douglas fir encroachment into mountain grasslands in southwestern Montana. *Journal of Range Management* 39, 272–276.
Arno, S.F. and Wilson, A.E. (1986) Dating past fires in curlleaf mountain mahogany communities. *Journal of Range Management* 39, 139–164.
Austin, M.P. (1980) An exploratory analysis of grassland dynamics: an example of lawn succession. *Vegetatio* 43, 87–94.
Bahre, C.J. (1991) *A Legacy Of Change: Historic Human Impact on Vegetation of the Arizona Borderlands*. University of Arizona Press, Tucson.
Bartolome, J.W., Klukkert, S.E. and Barry, W.J. (1986) Opal phytoliths as evidence for displacement of native Californian grasslands. *Madrōno* 33, 217–222.
Bender, M.M. (1968) Mass spectrometric studies of carbon 13 variations in corn and other grasses. *American Journal of Science, Radiocarbon Supplement* 10, 468–472.
Betancourt, J.L., Devender, T.R.V. and Martin, P.S. (1990) Synthesis and prospectus. In: Betancourt, J.L., Devender, T.R.V. and Martin, P.S. (eds) *Packrat Middens: The Last 40,000 Years of Biotic Change*. University of Arizona Press, Tucson, pp. 435–447.
Billings, W.D. (1994) Ecological impacts of cheatgrass and resultant fire on ecosystems in the western Great Basin. In: Monsen, S.B. and Kitchen, S.G. (eds) *Proceedings: Ecology and Management of Annual Rangelands*. General Technical Report INT-GTR-313, US Department of Agriculture/Forest Service, Intermountain Research Station, Ogden, Utah, pp. 22–30.

Blackburn, W.H. and Tueller, P.T. (1970) Pinyõn and juniper invasion in black sagebrush communities in eastcentral Nevada. *Ecology* 51, 841–848.

Blackmore, A.C., Mentis, M.T. and Scholes, R.J. (1990) The origin and extent of nutrientenriched patches within a nutrient-poor savanna in South Africa. *Journal of Biogeography* 17, 463–470.

Bolstad, P.V. (1992) Geometric errors in natural resource GIS data: tilt and terrain effects in aerial photographs. *Forest Science* 38, 367–380.

Bond, W.J. (1993) Keystone species. In: Schulze, E.-D. and Mooney, H.A. (eds) *Biodiversity and Ecosystem Function*. Springer-Verlag, New York, pp. 237–254.

Bond, W.J., Stock, W.D. and Hoffman, M.T. (1994) Has the karoo spread? A test for desertification using carbon isotopes from soils. *South African Journal of Science* 90, 391–397.

Borman, M.M. and Pyke, D.A. (1994) Successional theory and the desired plant community approach. *Rangelands* 16, 82–84.

Bosch, O.J.H. and Booysen, J. (1992) An integrative approach to rangeland condition and capability assessment. *Journal of Range Management* 45, 116–122.

Boutton, T.W., Harrison, A.T. and Smith, B.N. (1980) Distribution of biomass of species differing in photosynthetic pathway along an altitudinal transect in southeastern Wyoming grassland. *Oecologia* 45, 287–298.

Boutton, T.W., Nordt, L.C., Archer, S. and Casar, I. (1993) Stable carbon isotope ratios of soil organic matter and their potential use as indicators of paleoclimate. In: *Applications of Isotope Techniques in Studying Past and Current Environmental Changes in the Hydrosphere and Atmosphere*. International Atomic Energy Agency, Vienna, Austria, pp. 445–459.

Boutton, T.W., Archer, S.R. and Nordt, L.C. (1994) Climate, CO_2 and plant abundance. *Nature* 72, 625–626.

Brown, J.H. and Heske, E.J. (1990) Control of a desert–grassland transition by a keystone rodent guild. *Science* 250, 1705–1707.

Brown, J.R. and Archer, S. (1987) Woody plant seed dispersal and gap formation in a North American subtropical savanna woodland: the role of domestic herbivores. *Vegetatio* 73, 73–80.

Bryant, N.A., Johnson, L.F., Brazel, A.J., Balling, R.C., Hutchinson, C.F. and Beck, L.R. (1990) Measuring the effect of overgrazing in the Sonoran Desert. *Climatic Change* 17, 243–264.

Buffington, L.C. and Herbel, C.H. (1965) Vegetational changes on a semidesert grassland range. *Ecological Monographs* 35, 139–164.

Burrows, W.H., Beale, I.F., Silcock, R.G. and Pressland, A.J. (1985) Prediction of tree and shrub population changes in a semiarid woodland. In: Tothill, J.C. and Mott, J.J. (eds) *Ecology and Management of the World's Savannas*. Australian Academy of Science, Canberra, pp. 207–211.

Caldwell, M.M., White, R.S., Moore, T.T. and Camp, L.B. (1977) Carbon balance, productivity and water use of cold-winter desert shrub communities dominated by C_3 and C_4 species. *Oecologia* 29, 275–300.

Callaway, R.M. and Davis, F.W. (1993) Vegetation dynamics, fire, and the physical environment in coastal central California. *Ecology* 74, 1567–1578.

Caswell, H.A. (1989) *Matrix Population Models: Construction, Analysis and Interpretation*. Sinauer Associates, Sunderland, Massachusetts.

Coleman, D.C., Andrews, R., Ellis, J.E. and Singh, J.W. (1976) Energy flow and partitioning in selected manmanaged and natural ecosystems. *Agroecosystems* 3, 45–154.

Collins, S.L., Bradford, J.A. and Sims, P.L. (1987) Succession and fluctuation in *Artemisia* dominated grassland. *Vegetatio* 73, 89–99.

Coughenour, M.B. (1985) Graminoid responses to grazing by large herbivores: adaptations, exaptation, and interacting processes. *Annals of the Missouri Botanical Garden* 72, 852–863.

Dankwerts, J.E., O'Reagain, P.J. and O'Connor, T.G. (1993) Range management in a changing environment: a southern African perspective. *Rangeland Journal* 15, 133–144.

De Angelis, D.L. and Waterhouse, J.C. (1987) Equilibrium and non-equilibrium concepts in ecological models. *Ecological Monographs* 57, 1–21.

Detling, J.K. (1988) Grasslands and savannas: regulation of energy flow and nutrient cycling by herbivores. In: Pomeroy, L.R. and Alberts, J.J. (eds) *Concepts of Ecosystem Ecology*. Springer-Verlag, New York, pp. 131–148.

Dzurec, R.S., Boutton, T.W., Caldwell, M.M. and Smith, B.N. (1985) Carbon isotope ratios of soil organic carbon and their use in assessing community composition changes in Curlew Valley, Utah. *Oecologia* 66, 17–24.

Ellery, W.N. and Mentis, M.T. (1992) How old are South Africa's grasslands? In: Furley, P.A., Proctor, J. and Ratter, J.A. (eds) *Nature and Dynamics of Forest–Savanna Boundaries*. Chapman & Hall, London, pp. 283–292.

Ellis, J.E. (1992) Recent advances in arid land ecology. In: Valdivial, C. (ed.) *Sustainable Crop–Livestock Systems for the Bolivian Highlands*. University of Missouri Press, Missouri, pp. 1–14.

Ferguson, C.W. (1964) *Annual Rings in Big Sagebrush* Artemisia tridentata. University of Arizona Press, Tucson.

Fisher, R.F., Jenkins, M.J. and Fisher, W. (1986) Fire and the prairie–forest mosaic of Devils Tower National Monument. *American Midland Naturalist* 117, 250–257.

Flinn, R.C., Archer, S., Boutton, T.W. and Harlan, T. (1994) Identification of annual rings in arid land woody plant, *Prosopis glandulosa*. *Ecology* 75, 850–853.

Forman, R.T.T. and Russell, E.W.B. (1983) Evaluation of historical data in ecology. *Ecological Society of America Bulletin* 64, 5–7.

Foster, D.R. (1992) Landuse history (1730–1990) and vegetation dynamics in central New England, USA. *Journal of Ecology* 80, 753–772.

Fredlund, G.G. and Tieszen, L.L. (1994) Modern phytolith assemblages from the North American Great Plains. *Journal of Biogeography* 21, 321–335.

Friedel, M.H. (1991) Range condition assessment and the concept of thresholds: a viewpoint. *Journal of Range Management* 44, 422–426.

Friedel, M.H., Bastin, G.N. and Griffin, G.F. (1988) Range assessment and monitoring in arid lands: the derivation of functional groups to simplify vegetation data. *Journal of Environmental Management* 27, 85–97.

Fritts, H.C. and Swetnam, T.W. (1989) Dendroecology: a tool for evaluating variations in past and present forest environments. *Advances in Ecological Research* 19, 111–188.

Gadgil, M. and Meher-Homji, V.M. (1985) Land use and productive potential of Indian savannas. In: Tothill, J.C. and Mott, J.J. (eds) *Ecology and Management of the World's Savannas*. Australian Academy of Science, Canberra, pp. 107–113.

George, M.R., Brown, J.R. and Clawson, W.J. (1992) Application of non-equilibrium ecology to management of Mediterranean grasslands. *Journal of Range Management* 45, 436–440.

Grimm, E.C. (1983) Chronology and dynamics of vegetation change on the prairie–woodland region of southern Minnesota, USA. *New Phytologist* 93, 311–350.

Hall, F.G., Botkin, D.B., Strebel, D., Woods, K.D. and Goetz, S.J. (1991) Large scale patterns of forest succession as determined by remote sensing. *Ecology* 72, 628–640.

Hastings, J.R. and Turner, R.L. (1965) *The Changing Mile: An Ecological Study of Vegetation Change with Time in the Lower Mile of an Arid and SemiArid Region.* University of Arizona Press, Tucson.

Henry, J.D. and Swan, J.M.A. (1974) Reconstructing forest history from live and dead plant material – an approach to the study of forest succession in southwest New Hampshire. *Ecology* 55, 772–783.

Herbel, C.H., Ares, F.N. and Wright, R.A. (1972) Drought effects on a semidesert grassland. *Ecology* 53, 1084–1093.

Hibbard, K.A. (1995) Landscape patterns of carbon and nitrogen dynamics in a subtropical savanna: observations and models. PhD dissertation, Texas A&M University.

Hobbie, S. (1992) Effects of plant species on nutrient cycling. *Trends in Ecology and Evolution* 7, 336–339.

Hoffman, M.T. and Cowling, R.M. (1990) Vegetation change in the semi-arid eastern Karoo over the last two hundred years: an expanding Karoo – fact or fiction? *South African Journal of Science* 86, 286–294.

Huntsinger, L. and Bartolome, J.W. (1992) Ecological dynamics of *Quercus* dominated woodlands in California and southern Spain: a state–transition model. *Vegetatio* 100, 299–305.

Idso, S.B. (1992) Shrubland expansion in the American southwest. *Climate Change* 22, 85–86.

Jacoby, G.C. (1989) Overview of tree ring analysis in tropical regions. *International Association of Wood Anatomists Bulletin* 10, 99–108.

Janzen, D.H. (1984) Dispersal of small seed by big herbivores: foliage is the fruit. *American Naturalist* 123, 338–353.

Janzen, D.H. (1986) Chihuahuan Desert nopaleras: defaunated big mammal vegetation. *Annual Review of Ecology and Systematics* 17, 595–636.

Johnson, H.B., Polley, H.W. and Mayeux, H.S. (1993) Increasing CO_2 and plant–plant interactions: effects on natural vegetation. *Vegetatio* 104–105, 157–170.

Johnson, S.R. and Young, D.R. (1992) Variation in tree ring width in relation to storm activity for Mid-Atlantic Barrier Island populations of *Pinus taeda*. *Journal of Coastal Research* 8, 99–104.

Jones, R.M. (1992) Resting from grazing to reverse changes in sown pasture composition: application of the 'state-and-transition' model. *Tropical Grasslands* 26, 97–99.

Joyce, L.A. (1993) The life cycle of the range condition concept. *Journal of Range Management* 46, 132–138.

Kalisz, P.J. and Boettcher, S.E. (1990) Phytolith analysis of soils at Buffalo Beats, a small forest opening in southeastern Ohio. *Bulletin of Torrey Botanical Club* 117, 445–449.

Kalisz, P.J. and Stone, E.L. (1984) The longleaf pine islands of the Ocala National Forest, Florida: a soil study. *Ecology* 65, 1743–1754.

Keeley, J.E. (1993) Utility of growth rings in the age determination of chaparral shrubs. *Madroño* 40, 1–14.

Knight, C.L., Briggs, J.M. and Nelis, M.D. (1994) Expansion of gallery forest on Konza Prairie Research Natural Area, Kansas, USA. *Landscape Ecology* 9, 117–125.

Laycock, W.A. (1991) Stable states and thresholds of range condition on North American rangelands: a viewpoint. *Journal of Range Management* 44, 427–433.

Li, B.-L. (1995) Stability analysis of a nonhomogeneous Markovian landscape model. *Ecological Modelling* 82, 247–256.

Longman, K.A. and Jenik, J. (1992) Forest–savanna boundaries: general considerations. In: Furley, P.A., Proctor, J. and Ratter, J.A. (eds) *Nature and Dynamics of Forest–Savanna Boundaries*. Chapman & Hall, New York, pp. 3–18.

Ludwig, J.A. and Tongway, D.J. (1993) Monitoring the condition of Australian arid lands: linked plant–soil indicators. In: McKenzie, D.H., Hyatt, D.E. and McDonald, V.J. (eds) *Ecological Indicators*, Vol. 1. Elsevier Applied Science, New York, pp. 763–772.

Ludwig, J.A., Tongway, D.J. and Marsden, S.G. (1994) A flow-filter model of simulating the conservation of limited resources in spatially heterogeneous semi-arid landscapes. *Pacific Conservation Biology* 1, 209–213.

Lura, C.L. and Nyren, P.E. (1992) Some effects of white grub infestation on northern mixed-grass prairie. *Journal of Range Management* 45, 352–354.

McAuliffe, J.R. (1988) Markovian dynamics of simple and complex desert plant communities. *American Naturalist* 131, 459–490.

McKell, C.M. (1989) Management practices for shrubdominated lands to assure multi-pleuse benefits. In: McKell, C.M. (ed.) *The Biology and Utilization of Shrubs*. Academic Press, San Diego, California, pp. 575–592.

McNaughton, S.J. (1993) Biodiversity and function of grazing ecosystems. In: Schulze, E.-D. and Mooney, H.A. (eds) *Biodiversity and Ecosystem Function*. Springer-Verlag, New York, pp. 361–408.

McNaughton, S.J. (1994) Conservation goals and the configuration of biodiversity. In: Forey, P.L., Humphries, C.J. and Vane-Wright, R.I. (eds) *Systematics and Conservation Evaluation*. Clarendon Press, Oxford, pp. 41–62.

McPherson, G.R. and Wright, H.A. (1990) Effects of cattle grazing and *Juniperus pinchotii* canopy cover on herb cover and production in western Texas. *American Midland Naturalist* 123, 144–151.

McPherson, G.R., Boutton, T.W. and Midwood, A.J. (1993) Stable carbon isotope analysis of soil organic matter illustrates vegetation change at the grassland/woodland boundary in southeastern Arizona, USA. *Oecologia* 93, 95–101.

Madany, M.H. and West, N.E. (1983) Livestock grazing–fire regime interactions within montane forests of Zion National Park, Utah. *Ecology* 64, 661–667.

Magnuson, J.J. (1990) Long-term ecological research and the invisible present. *Bioscience* 40, 495–501.

Malin, J.C. (1953) Soil, animal and plant relations of the grasslands historically recorded. *Scientific Monthly* 76, 207–220.

Milchunas, D.G. and Lauenroth, W.K. (1993) Quantitative effects of grazing on vegetation and soils over a global range of environments. *Ecological Monographs* 63, 327–366.

Milchunas, D.G., Sala, O.E. and Lauenroth, W.K. (1988) A generalized model of the effects of grazing by large herbivores on grassland community structure. *American Naturalist* 132, 87–106.

Miller, R.F. and Wigand, P.E. (1994) Holocene changes in semiarid pinyōn–juniper woodlands. *Bioscience* 44, 465–474.

Milton, S.J., Dean, W.R.J., du Plessis, M.A. and Siegfried, W.R. (1994) A conceptual model of arid rangeland degradation. *Bioscience* 44, 70–76.

Neilson, R.P. (1986) High resolution climatic analysis and southwest biogeography. *Science* 232, 27–34.

Noble, J.C., MacLeod, N.D., Ludwig, J.A. and Grice, A.C. (1991) Integrated shrub control strategies in Australian semi-arid woodlands. In: *Proceedings of the 4th International Rangeland Congress*, Montpellier, France, pp. 846–849.

O'Connor, T.G. (1993) The influence of rainfall and grazing on the demography of some African savanna grasses: a matrix modelling approach. *Journal of Applied Ecology* 30, 119–132.

O'Connor, T.G. (1994) Composition and population responses of an African savanna grassland to rainfall and grazing. *Journal of Applied Ecology* 31, 155–171.

Owen-Smith, N. (1987) Pleistocene extinctions: the pivotal role of megaherbivores. *Paleobiology* 13, 351–362.

Palmer, A.R., Hobson, C.G. and Hoffman, M.T. (1990) Vegetation change in a semi-arid succulent dwarf shrubland in the eastern Cape. *South African Journal of Science* 86, 392–395.

Pastor, J. and Naiman, R. (1992) Selective foraging and ecosystem processes in boreal forests. *American Naturalist* 139, 690–705.

Pieper, R.D. (1994) Ecological implications of livestock grazing. In: Vavra, M., Laycock, W.A. and Pieper, R.D. (eds) *Ecological Implications of Livestock Herbivory in the West*. Society for Range Management, Denver, pp. 177–211.

Piperno, D.R. (1988) *Phytolith Analysis: An Archaeological and Geological Perspective*. Academic Press, San Diego, California.

Polley, H.W., Johnson, H.B. and Mayeux, H.S. (1992) Carbon dioxide and water fluxes of C_3 Annuals and C_3 and C_4 perennials at sub-ambient CO_2 concentrations. *Functional Ecology* 6, 693–703.

Redetzke, K.A. and Van Dyne, G.M. (1976) A matrix model of a rangeland grazing system. *Journal of Range Management* 29, 425–430

Remmert, H. (ed.) (1991) *The Mosaic–Cycle Concept of Ecosystems*. SpringerVerlag, New York.

Richardson, D.M. and Brown, P.J. (1986) Invasion of mesic mountain fynbos by *Pinus radiata*. *South African Journal of Botany* 52, 529–536.

Roughton, R.D. (1972) Shrub age structures on a mule deer winter range in Colorado. *Ecology* 53, 615–625.

Savage, M. and Swetnam, T.W. (1990) Early 19th-Century fire decline following sheep pasturing in a Navajo ponderosa pine forest. *Ecology* 71, 2374–2378.

Scanlan, J.C. and Archer, S. (1991) Simulated dynamics of succession in a North American subtropical *Prosopis* savanna. *Journal of Vegetation Science* 2, 625–634.

Schlesinger, W.H., Reynolds, J.F., Cunningham, G.L., Huenneke, L.F., Jarrell, W.M., Virginia, R.A. and Whitford, W.G. (1990) Biological feedback in global desertification. *Science* 247, 1043–1048.

Scholes, R.J. and Walker, B.H. (1993) *An African Savanna: Synthesis of the Nylsvley Study*. Cambridge University Press, Cambridge.

Scifres, C.J., Hamilton, W.T., Conner, J.R., Inglis, J.M., Rasmussen, G.A., Smith, R.P., Stuth, J.W. and Welch, T.G. (1985) *Integrated Brush Management Systems for South Texas: Development and Implementation*. Texas Agricultural Experiment Station, College Station.

Scott, D., Robertson, J.S. and Archie, W.J. (1990) Plant dynamics of New Zealand tussock grassland infested with *Hieracium pilosella*: transition matrices of vegetation change. *Journal of Applied Ecology* 27, 235–241.

Shugart, H.H., West, D.C. and Emanuel, W.R. (1981) Patterns and dynamics of forests: an application of simulation models. In: West, D.C., Shugart, H.H. and Botkin, D.B. (eds) *Forest Succession: Concepts and Applications*. Springer-Verlag, Heidelberg, pp. 74–94.

Silva, J.F., Raventos, J., Caswell, H. and Trevisan, M.C. (1991) Population responses to fire in a tropical savanna grass, *Andropogon semiberbis*: a matrix model approach. *Journal of Ecology* 79, 345–356.

Sinclair, A.R.E. (1979) Dynamics of the Serengeti ecosystem. In: Sinclair, A.R.E. and Norton-Griffiths, M. (eds) *Serengeti: Dynamics of an Ecosystem*. University of Chicago Press, Chicago, pp. 1–30.

Skarpe, C. (1991) Impact of grazing in savanna ecosystems. *Ambio* 20, 351–356.

Smeins, F.E. and Merrill, L.B. (1988). Long-term change in semi-arid grassland. In: Amos, B.B. and Gehlbach, F.R. (eds) *Edwards Plateau Vegetation*. Baylor University Press, Waco, Texas, pp. 101–114.

Smith, B.N. and Epstein, S. (1971) Two categories of $^{13}C/^{12}C$ ratios for higher plants. *Plant Physiology* 47, 380–384.

Stafford Smith, M. and Pickup, G. (1993) Out of Africa, looking in: understanding vegetation change. In: Behnke, R.H., Jr, Scoones, I. and Kerven, C. (eds) *Range Ecology at Disequilibrium*. Overseas Development Institute, Regent's College, London, pp. 196–226.

Steinauer, E.M. and Bragg, T.B. (1987) Ponderosa pine (*Pinus ponderosa*) invasion of Nebraska sandhills prairie. *American Midland Naturalist* 118, 358–365.

Steuter, A.A., Jasch, B., Ihnen, J. and Tieszen, L.L. (1990) Woodland/grassland boundary changes in the middle Niobrara valley of Nebraska identified by ^{13}C values of soil organic matter. *American Midland Naturalist* 124, 301–308.

Stewart, G.H. (1986) Population dynamics of a montane conifer forest, Western Cascade Range, Oregon, USA. *Ecology* 67, 534–544.

Stott, P. (1991) Stability and stress in the savanna forests of mainland SouthEast Asia. In: Werner, P.A. (ed.) *Savanna Ecology and Management: Australian Perspectives and Intercontinental Comparisons*. Blackwell Scientific Publications, Oxford, pp. 29–41.

Swetnam, T.W. and Lynch, A.M. (1989) A tree ring reconstruction of western spruce budworm history in the southern Rocky Mountains. *Forest Science* 35, 962–986.

Teeri, J.A. and Stowe, L.G. (1976) Climatic patterns and distribution of C_4 grasses in North America. *Oecologia* 23, 1–12.

Thurow, T.L. (1991) Hydrology and erosion. In: Heitschmidt, R.K. and Stuth, J.W. (eds) *Grazing Management: An Ecological Perspective*. Timber Press, Portland, Oregon, pp. 141–160.

Tieszen, L.L. and Archer, S. (1990) Isotopic assessment of vegetation changes in grassland and woodland systems. In: Osmond, C.B., Pitelka, L.F. and Hidy, G.M. (eds) *Plant Biology of the Basin and Range*. Springer-Verlag, New York, pp. 293–321.

Tieszen, L.L. and Boutton, T.W. (1989) Stable carbon isotopes in terrestrial ecosystem research. In: Rundel, P.W., Ehleringer, J.R. and Nagy, K.A. (eds) *Stable Isotopes in Ecological Research*. Springer-Verlag, New York, pp. 167–195.

Tieszen, L.L., Senyimba, M.M., Imbamba, S.K. and Troughton, J.H. (1979) The distribution of C_3 and C_4 grasses and carbon isotope discrimination along an attitudinal and moisture gradient in Kenya. *Oecologia* 37, 337–350.
Tongway, D.J. and Ludwig, J.A. (1994) Small-scale resource heterogeneity in semi-arid landscapes. *Pacific Conservation Biology* 1, 201–208.
Twiss, P.C., Suess, E. and Smith, R.M. (1969) Morphological classification of grass phytoliths. *Soil Science Society of America* 33, 109–115.
Van der Maarel, E. and Werger, M.J.A. (1978) On the treatment of succession data. *Phytocoenosis* 7, 257–278.
Van Vegten, J.A. (1983) Thornbush invasion in a savanna ecosystem in eastern Botswana. *Vegetatio* 56, 3–7.
Vasek, F.C. (1980) Creosote bush: long-lived clones in the Mojave Desert. *American Journal of Botany* 67, 246–255.
Villalba, R., Veblen, T.T. and Ogden, J. (1994) Climatic influences on the growth of subalpine trees in the Colorado front range. *Ecology* 75, 1450–1462.
Walker, B.H. (ed.) (1987) *Determinants of Tropical Savannas*. IRL Press, Oxford.
Walker, B.H. (1993) Rangeland ecology – understanding and managing change. *Ambio* 22, 80–87.
Wellington, A.B., Polach, H.A. and Noble, I.R. (1979) Radiocarbon dating of lignotubers from mallee forms of *Eucalyptus*. *Search* 10, 282–283.
West, N. (1993) Biodiversity of rangelands. *Journal of Range Management* 46, 2–13.
Westoby, M., Walker, B. and Noy-Meir, I. (1989) Opportunistic management for rangelands not at equilibrium. *Journal of Range Management* 42, 266–274.
Whalley, R.D.B. (1994) Successional theory and vegetation change. *Tropical Grasslands* 28, 195–205.
Williams, J. and Chartres, C.J. (1991) Sustaining productive pastures in the tropics. I. Managing the soils resource. *Tropical Grasslands* 25, 73–84.
Williams, K., Hobbs, R.J. and Hamburg, S.P. (1987) Invasion of an annual grassland in Northern California by *Baccharis pulularis* ssp. *consanguinea*. *Oecologia* 72, 461–465.
Wright, H.A. and Bailey, A.W. (1982) *Fire Ecology*. John Wiley, New York.
Wyant, J.E. and Reid, R.S. (1992) Determining the age of *Acacia tortilis* with ring counts for South Turkana, Kenya: a preliminary assessment. *African Journal of Ecology* 30, 176–180.
Yeaton, R.I. and Bond, W.J. (1991) Competition between two shrub species: dispersal differences and fire promote coexistence. *American Naturalist* 138, 328–421.
Yorks, T.P., West, N.E. and Capels, K.M. (1992) Vegetation differences in desert shrublands of western Utah's Pine Valley between 1933 and 1989. *Journal of Range Management* 45, 569–578.
Yorks, T.P., West, N.E. and Capels, K.M. (1994) Changes in pinyōn–juniper woodlands in western Utah's Pine Valley between 1933–1989. *Journal of Range Management* 47, 359–364.

Animals and Animal Populations

Foraging Strategies of Grazing Animals*

E.A. Laca[1] and M.W. Demment[2]

[1]*Department of Range and Wildlife Management, Texas Tech University, Lubbock, TX 79409-2125, USA;* [2]*Department of Agronomy and Range Science, University of California, Davis, CA 95616-8515, USA*

INTRODUCTION

Foraging strategies of grazing animals are of paramount ecological and agronomic importance. Rangelands are the most abundant type of land on earth and are characterized by a dominant role of herbivory, mostly by domestic animals (Holechek *et al.*, 1995). Grazing strategies used by large herbivores to cope with their environments define a cardinal link between primary and secondary productivity. Grazers respond to changes in their resources by means of a grazing strategy. In turn, this strategy affects the structure of the plant community, in a perennial feedback loop. Therefore, an understanding of grazing strategies is at the core of our ability to forecast the effects of local and global changes in grazed systems.

In this chapter we use the term 'strategy' to refer to relevant patterns of foraging behaviour. Strategies emerge as large herbivores face the challenge of obtaining enough energy and nutrients to survive and reproduce efficiently in an environment characterized by spatial and temporal variability (Provenza and Balph, 1990). Emergent patterns are strictly dependent on the observer's goals, point of view and scale of observation. For example, Jarman and Sinclair (1979) underscored evolutionary relationships between feeding strategies and morphology of African ungulates, and sought to explain resource partitioning and variation of diets among species. Perhaps at the other extreme, Provenza (1995) proposed a learning paradigm focused on the development of foraging strategies of ruminants, and sought to explain qualitative variation of diets among

---

* This is Technical article T-9-738 of the College of Agricultural Science and Natural Resources, Texas Tech University. This research was partially supported by NSF grant IBN-9311463 to E.A.L. and BARD grant IS-2331-93C to M.W.D. and E.A.L.

©1996 CAB INTERNATIONAL.
The Ecology and Management of Grazing Systems (eds J. Hodgson and A.W. Illius)

individuals. Thus, different approaches to the study of foraging strategies of grazing animals have resulted in interpretations that differ in which aspects of foraging strategies are emphasized.

In this chapter we argue that foraging theory (Stephens and Krebs, 1986; Mangel and Clark, 1988) offers the strongest theoretical basis and framework to study foraging strategies of grazing animals. Until recently (Table 5.1), theoreticians concerned with grazing animals adopted the modelling approaches of range science and animal husbandry (Hughes, 1990). In our view, a better understanding of grazing strategies is emerging from the combination of various approaches within the context of foraging theory (e.g. Thornley *et al.*, 1994). New foraging models are superior because they can: (i) incorporate aspects from all other approaches; (ii) make qualitative and quantitative predictions; (iii) incorporate known mechanisms of diet selection; and, (iv) incorporate behavioural and other constraints, even if the specific neural mechanisms are not known (e.g. Thornley *et al.*, 1994). Thus, we discuss foraging theory and delineate some of the special features of herbivores in relation to early foraging models. We briefly review recent models and tests of foraging by large herbivores, emphasizing their structure and the constraints on grazing intake.

APPROACHES TO THE UNDERSTANDING AND PREDICTION OF GRAZING BEHAVIOUR

There are three general approaches to understanding and predicting how grazers interact with the plant community. First, one can take a purely empirical approach. This approach can be valuable if one measures responses under a broad range of conditions, but may be inefficient and provides little understanding of underlying processes. Moreover, simple behavioural rules can result in extremely plastic quantitative relationships between plant community characteristics and behavioural patterns (Laca *et al.*, 1993, 1994a; Distel *et al.*, 1995).

Second, one can take a purely mechanistic approach by studying the processes that determine intake behaviour and dietary choice, elucidating phenomena at progressively lower levels of aggregation until a reasonably 'hard-wired' basis of behaviour is found. While this approach improves our understanding of grazing processes, explanations at very fine levels of resolution can become excessively complex for making practical predictions. In addition, some of the mechanisms that result in foraging strategies can be purely behavioural (Newman *et al.*, 1994a) or partly determined by the central nervous system, about which there is not enough knowledge to use a reductionist approach.

Third, one can assume general principles that organize feeding behaviour to produce models that combine understanding of mechanisms and predictive power. This alternative is represented by allocation of behaviour by maximization of 'utility', and by teleonomic models based on the maximization of some currency related to fitness. Malecheck and Balph (1987) refer to these last two approaches as 'principles of reinforcement' or 'causal mechanisms', and 'optimal foraging theory' (OFT) or 'functional approach' respectively.

The teleonomic model of grazing intake and diet selection by Thornley *et al.* (1994) illustrates the power of the OFT approach (Table 5.1). The model assumes that sheep select diets from a two-species mixture and regulate quantity of intake such as to maximize the balance between net energy (NE) benefits and costs of obtaining the forage. In simplified terms, the model assumes that: (i) the NE obtained per unit intake declines with increasing daily intake; and (ii) the specific costs of obtaining intake increase as the difference between diet and available forage composition increases and decrease as leaf area index increases. The first assumption is a widely accepted empirical result (ARC, 1980; NRC, 1985); the second one has some empirical support (Penning, 1986; Penning *et al.*, 1991) and is logical. The model predicts that in certain situations animals limit intake and grazing time behaviourally, without invoking specific morphological or digestive constraints. By reducing grazing time, the model sheep actually maintain a better energy balance than if they grazed for longer. This is a significant contribution in that the prediction of grazing time has been a fundamental but elusive goal of intake models. However, the validity of this model is limited to homogeneous mixtures or situations in which animals cannot increase the encounter rate of preferred plants by means of specialized search strategies.

LEVELS OF EXPLANATION OF FORAGING BEHAVIOUR

The concepts of causal and functional explanations of behaviour, also referred to as proximate and ultimate, have been useful but they have also generated some confusion (Armstrong, 1991). It should be clear that explanations at different causal levels are not mutually exclusive alternatives but complementary. For example, the following explanations of why sheep select green ryegrass patches, when offered a variety of alternatives, are not mutually exclusive: (i) the perception of green food elicits more intense ingestive behaviour than foods of other colours (i.e. behaviour-eliciting stimuli (Armstrong, 1991)); (ii) they learned through reinforcement that biting into green patches is associated with a better food reward (i.e. change in responsiveness to stimuli); (iii) earlier in life they consumed green ryegrass and developed grazing skills and a rumen microflora specially able to handle this kind of forage (i.e. ontogeny of the behaviour); and (iv) green patches provided better nutrition and enhanced the reproductive output of the ancestors of current sheep (i.e. evolution of the behaviour). Often, scientists propose explanations of behaviours as alternative hypotheses, while in fact the explanations are not mutually exclusive because they refer to different levels of explanation.

Table 5.1. Models and tests of foraging strategies of large herbivores.

Reference	Type of study	Animal	Plants	Currency	Decision variables	Constraints
Fryxell, 1991	Static, deterministic model	Ruminants; parameters for mule deer (*Odocoileus hemionus*) or Thomson's gazelle (*Gazella thomsoni*)	Grasslands; communities where quality declines with increasing herbage mass	Net daily energy intake	Herbage mass of patches selected. Membership in aggregated groups of conspecifics	Intake rate per unit body weight increases asymptotically with herbage mass per unit area. Forage quality declines linearly with herbage mass per unit area. Digestion rate increases asymptotically with increasing forage quality
Illius, 1986	Model & test by observational study	Cattle (*Bos taurus*)	Ryegrass (*Lolium perenne*)	Daily energy intake	Digestibility below which herbage is rejected	Decline in bite mass with decreasing sward height. Decline in bite rate with increasing sward height and decreasing sward quality. Maximum of 10 h grazing per day. Digestive capacity limited to a maximum of 9.5 g of indigestible DM kg^{-1} W day^{-1}
Illius *et al.*, 1992	Manipulative experiment	Sheep (*Ovis aries*)	*Lolium perenne* and *Trifolium repens*	Intake rate of DM	Number of bites taken from each of two alternative patches. Selection of species within patches	Effects of sward structure and composition on bite mass. Discrimination of patches by pre-consumption cues
Vivas *et al.*, 1991	Stochastic and deterministic, static models. Observational field test.	Moose (*Alces alces*)	Birch (*Betula pubescens*)	Intake rate of digestible energy	Diameter at point where twig is clipped	Time to consume and digest twigs
Jiang and Hudson, 1993	Observational study	Wapiti (*Cervus elaphus*)	Grasslands dominated by *Bromus pompellianus*, *Poa pratensis*, *Taraxacum officinale* and *Trifolium* spp.	Not specified. Argument about patch-departure rule leads to think that currency is average cropping rate	Residence time in feeding stations within patches and patch residence time	Not specified
Laca and Demment, 1991	Static, deterministic model	Cattle (*Bos taurus*)	Heterogeneous grassland with two nutritionally different components (leaf and stem or two species)	Intake rate of digestible DM	Time per bite or proportion of each potential bite consumed	Chewing requirements g^{-1} of leaf and stem consumed. Search rate. Effects of sward bulk density and height on bite weight
Laca *et al.*, 1993	Static, deterministic model and test with manipulative experiment	Cattle (*Bos taurus*)	Ryegrass (*Lolium perenne*)	DM intake rate	Patch residence time	Walking speed between patches. Patch depression curves. Effects of sward structure on bite mass

Reference	Type of study	Species	Environment/Food	Currency	Constraints/Assumptions
Langvatn and Hanley, 1993	Manipulative experiment	Cervus elaphus	Seeded Phleum pratense	Intake rate of digestible protein or DM	None specified. Implicit: intake rate is an asymptotic function of herbage mass per unit area and was not affected by other differences between patches
Murden and Risenhoover, 1993	Manipulative experiment	White-tailed deer (Odocoileus virginianus) and Angora goats (Capri hircus)	Native semiarid oak woodland savanna	Intake of digestible energy or crude protein	Foraging time
Murray, 1991	Static, deterministic model	Topi (Damaliscus lunatus, 100-kg free ranging ruminant)	Natural pastures in which small bite mass is associated with long distances between bites. Herbage is divided into two classes that differ in metabolizability and can be selected by the grazer	Intake rate of net energy	Distance travelled per unit mass of intake increases exponentially as bite mass declines (and as selectivity increases?)
Newman et al., 1995	Stochastic dynamic programming model	Sheep (Ovis aries)	Mixtures of ryegrass (Lolium perenne) and clover (Trifolium repens)	Fitness (probability of surviving predation and starvation until reproductive season)	Sequence and duration of alternative behaviours: eat clover, eat grass, eat whatever is encountered, rest, ruminate
Thornley et al., 1994	Static, deterministic model	Sheep (Ovies aries)	Ryegrass (Lolium perenne) and clover (Trifolium repens) swards	Total intake in monocultures; intake of clover and grass in mixtures	Maximum gut capacity. Maximum storage of energy
Verlinden and Wiley, 1989	Deterministic, static model	Herbivores	Any with a variety of food types	Benefit = net energy intake per day − net energy spent gathering diet	Energetic efficiencies from gross to net energy. Energetic costs of grazing
Ward and Saltz, 1994	Observational study and graphical static, deterministic model	Gazella dorcas	Subterranean stems and bulbs of Madonna lily (Pancratium sickenbergeri)	Long-term rate of digestion of energy	Total capacity of digestive tract. Maximum daily foraging time
Wilmshurst et al., 1995	Manipulative experiment to test model by Fryxell (1991)	Wapiti (Cervus elaphus)	Mixture of Poa pratensis, Bromus inermis, Trifolium sp., Taraxacum officinale and Cirsium arvense	Net energy intake rate	Amount of energy spent digging to expose the plant increases exponentially with depth. Amount of plant exposed per unit depth increases with depth
				Net daily energy intake	Intake rate increases asymptotically with herbage mass per unit area. Forage quality declines linearly with herbage mass per unit area. Digestion rate increases asymptotically with increasing forage quality

Table 5.1. (cont.)

Reference	Assumptions	Conclusions	Comments
Fryxell, 1991	No interference competition among conspecifics in groups. Individuals freely choose feeding location. Individuals are stationary once they choose a feeding location	Large non-selective grazers should prefer patches with intermediate levels of herbage mass, where they maximize net energy gains. Large non-selective grazers aggregate, in part, to maintain patches of intermediate herbage mass per unit area	Selectivity within patches was not considered. No effect of group size on grazing time was considered (see Penning et al., 1993). No experimental test was conducted (see Wilmshurst et al., 1995, below)
Illius, 1986	Intake from each patch is proportional to % of area covered by patch and rate of intake of DM or digestible DM offered by the patch	Excellent agreement between predicted and observed patch selection. Digestive constraint was inoperative, emphasizing behavioural constraints on intake	Model incorporated sward variability in digestibility within patches. More variable patches were preferred, in part because the model did not incorporate the effect of greater selectivity on searching time
Illius et al., 1992	None specified	Sheep preferred patches that offered higher intake rate. Small-scale heterogeneity within patches was not exploited. Discrimination can severely constrain intake rate in heterogeneous swards	Argue that sheep cannot maximize intake rate because of information and discrimination constraints
Vivas et al., 1991	Eating and digestion are mutually exclusive in time. Moose can select twig size precisely (only in deterministic model)	Both models predicted average twig size within 2 mm. The stochastic model was better. Even when food supply is very low, large amounts are rejected due to the trade-off between quality and intake rate	Implicit assumption that all twigs are identical and therefore all are consumed. Twigs are in fact 'patches'; choice of clip diameter maps into choice of 'residence time'. Ignores problem of scale. Branches are fractals, and thus, potentially exhibit patchiness at multiple scales; e.g. a branch is a twig in a limb
Jiang and Hudson, 1993	None specified. Implicitly, same as for marginal value theorem (Stephens and Krebs, 1986)	Feeding stations are left when neck angle reaches threshold. Patches are left when cropping rate within a feeding station fall below average for the season	Observational study; no model used. Weak evidence that feeding stations are left when neck angle reaches a threshold value. Weak evidence that patches are left when cropping rates fall below the average for the season. Departure rule compared cropping rate within FS with average cropping rate within FS. According to the marginal value theorem the average should have incorporated travel time between feeding stations
Laca and Demment, 1991	Random encounters of potential bites. Time per bite is a linear function of bite weight with a positive intercept	Sward heterogeneity strongly affects maximum intake rate of nutrients. Effects of heterogeneity depend on correlation between potential bite quality and mass	Two scales of selectivity are depicted: between and within potential bites
Laca et al., 1993	Same as for marginal value theorem	Excellent agreement between predicted and observed residence times. Larger differences in height between patches allowed higher intake rates	Model overestimated intake rate

Reference	Assumptions	Predictions	Comments
Langvatn and Hanley, 1993	None specified. Implicit: selection at the patch scale is based only on patch level phenomena. Small-scale selectivity is not the basis for habitat–patch selection	Patch preferences matched estimated intake rate of digestible protein. Maximization of intake rate of either currency was the worst predictor of patch use	Intake rate and diet quality were not measured. Other studies have demonstrated that intake rate is strongly affected by sward structure. There was significant within-patch variation in quantity and quality of forage. The experiment does not rule out maximization at a small scale (see Danell et al., 1991)
Murden and Risenhoover, 1993	None specified. Same as for prey model (Stephens and Krebs, 1986)?	Ad libitum supplementation increased selectivity and movement rates while foraging in both animals. Even when supplemented ad libitum, animals continued to consume significant amounts of native species with lower protein and energy content than the supplement	No explicit model used for qualitative predictions. Effects of supplementation were accurately predicted
Murray, 1991	Heat production increases exponentially with selectivity. Forage intake per day is a function of body weight and diet metabolizability; pasture availability does not affect intake rate	Optimum selectivity and net energy decline steeply with increments in the cost of locomotion. Larger grazers achieve maximum energy retention at lower levels of selection than smaller animals because of their greater costs of locomotion and ability to digest lower quality forages	No experimental tests performed
Newman et al., 1995	Intake rate is a constant characteristic of each plant species. Predation hazard is 0 while ruminating	Preference depends on intake rate of alternatives. Optimal diet should change over the day and may be sensitive to predation risk	State depends probabilistically on behaviours and choice of behaviour depends on state. Probability of survival is the a priori common currency to compare the value of alternative actions
Thornley et al., 1994	In monocultures, cost of foraging per unit intake declines exponentially with increasing leaf area index. In mixtures, sheep encounter potential bites randomly, and cost of grazing increases with degree of selectivity	Traditional asymptotic relationship between daily intake and herbage mass is explained mechanistically as a result of maximization of benefit, without resorting to morphological, physiological or ingestive constraints	Equation relating energy costs of grazing per unit intake implies that the grazer has morphological and physiological constraints on intake rate
Verlinden and Wiley, 1989	Different food types are digested independently. Simultaneous search of foods. Complete knowledge and discrimination ability	Herbivores should exhibit partial preference for lowest food type in diet. Selectivity and daily energy absorption increase with increasing foraging time	Model provides a basis to predict grazing time a priori; but this requires measures of the marginal value of activities other than grazing
Ward and Saltz, 1994	Negligible search time between plants. Random dispersion of plants of different sizes	Small plants were preferred because of their smaller cost : benefit ratio. Large plants were not totally consumed because costs exceed benefits for the deeper parts of the plant	Predictions based on complex gain function determined by digging costs and distribution of plant mass over depth
Wilmshurst et al., 1995	No interference competition among conspecifics in groups. Individuals freely choose feeding location	Wapiti preferred patches with intermediate levels of biomass (1200 kg ha^{-1}) as predicted by the model (1100 kg ha^{-1}). Animals spent significant time grazing in suboptimal patches	Grazing of suboptimal patches attributed to individual differences in parameters and imperfect information

DM, dry matter; W, weight; FS, feeding station.

SPECIAL CHARACTERISTICS OF LARGE HERBIVORES

Foraging theory was initially developed by scientists interested in explaining community structure from differences observed in the feeding behaviour of species (Emlen, 1966; MacArthur and Pianka, 1966; Schoener, 1987). In pursuit of this goal, researchers focused on species, primarily insectivorous birds and lizards, whose behaviours are easily quantifiable. The ecology and feeding behaviour of these groups shaped the development of early optimal foraging models. During this period the nutrition of herbivores was not well understood, complicating the modelling of value of foods for the herbivore. During much of the 1970s a considerable modelling effort in foraging ignored large herbivores. Not until a better concept of herbivore nutrition was developed in the agricultural literature and adapted to ecological problems (e.g. Owen-Smith, 1979; Belovsky, 1981) were optimality models structured to address the particular characteristics of feeding in large herbivores. Therefore, the characteristics of large herbivores that we label 'special' are unique only in relation to the early OFT models. There are no a priori reasons that make OFT inapplicable to large herbivores.

A critical element for formulating a foraging model for large herbivores is the correct perspective of their feeding process. For example, Stephens and Krebs (1986) stated that, relative to most carnivorous predators, herbivores spend little time searching for food. Contrary to carnivorous predators, herbivores face a relatively abundant and conspicuous food resource, and spend much of their time harvesting and digesting the food. We propose a slightly different view of the grazing process, in which searching time is an important component of grazing time. Searching, handling and ingestive behaviours of large herbivores are finely interspersed in a small temporal scale (Laca *et al.*, 1994b), whereas in predators these behaviours are distinctly separated in larger time-scales. Because potential food for grazers apparently forms a practically continuous layer, searching time is not fully determined by the effort to reach plants. A more relevant factor is the degree of selectivity of the grazing animal, as indicated by the difference between the composition of its diet and that of the available forage.

A key feature of grazing is the heterogeneity in quality of potential foods. It was this heterogeneity and complexity of the feeding process that originally discouraged application of models developed for predators that consume prey of uniformly high nutritional quality. Nutritional quality added a new dimension to the original foraging models. Once a clearer picture of the nutritional ecology of large herbivores emerged (Bell, 1970; Janis, 1976; Demment, 1982; Van Soest, 1982), OFT was applied. One of the most distinctive features of large generalist herbivores is the trade-off between quality and quantity of forage they face at multiple temporal and spatial scales (Senft *et al.*, 1987). This trade-off has been a major factor behind the evolutionary forces that shaped different herbivore feeding and social strategies (Jarman, 1974; Janis, 1976; Demment and Van Soest, 1985).

In general, forages are not available as discrete 'food items'. The lack of discrete food items allows the animal a greater behavioural flexibility to solve its nutritional requirements. The formation of bite size adds another level of complexity to the modelling exercise.

ELEMENTS AND STRUCTURE OF FORAGING MODELS

Evolutionary basis

OFT is a body of hypotheses, most of which are formalized as quantitative models that share a common axiom and structure. The axiom is that, because animals that forage more efficiently have greater reproductive output (fitness), present-day animals forage optimally as a result of natural selection. 'Optimum' does not mean 'the best conceivable' but 'the maximum or minimum subject to specific constraints' (Krebs and McCleery, 1984).

The common structure of OFT models consists of a currency to be optimized, decision variables, representing the behavioural choices available to the animal, and constraints. These constraints define the relationship between decision variables and currency, and the feasible values of the decision variables (Stephens and Krebs, 1986).

Currency

The currency is the metric or criterion used to evaluate different behavioural options according to their effects on performance (fitness). Many models have assumed that long-term average intake rate of net energy (e/t) is such a currency (Fryxell, 1991; Murray, 1991; Thornley *et al.*, 1994; Ward and Saltz, 1994; Wilmshurst *et al.*, 1995), although there have been various alternatives, such as intake rate of digestible energy (Verlinden and Wiley, 1989; Vivas *et al.*, 1991; Murden and Risenhoover, 1993) and dry matter (DM) intake rate (Illius *et al.*, 1992; Laca *et al.*, 1993). One of the major flaws of these currencies is that they only take into account the nutritional or energetic consequences of feeding. However, feeding has effects on other aspects of fitness (e.g. it may increase risk of predation) and is interdependent with other activities. As a result, no single surrogate indicator of fitness, such as e/t, will be maximized at all times.

Is rate maximization a valid assumption? Should grazing behaviour tend to maximize intake rate of any nutrient? Strictly speaking, the answer is no. We should not expect animals to maximize intake rate while (apparently) feeding, because it is frequently valuable for them to intersperse other behaviours, such as scanning for predators, staying with the herd and chewing, which reduce intake rate. Absolute intake rate maximization may be even less applicable to ruminants because the choice of a particular intake rate through regulation of ingestive chewing determines the immediate rates of digestion and the later rumination required (Greenwood and Demment, 1988). Yet rate maximization is a useful assumption if we understand and control the context in which it is applied. For example, by reducing chewing, steers are able to eat faster when grazing time is restricted than when they have free access to forage (Greenwood, 1989). Thus, one might conclude that they do not maximize intake rate when they have free access to forage. However, assuming that the optimal amount of chewing depends only on available grazing time, prediction of patch choice or diet selection by rate maximization is valid if chewing requirements are measured under the same grazing time restriction in which they will be applied. Animals may maximize intake rate with respect to patch choice within the context of grazing time restrictions.

The currency problem of traditional OFT was resolved in theory by Mangel and Clark (1986). Instead of surrogate currencies, their 'unified' foraging theory maximizes fitness itself by integrating the effects of feeding choices and alternative behaviours with probability of survival and reproduction over defined periods of time. This integration uses a state-variable approach to model animal state over time as a result of various behavioural choices, and requires a terminal reward function that relates Darwinian fitness with animal state at the end of the modelled period. Optimization of sequences of choices is achieved by a stochastic–dynamic programming (SDP) procedure. By bringing all behavioural options to the common denominator of their effects on fitness, SDP allows quantitative modelling of choices between foraging and non-foraging behaviours. A traditional rate-maximizing model can only predict behavioural choices during grazing, but cannot predict when bouts of different behaviours should begin or end.

The model by Newman *et al.* (1995) illustrates the advantages of the SDP approach. In this model, sheep behaviour is dependent on three state variables: stored energy, digestible gut fill and indigestible gut fill. The animal's fitness is proportional to the probability that it survives to reproduce, and therefore it is maximized by minimizing the probability of starvation and predation. At each time step, the model animal chooses among five alternatives: graze species i, graze species j, graze whichever species is encountered first, rest or ruminate. Each activity may entail an energetic gain, an energetic cost and a predation probability. Therefore, when optimizing its actions, the animal must discount the energetic effect of a behaviour by the probability that it will be killed while performing the chosen behaviour. The model predicts the optimal sequence of behaviours and dietary choice over the day, grazing time, diet preference and total intake per day. Like the model by Thornley *et al.* (1994), this one provides a non-mechanistic explanation of why animals limit intake per day. For example, when energy reserves are plentiful, the asymptotic terminal reward function implies little fitness gain from more food. The animal may maximize fitness gains by stopping grazing and engaging in non-foraging activities with lower predation risks. The model also predicts that dietary preferences should be sensitive to perceived predation hazard.

Although theoretically and heuristically superior to the e/t approach, the SDP approach has practical limitations. First, the output of these models is so detailed that it may be hard to derive general trends that truly promote our understanding. Second, the degree of realism of SDP models is severely constrained by their large computational demands.

Decision variables

Decision variables describe the types of decisions animals can make to affect their performance. Typical decision variables are quality of items accepted in the diet, grazing time, bite weight, time spent at each patch type, etc. (see Table 5.1). The success of a model is strictly dependent on the selection of decision variables. The choice must reflect dimensions which offer relevant opportunities for choice by the animals.

Constraints on diet selection

The constraints are the formal expression of the foraging abilities of the animal, and they define the relationship between decision variables and currency. In a more general way, constraints are all the factors that limit the acquisition of the currency. The constraints should synthesize all relevant characteristics of the forager that impinge on performance and should include more factors than those explicitly identified as constraints. For example, Belovsky (1986) used a linear programming approach to predict diets of a guild of generalist herbivores. The explicit constraints were: (i) digestive capacity; (ii) daily feeding time; (iii) energy requirements; and (iv) salt and protein requirements. Additionally, the model involved parameters such as cropping rates for each dietary choice, gut volumes and digestibilities. These parameters represent a suite of morphological, physiological, behavioural and environmental constraints that limit the foragers' performance.

Although the classification of constraints as intrinsic and extrinsic has been useful (Penry, 1993), ingestive and digestive constraints are interdependent and simultaneously affect diet choice. This is particularly true for large generalist herbivores foraging in grasslands with intermediate levels of quality and quantity of forage. In this case, the range of possible states cannot be neatly partitioned into two mutually exclusive sets, one in which digestion is limiting and another in which ingestion is the limiting factor. Spalinger and Hobbs (1992) provide a summary of ingestive constraints of large herbivores. Illius and Gordon (1990) reviewed the constraints on diet selection by herbivores and integrated, to some extent, ingestive and digestive constraints (see Illius, 1990, Figs 3b and 4).

Foraging models have frequently ignored factors that constrain optimization of diets. Table 5.1 lists some of the constraints typically considered in studies of foraging by large herbivores. Constraints on the mechanics of reaching and handling forages are usually considered (e.g. Illius, 1986; Laca and Demment, 1991; Spalinger and Hobbs, 1992), whereas constraints on acquisition and processing of information by the animal are routinely omitted. Yet information constraints may be very important in determining the dietary choices made by foragers. Although nutritional constraints imposed by secondary compounds have rarely been included in models, Belovsky and Schmitz (1991) clearly showed that the effects of plant allelochemicals on foraging strategies can be easily taken into account with a linear programming approach.

From agronomic or ecological standpoints one may ask, what are the most relevant constraints on intake of nutrients by grazers? However, the most limiting constraint probably changes with many factors, both internal and external to the animal. From an evolutionary standpoint, one can argue that all constraints should be equally limiting to fitness. A disproportionately limiting constraint would be subject to intense natural selection. Time available to forage is an important constraint, because foraging and other activities such as resting or ruminating are mutually exclusive. However, foraging time can also be viewed as a variable whose value emerges from maximization of fitness (Thornley *et al.*, 1994; Newman *et al.*, 1995). Other important constraints on grazing intake are: (i) digestion, passage and gut volume; within any time period intake minus absorption and passage cannot exceed change in gut fill, and gut fill cannot exceed a maximum; (ii) water balance and body temperature must be maintained within certain levels; (iii) time necessary to search for, reach, harvest and chew each gram of forage ingested; and (iv) maximum bite area.

Confusion about what is and what is not a constraint may arise because a given constraint may not always be directly involved in determining behaviour if there are other more constraining factors. Some constraints may be redundant and therefore do not affect the solution. Evidently, whether a constraint is redundant or not potentially depends on the values of all other variables and parameters of the model or system.

EFFECTS OF ANIMAL STATE ON FORAGING DECISIONS

Because fitness is not linearly related to amount of nutrients acquired or body reserves, a bite of food should be more valuable to a hungry animal in poor physiological condition than to one that is full or in good nutritional condition (Houston, 1993; Newman *et al.*, 1995). The expected increment in fitness or value (*sensu* Staddon, 1983, p. 222) derived from an action depends on the state of the animal making the decision. There is evidence from operant conditioning experiments that the marginal value of many behaviours declines with increasing rate of the behaviour (Staddon, 1983).

Until recently, quantitative effects of the state of the animal have been largely ignored in studies of diet selection of grazing herbivores. Yet research indicates that herbivores exhibit foraging strategies that depend on state. Arnold (1981, 1985) reviewed intake patterns presumably generated by changes in animal state. Arnold (1981) provides a concise review of early evidence of diurnal patterns, in which diet quality tends to be greater in afternoon than in morning meals, presumably because of the desire to eat quickly during the morning grazing period. Newman *et al.* (1994b) and Parsons *et al.* (1994) found that the proportion of clover in the diet of sheep that had free access to adjacent monocultures of grass and clover was greater in the morning and declined in the afternoon. Considering that the intake rate of clover was greater than that of grass and that sheep prefer to eat a mixed diet, these results seem to be in agreement with those reviewed by Arnold (1981). An SDP approach to dietary selectivity (Newman *et al.*, 1995) potentially explains complex temporal patterns of diet selection that depend on predation hazard, time of day and relative and absolute values of intake rate of alternative forages.

Perhaps the most frequently studied manipulation of animal state has been fasting. Jung and Koong (1985) found that overnight fasting did not influence quality of diet selected by sheep, but it increased intake rate from 47 to 124 mg min^{-1} kg body weight $(BW)^{-0.75}$. In the same experiment, hunger satiation by feeding a high-quality feed resulted in higher quality of diet selected and lower intake rates. Greenwood and Demment (1988) found that steers fasted for 36 h exhibited significantly different grazing patterns from non-fasted ones. Intake and biting rate increased during the morning after the fast (27 vs. 17 g DM min^{-1}); however, no effects of fasting on bite mass or diet quality were detected. Edwards *et al.* (1994) found that sheep increased intake rate and reduced selectivity for a preferred pellet type as a result of fasting. Newman *et al.* (1994b) detected increases in intake rate, grazing time and proportion of grass in the diet of sheep as a result of fasting. These latter results support state dependence of foraging decisions but defy explanation, because animals had free access to monospecific clover and grass swards and could easily have consumed only one species if desired. Moreover, intake rate of clover was greater than intake rate of grass, and sheep eating pure clover grow faster and produce more milk (Gibb and Treacher, 1983, 1984).

Reproductive and physiological status also affect dietary strategies and intake rate (Arnold, 1985). Shearing plus mild cold stress of sheep fed chopped hay increased intake rate from 12 to 14.7 g min^{-1} (Hogan *et al.*, 1987). Lactating animals consume more forage than non-lactating ones, typically by increasing grazing time (Arnold, 1985). This empirical result is not explained by energy-maximization approaches (why does the dry animal not eat as much as the lactating one?), and illustrates the role of behavioural constraints on intake. The longer grazing time and greater intake exhibited by lactating herbivores is predicted by the SDP approach (Newman *et al.*, 1995).

What are the relevant variables that define state and can potentially influence foraging decisions? State-variable models have considered body mass, energy reserves, gut fill and/or quantity of a series of dietary fractions in the gut (e.g. Illius and Gordon, 1992). Unfortunately, the most detailed state-variable models of ruminants (Baldwin et al., 1987a,b,c) have ignored the behavioural processes of diet selection. The following state variables are potentially relevant in determining animal decisions.

1. Gut fill.
2. Physiological/reproductive status.
3. Fat and other energy reserves.
4. Water balance.
5. Blood levels of metabolites and toxins.
6. Dietary experience (taste aversions and preferences).
7. Dietary experience (location of foods, temporal cycles).
8. Degree of acclimatization in terms of gut morphology, enzyme expression and rumen microflora.
9. Position relative to herd and resources.

Models considering variables 1–4 (e.g. Loza et al., 1992) or even 5 (Illius and Jessop, 1995) have been developed. Perhaps the current challenge is to devise practical quantitative models that include the rest of the variables.

By definition, state variables integrate the results of foraging and other behavioural choices. Because they integrate over different time and spatial scales, different variables should serve to regulate diet selection at different scales (Laca and Ortega, 1995). This hierarchical approach to the regulation of intake and dietary selection offers a powerful research tool, because small-scale processes (e.g. gut fill dynamics) nested in larger-scale ones (change in body mass) can be studied with a degree of independence.

VALIDITY OF THE OPTIMALITY PARADIGM

Criticisms of OFT and rebuttals are presented by Stephens and Krebs (1986), Schoener (1987), Pierce and Ollason (1987) and Stearns and Schmid-Hempel (1987). In this section we briefly review some of the criticisms of the application of OFT to ruminants.

Provenza and Cincotta (1993) considered that ecological models of feeding behaviour are inadequate for ruminants, because in their view, these models assume that animals are genetically 'pre-adapted to their environments and are thus unaffected by feedback mechanisms'. However, Schoener (1987) clearly indicated that behaviours involved in achieving the predictions of OFT included a great deal of learning and very little detailed genetic programming. Instead of feeding behaviour being tightly programmed genetically, it is more likely that a general homeostatic mechanism is genetically programmed (Schoener, 1987).

Traditional optimal foraging models emphasized evolutionary aspects, but did not exclude learning or short-term changes in the state of the animal as important mechanisms (e.g. Waage, 1979; Ollason, 1987).

A fundamental weakness of the application of OFT models is the inappropriate extrapolation of results from small to broader scales of space and time. Consider a continuum along which the scale of time varies from life history to minutes. Behaviour monitored over the life of the organisms will reflect the integration of many dimensions of the animal's ecology. The appropriate explanation of behaviour is some integrated measure that allows the trade-off between processes that have different currencies. This measure may be fitness. At the other end of the continuum, over short intervals of time, behaviour can be explained by relatively simple measures of performance if the context of the observations is understood or controlled.

Does OFT have a reasonable chance to explain the short-term behaviour of a hungry animal released in a pasture? Yes, because feeding is important relative to other activities. As the intervals of time and space are expanded, the context of feeding behaviour becomes more complex and models based on simple currencies become increasingly inappropriate. Historically, OFT models have been based on currencies related to ingested nutrients per unit time (Schoener, 1987). The insight gained from these studies was applied to larger scales of time and space to understand community structure and function. Often, models and assumptions were incorporated in this integration placing small-scale models in a long-term context (MacArthur and Pianka, 1966).

A major challenge for grazing researchers is to step beyond the short-term time-frame along axes of time and space. The SDP approach is highly integrated but is limited by its own complexities. Foraging models that occupy the middle ground of the continuum are needed – models addressing behaviour over several days. The integration of digestive and ingestive processes is a natural step towards bridging the gap between large- and small-scale models of grazing strategies.

WHY AND HOW TO USE FORAGING MODELS

The main goal of foraging models is to make testable quantitative predictions about feeding behaviour. However, models are useful in other ways. First, estimates of the maximum performance achievable could be used as a measure for the evaluation of habitats. Second, a model with details on the mechanisms that determine productivity could suggest animal breeding and environmental modifications to elicit productivity-maximizing behaviour. Third, if observations and predictions disagree, there is an opportunity to increase our understanding of the grazing process. For example, Greenwood (1989) found that steers compensated for reduced grazing time by increasing intake rate while maintaining bite weight. Animals achieved a greater intake rate by reducing ingestive chewing and increasing rumination time. Ingestive chewing is usually considered a constraint on intake rate, but the steers treated it as a decision variable, presumably to maximize daily intake under the restrictions of feeding time.

The rejection of a model approaches the 'strong inference' scheme (Platt, 1964), but poses the question of what to do next. There are basically three options, and the choice depends on the extent and pattern of disagreement between model and reality and on the philosophy of the modeller.

The first option is to question the selection of constraints and decision variables, or even the validity of parameters measured. The mechanisms of feeding and the relevant constraints should be reviewed and studied in greater detail, and the model reformulated and tested again. For example, Schmitz (1990) predicted that supplemented deer should eat only supplemental feed, whereas they consumed significant amounts of browse. The disagreement between prediction and observation led to a revision of the constraints in the model. The supplement was administered in a few large feeders, which probably caused competition among animals, forcing individuals to browse (Schmitz, 1990). Reformulation of the model in the light of refuted predictions is not *ad hoc* or circular, provided the same data are not used as confirmation of the validity of the new model (Brown, 1993).

Second, one can hypothesize that the currency used is not correct. An alternative currency may be proposed and tested. Some models propose alternative currencies a priori, before the experimental tests, in which case the result of the tests is the selection of one alternative. For example, Belovsky (1981, 1986) tested time minimization and energy maximization as alternatives, and concluded that in all species tested a strategy of energy maximization agreed very well with observed diets.

Third, one can conclude that, although evolution tends to maximize fitness, the *optimum optimorum* has not been reached. The population or the individual may be stuck at a local optimum, or environmental changes may have changed the optimum (Mangel and Clark, 1988; Provenza and Cincotta, 1993). This position questions the validity of assuming that fitness is maximized under the current constraints.

Brown (1993) proposed that optimal foraging models have a hierarchy of assumptions.

1. Statistical assumptions, such as independence of samples.
2. Methodological assumptions, such as proper functioning of laboratory equipment.
3. Objective function and environmental context, for example, net energy maximization and relative abundance of foods.
4. Feasible behaviours, such as ability to discriminate foods.
5. Optimization assumption, i.e. fitness is not maximized.

This hierarchy also prescribes an order in which assumptions should be questioned to explain the differences between predicted and observed behaviours. The 'hard-core' (Brown, 1993) assumption of optimality should be doubted last. Yet it is possible to test different hard-core assumptions. Alternatives to the assumption of optimality such as satisficing (Ward, 1992) and phylogenetic baggage (Provenza and Cincotta, 1993) have been proposed in arguments against the optimal foraging programme. However, to our knowledge, no strong tests have been performed in which one or more core assumptions have been considered as mutually exclusive alternatives.

CONCLUSIONS

The study of feeding strategies of grazing animals has revealed evolutionary and short-term patterns. Studies across taxa have unveiled broad dietary strategies that result in niche separation and the ability of many species to coexist in grasslands (Jarman, 1974; Janis, 1976; McNaughton, 1979; Demment and Van Soest, 1985). This knowledge is significantly useful for grazing management, particularly to match resources available with proper mixtures of herbivores. Within taxa, and even within individuals, grazing and browsing animals exhibit behavioural strategies to meet the challenges of gathering and processing enough forage to survive and reproduce (and produce). Although the behavioural options and constraints of grazing livestock have been extensively studied, our understanding is limited. The application of optimal foraging to grazing behaviour is a very promising trend towards a general framework for grazing strategies.

Optimal foraging models can be useful tools in the study of foraging behaviour of grazing animals. Most of the criticisms of the application of optimal foraging models to grazing animals can be addressed by simply refining the constraints and by integrating animal state with behaviours other than foraging. Such reformulations fall well within the scope of OFT and constitute an example of how these models identify areas of foraging behaviour that need more basic research.

Early models of grazing behaviour emphasized instantaneous aspects of ingestion and digestion. Spatial distribution of forages was fairly consistently ignored, together with the grazer's abilities and constraints in searching for and learning about the resources. When animals graze, they are simultaneously subjected to social interactions, predator avoidance, thermoregulation needs and thirst. As animals consume and digest forages, they simultaneously have an impact on the plants and learn about the nutritional value and location of foods.

In a way, purely empirical studies are in the past, and optimal foraging models are part of the present. What is in the future? First, models that integrate the interaction of foraging with non-foraging behaviours and the effect of animal state will improve our understanding of grazing strategies. Integration of these factors has the potential to generate a priori predictions of grazing time and daily intake. Second, more models and experiments will explicitly take into account the spatial dimensions of grazing. These models are facilitated by the current exponential development of our ability to gather and process spatial data. Third, we expect to see conceptual and formal models integrating processes of diet selection and intake over scales of space and time. These models should allow us to formally apply our knowledge of grazing strategies to predict grazing intake and animal impact on the plant communities in heterogeneous natural and intensively managed landscapes.

ACKNOWLEDGEMENTS

Karen Launchbaugh, Nancy Mathews, Chris Goguen, Andrew Illius and Eugene Ungar provided helpful comments on a draft of this chapter.

REFERENCES

ARC (1980) *The Nutrient Requirements of Ruminant Livestock*. Commonwealth Agricultural Bureaux, Farnham Royal, 368 pp.

Armstrong, D.P. (1991) Levels of cause and effect as organizing principles for research in animal behaviour. *Canadian Journal of Zoology* 69, 823–829.

Arnold, G.W. (1981) Grazing behaviour. In: Morley, F.H.W. (ed.) Grazing Animals. Elsevier Scientific Publishing Company, Amsterdam, pp. 79–104.

Arnold, G.W. (1985). Regulation of food intake. In: Hudson, R.J. and White, R.G. (eds) *Bioenergetics of Wild Herbivores*. CRC Press, Boca Raton, Florida, pp. 81–101.

Baldwin, R.L., France, J. and Gill, M. (1987a) Metabolism of the lactating cow. I. Animal elements of a mechanistic model. *Journal of Dairy Research* 54, 77–105.

Baldwin, R.L., Thornley, J.H.M. and Beever, D.E. (1987b) Metabolism of the lactating cow. II. Digestive elements of a mechanistic model. *Journal of Dairy Research* 54, 107–131.

Baldwin, R.L., France, J., Beever, D.E., Gill, M. and Thornley, J.H.M. (1987c) Metabolism of the lactating cow. III. Properties of mechanistic models suitable for evaluation of energetic relationships and factors involved in the partition of nutrients. *Journal of Dairy Research* 54, 133–145.

Bell, R.H.V. (1970) The use of the herb layer by grazing ungulates in the Serengeti. In: Watson, A. (ed.) *Animal Populations in Relation to their Food Resources*. Blackwell Scientific Publications, Oxford, pp. 111–123.

Belovsky, G.E. (1981) Food plant selection by a generalist herbivore: the moose. *Ecology* 62, 1020–1030.

Belovsky, G.E. (1986) Optimal foraging and community structure: implications for a guild of generalist grassland herbivores. *Oecologia* 70, 35–52.

Belovsky, G.E. and Schmitz, O.J. (1991) Mammalian herbivore optimal foraging and the role of plant defenses. In: Palo, R.T. and Robbins, C.T. (eds) *Plant defenses against Mammalian Herbivory*. CRC Press, Boca Raton, Florida, pp. 1–28.

Brown, J.S. (1993) Model verification: optimal foraging theory. In: Scheiner, S.M. and Gurevitch, J. (eds) *Design and Analysis of Ecological Experiments*. Chapman and Hall, New York, pp. 360–377.

Danell, K., Edenius, L. and Lundberg, P. (1991) Herbivory and tree stand composition: moose patch use in winter. *Ecology* 72, 1350–1357.

Demment, M.W. (1982) The scaling of ruminoreticulum size with body weight in East African ungulates. *African Journal of Ecology* 20, 43–47.

Demment, M.W. and Van Soest, P.J. (1985) A nutritional explanation for body-size patterns of ruminant and nonruminant herbivores. *American Naturalist* 125, 641–672.

Distel, R.A., Laca, E.A., Griggs, T.C. and Demment, M.W. (1995) Patch selection by cattle: maximization of intake rate in horizontally heterogeneous pastures. *Applied Animal Behaviour Science* 45, 11–21.

Edwards, G.R., Newman, J.A., Parsons, A.J. and Krebs, J.R. (1994) Effects of the scale and spatial distribution of the food resource and animal state on diet selection: an example with sheep. *Journal of Animal Ecology* 63, 816–826.

Emlen, J.M. (1966) The role of time and energy in food preference. *American Naturalist* 100, 611–617.

Fryxell, J.M. (1991) Forage quality and aggregation by large herbivores. *American Naturalist* 138, 478–498.

Gibb, M.J. and Treacher, T.T. (1983) The performance of lactating ewes offered diets containing different proportions of fresh perennial ryegrass and white clover. *Animal Production* 37, 433–440.

Gibb, M.J. and Treacher, T.T. (1984) The performance of weaned lambs offered diets containing different proportions of fresh perennial ryegrass and white clover. *Animal Production* 39, 412–420.

Greenwood, G.B. (1989) What options does a grazing ruminant have? Unpublished PhD dissertation, University of California, Davis.

Greenwood, G.B. and Demment, M.W. (1988) The effect of fasting on short-term cattle grazing behaviour. *Grass and Forage Science* 43, 377–386.

Hogan, J.P., Kenney, P.A. and Weston, R.H. (1987) Factors affecting the intake of feed by grazing animals. In: Wheeler, J.L., Pearson, C.J. and Robards, G.E. (eds) *Temperate Pastures, Their Production, Use and Management*. Australian Wool Corporation/CSIRO, Melbourne, pp. 317–327.

Holechek, J.L., Pieper, R.D. and Herbel, C.H. (1995) *Range Management: Principles and Practices*, 2nd edn. Prentice Hall, Englewood Cliffs, New Jersey.

Houston, A.I. (1993) The importance of state. In: Hughes, R.N. (ed.) *Diet Selection: An Interdisciplinary Approach to Foraging Behaviour*. Blackwell Scientific Publications, Oxford, pp. 10–31.

Hughes, R.N. (1990) Preface. In: Hughes, R.N. (ed.) *Behavioural Mechanisms of Food Selection*. Springer-Verlag, Berlin and Heidelberg, pp. v–vii.

Illius, A.W. (1986) Foraging behaviour and diet selection. In: Gudmundsson, O. (ed.) *Grazing Research on Northern Latitudes*. Plenum Press, New York, pp. 227–236.

Illius, A.W. and Gordon, I.J. (1990) Constraints on diet selection and foraging behaviour in mammalian herbivores. In: Hughes, R.N. (ed.) *Behavioral Mechanisms of Food Selection*. Springer-Verlag, Berlin, pp. 369–392

Illius, A.W. and Gordon, I.J. (1992) Modelling the nutritional ecology of ungulate herbivores: evolution of body size and competitive interactions. *Oecologia* 89, 428–434.

Illius, A.W. and Jessop, N.S. (1995) Modelling metabolic costs of allelochemical ingestion by foraging herbivores. *Journal of Chemical Ecology* 21, 693–719.

Illius, A.W., Clark, D.A. and Hodgson, J. (1992) Discrimination and patch choice by sheep grazing grass–clover swards. *Journal of Animal Ecology* 61, 183–194.

Janis, C. (1976) The evolutionary strategy of the Equidae and the origins of rumen and cecal digestion. *Evolution* 30, 757–774.

Jarman, P.J. (1974) The social organisation of antelope in relation to their ecology. *Behaviour* 48, 215–267.

Jarman, P.J. and Sinclair, A.R.E. (1979) Feeding strategy and the pattern of resource partitioning in ungulates. In: Sinclair, A.R.E. and Norton-Griffiths, M. (eds) *Serengeti: Dynamics of an Ecosystem*. University of Chicago Press, Chicago and London, pp. 130–163.

Jiang, Z. and Hudson, R.J. (1993) Optimal grazing of wapiti (*Cervus elaphus*) on grassland: patch and feeding station departure rules. *Evolutionary Ecology* 7, 488–498.

Jung, H.G. and Koong, L.J. (1985) Effects of hunger satiation on diet quality by grazing sheep. *Journal of Range Management* 38, 302–305.

Krebs, J.R. and McCleery, R.H. (1984) Optimization in behavioural ecology. In: Krebs, J.R. and Davies, N.B. (eds) *Behavioural Ecology: An Evolutionary Approach*. Sinauer Associates, Sunderland, Massachusetts, pp. 91–121.

Laca, E.A. and Demment, M.W. (1991) Herbivory: the dilemma of foraging in a spatially heterogeneous food environment. In: Palo, R.T. and Robbins, C.T. (eds) *Plant Defenses Against Mammalian Herbivory*. CRC Press, Boca Raton, Florida, pp. 29–44.

Laca, E.A. and Ortega, I.M. (1995) Integrating foraging mechanisms across spatial and temporal scales. *Fifth International Rangeland Congress*, Salt Lake City, Utah (in press).

Laca, E.A., Distel, R.A., Griggs, T.C., Deo, G.P. and Demment, M.W. (1993) Field test of optimal foraging with cattle: the marginal value theorem successfully predicts patch selection and utilization. In: *17th International Grassland Congress*, Palmerston North, Hamilton and Lincoln, New Zealand, Australian Society of Animal Production, Rockhampton, Queensland, pp. 709–710.

Laca, E.A., Distel, R.A., Griggs, T.C. and Demment, M.W. (1994a) Effects of canopy structure on patch depression by grazers. *Ecology* 75, 706–716.

Laca, E.A., Ungar, E.D. and Demment, M.W. (1994b) Mechanisms of handling time and intake rate of a large mammalian grazer. *Applied Animal Behaviour Science* 39, 3–19.

Langvatn, R. and Hanley, T.A. (1993) Feeding-patch choice by red deer in relation to foraging efficiency. *Oecologia* 94, 164–170.

Loza, H.J., Grant, W.E., Stuth, J.W. and Forbes, T.D.A. (1992) Physiologically based landscape use model for large herbivores. *Ecological Modelling* 61, 227–252.

MacArthur, R.H. and Pianka, E.R. (1966) On optimal use of a patchy environment. *American Naturalist* 100, 603–609.

McNaughton, S.J. (1979) Grazing as an optimization process: grass–ungulate relationships in the Serengeti. *American Naturalist* 113, 691–703.

Malechek, J.C. and Balph, D.F. (1987) Diet selection by grazing and browsing livestock. In: Hacker, J.B. and Ternouth, J.H. (eds) *The Nutrition of Herbivores*. Academic Press, Sydney, Australia, pp. 121–132.

Mangel, M. and Clark, C.W. (1986) Towards a unified foraging theory. *Ecology* 67, 1127–1138.

Mangel, M. and Clark, C.W. (1988) *Dynamic Modeling in Behavioral Ecology*. Princeton University Press, Princeton, New Jersey.

Murden, S.B. and Risenhoover, K.L. (1993) Effects of habitat enrichment on patterns of diet selection. *Ecological Applications* 3, 497–505.

Murray, M.G. (1991) Maximizing energy retention in grazing ruminants. *Journal of Animal Ecology* 60, 1029–1045.
Newman, J.A., Parsons, A.J. and Penning, P.D. (1994a) A note on the behavioural strategies used by grazing animals to alter their intake rates. *Grass and Forage Science* 49, 502–505.
Newman, J.A., Penning, P.D., Parsons, A.J., Harvey, A. and Orr, R.J. (1994b) Fasting affects intake behaviour and diet preference of grazing sheep. *Animal Behavior* 47, 185–193.
Newman, J.A., Parsons, A.J., Thornley, J.H.M., Penning, P.D. and Krebs, J.R. (1995) Optimal diet selection by a generalist grazing herbivore. *Functional Ecology* 9, 255–268.
NRC (1985) *Nutrient Requirements of Sheep*, 6th revised edn. National Academy Press, Washigton, DC.
Ollason, J.G. (1987) Learning to forage in a regenerating patchy environment: can it fail to be optimal? *Theoretical Population Biology* 31, 13–32.
Owen-Smith, N. (1979) Assessing the foraging efficiency of a large herbivore, the kudu. *South African Journal of Wildlife Research* 9, 102–110.
Parsons, A.J., Newman, J.A., Penning, P.D., Harvey, A. and Orr, R.J. (1994) Diet preference of sheep: effects of recent diet, physiological state and species abundance. *Journal of Animal Ecology* 63, 465–478.
Penning, P.D. (1986) Some effects of sward conditions on grazing behaviour and intake by sheep. In: Gudmundsson, O. (ed.) *Grazing Research at Northern Latitudes*. Plenum Publishing Corporation, New York, pp. 219–226.
Penning, P.D., Parsons, A.J., Orr, R.J. and Treacher, T.T. (1991) Intake and behaviour responses by sheep to changes in sward characteristics under continuous stocking. *Grass and Forage Science* 46, 15–28.
Penning, P.D., Parsons, A.J., Newman, J.A., Orr, R.J. and Harvey, A. (1993) The effects of group size on grazing time in sheep. *Applied Animal Behaviour Science* 37, 101–109.
Penry, D.L. (1993) Digestive constraints on diet selection. In: Hughes, R.N. (ed.) *Diet Selection: An Interdisciplinary Approach to Foraging Behaviour*. Blackwell Scientific Publications, Oxford, pp. 32–55.
Pierce, G.J. and Ollason, J.G. (1987) Eight reasons why optimal foraging theory is a complete waste of time. *Oikos* 49, 111–117.
Platt, J.R. (1964) Strong inference. *Science* 146, 347–353.
Provenza, F.D. (1995) Postingestive feedback as an elementary determinant of food selection and intake in ruminants. *Journal of Range Management* 48, 2–17.
Provenza, F.D. and Balph, D.F. (1990) Applicability of five diet-selection models to various foraging challenges ruminants encounter. In: Hughes, R.N. (ed.) *Behavioural Mechanisms of Food Selection*. Springer-Verlag, Berlin and Heidelberg, pp. 423–460.
Provenza, F.D. and Cincotta, R.P. (1993) Foraging as a self organizational learning process: accepting adaptability at the expense of predictability. In: Hughes, R.N. (ed.) *Diet Selection: An Interdisciplinary Approach to Foraging Behaviour*. Blackwell Scientific Publications, Oxford, pp. 78–101.
Schmitz, O.J. (1990) Management implications of foraging theory: evaluating deer supplement feeding. *Journal of Wildlife Management* 54, 522–532.
Schoener, T.W. (1987) A brief history of optimal foraging ecology. In: Kamil, A.C., Krebs, J.R. and Pulliam, H.R. (eds) *Foraging Behavior*. Plenum Press, New York, pp. 5–67.

Senft, R.L., Coughenour, M.B., Bailey, D.W., Rittenhouse, L.R., Sala, O.E. and Swift, D.M. (1987) Large herbivore foraging and ecological hierarchies: landscape ecology can enhance traditional foraging theory. *BioScience* 37, 789–799.

Spalinger, D.E. and Hobbs, N.T. (1992) Mechanisms of foraging in mammalian herbivores: new models of functional response. *American Naturalist* 140, 325–348.

Staddon, J.E.R. (1983) *Adaptive Behavior and Learning*. Cambridge University Press, New York.

Stearns, S.C. and Schmid-Hempel, P. (1987) Evolutionary insights should not be wasted. *Oikos* 49, 118–125.

Stephens, D.W. and Krebs, J.R. (1986) *Foraging Theory*. Princeton University Press, Princeton, New Jersey.

Thornley, J.H.M., Parsons, A.J., Newman, J. and Penning, P.D. (1994) A cost–benefit model of grazing intake and diet selection in a two-species temperate grassland sward. *Functional Ecology* 8, 5–16.

Van Soest, P.J. (1982) *Nutritional Ecology of the Ruminant: Ruminant Metabolism, Nutritional Strategies, the Cellulolytic Fermentation, and the Chemistry of Forages and Plant Fibers*, 1st edn. O & B Books, Corvallis, Oregon.

Verlinden, C. and Wiley, R.H. (1989) The constraints of digestive rate: an alternative model of diet selection. *Evolutionary Ecology* 3, 264–273.

Vivas, H.J., Saether, B.E. and Andersen, R. (1991) Optimal twig-size selection of a generalist herbivore, the moose *Alces alces*: implications for plant–herbivore interactions. *Journal of Animal Ecology* 60, 395–408.

Waage, J.K. (1979) Foraging for patchily distributed hosts by the parasitoid *Nemeritis canescens*. *Journal of Animal Ecology* 48, 353–371.

Ward, D. (1992) The role of satisficing in foraging theory. *Oikos* 63, 312–317.

Ward, D. and Saltz, D. (1994) Foraging at different spatial scales: dorcas gazelles foraging for lilies in the Negev Desert. *Ecology* 75, 48–58.

Wilmshurst, J.F., Fryxell, J.M. and Hudson, R.J. (1995) Forage quality and patch choice by wapiti (*Cervus elaphus*). *Behavioral Ecology* 6, 209–217.

Biochemical Aspects of Grazing Behaviour*

K.L. Launchbaugh

Department of Range and Management, Texas Tech University, Lubbock, TX 79409-2125, USA

INTRODUCTION

A vast array of plants contain allelochemicals that are potentially toxic to herbivores. The herbivore's challenge is to acquire sufficient nutrients to evade starvation and produce viable offspring and yet avoid the consumption of lethal doses of phytochemicals. Mammalian herbivores possess several adaptations for walking this biological tightrope. These adaptations consist of some innate avoidance patterns and mechanisms that alter the hedonic value of preingestive plant qualities (e.g. taste, odour, appearance) based on postingestive experiences (e.g. nutritional benefit or gastrointestinal distress). Plants become more palatable when their consumption has positive postingestive consequences and their palatability decreases when consumption results in gastrointestinal malaise. An understanding of grazed ecosystems will only be gained through a clear concept of how plants affect selective grazing patterns of herbivores. This chapter will first discuss ways in which plant allelochemicals alter the grazing behaviour and well-being of mammalian herbivores. Then, the focus will switch to ways in which herbivores protect themselves from overingestion of phytotoxins.

---

* This document is a contribution from the College of Agricultural Sciences and Natural Resources at Texas Tech University No. T-9-737.

©1996 CAB INTERNATIONAL.
The Ecology and Management of Grazing Systems (eds J. Hodgson and A.W. Illius)

PLANT ALLELOCHEMICALS: WHAT ARE THEY AND WHY DO THEY EXIST?

In the early 1800s a primary goal of plant biology was to understand the processes of plant growth and metabolism. As scientists probed, they gained a basic understanding of the chemicals involved in plant metabolism. They also found many chemicals that did not appear to be involved in metabolic processes supporting plant growth, development or reproduction and they termed these chemicals 'secondary metabolites'. Secondary compounds were viewed initially as waste products of primary metabolism. Then, in 1888 the German botanist E. Stahl suggested that plant secondary metabolites might play an important role in deterring herbivores. This idea attracted little scientific attention until 1959 when Fraenkel's landmark paper highlighted Stahl's work and argued strongly that plant secondary compounds play a primary role in mediating plant–animal interactions.

Subsequent research has revealed many important functions of secondary metabolites in plants. They act as regulators of plant growth or biosynthetic activities, transport facilitators and nutrient, energy or waste storage compounds (Rosenthal and Bell, 1979). Secondary compounds may also affect the plant's ability to survive abiotic conditions such as excessive radiation or moisture and temperature extremes (Harborne, 1988). Important biological interactions that are mediated by secondary compounds include: protection against plant pathogens; allelopathic effects to decrease competition with other plants; and, of course, herbivory resistance (Whittaker and Feeny, 1971; Rhoades and Cates, 1976; Lindroth, 1989). Why a particular secondary metabolite exists or evolved is an adaptationist's riddle. However, the fact that many of these compounds deter grazing cannot be disputed.

Not all compounds that deter grazing are secondary compounds and not all secondary compounds deter grazing. Therefore, the terms 'allelochemical' or 'phytotoxin' are generally preferred to describe the subset of plant chemicals that have a negative impact on the fundamental biochemical processes (McArthur et al., 1991), survival or growth (Reese, 1979) or selective behaviour (Provenza et al., 1988) of consuming herbivores (Whittaker and Feeny, 1971).

HOW DO PLANT ALLELOCHEMICALS AFFECT DIET SELECTION?

Understanding the role of plant allelochemicals in controlling plant/animal interactions is important for managing plant and animal populations in grazed ecosystems. Levels of plant defensive compounds are major determinants of herbivore production and defoliation patterns of plant communities. Understanding how allelochemicals affect herbivores and how herbivores respond to these compounds is critical to our understanding of the role of herbivory in grazed ecosystems.

Plant allelochemicals as grazing deterrents

It is common for herbivores to include small amounts of phytotoxins in their diet, but the norm is that selection of plants or plant parts is inversely related to their content of allelochemicals (Marten *et al.*, 1976; Provenza *et al.*, 1990; Freeland, 1991). It is therefore evident that animals possess grazing mechanisms to limit their exposure to phytotoxins. By understanding these grazing mechanisms we can predict ways in which the biochemical structure of plants influences their probability of being grazed. Successful antiherbivore strategies either decrease the attractiveness of the plant as animal forage or decrease the survival and reproduction of the herbivore (Fig. 6.1).

Not all chemicals are equally effective at deterring herbivory, and a chemical effective against one herbivore may not affect grazing by another. Furthermore, plants usually possess multiple defensive systems (Rhoades and Cates, 1976), which include several biochemical mechanisms, physical defences, growth forms or distribution patterns that decrease the probability or intensity of grazing.

Inherently deterrent plant flavours

Many authors consider the offensive flavour of some plant toxins to be an important defence against herbivory (Rhoades, 1979; Laycock *et al.*, 1988; Lindroth, 1989; Harborne, 1991). A significant number of plant allelochemicals are

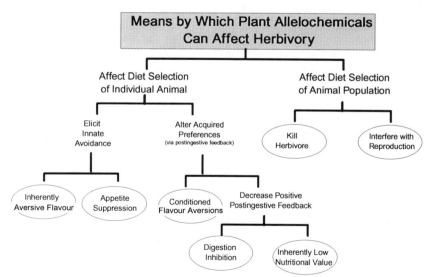

Fig. 6.1. Characteristics of plant allelochemicals that can reduce the probability or severity of grazing by mammalian herbivores.

bitter or otherwise unpleasant to the taste – at least to humans. Bitter-tasting plant compounds include many cyanogenic glycosides, alkaloids and flavonoids (Laycock *et al.*, 1988; Harborne, 1991; Molyneux and Ralphs, 1992) while unpalatable astringent taste qualities are ascribed to tannins (Harborne, 1991). Some argue that poisonous plants have evolved strong tastes and odours to advertise toxicity (Fraenkel, 1959; Rhoades, 1979; Harborne, 1988).

Most herbivores initially avoid foods with strong flavours (Provenza *et al.*, 1988; Launchbaugh and Provenza, 1994) or flavours described as bitter to humans (Bate-Smith 1972; Garcia and Hankins, 1975). The instinctive avoidance of bitter plants may have significant survival value since many plant toxins possess a bitter flavour (Garcia and Hankins, 1975). However, the inherent avoidance of bitter flavours is not universal (Nolte *et al.*, 1994). Furthermore, strong preferences can be formed for bitter-tasting foods when ingestion is followed by positive gastrointestinal consequences (Molyneux and Ralphs, 1992). Many foods eaten by humans in fact require a certain bitterness to confer palatability (e.g. coffee, beer, chocolate, cheddar cheese; Molyneux and Ralphs, 1992). Livestock have often been observed eating plants that are intensely bitter-smelling or tasting to humans. For example, cattle have been seen devouring the leaves and stems of a plant called stinky gourd (*Cucurbita foetidissima*), which has an acrid odour (E.A. Laca, Texas, 1994, personal communication). Although an offensive odour or flavour may decrease the probability that a plant will be eaten in great quantities when initially encountered, flavour *per se* will probably not deter experienced herbivores unless the tasting is followed by negative ingestive consequences.

Herbivore appetite suppression

Plant allelochemicals may act to decrease plant palatability or they may alternatively elicit in the herbivore the desire to stop eating (appetite suppression; Provenza, 1995). Garcia (1989) suggested that it is no accident that many phytochemicals, such as caffeine, nicotine and digitalis, cause a loss of appetite in vertebrates. This antiherbivore mechanism may limit the severity of grazing but there is no evidence that it affects plant palatability. On the contrary, preferences can be formed for foods that satisfy an animal's appetite (Booth, 1985). However, appetite suppression may effectively decrease herbivory of large clonal plants or extensive plant populations.

Conditioned aversions

The most effective means by which secondary compounds can alter plant palatability is by causing negative gastrointestinal consequences in the herbivore to elicit a conditioned food aversion. When a herbivore encounters a new plant, it generally eats small quantities of the plant (Chapple and Lynch, 1986; Provenza *et al.*, 1988). If gastrointestinal malaise does not follow ingestion, animals generally increase intake of the new plant over time (Chapple and Lynch, 1986;

Chapple *et al.*, 1987). However, if the herbivore becomes ill after eating a new plant it forms a dislike for the plant, termed a conditioned aversion (Burritt and Provenza, 1989a; Garcia, 1989; Provenza *et al.*, 1992). The formation of conditioned aversions involves both incentive modification (a decreased palatability of the flavour) and behaviour modification (avoidance behaviour by the animal based on visual, odour or taste cues; Fig. 6.2; Provenza *et al.*, 1992).

Strong conditioned aversion generally results when a phytotoxin stimulates the emetic centres of the midbrain and brainstem (Grant, 1987; Garcia, 1989; Provenza *et al.*, 1992, 1994a). In humans, compounds that stimulate the emetic system are described as causing nausea. Herbivores can readily form aversions to feeds that stimulate the emetic system, however they may not be able to avoid feeds that do not cause nausea (e.g. cause allergies, bloating or lower intestinal discomfort; Garcia, 1989; Provenza *et al.*, 1992). Further evidence for the important role of the emetic system is that antiemetic drugs can attenuate aversions induced by lithium chloride in sheep (Provenza *et al.*, 1994b). Some of the most dangerous toxic plants are therefore those that do not affect the emetic system because animals are generally unable to learn to avoid them through conditioned aversions (Provenza *et al.*, 1992).

The intensity of aversive feedback also affects the formation of food aversions. Doses of allelochemicals that make animals intensely ill cause stronger aversions than weak emetic stimulants (du Toit *et al.*, 1991; Launchbaugh and Provenza, 1994). The time required for the toxin to produce negative gastrointestinal effects is also important. Ruminant herbivores can readily associate novel feed flavours with gastrointestinal malaise given a delay of 8 hours between consumption and consequences (Burritt and Provenza, 1991). However, some

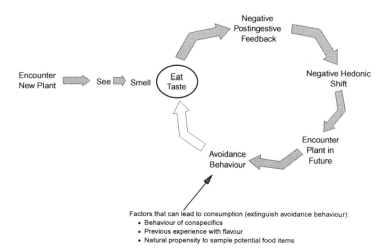

Fig. 6.2. Process by which conditioned aversions reduce the palatability of plants and the probability of future herbivory.

non-ruminant herbivores apparently require feedback within a few hours to form strong aversions (Provenza *et al.*, 1988). Thus, the intensity of a conditioned aversion may depend on how quickly a phytotoxin causes gastrointestinal malaise.

Very little is known about what a phytotoxin should taste like to be an effective aversive agent. Some researchers argue that foods with strong and salient flavours form stronger aversions than those with weak or mild flavours (Kalat and Rozin, 1970). However, when sheep were offered barley with 1 or 3% oregano (levels described as 'mild' and 'strong' by researchers) and subsequently given a dose of lithium chloride, a gastrointestinal poison known to cause aversions in sheep, the sheep formed equally strong aversions to the strong and mild flavoured barley (Launchbaugh and Provenza, 1994). Others have suggested that herbivores may more readily form aversions to bitter- than to sweet-flavoured feeds, since bitter flavours are more commonly associated with secondary compounds in natural settings (Garcia and Hankins, 1975). However, sheep formed equally strong aversions to sweet (sodium saccharinate)- and bitter (aluminium sulphate)-flavoured barley when tested with aversions induced by lithium chloride (Launchbaugh *et al.*, 1993). It is uncertain if the flavour of a plant can affect the strength or persistence of food aversions in mammalian herbivores.

Digestion inhibition

Chemical factors that reduce the digestibility of plant nutrients could also protect plants from herbivory by decreasing positive feedback in the formation of conditioned preferences. If an allelochemical reduces the digestive benefits of a plant, the palatability of that plant will decrease (Provenza *et al.*, 1992; Provenza, 1995). There are three ways in which plant chemicals can decrease the digestibility of plants. First, plant compounds can form insoluble complexes with nutrients, such as the protection of structural carbohydrates by lignin (Van Soest, 1994) and the precipitation of plant proteins by tannins (McLeod, 1974; Reese, 1979) and resinous compounds (Meyer and Karasov, 1991). Second, allelochemicals can inactivate digestive enzymes. For example, potatoes, soybeans, cottonseeds, peas, tomatoes and a wide variety of other plants contain trypsin and chymotrypsin inhibitors, which can interfere with protein digestion, particularly in monogastric animals (Reese, 1979; Slansky, 1992). Compounds that precipitate proteins, like gossypol (Calhoun *et al.*, 1991; Slansky, 1992) and tannins (McLeod, 1974; Reese, 1979), can also decrease enzyme activity by binding with enzyme proteins. Thirdly, allelochemicals can have antimicrobial effects that inhibit rumen or caecal/colon microbes (Nagy *et al.*, 1964; Oh *et al.*, 1967; Allison, 1978). Compounds that hinder the growth of microbial symbionts can reduce the amount of volatile fatty acids and amino acids available to the herbivore (Slansky, 1992).

The ability of tannins to render proteins indigestible is often cited as an example of digestion inhibition (Rhoades and Cates, 1976; Reese, 1979; Harborne, 1991). Tannins certainly have deleterious effects on vertebrate herbivores

(McLeod, 1974; Reese, 1979; Meyer and Karasov, 1991), but these effects are not necessarily symptomatic of digestion inhibition (Bernays *et al.*, 1989; Provenza, *et al.* 1990). Provenza *et al.* (1990), in a series of experiments with goats and condensed tannins in blackbrush (*Coleogyne ramosissima*), determined that the grazing deterrent characteristics of condensed tannins in blackbrush were not due to digestion inhibition but rather to negative gastrointestinal events that yielded conditioned food aversions. Erosion of gut epithelia and induced mineral imbalances in the animal have also been implicated in the aversive nature of tannins (Bernays *et al.*, 1989). These studies suggest caution when ascribing digestion reduction as a sole means by which a phytochemical evokes grazing avoidance.

Nutrient dilution

Plants may evade grazing either by possessing compounds that cause negative gastrointestinal feedback or by minimizing digestive benefits. The high content of cellulose, hemicellulose and waxes in some plants may represent mechanisms of nutrient dilution (Herms and Mattson, 1992). It is apparent that plants do not need to be completely repellent to herbivores; they only need to be less desirable to herbivores than neighbouring plants. Herbivores can quite effectively distinguish between plants that differ in digestible energy or nutrients (Villalba and Provenza, 1996a, b). Thus, digestion inhibition and low nutritive value may influence herbivory if more attractive forages are available to the herbivore. Herbivory may in fact encourage the evolution of plants with low nutritive quality through disproportional selection of the most nutritious members of a plant population (Herms and Mattson, 1992).

Plant allelochemicals that directly affect reproductive fitness

Rather than affecting the diet selection or appetite of herbivores some allelochemicals may reduce the reproductive fitness of herbivores to deter future herbivory (Rhoades and Cates, 1976). Many plant allelochemicals affect the reproductive system of herbivores. For example, gossypol in cottonseed meal can decrease reproductive success of monogastric females (Randel, 1991) and cause male infertility (Cheeke and Shull, 1985). Several mycotoxins (Cheeke and Shull, 1985) and isoflavones (Harborne, 1991) in clovers (*Trifolium* spp.) and alfalfa (*Medicago sativa*) have oestrogenic effects resulting in infertility. Phenolic acids in saltgrass (*Distichlis spicata*) trigger the cessation of breeding in some herbivorous rodents (Harborne, 1991). Teratogenic effects have been observed for a number of toxic plants including lupines (*Lupinus* spp.), poison hemlock (*Conium maculatum*), veratrum (*Veratrum californicum*) and crazyweeds (*Astragalus* and

Oxytropis spp; Panter *et al.*, 1992). Agents causing abortion have been identified in ponderosa pine needles (*Pinus ponderosa;* James *et al.*, 1992).

Decreasing reproductive success of herbivores could constitute an anti-herbivore device because there is a genetic basis for tolerance systems in herbivores (O'Reilly, *et al.* 1968; Vesell, 1968; Oliver *et al.*, 1979; Mead *et al.*, 1985). Thus, phytotoxins that affect the reproductive system of the ingesting herbivores could selectively remove traits from the gene pool that allow animals to detoxify or tolerate a particular phytotoxin. Additionally, young animals learn from their mothers what to eat and what to avoid (Galef and Beck, 1990; Mirza and Provenza, 1990, 1994; Thorhallsdottir *et al.*, 1990a, b). Therefore, phytotoxins could successfully deter future grazing if they prevented the herbivores that eat them from reproducing or if avoidance behaviour was taught to their offspring.

Ingestive benefits of plant allelochemicals

Herbivores possess several mechanisms to identify toxic compounds in forage and yet diet studies commonly reveal the consumption of toxic plants. This suggests that animals knowingly ingest phytotoxins (Freeland and Janzen, 1974). It is often assumed that animals eat plant toxins as a necessary consequence of acquiring adequate nutrients (Brattsten, 1979). An alternative explanation is that many phytotoxins benefit the animal when eaten at low doses. The distinction between toxin and nutrient is not always clear. Even nutritional components of forage (e.g. fats, starches, amino acids, vitamins and minerals) can be toxic at high doses (Reese, 1979; Masters *et al.*, 1992; Provenza, 1995). Paracelsus' axiom '*Sola dosis facit venenum*', only the dose makes the poison (stated in Molyneux and Ralphs, 1992), is particularly appropriate when applied to plant toxins.

As plant allelochemicals are degraded or detoxified they may yield compounds that benefit the herbivore. Some mammals can convert plant sterols to animal hormones or use terpenes as precursors of vitamin A and possibly vitamins E and K (Freeland and Janzen, 1974). Some vertebrates can metabolize phytic acid for phosphorus (Slansky, 1992). Within ruminants, tannins may bind plant proteins, sparing them from deamination by rumen microbes which can degrade their quality. When these tannin–protein complexes pass into the more acidic abomasum and duodenum they may dissociate and allow for digestion and absorption of the high quality plant protein (McLeod, 1974; Bernays *et al.*, 1989).

In an odd role reversal, some plant allelochemicals may aid in the detoxification of others. When some types of tannins and alkaloids are ingested together they form insoluble tannates that cannot be absorbed (Freeland and Janzen, 1974). Tannins may similarly complex with cyanogenic glycosides (Goldstein and Spencer, 1985) and saponins (Freeland *et al.*, 1985) to render them inert.

Phytotoxins may also be sought by animals for their biocidal or anti-parasite effects. For example, koalas, which feed almost exclusively on terpene-laden *Eucalyptus* plants, release some of the ingested essential oils through their lungs and skin (Eberhard *et al.*, 1975). This external release may effectively deter ectoparasites and insect predators (Slansky, 1992). The consumption of allelochemicals may also control internal parasites (Slansky, 1992).

Plant allelochemicals are common in the medicine cabinets and spice racks of human homes. Recent anecdotal evidence suggests that animals (primates, elephants, and bears) seek specific plants for their medicinal qualities (Cowen, 1990). It is clear that mammals form preferences for flavours eaten during recuperation from illness (Green and Garcia, 1971; Phy and Provenza, 1995). It is likely that animals select plants for medicinal value just as they do for positive nutritional value. The role of allelochemicals as spices for herbivores is more speculative (Bernays *et al.*, 1992). However, at least for humans, spices that can be quite toxic at high doses, due to their content of secondary compounds (e.g. vanilla, cinnamon, hot-pepper sauce, etc.), can have positive postingestive consequences at low doses.

HOW DO ANIMALS COPE WITH PLANT ALLELOCHEMICALS?

Generalist herbivores eat many plants in each feeding bout, and yet they successfully identify and avoid toxic plants in their environment and acquire diets that are more nutritious than the average level of nutrients on offer. Adaptive mechanisms for avoiding toxicosis include behavioural skills which limit the quantity of toxins ingested and internal systems that detoxify or tolerate the phytotoxins that are ingested.

Detection and avoidance of plant allelochemicals

Animals do not randomly select plants in their environment and rely solely on detoxification systems to save them from death by toxins. They rely also on behavioural avoidance systems to select the least toxic plants and plant parts (Provenza *et al.*, 1988; Provenza and Balph, 1990; Provenza, *et al.* 1992). How animals know which plants to avoid is a question of considerable debate (see Provenza and Balph, 1990). In the final analysis, animals must either be born knowing what to eat and avoid or they must learn appropriate dietary habits from other animals or through individual experiences.

Instinctive avoidance of plant allelochemicals

Some believe that animals instinctively recognize and avoid toxic plants (Bate-Smith, 1972; Rhoades, 1979). However, there is considerable evidence questioning the existence of an innate toxin recognition system (see Provenza and Balph,

1990). For example, herbivores can avoid toxins in environments in which they did not evolve (Provenza et al., 1988). An innate mechanism of food recognition, referred to as a specific hunger, is known for water and sodium (Rozin and Kalat, 1971; Rozin, 1976; Provenza and Balph, 1990). However, to apply a prewired recognition system for each nutrient or toxin encountered by a generalist herbivore is nearly inconceivable (Rozin and Kalat, 1971; Rozin, 1976). The investment in neurological machinery to avoid all possibly lethal toxins in a herbivore's environment is regarded as being simply too costly (Rozin, 1976).

There are, however, at least two innate mechanisms mentioned earlier that apply to the avoidance of plant allelochemicals. First, mammalian herbivores instinctively avoid plants with strong odours or tastes (Provenza et al., 1988). Since many plant allelochemicals have salient flavours, the initial avoidance of novel plants and particular avoidance of strong novel stimuli may prevent toxicosis when animals encounter new plants in their environment. This behaviour may allow animals to form flavour–consequence relationships through conditioned aversions at sublethal levels of toxin ingestion (Launchbaugh and Provenza, 1994). Secondly, when mammals become ill after eating a novel food they innately associate the gastrointestinal feedback with the flavour rather than the sight or texture of the food (Rozin, 1976; Garcia, 1989; Provenza et al., 1992). Since changes in plant biochemistry are more likely to be associated with changes in flavour than changes in sight or texture, this innate propensity of animals to form flavour–consequence relationships may improve their ability to avoid toxic plant allelochemicals.

Role of social models in directing selective behaviour

Livestock live in multigenerational groups, in which dietary information can be passed from experienced to inexperienced animals. Livestock may not need perfect dietary information at birth because they are born with a reliable model – their mother. Lambs quickly learned to avoid a harmful novel food their mothers were trained to avoid, and to consume a novel alternative selected by their mothers (Mirza and Provenza, 1990, 1994). Young livestock can also learn appropriate food choices from other adult animals and peers (Thorhallsdottir, 1990a, b).

Learning from mother may begin before young herbivores take their first bite. Flavours in the uterine fluid can influence food aversions (Smotherman, 1982). Mother's milk is also a source of information for young livestock. Nolte and Provenza (1991) found that orphan lambs raised on onion-flavoured milk later preferred onion-flavoured feed. Thus, before a herbivore begins to forage, it may have substantial information about the plants in its environment. As young herbivores gain individual experiences with plants, mother's influence diminishes. Mirza and Provenza (1990) found that lambs 6 weeks of age more closely matched the diets of their mothers than lambs that were 12 weeks old.

Individual learning about plant allelochemicals

Conditioned aversions are powerful mechanisms by which animals can learn to identify and avoid toxic plants (Rozin, 1976; Provenza and Balph, 1988, 1990; Provenza *et al.*, 1988, 1992; Garcia, 1989). As described above, aversions are formed when the ingestion of a plant is followed by negative gastrointestinal feedback. Aversions can be easily demonstrated by offering an animal a single novel food and inducing gastrointestinal illness immediately following ingestion. It is more difficult to explain how conditioned aversions allow generalist herbivores to avoid toxic plants in complex and variable foraging environments. Two questions are particularly relevant. First, how do herbivores determine the plant or plants that cause illness when each meal contains many plants? Second, how do animals learn to alter intake of plants in response to spatial and temporal variation in plant toxicity?

How herbivores determine which plants are toxic

Generalist herbivores construct meals containing many plant species. It is difficult to ascertain how herbivores attribute illness to specific plants rather than to all plants in the diet. One point is clear: herbivores view the edible world as either familiar or novel and form aversions to novel dietary items (Rozin, 1976; Burritt and Provenza, 1989b, 1991). When lambs were offered a non-toxic novel feed along with a familiar feed containing lithium chloride, to induce illness, the lambs formed aversions to the novel feed even though the familiar feed contained the toxin (Burritt and Provenza, 1989b). Furthermore, when lambs were given a capsule of lithium chloride after eating a meal containing four familiar and one novel feed they subsequently avoided only the novel feed (Burritt and Provenza, 1991).

Rats that ate two novel foods in one meal and subsequently experienced illness became averse to the food avoided by peers (Galef and Beck, 1990). Thus, social models may improve an animal's ability to discriminate between toxic and safe plants. Conspecifics may also play a role in ameliorating aversions when the toxicity of a plant has diminished. Ewes and lambs averse to a pelleted ration ingested more of the pellet when feeding with non-averse conspecifics than when feeding alone (Thorhallsdottir *et al.*, 1990b). Likewise, lambs formed weaker aversions to a palatable shrub when conditioned and tested with conspecifics that were not averse to the shrub than when exposed to the shrub with other averse lambs (Provenza and Burritt, 1991). Cattle also consumed more of a toxic plant that they had been conditioned to avoid when they were with non-averse peers (Lane *et al.*, 1990).

Factors other than novelty and the behaviour of conspecifics can be employed by herbivores to attribute illness to specific plants in the diet. These include: previous experience with a flavour (Provenza, 1995); the proportions of each novel feed eaten during a meal (Provenza *et al.*, 1994b); temporal contiguity between ingestion and gastrointestinal feedback (Provenza et al., 1993); and the salience or strength of flavours in a meal (Kalat and Rozin, 1970; Launchbaugh et al., 1993; Provenza et al., 1994b). Herbivores therefore possess several ways of identifying the most likely sources of gastrointestinal illness in their diets.

How herbivores avoid toxins in a variable environment

Plant toxicity varies depending on plant and animal characteristics (Launchbaugh et al., 1993). How do animals recognize changes in phytotoxicity and respond appropriately? When flavour and toxicity are highly correlated, herbivores apparently first regulate food intake based on postingestive feedback and then simply adjust intake on the basis of flavour intensity (Launchbaugh et al., 1993). This pattern was observed in sheep that were fed oats containing a particular concentration of lithium chloride. When the percentage of lithium chloride in the oats increased, the lambs decreased intake and responded to a decrease in lithium chloride concentration by increasing consumption (Launchbaugh et al., 1993).

However, flavour may not always be a good indicator of toxicity. For instance, some compounds may be present in such small amounts that they add little or nothing to overall plant flavour (Provenza and Balph, 1990). Also, minute changes in stereochemistry, which may be difficult for herbivores to detect through taste or smell, can dramatically change the toxicity of many phytochemicals (Provenza et al., 1992; Launchbaugh et al., 1993). Factors such as diet selection or an animal's nutritional state may also alter plant toxicity but not affect plant flavour (Freeland and Janzen, 1974; Brattsten, 1979). When lambs were offered a feed with a constant flavour and variable toxic feedback, they consumed an amount based on the maximum (rather than the average) dose of the toxin they had received (Launchbaugh et al., 1993). This response reduced the risk of lambs overingesting a toxin even when toxicity changes were not detectable through flavour changes.

Detoxification and tolerance of allelochemicals

Animals possess several mechanisms to negate or restrict the toxic effects of plant allelochemicals once they have been ingested. Detoxification pathways are designed either to metabolize an allelochemical in such a way that its components can be used by microorganisms or metabolic pathways or to convert the toxin into a metabolic product that can be easily excreted (Brattsten, 1992). Animals can also enhance their ability to tolerate the toxic effects of some plant allelochemicals (Slansky, 1992).

Gut-level detoxification

Mediating chemical reactions in the gut represents the first line of defence against the potential toxic effects of ingested plant allelochemicals. Animals exhibit several chemical and biological mechanisms to alter the toxicity of allelochemicals in the gut.

Environment modification

Environmental conditions such as temperature, pH and solution polarity can significantly affect the degradation or absorption of allelochemicals (McArthur *et al.*, 1991). The environment in the herbivore's digestive tract often changes in response to allelochemicals in such a way as to decrease their reactivity (McArthur *et al.*, 1991). The extent to which environmental modification allows grazing animals to deal with secondary compounds is unknown, but represents a potential mechanism to limit the negative impacts of plant allelochemicals.

Complex formation

When plant allelochemicals are ingested they may form insoluble complexes with other compounds in the gut (McArthur *et al.*, 1991; Provenza *et al.*, 1992). Many tannins, for example, can be inactivated through complex formation with nutrients (McLeod, 1974; Reese, 1979) or other allelochemicals (Freeland *et al.*, 1985). This complexing principle was observed by McNabb *et al.* (1993) when sheep fed *Lotus pedunculatus*, a plant containing condensed tannins, showed greater ruminal absorption of amino acids when polyethylene glycol was added to the rumen. The polyethylene glycol apparently complexed with ingested tannins to prevent them from binding with dietary and microbial protein. Salivary proteins have also been shown to complex with tannins in the diet, resulting in increased protein and dry-matter digestibility in mule deer and black bears fed tannin-rich diets (Robbins *et al.*, 1991).

Detoxification by gut microbes

Many chemical reactions which alter the toxicity of plant allelochemicals are mediated by gut microflora. Ruminants generally suffer fewer negative effects from poisonous plants than non-ruminants (Smith, 1992). This results from microbial fermentation occurring prior to mammalian enzyme digestion in ruminants, while the site of fermentation in non-ruminants is distal to the stomach and small intestine (Van Soest, 1994). Chemical interactions between plant allelochemicals and gut microbes are of two major classes: addition of functional groups primarily by hydrolysis or reduction; and degradations, such as dealkylation, deamination, decarboxylation, dehalogenation, and ring fusion (Fig. 6.3; Allison, 1978; Lindroth, 1989; McArthur *et al.*, 1991).

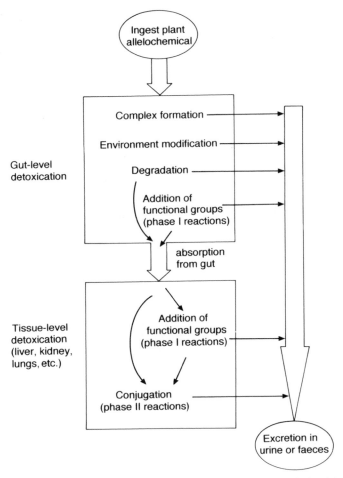

Fig. 6.3. Physiological biochemical mechanisms that detoxify and limit the deleterious effects of plant allelochemicals once ingested.

Microbial modifications of allelochemicals generally benefit the herbivore. For example, rumen microbes in sheep detoxify oxalates by forming insoluble calcium oxalates, which are excreted in the faeces (James and Cronin, 1974). Likewise, dietary selenium can be converted by gut microbes to a less soluble form of selenium (Allison, 1978). Gossypol in cottonseed meal, trypsin inhibitors in soybeans and glucosinolates in rapeseed meal are all secondary compounds in livestock feed known to be detoxified by rumen microbes (Cheeke and Shull, 1985). Other compounds known to lose their toxicity in the rumen include the amino acid mimosine (Hegarty *et al.*, 1964), monoterpene essential oils (Oh *et al.*, 1967; Allison, 1978), pyrrolizidine alkaloids (Lanigan and Smith, 1970), *Digitalis* alkaloids (Freeland and Janzen, 1974) and some mycotoxins (Cheeke and Shull, 1985). Kronberg and Walker (1993) have suggested that differences in leafy spurge (*Euphorbia esula*) consumption by sheep, goats and cattle can be explained by species differences in the degradation of a diterpene diester by rumen microbes. Likewise, greater consumption of Douglas fir (*Pseudotsuga menziesii*) needles by deer than sheep has been attributed to differences in rumen microbial populations between the two species (Oh *et al.*, 1967).

Microbial interactions with plant allelochemicals are not always beneficial, for they can lead to increased toxicity. Nitrate, for example, is converted to the more toxic nitrite by rumen microbes (Allison, 1978). Cyanogenic glycosides, which are not toxic to monogastric herbivores, are hydrolysed by gut microbes to form hydrogen cyanide, which is highly toxic (Conn, 1979). The phytoestrogen, formononetin, in subterranean clover (*Trifolium subterran*) can be degraded in the sheep rumen to a more potent estrogen (Cheeke and Shull, 1985).

Tissue-level detoxification

Plant allelochemicals and toxic products of microbial fermentation that are absorbed from the gut are usually metabolized by specific enzyme systems, located primarily in the liver but also in the kidneys, lungs, gut wall and other body tissues (McArthur *et al.*, 1991). The major enzyme reactions can be organized into two groups of reactions called phase I (or primary) and phase II (or secondary) reactions (Fig. 6.3; Freeland and Janzen, 1974; Brattsten, 1979; McArthur *et al.*, 1991).

Phase I reactions

If the absorbed compound contains a hydroxyl, carboxyl, amine or sulphate group, it can enter the conjugation phase (phase II); otherwise it must first have a functional group added in a phase I reaction (Freeland and Janzen, 1974). The addition of functional groups usually occurs through oxidation, reduction or hydrolysis (McArthur *et al.*, 1991). The most important phase I reactions are oxidation reactions catalysed by mixed function oxidases (MFOs; Brattsten, 1992). The actions of MFOs are not clearly understood but they are known to play a major role in the detoxification of plant allelochemicals.

Phase II reactions

The phase II reactions involve the conversion of lipophilic compounds to excretable products by conjugations with polar components to increase water solubility, a necessary requirement for excretion in mammalian systems (Freeland and Janzen, 1974). The polar products of conjugation are generally excreted in the urine, but larger products (molecular weight greater than 300) may return in bile to the gut and then be excreted in faeces (Freeland and Janzen, 1974; McArthur *et al.*, 1991). Common conjugates include sugars, amino acids, sulphates, phosphates and others (Brattsten, 1992; Smith, 1992).

Tolerance at target site

A plant allelochemical that fails to be detoxified or inactivated in the gut or body tissue may have a variety of deleterious effects on the animal. These effects include damage to muscles, nerves, deoxyribonucleic acid (DNA) and cell membranes, interference with basic metabolic processes, and other effects, many yet unknown (Slansky, 1992). The last line of defence against the deleterious effects of plant allelochemicals occurs when the tissues vulnerable to damage by an allelochemical become less sensitive or shielded in a process called physiological tolerance (Provenza *et al.*, 1992) or target site insensitivity (Slansky, 1992). Tolerance is observed as either a reduced effect in response to the same toxin dose or an increased dose required to maintain the same level of effect (Provenza *et al.*, 1992). The potential importance of tolerance in avoiding toxicosis is considerable, although little is known about this physiological mechanism in mammalian herbivores (Provenza *et al.*, 1992).

Influence of previous dietary experience on detoxification

When an animal is exposed to a toxic plant or feed component, it may initially eat a small amount. With continued consumption, the animal may gain an ability to survive high levels of an allelochemical (Freeland and Janzen, 1974). This increased immunity to the toxic effects of a particular plant may be due to physiological tolerance at the target site or enhanced detoxification. It is well known that enzyme systems in animal tissue can increase their detoxification capacity and efficiency in the presence of a toxic substrate (Freeland and Janzen, 1974; Harborne, 1988; Brattsten, 1992). These inducible defenses could explain increased tolerance of animals to plant allelochemicals with continued exposure.

Rumen microbes may also facilitate adaptation to diets high in phytotoxins. Microbial populations can change rapidly depending on the substrates available for degradation (Van Soest, 1994). For example, ruminants that have been adapted to oxalate-containing diets can tolerate levels of oxalate that would be lethal to non-adapted animals (James and Cronin, 1974; Allison and Cook, 1981). This protection from oxalates is attributed to oxalate-degrading rumen bacteria (Allison and Cook, 1981). One particular species of rumen bacteria, *Oxalobacter fomigenes*, is known to use oxalate as its only substrate for growth (Allison and Cook, 1981). Dietary experience leading to increased tolerance of ingested nitrates is attributed to increased populations of nitrite-degrading microbes (Smith, 1992). The inhibitory effects of essential oils in Douglas fir can likewise be diminished with dietary experience in deer (Oh *et al.*, 1967).

Changes in digestive morphology and physiology may also alter the toxicity of phytochemicals (Lindroth, 1989). Changes in diet quality often result in changes in rumen size, rate of passage, rumen surface characteristics and saliva production (Van Soest, 1994). Some have suggested that changes in digestive morphology and physiology improve an animal's feed efficiency and detoxification of plant allelochemicals (Provenza and Balph, 1988; Distel and Provenza, 1991). Of particular interest is the production of proline-rich saliva in response to high-tannin diets (Mehansho *et al.*, 1985; Robbins *et al.*, 1991). Changes in the composition and production of saliva may represent an inducible defence against plant tannins (Mehansho *et al.*, 1985; Distel and Provenza, 1991; Robbins *et al.*, 1991).

Influence of animal nutritional status on detoxification

Detoxification of plant allelochemicals comes with a price paid in animal nutrient and energy resources. Improving the nutritional state of the animal can often lead to increased rates of detoxification or decreased toxic effects (Freeland and Janzen, 1974; Boyd and Campbell, 1983). Nutrient and energy availability can influence plant toxicity by altering: rates of gastrointestinal absorption and enzymatic detoxification; availability of substrates for conjugation; environmental conditions in the gastrointestinal tract or body fluids; and target-site responsiveness (Boyd and Campbell, 1983; Slansky, 1992). Dietary nutrients and energy are also required to maintain healthy rumen microbial populations (Van Soest, 1994), which are important in the detoxification of many plant allelochemicals (Allison, 1978).

Supplementation of specific nutrients may decrease the toxic effects of allelochemicals. For example, increased dietary calcium often impedes the absorption of oxalates, which bind with calcium ions to form insoluble calcium oxalates that are then excreted (James and Cronin, 1974). Likewise, increases in soluble dietary protein can decrease the absorption of gossypol (Calhoun *et al.*, 1991) and some types of tannin (Bernays *et al.*, 1989) through the formation of insoluble complexes. Vitamin A supplementation can decrease the deleterious effects of gossypol from cottonseed meal (Calhoun *et al.*, 1991). Increased available sulphur (Conn, 1979) and vitamin B_{12} (Brattsten, 1979) have both been shown to decrease the formation of hydrogen cyanide from cyanogenic glycosides. The administration of ferrous salts and sulphates can mitigate the negative effects of gossypol (Reese, 1979) and mimosine (Rosenthal and Bell, 1979).

Selection of varied diets

Mammalian herbivores possess many ways of detoxifying or tolerating phytotoxins and these various detoxification pathways are generally specific to a single phytotoxin or group of phytotoxins. Some have suggested that herbivores can eat a variety of plants to spread the ingestion of toxins over many detoxification pathways (Freeland and Janzen, 1974; Freeland, 1991; Bernays *et al.*, 1992). In this way, herbivores might avoid overloading a particular detoxification pathway and thus avoid toxicosis. Furthermore, interactions between allelochemicals in the diet can reduce toxicity and herbivores can learn to select diets to take advantage of these beneficial interactions (Freeland *et al.*, 1985). Thus, selection of a varied diet constitutes a mechanism by which herbivores can significantly lower their risk of toxicosis.

HERBIVORE MANAGEMENT IN RELATION TO PLANT CHEMICAL DEFENCES

Plants heavily laden with defensive chemicals are generally considered undesirable in ecosystems managed for livestock production or wildlife management. Traditional rangeland management approaches are based on reducing the abundance of chemically defended plants with fire, mechanical or herbicidal methods. These techniques are often prohibitively costly and may have detrimental environmental impacts. Alternatively, livestock management procedures based on how animals cope with plant allelochemicals can be developed to manage ecosystems abundant with chemically defended plants.

Avoid overstocking

As preferred forages become limited, due to high stocking rates or drought, animals generally increase their consumption of less preferred plants which often contain allelochemicals. As a result, deaths attributed to poisonous plants generally increase when nutritious forage becomes limited (Meyer and Karasov, 1991; Slansky, 1992; Taylor and Ralphs, 1992). Deaths may reflect increased consumption of toxins, a decreased ability to detoxify ingested toxins due to poor nutritional state or starvation because of restricted availability of nutritious plants (Meyer and Karasov, 1991).

There are situations when a manager may choose to overstock to make greater use of available forages and avoid the financial costs of destocking. A high stocking rate can effectively increase the consumption of chemically defended plants if detoxification pathways can be induced to deal with the increased intake of phytotoxins. Careful management of stocking rates can therefore be used to initiate animal adaptation to toxic plants (Taylor and Ralphs, 1992). Anecdotal evidence indicates that there are some plants which animals will not eat even when hungry. For example, hunger apparently cannot entice animals to eat African rue (*Peganum harmula*; Mathews, 1941) or drymary (*Drymaria pachyphylla*; Lantow, 1923). Mammalian herbivores limit their consumption of feeds based on the content of toxins (Launchbaugh *et al.*, 1993). Heavy stocking rates will probably not significantly increase the consumption of toxic plants per animal unless physiological, neurological or morphological changes lead to greater detoxification or tolerance. Also, as mentioned above, imposing high stocking rates in order to increase the consumption of chemically defended plants can have negative impacts on animal performance.

Offer nutrient resources for detoxification

If increasing the consumption of chemically defended plants is desired, then starving animals on to these plants may not be the answer. Nutrient deprivation often decreases the rate of detoxification and increases an animal's toxic response (Boyd and Campbell, 1983). Conversely, supplementation of vitamins, minerals, amino acids and carbohydrates often enhances the ability of herbivores to detoxify or tolerate phytotoxins (Freeland and Janzen, 1974; Brattsten, 1979; Conn, 1979; Boyd and Campbell, 1983). Many research and management opportunities exist to identify compounds that complex and inactivate allelochemicals in the diet (McNabb *et al.*, 1993). A detailed understanding of the detoxification pathways of specific compounds can therefore lead to supplementation programmes designed to entice rather than coerce livestock into eating chemically defended plants.

Select proper animal

Animal species differ in their ability to tolerate or detoxify secondary plant metabolites (Oh *et al.*, 1967, Taylor and Ralphs, 1992; Kronberg and Walker, 1993). Therefore, management based on using chemically defended plants begins with the selection of the proper animal species for the vegetation at hand (Taylor and Ralphs, 1992). Individuals within a species also vary in their consumption of defended plants. In a feeding trial with domestic goats, certain individuals consistently ate more juniper than the herd average, while others consistently ate less than average (Pritz and Launchbaugh, 1995). These inter- and intraspecies differences result from differences in digestive morphology, detoxification capacities or physiological tolerances. Selecting a species, breed or individual within a herd or flock with desired dietary preferences could significantly alter consumption of specific chemically defended plants.

Breed animals with desired diet characteristics

Many of the differences in diet preferences between individuals can be traced to inherited physiological, neurological or morphological characteristics. For example, enzyme systems for detoxification of fluoroacetates are inherited and resistance to fluoroacetate poisoning has in fact been used as a genetic population marker in native Australian mammals (Oliver *et al.*, 1979; Mead *et al.*, 1985).

Breeding programmes may be implemented to influence the consumption of chemically defended plants if a significant portion of inherited characteristics account for consumption of the plants.

Give animals proper early-life experiences

There is growing evidence that previous dietary experiences can influence the flavour preferences (Nolte and Provenza, 1991) of animals and the ability of animals to digest (Distel *et al.*, 1994), detoxify (Distel and Provenza, 1991; Robbins *et al.*, 1991) and harvest (Ortega-Reyes and Provenza, 1993) certain plants. Furthermore, experiences early in life often have a more lasting effect on consumption patterns than experiences later in life (Distel *et al.*, 1994). Implementing diet programmes for young animals may be necessary to increase the mature animal's consumption of a particular chemically defended plant.

SUMMARY

There are only a few avenues by which plant allelochemicals can have an impact on animal behaviour (Fig. 6.1). Likewise, only a few strategies are employed by herbivores to mitigate the negative impacts of allelochemicals. Yet, from these basic defence approaches, a tremendous complexity of interactions between plants and herbivores is born. The key to management of grazed ecosystems therefore lies in understanding the many variations on the simple themes of plant–herbivore interactions.

REFERENCES

Allison, M.J. (1978) The role of ruminal microbes in the metabolism of toxic constituents of plants. In: Keeler, R.F., VanKampen, K.R and James, L.F. (eds) *Effects of Poisonous Plants on Livestock.* Academic Press, New York, pp. 101–120.

Allison, M.J. and Cook, H.M. (1981) Oxalate degradation by microbes of the large bowel of herbivores: the effect of dietary oxalate on ruminant adaptation. *Science* 212, 675–676.

Bate-Smith, E.C. (1972) Attractants and repellents in higher animals. In: Harborne, J.B. (ed.) *Phytochemical Ecology.* Academic Press, New York, pp. 45–56.

Bernays, E.A., Cooper-Driver, G. and Bilgener, M. (1989) Herbivores and plant tannins. *Advances in Ecological Research* 19, 263–302.

Bernays, E.A., Bright, K., Howard, J.J., Raubenheimer, D. and Champagne, D. (1992) Variety is the spice of life: frequent switching between foods in the polyphagous grasshopper *Taeniopoda eques* Bermeister. *Animal Behavior* 44, 721–731.

Booth, D.A. (1985) Food-conditioned eating preferences and aversions with interoceptive elements: conditioned apetites and satieties. In: Braveman, N.S. and Bronstein, P.

(eds) *Experimental Assessments and Clinical Applications of Conditioned Food Aversions. Annals of the New York Academy of Science* 443, 22–41.

Boyd, J.N. and Campbell, T.C. (1983) Impact of nutrition on detoxication. In: Caldwell, J. and Jakoby, W.B. (eds) *Biological Basis for Detoxication.* Academic Press, New York, pp. 287–306.

Brattsten, L.B. (1979) Biochemical defense mechanisms in herbivores against plant allelochemicals. In: Rosenthal, G.A. and Janzen, D.H. (eds) *Herbivores: Their Interaction with Secondary Plant Metabolites.* Academic Press, New York, PP. 200–270.

Brattsten, L.B. (1992) Metabolic defenses against plant allelochemicals. In: Rosenthal, G.A. and Borenbaom M.R. (eds) *Herbivores: Their Interaction with Secondary Plant Metabolities. Vol. 2: Ecological and Evolutionary Processes.* Academic Press, New York, pp. 176–242.

Burritt, E.A. and Provenza, F.D. (1989a) Food aversion learning: conditioning lambs to avoid a palatable shrub *(Cerocarpus montanus). Journal of Animal Science* 67, 650–653.

Burritt, E.A. and Provenza, F.D. (1989b) Food aversion learning: ability of lambs to distinguish safe from harmful foods. *Journal of Animal Science* 67, 1732–1739.

Burritt, E.A. and Provenza, F.D. (1991) Ability of lambs to learn with a delay between food ingestion and consequences given meals containing novel and familiar foods. *Applied Animal Behaviour Science* 32, 179–184.

Calhoun, M.C., Huston, J.E., Calk, C.B., Baldwin, B.C., Jr and Kuhlman, S.W. (1991) Effects of gossypol on digestive metobolic function of domestic livestock. In: Jones, L.A., Kinard, D.H. and Mills, J.S. (eds) *Cattle Research With Gossypol Containing Feeds.* Special Session, Annual Meeting of the American Society of Animal Science. National Cottonseed Producers Association, Memphis, Tennessee.

Chapple, R.S. and Lynch, J.J. (1986) Behavioural factors modifying acceptance of supplementary foods by sheep. *Research and Development in Agriculture* 3, 113–120.

Chapple, R.S., Wadzicka-Tomanszewska, M. and Lynch, J.J. (1987) The learning behaviour of sheep when introduced to wheat. I. Wheat acceptance by sheep and the effects of trough familiarity. *Applied Animal Behaviour Science* 18, 157–162.

Cheeke, P.R. and Shull, L.R. (1985) Natural toxicants and their general biological effects. In: Cheeke, P.R. and Shull, L.R. (eds) *Natural Toxicants in Feeds and Poisonous Plants.* AVI Publication Company, Westport, Connecticut.

Conn, E.E. (1979) Cyanids and cyanogenic glycosids. In: Rosenthal, G.A. and Janzen, D.H. (eds) *Herbivores: Their Interaction with Secondary Plant Metabolites.* Academic Press, New York, pp. 387–412.

Cowen, R. (1990) Medicine on the wild side. *Science News* 138, 280–282.

Distel, R.A. and Provenza, F.D. (1991) Experience early in life affects voluntary intake of blackbrush by goats. *Journal of Chemical Ecology* 17, 431–450.

Distel, R.A., Villalba, J.J. and Laborde, H.E. (1994) Effects of early experience on voluntary intake of low quality roughage by sheep. *Journal of Animal Science* 72, 1191–1195.

du Toit, J.T., Provenza, F.D. and Nastis, A.S. (1991) Conditioned food aversions: How sick must a ruminant get before it detects toxicity in foods? *Applied Animal Behaviour Science* 30, 35–46.

Eberhard, I.H., McNamara, J., Pearse, R.J. and Southwell, I.A. (1975) Ingestion and excretion of *Eucalyptus punctata* (D.C.) and its essential oils by the koala, *Phascolarctos cinereus* (Goldfuss). *Australian Journal of Zoology* 23, 169–179.

Fraenkel, G.S. (1959) The raison d'être of secondary plant substances. *Science* 129, 1466–1470.
Freeland, W.J. (1991) Plant secondary metabolites: biochemical coevolution with herbivores. In: Palo, R.T. and Robbins, C.T. (eds) *Plant Defenses Against Mammalian Herbivores*. CRS Press, Boca Raton, Florida, pp. 61–82.
Freeland, W.J. and Janzen, D.H. (1974) Strategies in herbivory by mammals: the role of plant secondary compounds. *American Naturalist* 108, 269–289.
Freeland, W.J., Calcott, P.H. and Anderson, L.R. (1985) Tannins and saponins: interaction in herbivore diets. *Biochemical Systematic Ecology* 13, 189–193.
Galef, B.G., Jr and Beck, M. (1990) Diet selection and poison avoidance by mammals individually and in social groups. In: Stricker, E.M. (ed.) *Handbook of Behavioral Neurobiology*, Vol. 10: *Neurobiology of Food and Fluid Intake*. Plenum Press, New York, pp. 329–352.
Garcia, J. (1989) Food for Tolman: cognition and cathexis in concert. In: Archer, T. and Nilsson, L. (eds) *Aversion, Avoidance and Anxiety*. Lawrence Erlbaum and Associates, Hillsdale, New Jersey, pp. 45–85.
Garcia, J. and Hankins, W.G. (1975) The evolution of bitter and the acquisition of toxiphobia. In: Denton, D. and Coghlan, J. (eds) *Olfaction and Taste*, Vol. 5. Academic Press, New York, pp. 39–41.
Goldstein, W. and Spencer, K.C. (1985) Inhibition of cyanogenesis by tannin. *Journal of Chemical Ecology* 11, 845–847.
Grant, V.L. (1987) Do conditioned taste aversions result from activation of emetic mechanisms? *Psychopharmacology* 93, 405–415.
Green, K.F. and Garcia, J. (1971) Recuperation from illness: flavor enhancement for rats. *Science* 173, 744–751.
Harborne, J.B. (1988) *Introduction to Ecological Biochemistry*, 3rd edn. Academic Press, San Diego, California.
Harborne, J.B. (1991) The chemical basis of plant defense. In: Palo, R.T. and Robbins, C.T. (eds) *Plant Defenses Against Mammalian Herbivory*. CRC Press, Boca Raton, Florida, pp. 45–60
Hegarty, M.P., Schinckel, P.G. and Court, R.D. (1964) Reaction of sheep to the consumption of *Leucaena glauca* Benth. and to its toxic principle mimosine. *Australian Journal of Agricultural Research* 15, 153–167.
Herms, D.A. and Mattson, W.J. (1992) The dilemma of plants: to grow or defend. *Quarterly Review of Biology* 67, 283–313.
James, L.F. and Cronin, E.H. (1974) Management practices to minimize death losses of sheep grazing halogeton-infested rangeland. *Journal of Range Management* 27, 424–426.
James, L.F., Nielson, D.B. and Panter, K.E. (1992) Impact of poisonous plants on the livestock industry. *Journal of Range Management* 45, 3–8.
Kalat, J.W. and Rozin, P. (1970) 'Salience': a factor which can override temporal contiguity in taste-aversion learning. *Journal of Comparative Physiology and Psychology* 71, 192–197.
Kronberg, S.L. and Walker, J.W. (1993) Ruminal metabolism of leafy spurge in sheep and goats: a potential explanation for differential foraging on spurge by sheep, goats, and cattle. *Journal of Chemical Ecology* 19, 2007–2017.

Lane, M.A., Ralphs, M.A., Olsen, J.D., Provenza, F.D and Pfister, J.A. (1990) Conditioned taste aversions: potential for reducing cattle loss to larkspur. *Journal of Range Management* 43, 127–131.

Lanigan, G.W. and Smith, L.W. (1970) Metabolism of pyrrolizidine alkaloids in the ovine rumen. *Australian Journal Agricultural Research* 21, 493–500.

Lantow, J.L. (1923) The poisoning of livestock by *Drymaria pachyphylla*. *New Mexico Agricultural Experiment Station Bulletin* 173, 113.

Launchbaugh, K.L. and Provenza, F.D. (1994) The effect of flavor concentration and toxin dose on the formation and generalization of flavor aversions in lambs. *Journal of Animal Science* 72, 10–13.

Launchbaugh, K.L., Provenza, F.D. and Burritt, E.A. (1993) How herbivores track variable environments: response to variability of phytoxins. *Journal of Chemical Ecology* 19, 1047–1056.

Laycock, W.A., Young, J.A. and Ueckert, D.N. (1988) Ecological status of poisonous plants on rangelands. In: James, L.F., Ralphs, M.H. and Nielson, D.B. (eds) *The Ecology and Economic Impact of Poisonous Plants on Livestock Production*. Westview Press, Boulder, Colorado, pp. 27–42.

Lindroth, R.L. (1989) Mammalian herbivore–plant interactions. In: Abrahamson, W.G. (ed.) *Plant–Animal Interactions*. McGraw-Hill Book Co., New York, pp. 163–205.

McArthur, C., Hagerman, A. and Robbins, C.T. (1991) Physiological strategies of mammalian herbivores against plant defenses. In: Palo, R.T. and Robbins, C.T. (eds) *Plant Defenses Against Mammalian Herbivory*. CRC Press, Boca Raton, Florida, pp. 103–114.

McLeod, M.N. (1974) Plant tannins – their role in forage quality. *Nutrition Abstracts and Reviews* 44, 803–815.

McNabb, W.C., Waghorn, G.C., Barry, T.N. and Shelton, I.D. (1993) The effect of condensed tannins in *Lotus pedunculatus* on the digestion and metabolism of methionine, cystine and inorganic sulphur in sheep. *British Journal of Nutrition* 70, 647–661.

Marten, G.C., Jordon, R.M. and Hovin, A.W. (1976) Biological significance of reed canarygrass alkaloids and associated palatability variation to grazing sheep and cattle. *Agronomy Journal* 68, 909–914.

Masters, D.G., White, C.L., Peter, D.W., Porser, D.B., Roe, S.P. and Barnes, M.J. (1992) A multi-element supplement for grazing sheep. II. Accumulation of trace elements in sheep fed different levels of supplement. *Australian Journal of Agricultural Research* 43, 809–817.

Mathews, F.P. (1941) Poisonous plants in the Davis Mountains. *Texas Agricultural Experiment Station Annual Report* 54, 93.

Mead, R.J., Oliver, A.J., King, D.R. and Hubach, P.H. (1985) The co-evolutionary role of fluoroacetate in plant–animal interactions in Australia. *Oikos* 44, 55–60.

Mehansho, H., Clements, S., Shears, B.T., Smith, S. and Carlson, D.M. (1985) Induction of proline-rich glycoprotein synthesis in mouse salivary glands by isoproternol and by tannins. *Journal of Biological Chemistry* 260, 4418–4423.

Meyer, M.W. and Karasov, W.H. (1991) Chemical aspects of herbivory in arid and semiarid habitats. In: Palo, R.T. and Robbins, C.T. (eds) *Plant Defenses Against Mammalian Herbivory*. CRC Press, Boca Raton, Florida, pp. 167–188.

Mirza, S.N. and Provenza, F.D. (1990) Preference of the mother affects selection and avoidance of foods by lambs differing in age. *Applied Animal Behaviour Science* 28, 255–263.

Mirza, S.N. and Provenza, F.D. (1994) Socially induced food avoidance in lambs: direct or indirect material influence. *Journal of Animal Science* 72, 899–902.

Molyneux, R.J. and Ralphs, M.H. (1992) Plant toxins and palatability to herbivores. *Journal of Range Management* 45, 13–18.

Nagy, J.G., Steinhoff, H.W. and Ward, G.M. (1964) Effects of essential oils of sagebrush on deer rumen microbial function. *Journal of Wildlife Management* 28, 785–790.

Nolte, D.L. and Provenza, F.D. (1991) Food preferences in lambs after exposure to flavors in milk. *Applied Animal Behaviour Science* 32, 381–389.

Nolte, D.L., Mason, J.R. and Lewis, S.L. (1994) Tolerance of bitter compounds by an herbivore, *Cavia porcellus*. *Journal of Chemical Ecology* 20, 303–308.

Oh, H.K., Sakai, T., Jones, M.B. and Longhurst, W.M. (1967) Effects of various essential oils isolated from douglas fir needles upon sheep and deer microbial activity. *Applied Microbiology* 15, 777–784.

Oliver, A.J., King, D.R. and Mead, R.J. (1979) Fluoroacetate tolerance, a genetic marker in some Australian mammals. *Australian Journal of Zoology* 27, 363–372.

O'Reilly, R.A., Pool, J.G. and Aggler, P.M. (1968) Hereditary resistance to coumarin anticoagulent drugs in man and rat. *Annals of the New York Academy of Science* 151, 913–931.

Ortega-Reyes, L. and Provenza, F.D. (1993) Amount of experience and age affect the development of foraging skills of goats browsing blackbrush (*Coleogyne ramosissima*). *Applied Animal Behaviour Science* 36, 169–183.

Panter, K.E., Keeler, R.K., James, L.F. and Bunch, T.D. (1992) Impact of plant toxins on fetal and neonatal development: a review. *Journal of Range Management* 45, 52–57.

Phy, T. and Provenza, F.D. (1995) Are sheep's preferences for foods that rectify lactic acidosis caused by hedonic shifts or anticipated consequences? Abstract. *Annual Meeting of the Society of Range Management* 48, 50.

Pritz, R.K. and Launchbaugh, K.L. (1995) Effects of goat breed and early dietary experience on juniper consumption. Abstract. *Annual Meeting of the Society of Range Management* 48, 51.

Provenza, F.D. (1995) Postingestive feedback as an elementary determinant of food preference and intake in ruminants. *Journal of Range Management* 48, 2–17.

Provenza, F.D. and Balph, D.F. (1988) Development of dietary choice in livestock on rangeland and its implication for management. *Journal of Animal Science* 66, 2356–2368.

Provenza, F.D. and Balph, D.F. (1990) Applicability of five diet selection models to various foraging challenges ruminants encounter, In: Hughes, R.N. (ed.) *Behavioural Mechanisms of Food Selection*, Vol. 20. NATO ASI Series G: Ecological Sciences. Springer-Verlag, Heidelberg, pp. 423–459.

Provenza, F.D. and Burritt, E.A. (1991) Socially induced diet preference ameliorates conditioned food aversion in lambs. *Applied Animal Behaviour Science* 31, 229–236.

Provenza, F.D., Balph, D.F., Olsen, J.D., Dwyer, D.D., Ralphs, M.H. and Pfister, J.A. (1988) Toward understanding the behavioral responses of livestock to poisonous plants. In: James, L.F., Ralphs, M.H. and Nielson, D.B. (eds) *The Ecology and Economic Impact of Poisonous Plants on Livestock Production*. Westview Press, Boulder, Colorado, pp. 407–424.

Provenza, F.D., Burritt, E.A., Clausen, T.P., Bryant, J.P., Reichardt, P.B. and Distel, R.A. (1990) Conditioned flavor aversions: a mechanism for goats to avoid condensed tannins in blackbrush. *American Naturalist* 136, 810–828.

Provenza, F.D., Pfister, J.A. and Chaney, C.D. (1992) Mechanisms of learning in diet selection with reference to phytotoxicosis in herbivores. *Journal of Range Management* 45, 36–45.

Provenza, F.D., Nolan, J.V. and Lynch, J.J. (1993) Temporal contiguity between food ingestion and toxicosis affects the acquisition of food aversions in sheep. *Applied Animal Behaviour Science* 38, 269–281.

Provenza, F.D., Ortega-Reyes, L., Scott, C.B., Lynch, J.J. and Burritt, E.A. (1994a) Antiemetic drugs atttenuate food aversions in sheep. *Journal of Animal Science* 72, 1989–1994.

Provenza, F.D., Lynch, J.J., Burritt, E.A. and Scott, C.B. (1994b) How goats learn to distinguish between novel foods that differ in postingestive consequences. *Journal of Chemical Ecology* 20, 609–624.

Randel, R.D. (1991) Effects of gossypol on reproductive performance of domestic livestock. In: Jones, L.A., Kinard, D.H. and Mills, J.S. (eds) *Cattle Research With Gossypol Containing Feeds*. Special Session, Annual Meeting of the American Society of Animal Science. National Cottonseed Producers Association, Memphis, Tennessee.

Reese, J.C. (1979) Interactions of allelochemicals with nutrients in herbivore food. In: Rosenthal, G.A. and Janzen, D.H. (eds) *Herbivores: Their Interactions With Secondary Plant Metabolites*. Academic Press, New York, pp. 309–330.

Rhoades, D.F. (1979) Evolution of plant chemical defense against herbivores. In: Rosenthal, G.A. and Janzen, D.H. (eds) *Herbivores: Their Interaction with Secondary Plant Metabolites*. Academic Press, New York, pp. 4–55.

Rhoades, D.F. and Cates, R.G. (1976) Toward a general theory of plant antiherbivore chemistry. *Recent Advances in Phytochemistry* 10, 168–213.

Robbins, C.T., Hagerman, A.E., Austin, P.J., McArthur, C. and Hanley, T.A. (1991) Variation in mammalian physiological responses to a condensed tannin and its ecological implications. *Journal of Mammology* 72, 480–486.

Rosenthal, G.A. and Bell, E.A. (1979) Naturally occurring, toxic nonprotein amino acids. In: Rosenthal, G.A. and Janzen, D.H. (eds) *Herbivores: Their Interactions with Secondary Plant Metabolites*. Academic Press, New York, pp. 353–386.

Rozin, P. (1976) The selection of food by rats, humans, and other animals. In: Rosenblatt, J.S., Hinde, R.A., Beer, C. and Shaw, E. (eds) *Advances in the Study of Behaviour*. Academic Press, New York, pp. 21–76.

Rozin, P. and Kalat, J.W. (1971) Specific hungers and poison avoidance as adaptive specialization of learning. *Psychological Reviews* 78, 459–487.

Slansky, F. (1992) Allelochemical–nutrient interactions in herbivore nutritional ecology. In: Rosenthal, G.A. and Berenbaum, M.R. (eds) *Herbivores: Their Interactions with Secondary Plant Metabolites*, Vol. 2: *Evolutionary and Ecological Processes*. Academic Press, New York, pp. 135–175.

Smith, G.S. (1992) Toxification and detoxification of plant compounds by ruminants: an overview. *Journal of Range Management* 45, 25–30.

Smotherman, W.P. (1982) Odor aversion learning by the rat fetus. *Physiology and Behavior* 29, 769–771.

Taylor, C.A., Jr, and Ralphs, M.M. (1992) Reducing livestock losses from poisonous plants through grazing management. *Journal of Range Management* 45, 9–12.

Thorhallsdottir, A.G., Provenza, F.D. and Balph, D.F. (1990a). Ability of lambs to learn about novel foods while observing or participating with social models. *Applied Animal Behaviour Science* 25, 25–33.

Thorhallsdottir, A.G., Provenza, F.D. and Balph, D.F. (1990b) Social influence on conditioned food aversions in sheep. *Applied Animal Behaviour Science* 25, 45–50.

Van Soest, P.J. (1994) Plant defensive chemicals. In: Van Soest, P.J. (ed.) *Nutritional Ecology of the Ruminant*, 2nd edn. Cornell University Press, Ithaca, New York, pp. 196–212.

Vesell, E.S. (1968) Genetic and environmental factors affecting hexobarbital metabolism in mice. *Annals of the New York Academy of Science* 151, 900–913.

Villalba, J.J. and Provenza, F.D. (1996a) Preference for wheat straw by lambs conditioned with intraruminal infusions of sodium propionate. *Journal of Animal Science* (submitted).

Villalba, J.J. and Provenza, F.D. (1996b) Preference for wheat straw by lambs conditioned with intraruminal infusions of starch. *British Journal of Nutrition* (submitted).

Whittaker, R.H. and Feeny, P.P. (1971) Allelochemicals: chemical interactions between species. *Science* 171, 757–770.

Ingestive Behaviour 7

E.D. Ungar

Department of Agronomy and Natural Resources, Institute of Field and Garden Crops, Agricultural Research Organization, PO Box 6, Bet Dagan, Israel

INTRODUCTION

Context and significance

The gathering of food is the central objective of ruminants at pasture, and one of obvious importance to animal nutrition and vegetation response. For more than half a century, the subject has been the focus of considerable scientific attention (e.g. Johnstone-Wallace and Kennedy, 1944), and many excellent reviews of various aspects of this broad field are available (Tribe, 1950; Arnold, 1964, 1981, 1985; Fontenot and Blaser, 1965; Arnold and Dudzinski, 1978; Cordova *et al.*, 1978; Mannetje and Ebersohn, 1980; Van Dyne *et al.*, 1980; Freer, 1981; Hodgson 1982, 1984, 1985; Allison, 1985; Provenza and Balph, 1987; Forbes, 1988; Coleman *et al.*, 1989; Milne, 1991; Krysl and Hess, 1993).

The ambit of the plant–animal interface – intake, ingestive behaviour and diet selection (which will be referred to simply as intake) – is of interest for various reasons. Firstly, it is of ecological interest as a fundamental interaction between trophic levels. As a central theme in ecology, the gathering of food draws much experimental and theoretical attention (Crawley, 1983; Stephens and Krebs, 1986; Hughes, 1990). Secondly, it is of interest in the area of natural resource management. Grazing by wild or domesticated herbivores in areas whose primary use is not agricultural or economic (e.g. nature reserves, wilderness areas, wildlife refuges, national parks and forests) can have an impact on the landscape and vegetation. One would expect a better knowledge and understanding of how animals exploit their environment to be relevant to the management of these

©1996 CAB INTERNATIONAL.
The Ecology and Management of Grazing Systems (eds J. Hodgson and A.W. Illius)

resources (Van Dyne *et al.*, 1980). Thirdly, it is of agricultural interest. Grazing by domesticated ruminants for the production of animal products is a vast activity both geographically and economically. The plant–animal interface is the central feature of these systems (Forbes, 1988). Intake is a major determinant of animal production and, through its effect on sward structure, of plant production. Ultimately, better knowledge and understanding of intake should facilitate better management.

Beyond such generalities, the significance of intake can be expressed in considerably more concrete terms. For example, intake is the most appropriate criterion on which to base many within-season management decisions which are best taken in response to current conditions. These include the following.

1. Supplementary feeding of grazing livestock.
2. The allocation of area in intensive methods of grazing.
3. The timing of weaning.
4. The termination of grazing on seasonal pastures.
5. The movement of young stock from pasture to a fattening unit.
6. Animal density adjustments when they are based on the balance between short-term animal consumption and pasture growth.

Approaches

The many approaches to studying intake are difficult to categorize. They are partly defined by the balance between management-orientated problem-solving, hypothesis-testing, the provision of factual information and the elucidation of causal relationships. The methodology may range from the comparison of controlled experimental treatments to the gathering of data from a single natural or managed study area. The emphasis may be descriptive ('just give me the facts') or on modelling ('don't confuse me with the facts').

Both plant- and animal-orientated grazing studies can be conducted over a very wide range of temporal and spatial scales (Senft *et al.*, 1987; Coleman *et al.*, 1989). Different scales of study can entail different disciplines and a terminological shift. The scales are nevertheless related, and it might be argued that the relationship between them is the essence of scientific explanation. Shipley *et al.* (1994) expressed clearly the importance of time-scale in intake studies: at the longest time-scale (lifetime), nutrient intake is regulated to meet the costs of maintenance and production. On daily scales, intake rate is limited by the post-ingestive processes of digestion and excretion, by available grazing time and by the rate of intake during active grazing. Within periods of active grazing (minutes), consumption rate is limited by the spatial and morphological properties of the vegetation and by the ingestive apparatus of the animal. This latter aspect of ruminant grazing is at the shortest/smallest scale at which it is studied. It is the focus of this chapter since there have been important developments in this area of research over the last decade which have not been reviewed widely. Diet selection in the sense of choice between species or plant parts is regarded as being dependent on processes at greater temporal and spatial scales, and is not treated here. Examples of recent studies that discuss the complexities of diet selection, representing quite different conceptual approaches, are Belovsky and Schmitz (1994), Parsons *et al*. (1994) and Provenza (1995).

Demment and Laca (1993, 1994) argued the case for reductionism and synthesis in the grazing sciences generally, and outlined such a research agenda for the study of intake. This chapter follows a similar conceptual approach. Although the gathering of food deals with the interface between plant and animal, the subject is approached here with an emphasis on the animal: how does the animal harvest plant material and how does the harvesting process respond to features of the vegetation.

THE PROBLEM OF MEASUREMENT

The rate of intake and the quality of herbage selected by grazing animals are the key variables that the study of intake seeks to understand. They are both notoriously difficult to measure. Intake rate cannot be determined accurately on a daily time-scale at all. 'Without an absolute measure of herbage consumption by grazing animals there is doubt as to the accuracy of any technique' (Chacon *et al*., 1976). 'No method has been devised by which intake of grazing livestock can be accurately quantified... present methods for estimating intake by grazing livestock lack precision and are often tedious, expensive, and time consuming' (Cordova *et al*., 1978). The current difficulty, cost and uncertainty entailed in the measurement and prediction of daily intake rate by grazing ruminants is a major constraint to understanding and managing grazing systems because of the centrality of this variable.

Study of the factors that determine or correlate well with intake rate and quality is further motivated by the problem of measurement, since an understanding of these factors may yield predictive tools that are useful for the management of grazing systems. Such predictive tools may focus on characteristics of the grazed vegetation (e.g. height) or on behavioural components of intake that can be monitored (e.g. biting rate). The problem of measurement has spurred much effort in the development of methodologies, primarily in the management-orientated agricultural context where the desire for accuracy is greatest. Some of these are discussed below.

THE FUNCTIONAL RESPONSE

The rate at which an animal can gather food during active foraging bouts is a function of the capacity of its gathering apparatus and of the abundance of the food. The relationship between intake rate and abundance is termed the functional response (Solomon, 1949; Holling, 1959a, b). This is a state–rate relationship in which the state variable is usually defined as mass per unit area and the rate variable as mass ingested per unit time. It has been shown empirically in numerous studies that the shape of this function for small and large ruminants is a saturation curve (e.g. Arnold and Dudzinski, 1969; Allden and Whittaker, 1970; Young et al., 1980; Birrell, 1981). The functionl response has received ample theoretical treatment too. For example, Noy-Meir (1978a) examined four forms of the functional response (ramp, inverted exponential, Michaelis–Menten and sigmoid consumption functions), giving a theoretical derivation for each on the basis of assumptions about how a herbivore grazes. Spalinger and Hobbs (1992) derived three models of functional response to account for foraging in environments characterized by different patterns of plant availability.

When intake is defined on a daily time-scale, the slope of the initial ascending section of the functional response has units of area per animal per day. It represents the area of pasture from which all herbage is removed per day (allowing for an ungrazable residual quantity). The steepest section of the functional response is the maximum harvesting capacity of the gathering apparatus of the animal, in terms of the area that can be harvested per day. So, if the daily requirements of the animal are spread over an area greater than this maximum, the animal will not be able to meet that requirement. Expressed differently, a grazing ruminant can be allocated a total area containing a vastly greater quantity of high-quality herbage than its daily requirements and yet be unable to harvest that requirement because of the constraining functional response.

The superimposition of the functional response and the relationship between herbage growth rate and mass per unit area has yielded penetrating analyses of the overall stability properties of grazing systems (Noy-Meir, 1975, 1976, 1978a, b). Nevertheless, this form of the functional response is not intended to be an accurate predictive tool. The limitation is clear: mass per unit area is a two-dimensional description of the sward while bites are taken in three-dimensions.

> The measurement of pasture conditions by total yield inadequately describes any pasture, for the same yield can be presented . . . in many different ways. The relevance of pasture structure is in the mechanics of harvesting by the sheep. It is logical to assume that intake will be limited more by the length of herbage than by the yield per acre, because this will influence the amount that can be taken at each bite.
> (Arnold and Dudzinski, 1967)

This point was demonstrated vividly in experiment B of Allden and Whittaker (1970), in which herbage was removed by cultivation to create pasture strips. Bites and intake need to be functionally related to the spatial structure of a sward.

THE BITE

The bite is the fundamental unit of intake. Biting, or the taking of a bite, is defined here as the series of head and mouth-part movements that precede and include the severance and bringing into the mouth of herbage. The bite is the herbage thus ingested. A bite can be viewed as having an effective volume. This is not the volume of the buccal cavity but rather the equivalent volume of sward in its natural spatial arrangement that is occupied by the herbage ingested in a bite. The simplest characterization of bite volume is a cylinder of defined surface area and depth. Bite weight is the product of bite volume and the bulk density of herbage where the bite was removed (Hodgson, 1985).

From bite to intake rate

The connection between the individual bite and intake rate can be defined arithmetically or functionally. The arithmetic definition states that intake rate over a defined period is the product of the mean bite weight and the mean rate of biting over that period. Grazing over time-scales of a few hours to a day comprises bouts of active grazing (Birrell, 1991) and periods of rumination and other activities. Hence the intake equation for such time-scales can be defined as the product of the mean bite weight, the mean rate of biting during active grazing bouts and the active grazing time (Spedding *et al.*, 1966; Allden and Whittaker, 1970; Chacon *et al.*, 1976).

This model of intake rate imparts no information about the functional relationship between its components, and the research task is to establish the empirical relationship between them. The response of bite rate and grazing time to bite weight is generally negative. One interpretation has been that the animal is 'compensating' in an attempt to maintain daily intake rate (Chacon *et al.*, 1978; Penning *et al.*, 1991a; Morris *et al.*, 1993). While this may be true for grazing time, the inverse relation between bite rate and bite weight could be explained in terms of the mechanics of grazing rather than as a volitional process (Hodgson, 1986; Spalinger and Hobbs, 1992). In one of the most widely cited studies of its kind, Allden and Whittaker (1970, experiment A) examined the effect of a wide range of herbage mass levels on short-term intake rate and daily grazing time. The response of grazing behaviour could be divided into three zones. In the high range of herbage mass, short-term intake rate and daily grazing time are unaffected by mass. In the medium range of herbage mass, short-term intake rate declines and daily grazing time increases in compensatory fashion in response to declining mass. In the low range of herbage mass, daily grazing time cannot fully compensate for further decline in short-term intake rate. Allden and Whittaker (1970, experiment C) demonstrated the compensatory relationship between bite weight and biting rate in maintaining short-term intake rate over a wide range of tiller length. In some studies, neither grazing time nor rate of biting fully compensated for declining bite weight (e.g. Chacon and Stobbs, 1976; Jamieson and Hodgson, 1979b).

Ecological foraging theory provides a more functionally orientated framework by which intake rate can be defined. Even though the basic concepts were developed in the context of predator–prey relations, they have been applied with relative ease to herbivory. The functional definition of intake rate is based on a time budget, which defines the time required to perform each of the behavioural components associated with taking a bite, and bite weight. Intake rate is the ratio of bite weight to the time required to take a bite.

THE GRAZING PROCESS

The behavioural components associated with taking a bite are searching and handling. Searching includes the movement of the animal through its foraging environment, and all cognitive and sensory processes entailed in the decision to take a bite at a specific spot in the sward. These processes are not simple to quantify because they entail assumptions about how the animal moves through and perceives its environment. The only aspect of searching that lends itself to direct measurement is the speed of walking. Handling, however, is an easier process to study experimentally. Handling comprises biting (manipulative movements of head and mouth parts to gather herbage into the mouth, and severance of the herbage), chewing and swallowing. Not all the above behavioural components are mutually exclusive. For example, direct observation shows that an animal can chew and walk (search) at the same time. If the animal is grazing an abundant sward of high quality, searching may not limit the grazing process during bouts of active grazing and the intake equation reduces to the ratio of bite weight and handling time. The latter are inextricably linked, most obviously by a positive relation between chewing time and bite weight. (For a discussion of the role of ingestive mastication in the fragmentation of forages, see Pond *et al.*, 1984; Welch and Hooper, 1988; Beauchemin, 1992. For an introduction into the mechanics of food comminution, see Shipley *et al.*, 1994.)

The next step in developing an explanatory model of short-term intake rate is to bring together sward structure, bite dimensions, bite weight and handling time. Once again, the problem of measurement has hampered progress. In the field, it is difficult to observe processes at the bite level, and to precisely and independently control features of sward structure, such as height and bulk density.

> Ideally the separation of plant height from yield per unit area would require different densities of plants, but any attempt to attain such differences without affecting the size, growth habit, leaf number, and structure of the individual plants at different densities would pose many technical problems.
>
> (Allden and Whittaker, 1970)

It is not possible to devise experiments which alter sward structure without affecting some other characteristics of the sward such as feed quality.

(Stobbs, 1975)

MICROSWARDS

An extreme response to this problem has been the use of microswards of one form or another. This approach is radical because it seeks to understand what happens in nature, using an unnatural experimental arena: (i) the animal is removed, sometimes completely, from its normal foraging context; (ii) the forage presented to the animal often deviates strongly from anything the animal is likely to encounter naturally; and (iii) since the microsward is small, there is a severe restriction on the duration of an observation. Yet the appeal to the experimenter is strong: the technique povides an unparalleled degree of control over sward structure, including its uniformity and the ability to separate ordinarily confounding effects, and foraging behaviour can be scrutinized at a level of detail that is impossible to match under normal field conditions.

Various microsward techniques have been developed over the last decade. Table 7.1 summarizes some of the salient features of these studies. Seven of the studies cited used artificial swards constructed by attaching plant units to a board of some kind in a precise spatial arrangement. Burlison *et al.* (1991) created the microswards in the field, Illius *et al.* (1992) sowed the microswards in trays, and Hughes *et al.* (1991) and Newman *et al.* (1992) cut turfs from the field. In some cases, construction of the foraging arena was modular. Measurement of bite weight, biting rate and intake rate was common to almost all studies, although some included more detailed observations related to the mechanics of grazing. Many of the studies reported the overwhelming importance of bite weight in determining intake rate. Much of the remainder of this chapter is based on results and insights gained from such studies. The microsward methodology is described in more detail below, using two of the above studies which are of greatest relevance to this chapter.

The microsward technique was pioneered by Black and Kenney (1984) for sheep. Rows of holes were drilled through a hardwood board measuring 45 by 30 cm. Tillers were threaded through the holes and attached to the underside of the board. Sward bulk density was controlled by the number of tillers threaded through each hole and by the distance between the rows of holes. Sward height was controlled by careful preparation and threading of the tillers. The board was bolted to the pen floor and a sheep allowed to graze for 30 s. Measurements included the number of prehending bites and 'chewing' rate (actually the number of cycles of opening and closing the jaws per unit time).

The achievement of this technique was that it allowed the investigators to separate the effects of sward height and bulk density over the vast range of

Table 7.1. Studies of intake based on microswards.

Source	Animal	Herbage	Area grazed (m^2)	Emphasis
Black and Kenney, 1984 (series I)	Sheep	Ryegrass	0.14	Relation between sward height and plant density and intake
Spalinger et al., 1988	Deer	Three forb (whole), four shrub (leaves, stems) and one tree species (stems, needles)	23.04	Relation between forage biomass availability and bite size and intake
Burlison et al., 1991	Sheep	Oats, ryegrass, timothy, browntop bent	0.26	Relation between sward height, bulk density and bite weight
Hughes et al., 1991	Sheep	Ryegrass	0.10	Force exerted to sever bites
Illius et al., 1992	Sheep	Ryegrass–clover mixtures	0.26	Patch choice
Laca et al., 1992b	Cattle	Dallisgrass, alfalfa	c. 0.5	Relation between sward height and bulk density on intake; mechanics of foraging
Newman et al., 1992	Sheep	Ryegrass, clover	0.28 or 0.55	Patch choice
Shipley and Spalinger, 1992	(1) Mouse, deer, caribou (2) Hare	1) Maple leaves or shoots 2) Red clover	(1) Multiples of 0.09 (2) 3.55	Relation between bite size and intake
Gross et al., 1993a	Lemming	Alfalfa	0.43–2.23	Relation between plant size, plant density, biomass density and intake
Gross et al., 1993b	(1) Lemming (2) Prairie dog (3) Rabbit, peccary, deer, caribou, bear, elk, moose, horse, cow	Alfalfa	(1) 0.05 (2) 0.38 (3) 2.88	Relation between bite size and intake
Ginnett and Demment, 1995	Giraffe	*Acacia melanoxylon*	c. 0.5 or 1.0	Relation between bite size and intake; mechanics of foraging

herbage mass per unit area of 40 to 7610 kg ha^{-1}, without confounding effects of season, herbage quality or animal status. It was found that, given high quality, abundant, immediately available herbage, a hungry sheep can achieve a maximum intake rate over 30 s of 6–7 g dry matter (DM) min^{-1}. The maximum bite weight was ≈0.2 g DM. The study confirmed the findings of numerous field experiments that, while each of sward height, bulk density and mass per unit area may correlate well with intake rate, no one of these is an adequate explanatory variable. Black and Kenney (1984) reasoned that the herbage mass covered by a bite should be more directly related to intake rate. Although the experiments were not designed to estimate bite area (or its relation to sward structure), it was possible to derive a representative value of 8.4 cm^2. A close-fitting saturation-type relationship was indeed found between intake rate and the herbage mass effectively covered by one bite of 8.4 cm^2 in each treatment. Similar relationships were also found between intake rate and bite weight, and between bite weight and the herbage mass effectively covered by one bite.

Inspired by the study of Black and Kenney (1984), the hand-constructed sward technique was developed further by Laca *et al.* (1992a). The researchers worked with steers weighing about 750 kg. This required a completely new design of boards. The experimental protocol was also expanded. In its fullest form, a hand-constructed sward of about 0.5 m^2 (composed of 10 × 10 cm modules) was bolted to a force plate that recorded the forces applied to the herbage in three dimensions. The purpose of the force plate was to enable testing of the hypothesis that the force required to effect severance is a limiting factor in the determination of bite area. (A force plate was also used by Hughes *et al.*, 1991.) Grazing sessions were recorded on video, with the sound source coming from a small wireless microphone attached to the forehead of the animal. It was found that sound recordings obtained this way enable clear, unequivocal identification of prehending bites and chews. It was also found that visual counts of chews are not accurate, even when based on slow-motion play-back. Although the weight of individual bites cannot be determined, the video and sound records revealed much about the process of bite formation that cannot be inferred from measurements taken at the beginning and end of a grazing session. Each bolus swallowed was caught from an oesophageal fistula. The protocol was designed to enable detailed study of the relation between sward structure and bite dimensions and intake rate.

Measurements of bite dimensions were based on an initial short series of about six bites. The animal rarely placed a bite on a previously grazed area within the first six bites, and so sward structure at the bite site can be viewed as constant and equal to the initial structure as defined by the treatment. By counting the total number of plant units grazed and measuring the residual height of each, the average equivalent surface area and grazing depth of a bite can be computed. The six-bite technique for cattle is comparable to the 20-bite technique of Burlison *et al.* (1991).

Following the six-bite measurements, the board was reinitialized by removing all herbage from the grazed subsection of the board, and the animal was allowed to graze the board a second time for the determination of intake rate. It is obvious that the intake rate obtained from the board technique will, in part, be a function of the length of time the animal is given access to the board, simply because the biomass is continually being depleted. The functional response is a state–rate relationship and is therefore strictly instantaneous. Thus the criterion for terminating a grazing session in such experiments is a critical feature of the protocol.

BITE FORMATION

Jaw movements

Behaviours relevant to bite formation are jaw movements, swallowing and head movements. A jaw movement – a complete cycle of opening and closing of the jaws – is a basic behavioural component of bite formation and processing in the mouth. A jaw movement is required to chew, sweep the tongue beyond the gape of the mouth to gather herbage into the mouth (at least in cattle) and position the mouth to enclose herbage. The latter two are termed manipulative or prehending jaw movements. Head movements are for reaching new bite sites and for severing herbage gathered into the mouth and clamped between the incisors and the dental pad. The taking of a bite comprises one or more manipulative jaw movements and a severing head movement. The jaws briefly remain closed during the latter.

The microsward study of Black and Kenney (1984) suggests that the number of jaw movements per unit time is fairly constant in sheep. The allocation between chewing and prehending jaw movements will depend on bite weight. The heavier the bite, the more chewing is required, and hence the rate of prehensive biting declines. The idea that 'cropping' and chewing can be viewed as competing activities when herbivores forage in food-saturated environments is developed at length by Spalinger and Hobbs (1992). If we assume that there is a constant chewing requirement per unit weight ingested and that one prehending jaw movement is performed per bite, then this description can be expressed symbolically. Let:

I = rate of intake (mass time^{-1})
w = bite weight (mass bite^{-1})
j_i = rate of prehending jaw movements (time^{-1})
j_c = rate of chewing jaw movements (time^{-1})
j_t = rate of total jaw movements (time^{-1})
q = chewing jaw movements per unit mass ingested (mass^{-1})

Then:

$I = w\, j_i$
$j_t = j_i + j_c$
$j_c = I\, q$
$j_i = j_t - I\, q$

$$I = w(j_t - I\,q)$$
$$I = wj_t/(1 + w\,q) \tag{7.1}$$

This yields a saturation functional form between I and w, with j_t and q as constants. Rough estimates of these constants, derived from data presented by Black and Kenney (1984), are: $j_t = 100$ jaw movements per minute; $q = 12$ chewing jaw movements per gram ingested. Using these values, equation (1) above is almost identical to the empirical relationship between intake rate and bite weight shown by Black and Kenney (1984).

Penning et al. (1991a) also found the rate of jaw movements of sheep grazing swards ranging in height from 30 to 120 mm to be 'remarkably similar across all treatments', although the rates were higher: 149 and 157 jaw movements $\min^{-1}$ in the spring and autumn experiments respectively. Penning et al. (1991b) reported rates of jaw movement of 144 and 154 $\min^{-1}$ for clover and grass. Derrick et al. (1993) fed six forage species in chopped and unchopped form to sheep. The rate of chewing jaw movements ranged fom 115 to 134 $\min^{-1}$. Domingue et al. (1991) counted 154 and 128 jaw movements $\min^{-1}$ for goats and sheep respectively, when fed lucerne hay. For cattle, the data-logging system developed by Matsui and Okubo (1991) for studying jaw movements (chewing and prehensive) recorded a rate of jaw movements during grazing of about 80 $\min^{-1}$. The frequency distribution of the interval between two successive jaw movements shows that 93% of such intervals lie between 0.6 and 1.2 s. Matsui et al. (1994) found the rate of jaw movements of a Holstein heifer grazing (and depleting) a ryegrass sward to range between 60 and 102 $\min^{-1}$. The modal frequency of jaw movements during grazing periods increased from ~72 $\min^{-1}$ on day 2 to ~84 $\min^{-1}$ on days 5 to 10. Rook et al. (1994) used a gnathometer (Penning et al., 1984; Huckle et al., 1989) to record jaw movements of grazing cows. The rate of jaw movements during grazing ranged from 74 to 94 $\min^{-1}$ over all treatments examined.

Equation (1) is similar to other derivations of a functional response formula based on an assumption of competing cropping and chewing activities (e.g. Shipley and Spalinger, 1992; Spalinger and Hobbs, 1992; Gross et al., 1993b). One of the assumptions of the model is that these two components of ingestive behaviour are mutually exclusive. Microsward studies have recently shown that this assumption is not always true.

Compound jaw movements

The grinding sound of a chew and the tearing sound of herbage severance were clearly distinguishable in the sound recordings obtained from the microsward studies of Laca and coworkers. Together with the video image, these recordings revealed that cattle can perform a manipulative jaw movement and chew herbage already in the mouth within a single jaw movement (Laca et al., 1994b). When this compound type of jaw movement immediately preceded severance, it generated a distinctive chew–sever sound sequence. Ginnett and Demment (1995) found that giraffes also perform compound jaw movements.

The tongue

Tongue sweeping is an extremely important mechanism for enlarging the effective bite area in cattle (Laca *et al.*, 1992b). The potential effectiveness of the tongue depends primarily on the height of the herbage being grazed. Plant structural units outside the area covered by the gape of the mouth but within the sweep of the tongue can be grazed if their length is sufficient to reach the mouth of the animal. However, characteristics of the herbage, such as stiffness and vertical differentiation, will modify the response to and hence the effectiveness of tongue sweeping. The tongue is not used indiscriminately; the amplitude of tongue sweeps increases with sward height and the number of tongue sweeps per bite decreases with sward density. The tongue may also be important in anchoring the herbage in the mouth during severance. In a grazing trial using a heterogeneous sward surface, comprising interspersed tall and short leaves differing in length by a factor of two (Ungar *et al.*, 1991), initially short leaves (4 cm) would sometimes emerge intact from the mouth following the severance action of the head, while initially long leaves (8 cm) in the same bite were severed at a height less than 4 cm. This suggests that the clamping of the herbage between the incisors and the dental pad alone permits slippage.

The direct upshot of the gathering action of the tongue and, to a lesser extent, of the curvature of the dental arcade is that the residual sward surface shape following a single bite is almost hemispherical. The surface of a grazed sward may not look as if hemispherically shaped bites have been removed from it because the areas swept by the tongue in successive bites usually overlap on homogeneous swards. The residual shape of a bite forces us to clarify the definition of bite depth. There may clearly be a difference between the mean residual height of grazed plant structural units and the height above the ground of the incisors at severance. Incisor height can be close to the physical limit; at an initial sward height of 4 cm, the residual height at the centre of the bite area is approximately 0.5 cm (Ungar *et al.*, 1991).

THE TIME BUDGET OF BITE FORMATION

If the jaw movement is the basic behavioural component of biting, then it should be possible to explain the average time required to take a bite (in the absence of searching) in terms of the average number of jaw movements per bite. On the basis of 42 microsward grazing sessions with cattle, Laca *et al.* (1994b) found that linear regression of time per bite on jaw movements per bite explained 96% of the variance. The slope of the regression function represents the mean time to complete a jaw movement and was 0.68 s. The statistically significant intercept of 0.43 s represents a time cost per bite required for behavioural components of grazing other than jaw movements, such as head movements and swallowing.

It is now necessary to explain how many jaw movements are taken per bite. Bite weight is one of the main determinants of chewing requirement per bite, although other characteristics of the forage, such as fibrousness and length of prehended plant unit, may also influence the chewing requirement. For example, Newman *et al.* (1994) found that sheep performed 26 mastications g^{-1} DM when grazing grass and 15 on clover. (See also Beauchemin and Iwaasa, 1993; Derrick *et al.*, 1993.) Laca *et al.* (1994b) found a linear relationship between the number of chewing jaw movements (including compound jaw movements) and bite weight. The significant non-zero intercept obtained implies a decline in the chewing requirement per unit mass as bite weight increases. The number of manipulative jaw movements (pure or compound) required to take a bite may be expected to be more directly related to sward structure than to bite weight. It was indeed found that the total number of manipulative jaw movements was not significantly related to bite weight, but increased with decreasing sward height. This was assumed to reflect the greater difficulty in prehending short tillers and the animal's compensatory response.

Chambers *et al.* (1981) developed a bite-meter for use on cattle and sheep, incorporating a microswitch to monitor jaw movements, an accelerometer to record head movements, and a mercury switch to determine if the head is up or down (grazing). This equipment enabled jaw movements to be distinguished from harvesting bites, and ingestive mastication to be distinguished from rumination. For sheep there was evidence that the ratio of manipulative jaw movements to true bites increased with increasing sward height. The authors found that the ratio of jaw movements to head movements (i.e. bite severances) was consistently greater in sheep than in cattle at any given sward height. Based on their data, the number of jaw movements per minute for sheep was calculated to be in the range 77–90, close to the value assumed above on the basis of the Black and Kenney (1984) microsward experiments. For cattle, the range in jaw movements per minute was 48–99. In the experiment of Laca *et al.* (1994b), the rate of jaw movement for cattle was fairly constant at ~60 min^{-1}.

In the absence of the compound jaw movement, the total number of jaw movements required per bite in cattle would be the sum of the chewing and manipulative requirements. However, the animal exploits the ability to perform a compound jaw movement quite fully; at high bite weights, about 90% of total manipulative jaw movements are of the compound type. Thus the total number of jaw movements required per bite is closer to the greater of the chewing and manipulative requirements than to their sum, and the relationship with bite weight is well described by an initial horizontal section (manipulative requirements limiting) followed by a linearly increasing section (chewing requirements limiting). Consequently, the relationship between time per bite and bite weight is of the same form (Fig. 7.1). Various studies have failed to detect a 'compensatory'

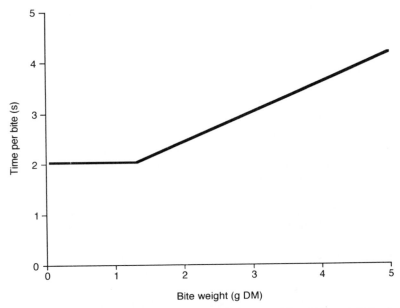

Fig. 7.1. The relationship between time per bite (T, in seconds) and bite weight (w, in g dry matter (DM)) for cattle. Time per bite includes all head and jaw movements used to prehend, sever and masticate herbage. Based on microsward trials using alfalfa and dallisgrass. Figure generated from the equation given in Laca *et al.* (1994b): T = max[2.03, 1.29 + 0.58 w]. Non-linearity derives from compound jaw movements, which both manipulate and chew.

increase in biting rate in response to declining bite weight (e.g. Arias *et al.*, 1990), and the above effect of the compound jaw movement on the time budget of grazing may be the simplest explanation.

BITE DIMENSIONS AND WEIGHT

The following account of bite dimensions is based on measurements by Laca *et al.* (1992b) for cattle. We shall assume that the dimensions of a bite can be summarized in terms of bite area and bite depth, even though the true shape of the sward volume swept or actually removed in the course of a bite may be more complex. Both bite area and bite depth are defined on the basis of sward rather than animal measurements. Bite area is the total area of plant structural units grazed to any extent divided by the number of bites taken. Bite depth is the difference between the initial and mean residual height of the same plant structural units. In relating bite area and bite depth to sward structure, measurements are based on an initially uniform leafy sward from which a short series of bites is taken without regrazing, but with some degree of overlap of the areas swept by the tongue between bites. In other words, bites are removed from the same horizon of the sward, and it is assumed that each bite is responding to the same initial sward structure, even though the microsward as a whole is being depleted with each bite removed.

On swards of low bulk density, bite area shows a steep curvilinear response to sward height. The steepness of the response declines with increasing bulk density (Fig. 7.2). At a sward height of 30 cm, maximum mean bite area is approximately 170 cm^2. (Based on close-up images on the video record, individual bites have been observed to reach 220 cm^2.)

Bite depth is primarily a function of sward height, but here too there is a negative interaction with bulk density (Fig. 7.3). At low bulk density, the slope of bite depth as a function of sward height is approximately 0.5, meaning that the animal removes about half the height. This result is contrary to the functional form assumed for bite depth in some theoretical studies of intake (e.g. Ungar and Noy-Meir, 1989) and, although possible explanations have been suggested, the reason for this 'take-half' behaviour remains unclear.

A comparison of bite dimension relationships across studies reveals large differences, although such comparisons are complicated by seemingly small

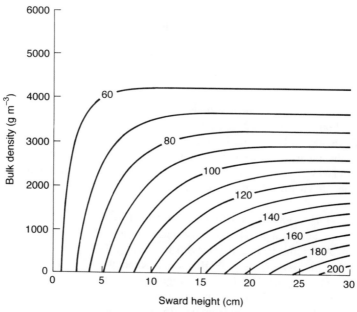

Fig. 7.2. Contour diagram showing the response of bite area (cm^2) to sward height (cm) and bulk density (g m^{-3}) for cattle. Bite area (contours) is the mean surface area of sward from which herbage is severed in the course of taking a bite. Based on microsward trials using alfalfa and dallisgrass. Figure generated from the equation given in Laca *et al.* (1992b).

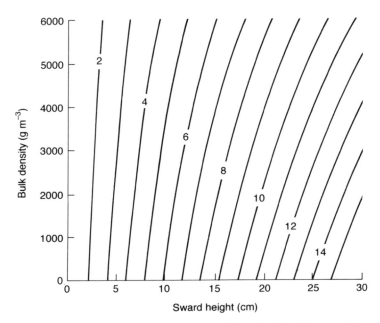

Fig. 7.3. Contour diagram showing the response of bite depth (cm) to sward height (cm) and bulk density (g m^{-3}) for cattle. Bite depth (contours) is the difference between initial sward height and the average height of residual (grazed) herbage for a single grazing horizon. Based on microsward trials using alfalfa and dallisgrass. Figure generated from the equation given in Laca *et al.* (1992b).

differences in definition and methodology. For example, sward height can be determined on the basis of the natural presentation of the sward surface or on the basis of extended tiller height. Postgrazing measurements can represent the field mean or the mean of grazed plant units only. As pointed out by Forbes and Coleman (1993), overall mean sward height includes both grazed and ungrazed areas, whereas sward influence on ingestive behaviour occurs only from grazed areas.

Wade *et al.* (1989) found that bite depth of cattle operated on a constant fraction basis, averaging 0.34 of initial extended tiller height over the height range 12–39 cm. A decline in bite depth with increasing bulk density was found for sheep in the microsward study of Black and Kenney (1984). In the microsward study of Burlison *et al.* (1991), linear regression of bite depth on initial sward height (of range 6 to 55 cm) showed that sheep removed 37% of sward height above 2.7 cm. Bite area ranged from 9 to 36 cm^2, but variance in bite area was not explained well by measured features of the sward. Maximum bite weight was 326 mg DM. In the microsward study of Illius *et al.* (1992) on pure ryegrass swards, sheep removed 71% of sward height above 2.9 cm, assuming minimal depletion. In a field experiment, Milne *et al.* (1982) determined the relationship between initial sward height and grazing depth in sheep. Based on their regression equation, sheep removed a constant one-third of the sward height above 2.4 cm. Data given by Curll and Wilkins (1982) in a study of the frequency and severity of defoliation of grass and clover plant units by sheep yields a curvilinear relationship between sward height and grazing depth (Fig. 7.4), and a proportion removed that declined from 60 to 45% with increasing sward height.

Allometric relationships

The relationship between the size of an animal and its harvesting apparatus is of considerable interest to ecologists. Illius and Gordon (1987) derived a general relationship between incisor breadth and body weight (W; in kg) from data on 32 grazing ruminant species. The range in body weight was 30–1200 kg. Incisor breadth, in mm, was $8.6\ W^{0.36}$. Burlison *et al.* (1991) measured sheep incisor breadths of 31 to 34 mm at live weights, for which the corresponding prediction is

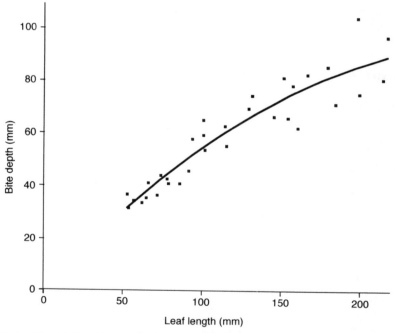

Fig. 7.4. The relationship between bite depth (D, in mm) and initial leaf length (L, in mm) of ryegrass tillers for grazing sheep. Bite depth calculated from initial leaf length and proportion removed given in Table 5 of Curll and Wilkins (1982). Points include all stocking rates, observation periods and levels of fertilizer application. Regression equation: $D = 0.68 + 0.65\ L - 0.0011\ L^2$; $r^2 = 0.87$.

37 to 42 mm. Penning *et al.* (1991b) measured sheep incisor breadth of 38 mm; the equation predicts 41 mm.

Significant within-species relationships between bite weight and animal size were detected by Illius (1989), in a study of different grazing systems with cattle. Bite weight, in mg organic matter (OM), was $3.4\ W^{0.76}$. This study includes a detailed examination of the allometric relations of intake rate and bite weight with live weight, sex, mature size and degree of maturity. The effect of sward height on the allometric coefficient for bite weight is also discussed.

Shipley *et al.* (1994) present a broad-ranging and yet penetrating analysis of the scaling of intake rate in mammalian herbivores. The authors proposed a simple model of short-term intake rate when food is spatially concentrated (i.e. in the absence of searching processes) and tested hypotheses on the scaling of the model parameters in 12 species of mammalian herbivores of body weight ranging from 0.05 (lemmings) to 547 kg (cattle). The parameters examined were the maximum rate of food processing in the mouth, the maximum bite size and the time expended to crop a bite. The authors found that maximum bite size scaled with $W^{0.72}$, rate of processing scaled with $W^{0.70}$, and cropping time did not scale with body weight. Maximum short-term dry-matter intake rate was $\sim 0.63\ W^{0.71}$. The study includes 39 published observations of the maximum short-term intake rate of mammalian herbivores (from voles to elephants).

The summit force hypothesis

Grazing animals must expend energy in order to sever and chew herbage. The amount of energy expended is a function of the mechanical properties of the plant material, which vary considerably between plant parts and species (Wright and Illius, 1995). It has been suggested that the force required to sever a bite may be a factor determining bite weight (Hodgson, 1985). This force would be expected to be related to the cross-sectional area of herbage material along the line of bite severance and hence to bite area. Bite depth *per se* would not be expected to influence the force required to sever a bite, although it may have an indirect effect if there is sward structural differentiation in the vertical plane. Hendricksen and Minson (1980) found that the shear load required to break stems of *Lablab purpureus* increased greatly with distance from the shoot apex. According to the summit force hypothesis, there is a limit to the force animals exert to sever the herbage encompassed within the bite area. Once this limit is reached, animals should respond to further increases in sward density or strength of the herbage by reducing bite area to maintain a constant maximum force of severance. The microsward study of Hughes *et al.* (1991) with sheep was the first attempt to measure the force exerted by a grazing ruminant directly. It did not yield supporting evidence for the above hypothesis. In the microsward study of Laca *et al.* (1992a) with cattle, which also used a force plate, it was found that bite area decreases with sward bulk density, as discussed earlier, but the force exerted in bite severance did not appear to reach a maximum or constant value. However, in an experiment designed specifically to test the summit force hypothesis, bite area was lower on the species with higher tensile strength of leaf blades at a given bulk density (Laca *et al.* 1993).

Bite volume and weight

Bite volume is computed as the product of bite area and bite depth. Bite weight is the product of bite volume and bulk density in the grazed horizon. The response of bite weight to sward height and bulk density for cattle, based on the microsward trials of Laca *et al.* (1992b), is shown in Fig. 7.5. A number of herbivores, including cattle, can achieve very high bite weights by a kind of 'spaghetti effect', whereby a long plant unit is severed and gradually drawn into the mouth during mastication. Similarly, cattle can use a long series of manipulative jaw movements in conjunction with tongue actions to gradually work their way down a very tall tiller, folding it into the mouth from the top down and severing it near the base.

The above account of ingestive behaviour at the bite level and its relation to intake rate is summarized in Fig. 7.6.

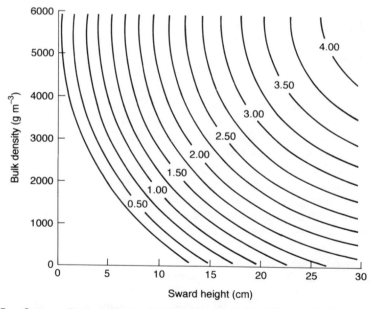

Fig. 7.5. Contour diagram showing the response of bite weight (g dry matter (DM)) to sward height (cm) and bulk density (g m^{-3}) for cattle. A bite is defined by severance of herbage, irrespective of the number of manipulative jaw movements entailed in gathering the herbage into the mouth. Based on microsward trials using alfalfa and dallisgrass. Figure generated from the basic data on which Laca *et al.* (1992b) was based.

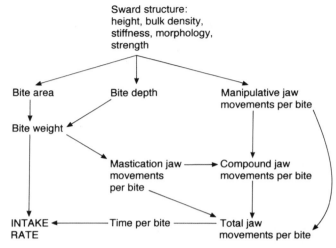

Fig. 7.6. The components of ingestive behaviour that mediate between sward structure and short-term intake rate.

FROM BITE TO PATCH

Depletion

Experiments with microswards have indicated that bite dimensions are determined by sward structure at the bite site rather than the average structure of the microsward as a whole. So, if we wish to predict intake rate for a series of bites, it is important to know where the animal places its bites. The extreme options are to place bites randomly or systematically on the face of the sward surface. Since a bite does not remove herbage to the ground, the residual herbage becomes a new potential bite site. If we extend this idea to the sward as a whole (assuming uniform initial height), we obtain a second bite horizon. The number of potential bite horizons will depend on the initial sward height. Thus systematic depletion could be by horizon grazing (first across, then down) or by profile grazing (first down, then across).

In general, steers grazing microswards place bites systematically by horizon, and the probability of regrazing (i.e. removal of a bite from a lower horizon) increases with degree of depletion of the higher horizon (Laca *et al.*, 1994a). Even within a horizon, bite weight tends to decline somewhat, because there is some degree of overlap in the area swept by adjacent bites and because grazing is never absolutely systematic. Small islands of ungrazed herbage which are of smaller area than the potential bite area tend to appear in a horizon. Thus average bite weight and intake rate decline with the degree of depletion of a microsward. The depletion curve defines the relationship between cumulative intake and grazing ('residence') time (or number of bites removed).

Laca *et al.* (1994a) determined depletion curves for steers grazing hand-constructed microswards of wheat leaf lamina with two contrasting structures but equal herbage mass. The depletion curves were constructed by allowing the animal to remove a different number of bites from each microsward over the range 10 to 60 bites. It was found that bite weight was initially larger and decreased more steeply with number of bites removed in the tall sparse swards than in the short dense ones (Fig. 7.7). Grazing style was horizon-orientated, resulting in an abrupt change in instantaneous intake rate when the top horizon was depleted. Similar depletion curves were obtained on the basis of bite sequences taken from ryegrass patches created in the field – an encouraging result in terms of the relevance of the hand-constructed, microsward technique.

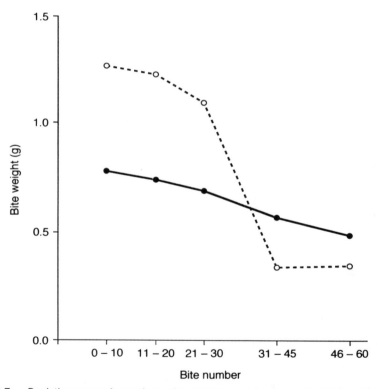

Fig. 7.7. Depletion curves for cattle grazing microswards of wheat leaf laminae. Sward structure: height = 20 cm, bulk density = 660 g m^{-3} (open symbol, dashed line); height = 10 cm, bulk density = 1300 g m^{-3} (closed symbol, solid line). Note that the two sward structures were of approximately equal mass per unit area. Based on Fig. 1(a) in Laca *et al.* (1994a).

Microswards and feeding stations

The concept of the feeding station is the critical link between the microsward methodology and the field. A number of investigators have noted that the typical pattern of grazing behaviour on many rangelands is for the grazing ruminant to walk (search), stop, lower the head and graze a semicircular area of herbage within the reach of the neck and head. The area of herbage thus depleted is termed a feeding station (Ruyle and Dwyer, 1985; Pfister, *et al.*, 1988; El Aich *et al.*, 1989). The microsward methodology of Laca *et al.* (1992a) uses an area of close to 0.5 m^2, which is about the largest area than can be reached by a mature steer without moving its forelegs. Thus the microsward (be it hand-constructed or in the field) represents a feeding station. Depletion trials can tell us all we want to know about grazing down a single feeding station. The big question now is: how much does, or should, a grazing animal deplete a feeding station before moving on?

Feeding stations and patches

One approach to answering this question is to make the small terminological switch from a feeding station to a patch. In foraging theory patchiness usually has two implications. Firstly, beyond some level of depletion, the animal experiences diminishing marginal returns to continued foraging. Secondly, a time cost is incurred in moving to another (undepleted) patch. So the question is how much should the organism deplete a patch before moving to a new one? The answer depends on the goal of the foraging organism. A common assumption is that the animal seeks to maximize intake rate. The marginal value theorem states that overall intake rate is maximized by leaving patches when intake rate within a patch falls to the average for the whole habitat (Charnov, 1976). The marginal value theorem can also be used to determine the optimal depletion level in a heterogeneous environment, i.e. where the patches are different.

This brings us to the portals of optimal foraging theory and questions of foraging strategies at longer time scales than those considered thus far. Approaches based on optimality are discussed in detail in Chapter 5.

THE FIELD LEVEL

Non-patchy swards

It is not absolutely clear that the patch paradigm is the most appropriate for all grazing situations. Are all swards patchy? Do foraging animals behave on uniform swards as if they were patchy? These questions have not been addressed experimentally. However, models of the grazing process have been developed that do not use the patch concept, as seen in the following two examples.

1. Consider a steady-state situation in which the sward is not being depleted. The sward can be viewed as a population of bites of different weights (determined by local sward structure), randomly distributed in the horizontal plane. Which bites should the animal select from this population of bites? The trade-off is clear: the greater the selectivity for heavier bites, the longer the time interval between selected bites. It can be shown that, in order to maximize intake rate, the animal should select all bites above a threshold bite weight and pass over all bites below it (Ungar and Noy-Meir, 1989). Furthermore, when the animal forages in this optimal way, its instantaneous intake rate will be greater on a sward that has a wider variation in bite weights, for swards offering the same mean bite weight.

2. Consider a sward of relatively uniform structure in the horizontal plane, which is being depleted by grazing. As discussed earlier, such swards can be viewed as comprising discrete grazing horizons. Within certain logical constraints, the animal can allocate its bites between these horizons in a very large number of permutations between the extreme strategies of strict horizon and profile grazing. The consequences for intake rate will depend on the characteristic bite dimensions and bite weight of each horizon. This will depend on the initial height of the sward and the vertical profile of bulk density. Thus the heaviest bites may be in the top horizon for one sward and in the bottom horizon for another. A conceptual model based on these ideas was explored by Ungar *et al.* (1992). This highlighted the need to understand the allocation of bites between grazing horizons and the factors influencing it.

Area allocation

It is clear from the above discussion that the fact that each bite taken by an animal depletes the sward and changes the mean sward structure does not necessarily mean that bite weight or instantaneous intake rate must change with each bite. As long as the animal selects bites from previously ungrazed (unsearched) areas (i.e. no revisitation, as if the path of the animal can be unwound on a straight line; Arditi and Dacorogna, 1988), bite weight and intake rate can be maintained. (The same applies on patchy swards at the between-patch scale.) If an animal is allocated a sufficiently large area for it to encounter 'initial condition' sward for an entire day, then it is valid to predict daily intake rate on the basis of a state–rate functional response in which the state is the initial condition of the sward.

In set-stocking grazing systems an animal is rarely allocated an area less than 20 times its total daily bite area, and the ratio is more commonly closer to 100. Thus a state–rate functional response for the prediction of daily intake rate is conceptually valid. However, in some grazing systems (e.g. rotational or strip-grazing) the foraging area available to the animal is closer to or even less than the potential total daily bite area. Consequently, the initial conditions of the sward are not relevant to what the animal experiences on a daily time-scale and a state–rate functional response that attempts to predict daily intake rate does not hold.

For this reason, intake studies on such high-depletion systems attempt to relate daily intake rate to herbage allowance expressed in units of mass per animal per unit time (Hodgson, 1976). Unlike mass per unit area, herbage allowance tells us absolutely nothing about sward structure. It is therefore not surprising that intake–allowance relationships are of low generality (e.g. Fig. 1 in Gibb and Treacher, 1976; Fig. 2 in Le Du *et al.*, 1979). In one high-depletion study, intake rate was more closely related to postgrazing sward height than to herbage allowance (Baker *et al.*, 1981).

The area allocation at which a state–rate functional response ceases to apply is difficult to define. The study of Jamieson and Hodgson (1979a), using continuous stocking management, allocated approximately 500 m^2 per calf, and this may be in the borderline area. This low area allocation for a set-stocking study may explain the fact that a clear asymptote in the functional response was not obtained.

Area grazed and horizon grazing

It has been argued that, since an animal removes bites from a certain area of sward each day, the average frequency of defoliation of a plant under set stocking and rotational management may be similar. Hence there is no essential difference in defoliation pattern and this may explain why no consistent difference between grazing methods has been found (Spedding, 1965; Allden and Whittaker, 1970). This line of reasoning holds as long as the area allocated per animal does not impinge on bite selection. If initial sward structure is uniform in the horizontal plane and the unconstrained animal removes bites predominantly from the top horizon, the critical zone of area allowance can be estimated simply. Bite area for large mature cattle on swards above 10 cm height (of relatively low bulk density) is at least 100 cm^2 (Fig. 7.2) and daily bite number is at least 25,000 if there are no motivational constraints on grazing behaviour (Hodgson *et al.*, 1991). Thus the animal removes bites from an area of 250 m^2 day^{-1}. (Dairy cows in the experiment of Wade *et al.* (1989) were estimated to have grazed 230 m^2 each on the first day of grazing a fresh paddock.) Let us assume that herbage mass is 2500 kg DM ha^{-1} and animal weight is 500 kg. Each unit of herbage allowance (g kg^{-1} live weight (LW) day^{-1}) requires an area of sward that contains 500 g herbage, i.e. 2 m^2. Since unconstrained top-horizon grazing covers 250 m^2 day^{-1}, an allowance level less than 125 g kg^{-1} LW day^{-1} will result in bites being taken from the second horizon. Note that this allowance threshold increases with increasing herbage mass per unit area. It follows that, if bite weight declines with depth of grazing horizon, intake rate would be expected to decline with increasing mass per unit area at a constant allowance level, as found by Reardon (1977).

An estimate of the total daily bite area ('grazed area') can be derived from data on defoliation intervals. The grazed area can be estimated as the area allocated per animal divided by the defoliation interval. A representative value for sheep based on the defoliation intervals of Curll and Wilkins (1982) for ryegrass tillers is 40 m^2 day^{-1}. There were clear trends in grazed area with stocking rate and observation period, but interpretation is difficult given the simultaneous changes that were occurring in tiller density, sward height, mass per unit area and herbage quality. An additional estimate for sheep of 35 m^2 day^{-1} can be derived from data presented by Mazzanti and Lemaire (1994). This estimate is based on an average stocking density of 32 sheep ha^{-1} and an average defoliation interval of tillers of 9 days.

The entire range of allowances examined in strip-grazing experiments is generally below the threshold that would allow top-horizon grazing only. Where increases in area allocation have not resulted in changes in intake rate (e.g. treatments II and III of Dougherty et al., 1989c; experiment 1 of Dougherty et al., 1992), pure top-horizon grazing could fit the available data. As heterogeneity of sward structure increases, the area allocation that impinges on grazing behaviour will increase because of bite selectivity. In other words, the 250 m^2 of grazed area per day is less than the area scanned by the animal. In an analysis of strip-grazing experiments, Hodgson (1981) suggested that 'herbage allowance may be assumed to influence intake through its effect upon the rate at which animals graze down through successive horizons in the sward'. Stobbs (1977) found that the proportion of herbage removed by grazing cows from successive horizons of 10 cm differed between levels of herbage allowance. At the highest allowance, approximately half the herbage was removed from layers above 20 cm height, and only 14% of the herbage in the 10–20 cm layer was grazed. At the lowest allowance, over 80% of the herbage was removed from layers above 20 cm height, and about half the herbage was removed from the 10–20 cm layer. These results are consistent with a horizon-orientated rather than a profile-orientated grazing style.

A sward would not be expected to be grazed down by horizon if the rate of depletion is relatively slow and herbage quality tends to be higher on previously grazed areas. For example, Illius et al. (1987) found that a greater proportion of bites were taken from patches with the shortest herbage (and highest quality) than would be expected on the basis of their fractional cover. The opposite was found for patches with tall, lower-quality herbage. Similarly, if plant differentitation creates a non-uniform vertical profile of herbage quality or preference, with a relatively open architecture, lower horizons can be accessed and grazed preferentially and grazing style is not horizon-orientated (L'Huillier et al., 1984).

The allocation of bites between grazing horizons may be influenced by the perception of the animal of the available grazing area. In other words, the allocation of bites may differ between herbage allowance levels for the same sward structure from the time a fresh grazing area is allocated. This is suggested by the results of Jamieson and Hodgson (1979a), who found that the initial bite weight on a new strip of herbage declined with increasing allowance and that the terminal bite weight increased with increasing allowance. The relation between grazing style and area allocation on high depletion systems has not yet been examined closely, although related studies have been conducted (Dougherty *et al.*, 1989a). There may also be effects on grazing behaviour of spatial proximity between grazing animals, independently of the overall area allocation per animal. One such effect that has been studied is social facilitation (e.g. Benham, 1982; Rook and Penning, 1991).

If there are indeed grazing horizons, then the frequency distribution of sward height measurements needs to be considered as well as the field mean. Few studies have emphasized this aspect of sward structure. It seems plausible that the double-normal frequency distributions of sward height used by Gibb and Ridout (1986, 1988) and Gibb *et al.* (1989) are related to grazing horizons. In these studies, the two peaks of bimodal height distributions were associated with the shorter, frequently grazed and the taller, infrequently grazed components of the sward. The height measurements were based on a plate meter of 930 cm^2 surface area, which is considerably larger than the bite scale. Given this limitation, it might be suggested that the greater of the two means (m2) fitted to each double-normal frequency distribution corresponds to the ungrazed horizon and that the lesser of the two means (m1) corresponds to the first grazing horizon. It would then be of interest to look at the relationship between ungrazed ('initial') sward height (m2) and grazing depth of the first horizon (m2 − m1). This is shown in Fig. 7.8. The regression equation means that the sheep removed two-thirds of the initial height above 1.7 cm.

CONCLUDING REMARKS

Research on ingestive behaviour over the last decade or so has seen the introduction of new methodologies based on microswards and a strengthening of the overlap between the ecological and agricultural disciplines. From the agricultural point of view, it is too early to say whether this will translate into concrete improvements in management. If it does, the benefits will almost certainly be felt first in the domain of intensively grazed, temperate, monocultural grasslands and last in the domain of extensive, natural rangelands.

The elementary components of intake outlined in this chapter are relevant to all grazing situations, but the quantitative relationships are not universal. Numerous factors beyond those discussed here, from the internal state of the animal (Dougherty *et al.*, 1987, 1989b; Greenwood and Demment, 1988; Newman *et al.*, 1994) to face flies (Dougherty *et al.*, 1993), influence short-term ingestive behaviour. It is by no means clear that delving deeper into the processes will yield a mechanistic reduction in the sense of quantitatively universal relationships; subtle complexity is present at all levels of organization and ultimately

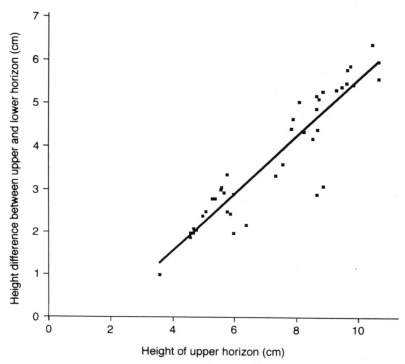

Fig. 7.8. The relationship between the height difference between the upper and lower horizons of a sward (cm) and the height of the upper horizon (cm). Derived from data presented by Gibb and Ridout (1988, Table 1) and Gibb *et al.* (1989, Table 3), giving the maxima of double-normal frequency distributions fitted to measurements of sward height under grazing. It was assumed here that these maxima correspond to grazing horizons. Hence the *y* axis represents bite depth (D, cm), and the *x* axis represents sward height (H, cm). Regression equation: $D = -1.1 + 0.66 H$; $r^2 = 0.86$.

abuts on animal psychology (Provenza and Balph, 1990). While scientifically fascinating, with implications for animal physiology and evolutionary theory, it remains to be seen to what extent the descent down spatial and temporal scales in the study of ingestive behaviour will enable a more agriculturally useful synthesis.

REFERENCES

Allden, W.G. and Whittaker, I.A.McD. (1970) The determinants of herbage intake by grazing sheep: the interrelationship of factors influencing herbage intake and availability. *Australian Journal of Agricultural Research* 21, 755–766.

Allison, C.D. (1985) Factors affecting forage intake by range ruminants: a review. *Journal of Range Management* 38, 305–311.
Arditi, R. and Dacorogna, B. (1988) Optimal foraging on arbitrary food distributions and the definition of habitat patches. *American Naturalist* 131, 837–846.
Arias, J.E., Dougherty, C.T., Bradley, N.W., Cornelius, P.L. and Lauriault, L.M. (1990) Structure of tall fescue swards and intake of grazing cattle. *Agronomy Journal* 82, 545–548.
Arnold, G.W. (1964) Factors within plant associations affecting the behaviour and performance of grazing animals. In: Crisp, D.J. (ed.) *Grazing in Terrestrial and Marine Environments*. Blackwell Scientific Publications, Oxford, UK, pp. 133–154.
Arnold, G.W. (1981) Grazing behaviour. In: Morley, F.H.W. (ed.) *Grazing Animals*. Elsevier Scientific Publishing Company, Amsterdam, The Netherlands, pp. 79–104.
Arnold, G.W. (1985) Ingestive behaviour. In: Fraser, A.F. (ed.) *Ethology of Farm Animals*. Elsevier, Amsterdam, The Netherlands, pp. 183–200.
Arnold, G.W. and Dudzinski, M.L. (1967) Studies on the diet of the grazing animal II. The effect of physiological status in ewes and pasture availability on herbage intake. *Australian Journal of Agricultural Research* 18, 349–359.
Arnold, G.W. and Dudzinski, M.L. (1969) The effect of pasture density and structure on what the grazing animal eats and on animal productivity. In: James, B.J.F. (ed.) *Intensive Utilization of Pastures*. Angus and Robertson, Sydney, Australia, pp. 42–48.
Arnold, G.W. and Dudzinski, M.L. (1978) *Ethology of Free-ranging Domestic Animals*. Elsevier Scientific Publishing Company, Amsterdam, The Netherlands.
Baker, R.D., Alvarez, F. and Le Du, Y.L.P. (1981) The effect of herbage allowance upon the herbage intake and performance of suckler cows and calves. *Grass and Forage Science* 36, 189–199.
Beauchemin, K.A. (1992) Effects of ingestive and ruminative mastication on digestion of forage by cattle. *Animal Feed Science and Technology* 40, 41–56.
Beauchemin, K.A. and Iwaasa, A.D. (1993) Eating and ruminating activities of cattle fed alfalfa or orchardgrass harvested at two stages of maturity. *Canadian Journal of Animal Science* 73, 79–88.
Belovsky, G.E. and Schmitz, O.J. (1994) Plant defenses and optimal foraging by mammalian herbivores. *Journal of Mammalogy* 75, 816–832.
Benham, P.F.J. (1982) Synchronisation of behaviour in grazing cattle. *Applied Animal Ethology* 8, 403–404.
Birrell, H.A. (1981) Some factors which affect the liveweight change and wool growth of adult Corriedale wethers grazed at various stocking rates on perennial pasture in Southern Victoria. *Australian Journal of Agricultural Research* 32, 353–370.
Birrell, H.A. (1991) The effect of stocking rate on the grazing behaviour of Corriedale sheep. *Applied Animal Behaviour Science* 28, 321–331.
Black, J.L. and Kenney, P.A. (1984) Factors affecting diet selection by sheep. II. Height and density of pasture. *Australian Journal of Agricultural Research* 35, 565–578.
Burlison, A.J., Hodgson, J. and Illius, A.W. (1991) Sward canopy structure and the bite dimensions and bite weight of grazing sheep. *Grass and Forage Science* 46, 29–38.
Chacon, E. and Stobbs, T.H. (1976) Influence of progressive defoliation of a grass sward on the eating behaviour of cattle. *Australian Journal of Agricultural Research* 27, 709–727.

Chacon, E., Stobbs, T.H. and Sandland, R.L. (1976) Estimation of herbage consumption by grazing cattle using measurements of eating behaviour. *Journal of the British Grassland Society* 31, 81–87.

Chacon, E.A., Stobbs, T.H. and Dale, M.B. (1978) Influence of sward characteristics on grazing behaviour and growth of Hereford steers grazing tropical grass pastures. *Australian Journal of Agricultural Research* 29, 89–102.

Chambers, A.R.M., Hodgson, J. and Milne, J.A. (1981) The development and use of equipment for the automatic recording of ingestive behaviour in sheep and cattle. *Grass and Forage Science* 36, 97–105.

Charnov, E.L. (1976) Optimal foraging, the marginal value theorem. *Theoretical Population Biology* 9, 129–136.

Coleman, S.W., Forbes, T.D.A. and Stuth, J.W. (1989) Measurements of the plant–animal interface in grazing research. In: Marten, G.C. (ed.) *Grazing Research: Design, Methodology, and Analysis*. Crop Science Society of America, American Society of Agronomy, Madison, Wisconsin, USA, pp. 37–51.

Cordova, F.J., Wallace, J.D. and Pieper, R.D. (1978) Forage intake by grazing livestock: a review. *Journal of Range Management* 31, 430–438.

Crawley, M.J. (1983) *Herbivory: The Dynamics of Animal–Plant Interactions*. University of California Press, Berkeley, California, USA.

Curll, M.L. and Wilkins, R.J. (1982) Frequency and severity of defoliation of grass and clover by sheep at different stocking rates. *Grass and Forage Science* 37, 291–297.

Demment, M.W. and Laca, E.A. (1993) The grazing ruminant: models and experimental techniques to relate sward structure and intake. *Proceedings 7th World Conference on Animal Production*, 439–460.

Demment, M.W. and Laca, E.A. (1994) Reductionism and synthesis in the grazing sciences: models and experiments. *Proceedings of the Australian Society of Animal Production* 20, 6–16.

Derrick, R.W., Moseley, G., Wilman, D. (1993) Intake, by sheep, and digestibility of chickweed, dandelion, dock, ribwort and spurrey, compared with perennial ryegrass. *Journal of Agricultural Science (Cambridge)* 120, 51–61.

Domingue, B.M.F., Dellow, D.W. and Barry, T.N. (1991) The efficiency of chewing during eating and ruminating in goats and sheep. *British Journal of Nutrition* 65, 355–363.

Dougherty, C.T., Bradley, N.W., Cornelius, P.L. and Lauriault, L.M. (1987) Herbage intake rates of beef cattle grazing alfalfa. *Agronomy Journal* 79, 1003–1008.

Dougherty, C.T., Bradley, N.W., Cornelius, P.L. and Lauriault, L.M. (1989a) Accessibility of herbage allowance and ingestive behavior of beef cattle. *Applied Animal Behaviour Science* 23, 87–97.

Dougherty, C.T., Cornelius, P.L., Bradley, N.W. and Lauriault, L.M. (1989b) Ingestive behavior of beef heifers within grazing sessions. *Applied Animal Behaviour Science* 23, 341–351.

Dougherty, C.T., Lauriault, L.M., Cornelius, P.L. and Bradley, N.W. (1989c) Herbage allowance and intake of cattle. *Journal of Agricultural Science (Cambridge)* 112, 395–401.

Dougherty, C.T., Bradley, N.W., Lauriault, L.M., Arias, J.E. and Cornelius, P.L. (1992) Allowance-intake relations of cattle grazing vegetative tall fescue. *Grass and Forage Science* 47, 211–219.

Dougherty, C.T., Knapp, F.W., Burrus, P.B., Willis, D.C. and Bradley, N.W. (1993) Face flies (*Musca autumnalis* De Geer) and the behavior of grazing beef cattle. *Applied Animal Behaviour Science* 35, 313–326.

El Aich, A., Moukadem, A. and Rittenhouse, L.R. (1989) Feeding station behavior of free-grazing sheep. *Applied Animal Behaviour Science* 24, 259–265.

Fontenot, J.P. and Blaser, R.E. (1965) Symposium on factors influencing the voluntary intake of herbage by ruminants: selection and intake by grazing animals. *Journal of Animal Science* 24, 1202–1208.

Forbes, T.D.A. (1988) Researching the plant–animal interface: the investigation of ingestive behavior in grazing animals. *Journal of Animal Science* 66, 2369–2379.

Forbes, T.D.A. and Coleman, S.W. (1993) Forage intake and ingestive behavior of cattle grazing Old World bluestems. *Agronomy Journal* 85, 808–816.

Freer, M. (1981) The control of food intake by grazing animals. In: Morley, F.H.W. (ed.) *Grazing Animals*. Elsevier Scientific Publishing Company, Amsterdam, The Netherlands, pp. 105–124.

Gibb, M.J. and Ridout, M.S. (1986) The fitting of frequency distributions to height measurements on grazed swards. *Grass and Forage Science* 41, 247–249.

Gibb, M.J. and Ridout, M.S. (1988) Application of double normal frequency distributions fitted to measurements of sward height. *Grass and Forage Science* 43, 131–136.

Gibb, M.J. and Treacher, T.T. (1976) The effect of herbage allowance on herbage intake and performance of lambs grazing perennial ryegrass and red clover swards. *Journal of Agricultural Science (Cambridge)* 86, 355–365.

Gibb, M.J., Baker, R.D. and Sayer, A.M.E. (1989) The impact of grazing severity on perennial ryegrass/white clover swards stocked continuously with beef cattle. *Grass and Forage Science* 44, 315–328.

Ginnett, T.F. and Demment, M.W. (1995) The functional response of herbivores: analysis and test of a simple mechanistic model. *Functional Ecology* 9, 376–384.

Greenwood, G.B. and Demment, M.W. (1988) The effect of fasting on short-term cattle grazing behaviour. *Grass and Forage Science* 43, 377–386.

Gross, J.E., Hobbs, N.T. and Wunder, B.A. (1993a) Independent variables for predicting intake rate of mammalian herbivores: biomass density, plant density, or bite size? *Oikos* 68, 75–81.

Gross, J.E., Shipley, L.A., Hobbs, N.T., Spalinger, D.E. and Wunder, B.A. (1993b) Functional response of herbivores in food-concentrated patches: tests of a mechanistic model. *Ecology* 74, 778–791.

Hendricksen, R. and Minson, D.J. (1980) The feed intake and grazing behaviour of cattle grazing a crop of *Lablab purpureus* cv. Rongai. *Journal of Agricultural Science (Cambridge)* 95, 547–554.

Hodgson, J. (1976) The influence of grazing pressure and stocking rate on herbage intake and animal performance. In: Hodgson, J. and Jackson, D.K. (eds) *Pasture Utilization by the Grazing Animal*. Occasional Symposium No. 8, British Grassland Society, Hurley, UK, pp. 93–103.

Hodgson, J. (1981) Variations in the surface characteristics of the sward and the short-term rate of herbage intake by calves and lambs. *Grass and Forage Science* 36, 49–57.

Hodgson, J. (1982) Influence of sward characteristics on diet selection and herbage intake by the grazing animal. In: Hacker, J.B. (ed.) *Nutritional Limits to Animal Production from Pasture*. Commonwealth Agricultural Bureaux, Farnham Royal, UK, pp. 153–166.

Hodgson, J. (1984) Sward conditions, herbage allowance and animal production: an evaluation of research results. *Proceedings of the New Zealand Society of Animal Production* 44, 99–104.

Hodgson, J. (1985) The control of herbage intake in the grazing ruminant. *Proceedings of the Nutrition Society* 44, 339–346.

Hodgson, J. (1986) Grazing behaviour and herbage intake. In: Frame, J. (ed.) *Grazing*. Occasional Symposium No. 19, British Grassland Society, Hurley, UK, pp. 51–64.

Hodgson, J., Forbes, T.D.A., Armstrong, R.H., Beattie, M.M. and Hunter, E.A. (1991) Comparative studies of the ingestive behaviour and herbage intake of sheep and cattle grazing indigenous hill plant communities. *Journal of Applied Ecology* 28, 205–227.

Holling, C.S. (1959a) The components of predation as revealed by a study of small-mammal predation of the European Pine Sawfly. *Canadian Entomologist* 91, 293–320.

Holling, C.S. (1959b) Some characteristics of simple types of predation and parasitism. *Canadian Entomologist* 91, 385–398.

Huckle, C.A., Clements, A.J. and Penning, P.D. (1989) A technique to record eating and ruminating behaviour in dairy cows. In: Phillips, C.J.C. (ed.) *New Techniques in Cattle Production*. Butterworths, London, pp. 236–237.

Hughes, R.N. (1990) *Behavioural Mechanisms of Food Selection*. NATO ASI Series, Vol. G 20. Springer-Verlag, Berlin, Germany.

Hughes, T.P., Sykes, A.R., Poppi, D.P. and Hodgson, J. (1991) The influence of sward structure on peak bite force and bite weight in sheep. *Proceedings of the New Zealand Society of Animal Production* 51, 153–158.

Illius, A.W. (1989) Allometry of food intake and grazing behaviour with body size in cattle. *Journal of Agricultural Science (Cambridge)* 113, 259–266.

Illius, A.W. and Gordon, I.J. (1987) The allometry of food intake in grazing ruminants. *Journal of Animal Ecology* 56, 989–999.

Illius, A.W., Wood-Gush, D.G.M. and Eddison, J.C. (1987) A study of the foraging behaviour of cattle grazing patchy swards. *Biology of Behaviour* 12, 33–44.

Illius, A.W., Clark, D.A. and Hodgson, J. (1992) Discrimination and patch choice by sheep grazing grass–clover swards. *Journal of Animal Ecology* 61, 183–194.

Jamieson, W.S. and Hodgson, J. (1979a) The effect of daily herbage allowance and sward characteristics upon the ingestive behaviour and herbage intake of calves under strip-grazing management. *Grass and Forage Science* 34, 261–271.

Jamieson, W.S. and Hodgson, J. (1979b) The effects of variation in sward characteristics upon the ingestive behaviour and herbage intake of calves and lambs under a continuous stocking management. *Grass and Forage Science* 34, 273–282.

Johnstone-Wallace, D.B. and Kennedy, K. (1944) Grazing management practices and their relationship to the behaviour and grazing habits of cattle. *Journal of Agricultural Science (Cambridge)* 34, 190–197.

Krysl, L.J. and Hess, B.W. (1993) Influence of supplementation on behavior of grazing cattle. *Journal of Animal Science* 71, 2546–2555.

Laca, E.A., Ungar, E.D., Seligman, N.G., Ramey, M.R. and Demment, M.W. (1992a) An integrated methodology for studying short-term grazing behaviour of cattle. *Grass and Forage Science* 47, 81–90.

Laca, E.A., Ungar, E.D., Seligman, N.G. and Demment, M.W. (1992b) Effects of sward height and bulk density on bite dimensions of cattle grazing homogeneous swards. *Grass and Forage Science* 47, 91–102.

Laca, E.A., Demment, M.W., Distel, R.A. and Griggs, T.C. (1993) A conceptual model to explain variation in ingestive behaviour within a feeding patch. *Proceedings 17th International Grassland Congress*, 710–712.

Laca, E.A., Distel, R.A., Griggs, T.C. and Demment, M.W. (1994a) Effects of canopy structure on patch depression by grazers. *Ecology* 75, 706–716.

Laca, E.A., Ungar, E.D. and Demment, M.W. (1994b) Mechanisms of handling time and intake rate of a large mammalian grazer. *Applied Animal Behaviour Science* 39, 3–19.

Le Du, Y.L.P., Combellas, J., Hodgson, J. and Baker, R.D. (1979) Herbage intake and milk production by grazing dairy cows 2. The effects of level of winter feeding and daily herbage allowance. *Grass and Forage Science* 34, 249–260.

L'Huillier, P.J., Poppi, D.P. and Fraser, T.J. (1984) Influence of green leaf distribution of diet selection by sheep and the implications for animal performance. *Proceedings of the New Zealand Society of Animal Production* 44, 105–107.

Mannetje, L.'t and Ebersohn, J.P. (1980) Relations between sward characteristics and animal production. *Tropical Grasslands* 14, 273–280.

Matsui, K. and Okubo, T. (1991) A method for quantification of jaw movements suitable for use on free-ranging cattle. *Applied Animal Behaviour Science* 32, 107–116.

Matsui, K., Kurokawa, Y. and Okubo, T. (1994) Changes of heart rate, grazing and rumination time and jaw movements in cattle with grazing days. *Animal Science and Technology (Japan)* 65, 16–21.

Mazzanti, A. and Lemaire, G. (1994) Effect of nitrogen fertilization on herbage production of tall fescue swards continuously grazed by sheep. 2. Consumption and efficiency of herbage utilization. *Grass and Forage Science* 49, 352–359.

Milne, J.A. (1991) Diet selection by grazing animals. *Proceedings of the Nutrition Society* 50, 77–85.

Milne, J.A., Hodgson, J., Thompson, R., Souter, W.G. and Barthram, G.T. (1982) The diet ingested by sheep grazing swards differing in white clover and perennial ryegrass content. *Grass and Forage Science* 37, 209–218.

Morris, S.T., Parker, W.J., Blair, H.T. and McCutcheon, S.N. (1993) Effect of sward height during late pregnancy on intake and performance of continuously stocked June- and August-lambing ewes. *Australian Journal of Agricultural Research* 44, 1635–1651.

Newman, J.A., Parsons, A.J. and Harvey, A. (1992) Not all sheep prefer clover: diet selection revisited. *Journal of Agricultural Science (Cambridge)* 119, 275–283.

Newman, J.A., Penning, P.D., Parsons, A.J., Harvey, A. and Orr, R.J. (1994) Fasting affects intake behaviour and diet preference of grazing sheep. *Animal Behaviour* 47, 185–193.

Noy-Meir, I. (1975) Stability of grazing systems: an application of predator–prey graphs. *Journal of Ecology* 63, 459–481.

Noy-Meir, I. (1976) Rotational grazing in a continuously growing pasture: a simple model. *Agricultural Systems* 1, 87–112.

Noy-Meir, I. (1978a) Stability in simple grazing models: effects of explicit functions. *Journal of Theoretical Biology* 71, 347–380.

Noy-Meir, I. (1978b) Grazing and production in seasonal pastures: analysis of a simple model. *Journal of Applied Ecology* 15, 809–835.

Parsons, A.J., Newman, J.A., Penning, P.D., Harvey, A. and Orr, R.J. (1994) Diet preference of sheep: effects of recent diet, physiological state and species abundance. *Journal of Animal Ecology* 63, 465–478.

Penning, P.D., Steel, G.L. and Johnson, R.H. (1984) Further development and use of an automatic recording sysem in sheep grazing studies. *Grass and Forage Science* 39, 345–351.

Penning, P.D., Parsons, A.J., Orr, R.J. and Treacher, T.T. (1991a) Intake and behaviour responses by sheep to changes in sward characteristics under continuous stocking. *Grass and Forage Science* 46, 15–28.

Penning, P.D., Rook, A.J. and Orr, R.J. (1991b) Patterns of ingestive behaviour of sheep continuously stocked on monocultures of ryegrass or white clover. *Applied Animal Behaviour Science* 31, 237–250.

Pfister, J.A., Malechek, J.C. and Balph, D.F. (1988) Foraging behaviour of goats and sheep in the Caatinga of Brazil. *Journal of Applied Ecology* 25, 379–388.

Pond, K.R., Ellis, W.C. and Akin, D.E. (1984) Ingestive mastication and fragmentation of forages. *Journal of Animal Science* 58, 1567–1574.

Provenza, F.D. (1995) Postingestive feedback as an elementary determinant of food preference and intake in ruminants. *Journal of Range Management* 48, 2–17.

Provenza, F.D. and Balph, D.F. (1987) Diet learning by domestic ruminants: theory, evidence and practical implications. *Applied Animal Behaviour Science* 18, 211–232.

Provenza, F.D. and Balph, D.F. (1990) Applicability of five diet-selection models to various foraging challenges ruminants encounter. In: Hughes, R.N. (ed.) *Behavioural Mechanisms of Food Selection*. NATO ASI Series, Volume G 20, Springer-Verlag, Berlin, Germany, pp. 423–459.

Reardon, T.F. (1977) Effect of herbage per unit area and herbage allowance on dry matter intake by steers. *Proceedings of the New Zealand Society of Animal Production* 37, 58–61.

Rook, A.J. and Penning, P.D. (1991) Synchronisation of eating, ruminating and idling activity by grazing sheep. *Applied Animal Behaviour Science* 32, 157–166.

Rook, A.J., Huckle, C.A. and Penning, P.D. (1994) Effects of sward height and concentrate supplementation on the ingestive behaviour of spring-calving dairy cows grazing grass–clover swards. *Applied Animal Behaviour Science* 40, 101–112.

Ruyle, G.B. and Dwyer, D.D. (1985) Feeding stations of sheep as an indicator of dimished forage supply. *Journal of Animal Science* 61, 349–353.

Senft, R.L., Coughenour, M.B., Bailey, D.W., Rittenhouse, L.R., Sala, O.E. and Swift, D.M. (1987) Large herbivore foraging and ecological hierarchies. *BioScience* 37, 789–799.

Shipley, L.A. and Spalinger, D.E. (1992) Mechanics of browsing in dense food patches: effects of plant and animal morphology on intake rate. *Canadian Journal of Zoology* 70, 1743–1752.

Shipley, L.A., Gross, J.E., Spalinger, D.E., Hobbs, N.T. and Wunder, B.A. (1994) The scaling of intake rate in mammalian herbivores. *American Naturalist* 143, 1055–1082.

Solomon, M.E. (1949) The natural control of animal populations. *Journal of Animal Ecology* 18, 1–35.

Spalinger, D.E. and Hobbs, N.T. (1992) Mechanisms of foraging in mammalian herbivores: new models of functional response. *American Naturalist* 140, 325–348.

Spalinger, D.E., Hanley, T.A. and Robbins, C.T. (1988) Analysis of the functional response in foraging in the Sitka black-tailed deer. *Ecology* 69, 1166–1175.

Spedding, C.R.W. (1965) The physiological basis of grazing management. *Journal of the British Grassland Society* 20, 7–14.

Spedding, C.R.W., Large, R.V. and Kydd, D.D. (1966) The evaluation of herbage species by grazing animals. In: Hill, A.G.G. (ed.) *Proceedings of the 10th International Grassland Congress*. Valtioneuvoston Kirjapaino, Helsinki, Finland, pp. 479–483.

Stephens, D.W. and Krebs, J.R. (1986) *Foraging Theory*. Princeton University Press, Princeton, New Jersey, USA.

Stobbs, T.H. (1975) The effect of plant structure on the intake of tropical pasture. III influence of fertilizer nitrogen on the size of bite harvested by Jersey cows grazing *Setaria anceps* cv. Kazungula swards. *Australian Journal of Agricultural Research* 26, 997–1007.

Stobbs, T.H. (1977) Short-term effects of herbage allowance on milk production, milk composition and grazing time of cows grazing nitrogen-fertilized tropical grass pasture. *Australian Journal of Experimental Agriculture and Animal Husbandry* 17, 892–898.

Tribe, D.E. (1950) The behaviour of the grazing animal – a critical review of present knowledge. *Journal of the British Grassland Society* 5, 209–214.

Ungar, E.D. and Noy-Meir, I. (1989) Herbage intake in relation to availability and sward structure: grazing processes and optimal foraging. *Journal of Applied Ecology* 25, 1045–1062.

Ungar, E.D., Genizi, A. and Demment, M.W. (1991) Bite dimensions and herbage intake by cattle grazing short hand-constructed swards. *Agronomy Journal* 83, 973–978.

Ungar, E.D., Seligman, N.G. and Demment, M.W. (1992) Graphical analysis of sward depletion by grazing. *Journal of Applied Ecology* 29, 427–435.

Van Dyne, G.M., Brockington, N.R., Szocs, Z., Duek, J. and Ribic, C.A. (1980) Large herbivore subsystem. In: Breymeyer, A.I. and Van Dyne, G.M. (eds) *Grasslands, Systems Analysis and Man*. Cambridge University Press, Cambridge, UK, pp. 269–537.

Wade, M.H., Peyraud, J.L., Lemaire, G. and Comeron, E.A. (1989) The dynamics of daily area and depth of grazing and herbage intake of cows in a five day paddock system. *Proceeedings 16th International Grassland Congress*, Nice, France, pp. 1111–1112.

Welch, J.G. and Hooper, A.P. (1988) Ingestion of feed and water. In: Church, D.C. (ed.) *The Ruminant Animal*. Prentice Hall, Englewood Cliffs, New Jersey, USA, pp. 108–116.

Wright, W. and Illius, A.W. (1995) A comparative study of the fracture properties of five grasses. *Functional Ecology* 9, 269–278.

Young, N.E., Newton, J.E. and Orr, R.J. (1980) The effect of a cereal supplement during early lactation on the performance and intake of ewes grazing perennial ryegrass at three stocking rates. *Grass and Forage Science* 35, 197–202.

The Ruminant, the Rumen and the Pasture Resource: Nutrient Interactions in the Grazing Animal

H. Dove

CSIRO Division of Plant Industry, GPO Box 1600, Canberra, ACT 2601, Australia

INTRODUCTION

The ruminant animals of the world provide humans with most of their animal protein, a significant proportion of the fibre used in clothing and, in developing countries, a major source of draught power. None of this would be possible without the ruminant's capacity to digest plant fibre. In this chapter, I propose to review what might be termed 'digestive behaviour' or 'digestive ecology' in the grazing animal, that is, the way in which the animal and the pasture resource interact from a nutritional point of view. My starting-point is the concept that efficient production from ruminants represents a complex balance between the changing nutrient requirements of the ruminant itself (e.g. growth vs. pregnancy vs. lactation), the requirements of the rumen microbial ecosystem for nutrient input and removal (intake, comminution of particles, absorption) and the changing external supply of herbage nutrients (pasture growth, maturity and senescence). Successful grazing management demands an appreciation of the interplay between these components, since all three can impose constraints upon production.

NUTRIENT REQUIREMENTS AND TRANSACTIONS IN THE RUMEN

A major advance in this area was the development and application of marker-based techniques for monitoring the fate of consumed nutrients in the rumen. In a series of classic papers, Hogan, Weston and colleagues used this approach to

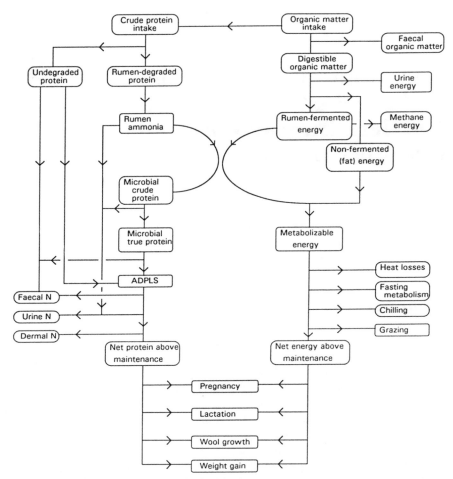

Fig. 8.1. Diagrammatic representation of the flows of energy and protein during ruminant digestion (adapted from Freer *et al.*, 1996). ADPLS, apparently digested protein leaving the stomach; N, nitrogen.

investigate the nitrogen (N) and energy transactions in the rumen, the relationship between these and the resultant flow of nutrients from the rumen (see Hogan, 1982; Weston and Hogan, 1986). These and similar studies provided the much needed link between the nutrient content of herbage, on the one hand, and the nutrient requirements of the animal, on the other. They also formed the basis of the various schemes representing the nutrient requirements of ruminants, subsequently developed in many parts of the world (e.g. ARC, 1980, 1984; SCA, 1990; AFRC, 1992). A diagrammatic representation of dietary protein and energy flows through the ruminant, based substantially on SCA (1990) and AFRC (1992), is shown in Fig. 8.1. As Beever (1993) has recently stressed, it is the synchrony between the protein and energy transactions which is of pivotal importance in the conversion of pasture nutrients into animal product.

The intake of the animal can be viewed as comprising an indigestible component, (ultimately excreted) and a potentially digestible component which may be fermented in the rumen into energy substrates (ARC, 1980; SCA, 1990). In the ARC (1980) and SCA (1990) systems, after allowance was made for the inevitable losses of energy in urine, all the remaining digestible material was fermentable. More recently, AFRC (1992) refined this approach to allow for the fact that dietary lipid is not fermented, although it does contribute to the total metabolizable energy (ME) supply. The fermentation of digestible material to energy substrates is frequently a rate-limiting step, especially in relation to protein transactions.

A feature common to the schemes listed above is the separation of dietary crude protein (CP = N × 6.25) into 'rumen-degradable protein' (RDP), which can contribute to rumen microbial demands for N, and 'undegradable protein' (UDP), which represents dietary protein passing through the rumen unaffected. In the AFRC (1992) scheme, RDP is further broken down conceptually into 'quickly degrading' and 'slowly degrading' fractions, on the basis of the manner in which dietary N disappears *in sacco*, that is, from nylon or equivalent bags suspended in the rumen. Ørskov and McDonald (1979) described this mathematically as

$$N \text{ disappearance} = a + b(1 - e^{-ct})$$

The y-intercept a is taken to be the quickly degrading fraction, while b is the slowly degrading fraction, which disappears at rate c. This concept is based on the idea that protein which disappears from the bag has been degraded. However, the results of a number of studies have shown that protein can be soluble in the rumen (so that it moves out of the bag) but remain substantially undegraded or at least slowly degrading, in the rumen (e.g. Mangan, 1972; Spencer *et al.*, 1988). For this reason, the distinction between quickly degrading and slowly degrading fractions is not made in Fig. 8.1.

Dietary RDP is eventually degraded to ammonia, which has the potential to be converted into microbial protein, the main source of protein for the grazing ruminant. The extent to which it is so converted depends on the rate at which adenosine triphosphate (ATP) is supplied by the fermentation of plant structural and non-structural carbohydrates (NSC) (see SCA, 1990, for further discussion). If fermentation is slow and the rate of energy supply is reduced, then ammonia will be diverted to urinary excretion as urea. Given the rapidity with which the protein of green herbage is degraded to ammonia, it is normal for there to be a substantial loss of protein as urinary urea, during passage through the rumen (e.g. Ulyatt and Egan, 1979). In contrast, when animals consume low-quality roughages with low N contents (e.g. < 1%), fermentation rate can be reduced because the supply of RDP in the rumen is inadequate for microbial growth and activity. Hence, one of the key elements in the efficient conversion of herbage into animal product is the synchrony between the supply of rumen ammonia (from herbage RDP) and the supply of energy-yielding substrates (from herbage carbohydrates).

Taken together, microbial protein and UDP make up the component referred to as 'apparently digestible protein leaving the stomach' (ADPLS in Fig. 8.1) in the ARC (1984) and SCA (1990) schemes or, with some modification, the 'truly absorbed amino N' of the AFRC (1992) scheme. After obligatory losses of ADPLS and ME, the net supplies of protein and energy can contribute to productive functions in the animal, and they do so with efficiencies which themselves are functions of the quality of the diet (for detailed discussion, see SCA, 1990, and Oldham, 1993).

Figure 8.1 represents a substantial oversimplification of the concepts involved in the current feeding standards and, for more detailed treatments, the reader is referred to SCA (1990), AFRC (1992), Webster (1992) or Oldham (1993). Nevertheless, the scheme as outlined in the figure does provide a framework within which one can assess the likely impact of changes in the nutrient content of or the nutrient supply from the grazed pasture.

THE FEEDING VALUE OF HERBAGE

The 'feeding value' of herbage can be viewed as the product of the nutritive value of the herbage itself, the herbage intake by the animal and the efficiencies with which consumed nutrients are utilized (Ulyatt, 1973). Assuming for the moment that the supply of herbage is not limiting the grazing animal, the nutrient content of that herbage can influence the eventual supply of nutrients to the animal in several ways.

1. The physical and chemical nature of the herbage can provide cues that influence diet choice and short-term eating rate (Arnold, 1981).
2. The cell wall content and particularly the anatomical distribution of cell wall constituents can influence the susceptibility of plant material to microbial colonization and digestion (Wilson, 1994).
3. The content of cell walls and N and their interaction will influence the rate of particle breakdown in and flow from the rumen, under the combined effects of digestion and rumination. The slower the rate of outflow, the lower the intake must be.

For the purposes of this discussion, I shall assume that herbage vitamin and mineral contents are not limiting. The major factors which then influence nutritive value are the contents of structural carbohydrates and NSC and the protein content, but these should never be viewed in isolation from their likely interactions in the rumen. For detailed treatments of the effects of plant species, plant development, plant anatomy and environmental influences on herbage nutritive value, the reader is referred to reviews by Norton (1982), Wilson (1982), Weston and Hogan (1986), Jones and Wilson (1987) and Wilson (1994). The distinction frequently drawn in such reviews between the nutrients coming from cell walls and those from cell contents is nutritionally sensible, since both the events in the rumen and the ultimate herbage intake can be related to the herbage levels of cellulose and lignin from cell walls and of NSC and protein from cell contents. This is discussed further in the sections which follow.

While the distinction between cell walls and cell contents is useful and lends itself to simple chemical analyses, such as neutral detergent fibre (NDF; Goering and Van Soest, 1970) or water soluble carbohydrates (WSC; Yemm and Willis, 1954), inadequate attention to analytical procedures can cause confusion. For example, NDF can underestimate cell wall material in legumes due to the solubilization of pectic material from the cell walls. Conversely, the NDF fraction may be overestimated if contaminated by N or starch when these are present in large amounts in the sample (Wilson, 1994).

Nutrients from plant cell walls

Depending upon plant species and stage of maturity, the cell wall constituents comprise 25–85% of the dry weight of forage. The main components are the polymers of glucose, galactose, xylose, arabinose and mannose. Cellulose, the major polymer, consists of glucose units linked by 1–4-β-glycosidic bonds into long, straight chains, which are themselves aggregated into a partly crystalline, partly amorphous complex. The remaining cell wall polysaccharides are generically referred to as hemicellulose. Cell walls between adjacent cells are 'cemented' by pectic material, consisting of long chains of galacturonic acid units. In grasses, the contribution of pectins to cell walls is only of the order of 1–2% but, in legumes, may be ten times as high (Wilson, 1994).

The remaining cell-wall component is lignin, which is an indigestible phenolic polymer comprising 2–20% of plant dry weight. However, the effect of lignin on digestibility may be out of proportion to its actual content in herbage, because it bonds to cell wall polysaccharides and restricts microbial access during digestion (Wilson, 1994). As forage plants mature, the increasing crystallization of cellulose and the increased bonding between cellulose and lignin are essential for plant strength, but are also a major constraint to the ruminant in terms of both physical resistance to chewing and rumination and susceptibility to microbial colonization and digestion. The effects of maturation are compounded by the increasing stem : leaf ratio, since stem usually contains more cell wall material. Both effects on plant cell wall content can be delayed by defoliation, providing an obvious link between nutritive value and grazing management.

To be useful in feeding standard schemes, such as those outlined above and in Fig. 8.1, components of nutritive value, such as the content of cell walls, must ultimately be expressed in ME terms. This is usually achieved via the effects of forage cell wall content on digestibility, and the use of well-established empirical relationships between ME and digestibility (ARC, 1980; SCA, 1990) or between

ME and digestibility plus fat content (AFRC, 1992; Freer *et al.*, 1996). The digestibility of the forage influences both the voluntary intake and the efficiency of utilization of ME.

Cell wall components and physical limitation of voluntary intake

The mechanisms which control voluntary intake in ruminants have been reviewed by previous authors (e.g. Freer, 1981; Weston and Hogan, 1986; Kennedy, 1990), with general agreement that, in the grazing animal, intake control is usually dominated by the effects of plant cell wall material in the digestive tract, especially the rate at which digesta particles can leave the rumen. Herbage particles will only leave the rumen when reduced to a size which permits passage to the omasum. In sheep, for example, this critical size is taken to be 1 mm (Poppi *et al.*, 1980). The relative effects of mastication, rumination, ease of microbial colonization and rate of enzymic digestion on particle outflow have been discussed by Weston and Hogan (1986) and by Kennedy (1990).

Many of the factors influencing particle outflow are difficult to measure and thus difficult to use routinely for predicting voluntary intake. Fortunately, the cell wall characteristics which influence the rate of disappearance of digesta from the rumen are also closely reflected in the digestibility of the herbage. As a consequence, there are many reports in the literature in which voluntary feed consumption (VFC) has been related to herbage digestibility. In his review of the topic, Minson (1982) noted that the relationship was influenced by plant species (including grass vs. legume), by cultivar within species and by fertilizer regime. A typical example of the relationship between digestibility and VFC is the report by Freer and Jones (1984), working with young sheep (27 kg live weight) fed phalaris, annual ryegrass, lucerne or subterranean clover (Fig. 8.2). These authors found a significantly different relationship between digestibility and VFC when the grasses and clovers were compared, but there were no differences between the two grasses or between the two legumes. Depending on the level of digestibility, the VFC from legumes was 100–200 g organic matter (OM) day^{-1} more than from grasses, while the intake of both classes of plant increased by 16–19 g OM day^{-1} for each unit increase in digestibility.

The effect of herbage cell wall content on VFC can be modified by the energy demand of the particular class of animal being considered (Freer, 1981; Weston and Hogan, 1986) since, in theory, this sets the upper limit to VFC. Attempts have been made to accommodate this effect by relating VFC to various exponents of live weight, but there is no general agreement about which exponent should be used.

A more promising and potentially useful attempt to relate VFC to features of both the herbage and the energy demand of the animal is the concept of 'forage consumption constraint' (FCC), first discussed by Weston and Davis (1991). The value of this approach is in the integration of the concept of physical limitation of intake with the concept of the animal's capacity to use energy. It is based on the

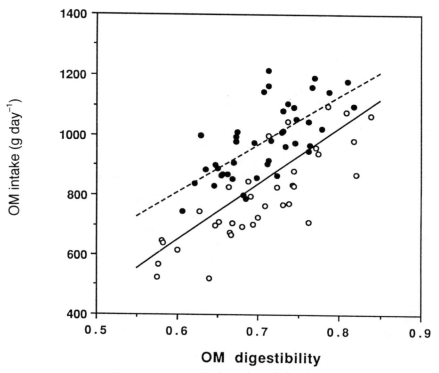

Fig. 8.2. Effect of organic matter (OM) digestibility of herbage on the voluntary OM intake of young sheep consuming grass (o) or legume (●) diets (adapted from Freer and Jones, 1984).

idea that, in feeds which contain no toxic compounds or deficiencies of required nutrients, VFC is regulated by an interplay between the rate of clearance of dry matter (DM) from the rumen, and the amount of energy available to the animal relative to its capacity to use energy (Weston, 1982; Weston and Davis, 1991). The FCC is defined as the difference between the VFC actually achieved and that which the animal would need to consume, theoretically, to satisfy its capacity to use energy. For mature wethers, this capacity was taken to be 60 g digestible OM (DOM) kg^{-1} body weight$^{0.75}$ (metabolic body weight (MBW)). The maximum theoretical intake would thus be 60/OM digestibility. This relationship is shown in Fig. 8.3 (solid line), together with the equivalent relationship for a 30 kg weaner (broken line), for which the equivalent capacity was taken to be 90 g DOM kg$^{-0.75}$. The data points are the actual levels of VFC achieved by sheep in the studies of Minson *et al.* (1964), Troelsen and Campbell (1969) and Freer and Jones (1984).

In all three of these studies and with substantial overlap between them, VFC was positively related to digestibility. At digestibilities above 0.75, the observed

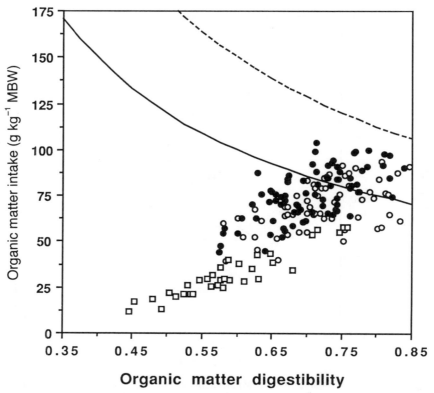

Fig. 8.3. Effects of herbage digestibility on voluntary food intake per kg metabolic body weight (MBW = $kg^{0.75}$), compared with the maximum theoretical intake of sheep if digestibility was not a constraint (Weston and Davis, 1991). ○, Data of Minson et al. (1964); □, data of Troelsen and Campbell (1969); ●, data of Freer and Jones (1984); ———, maximum theoretical intake of mature wether; – – – –, maximum theoretical intake of 30 kg weaner.

VFC approached or were in the zone between the maximum theoretical intakes for mature or 30 kg sheep. This was particularly so for the 27 kg young sheep in the study by Freer and Jones (1984). Two points can be made from Fig. 8.3. First, the difference between the observed VFC and the maximum theoretical intake (i.e. FCC) increases markedly as digestibility decreases. Second, at any given digestibility, the FCC for weaners is greater than that of mature sheep, even though their VFC may be the same. This reflects the greater capacity of the young animals to use energy. Weston and Davis (1991) have suggested that the importance of FCC as a concept is that it takes into account this capacity to use energy, and can be more closely related than VFC to forage characteristics such as NDF content or comminution energy. While Weston and Davis (1991) found this to be so, later studies show equally good relationships between either VFC or FCC and the energy required to shear herbage (e.g. Baker et al., 1993).

The relevance of the concept of FCC under field conditions is that, once FCC : digestibility relationships are established for 'palatable, nutrient-balanced' forages, the adverse effects of secondary compounds or nutrient imbalances can be quantified in terms of an increase in the FCC above that which would be expected from the digestibility alone. Clearly, to be used in that way, this approach will require accurate techniques for estimating both digestibility and intake in the field. Recent work has shown that, by using a combination of the natural alkanes of plant cuticular wax plus dosed, synthetic alkanes as faecal markers, the diet composition, digestibility and intake can be estimated accurately in individual grazing animals (Dove and Mayes, 1991). This advance, coupled with the use of new telemetric techniques for monitoring chewing and rumination activity in the field (Klein *et al.*, 1994), should permit the application of the FCC concept under field conditions. This would, in turn, provide a much needed link between the many indoor studies relating VFC to digestibility and the study of the factors limiting intake in the grazing animal.

Nutrients from plant cell contents

The main components of the cell contents are the NSC (sugars and storage polysaccharides), protein and non-protein N, lipids, organic acids, vitamins and minerals. As a group, these are frequently described as 'neutral detergent solubles' (NDS = (1 − NDF)) and, based on Van Soest (1967), are generally regarded as close to 100% digestible. However in grasses at least, there is mounting evidence that the NDS fraction can have a much lower digestibility. For example, in some species/plant parts, the digestibility of NDS was as low as 35–40% (Table 8.1). Since the values in Table 8.1 are, of necessity, *in vitro* digestibility figures, they should be treated with some caution. However, they do suggest that equal caution should be applied to the frequent assumption that the NDS fraction is completely digestible. For discussion of the factors influencing organic acid, vitamin and mineral contents of herbage, the reader is referred to Jones and Wilson (1987). The current discussion will be limited to the nutritional significance of plant lipids, protein and NSC.

Table 8.1. The digestibility of neutral detergent solubles (NDS) in a range of grass species (adapted from Ballard, 1988).

Species	Digestibility	Reference
Dactylis	0.90	Colburn et al. (1968)
Triticum	0.83	Moir (1971)
Panicum	0.46–0.67	Minson (1971)
Digitaria	0.35–0.61	Minson (1984)
Lolium rigidum	0.39–0.95	Ballard *et al.* (1990)
Lolium rigidum	0.40–1.00	Armstrong *et al.* (1992)
Lolium perenne	0.72–0.92	Radojevic *et al.* (1994)

Lipids

The lipids of herbage are mostly associated with the chloroplasts and constitute about 3–10% of herbage. The factors which influence lipid content have been reviewed by Hawke (1973). Lipids from cell contents can contribute significantly to the energy content of herbage and should be taken into account when computing ME content (Freer et al., 1996). However, they are not fermented in the rumen and thus do not contribute to the supply of energy for microbial protein synthesis (AFRC, 1992).

Protein

Although up to 10% of the dry weight of cell wall material consists of protein (Harris, 1983), most of the plant protein is associated with cell contents. The CP content of herbage can vary from 3 to 30% of the DM and, in general, leaves have more protein than stem (Sheehan et al., 1985) because of their chloroplast content (Brady, 1976). About half of the plant protein is water-soluble and, of this, about 50% is the enzyme ribulose biphosphate carboxylase (Brady, 1976; Peoples et al., 1980), sometimes referred to as 'fraction I' protein. This is released rapidly from cells during chewing and is quickly and extensively degraded in the rumen (Mangan, 1972; Nugent and Mangan, 1981). The remainder of the soluble protein is the so-called 'fraction II' protein, derived from chloroplasts and cytoplasm, while the insoluble protein is from chloroplast membranes and is associated with lipids. The various protein fractions would seem to have different rates of proteolysis in the rumen, and yet *in sacco* measurements suggest that the degradability of forage protein is high (Nugent and Mangan, 1981; Dove and McCormack, 1986). However, such results should be interpreted with caution since they do not measure proteolysis but only N disappearance from a fibre bag. As Thomson (1982) has emphasized, there is a large gap between the assessment of forage in terms of total N or CP and the detailed understanding we have of the nature of chloroplast and cytoplasmic proteins. Without an understanding of the latter, attempts to breed pasture plants with 'reduced protein degradability' may prove frustrating.

The factors which influence the protein content of green herbage have been discussed by Minson (1976), Norton (1982) and Jones and Wilson (1987). In general, legumes tend to have higher protein contents than grasses (Lyttleton, 1973), while tropical species, especially tropical grasses, have lower levels than temperate species (Minson, 1976; Norton, 1982). Since the protein content of leaves is higher, herbage protein content is also strongly influenced by the stage of growth, as it influences the leaf : stem proportions in the sward (Lyttleton, 1973; Norton, 1982; Ballard et al., 1990). The N status of the soil, arising either from N fertilizer or from biological N fixation, also markedly influences herbage N content.

Protein is a key nutrient for both the rumen and the ruminant, and several key points about protein supply can be made (see also Fig. 8.1).

1. When animals graze vegetative, temperate pastures, the protein content of the pasture is itself unlikely to limit production, so that breeding for higher protein content alone may not be useful. However, the rapid degradation of forage protein in the rumen can result in inefficient N capture into microbial protein synthesis, if the supply of energy from carbohydrate fermentation lags behind the supply of N from protein degradation. Since it nearly always does, some loss of herbage protein during rumen digestion is normal. For example, it can be calculated from the data of Ulyatt and Egan (1979) that, in an animal consuming 1500 g DM containing 22% CP, the protein intake is 330 g CP but the estimated protein flow into the duodenum is only 200 g CP. However, there are also occasions on which the supply of readily fermentable substrates in the rumen is so reduced as to severely limit microbial protein production in animals grazing green herbage (Dove and Milne, 1994).

2. Deficiencies of protein as such are a major limitation to the intake and utilization of tropical forages (see Minson, 1976; Weston and Hogan, 1986).

3. In general, ruminant nutritionists have not paid much attention to the amino acid pattern of the protein leaving the rumen, beyond the acceptance of the fact that, relative to the amino acid requirements for wool growth, rumen protein is deficient in the sulphur amino acids (Reis and Schinkel, 1963). This latter point has formed the basis of recent attempts to identify plant proteins high in sulphur amino acids and resistant to rumen degradation (Spencer *et al.*, 1988) and to transfer the genes encoding for these proteins into pasture legumes (see McNabb *et al.*, 1993, for discussion). However, Oldham (1993) has suggested that ruminant nutritionists need to pay more attention in general to the amino acid composition of the material flowing from the rumen, as it relates to the amino acid requirements of the host animal. Certainly, there are enough data from studies involving rumen-protected proteins or post-ruminal infusions of proteins or amino acid mixtures to suggest that, quite frequently, amino acid absorption from the small intestine limits production (e.g. Barry, 1981; Fraser *et al.*, 1991). The problem is not so much showing that this is the case as knowing what then to do about it within a grazing system. This issue remains unresolved.

4. When animals consume low-quality roughages such as dead pasture, the protein content may be so low (< 3–5%) as to be not only deficient relative to the protein requirements of the host animal but also low enough to limit the supply of RDP for microbial growth and fermentative acitivity. As a result, the rate of cell wall digestion drops dramatically, material leaves the rumen more slowly and intake falls. Supplements of RDP can readily redress this problem with a resultant increase in roughage intake (see below).

Non-structural carbohydrates

The NSC content of plant tissues consists principally of a mixture of sugars and storage polysaccharides. Legumes generally have higher levels of storage carbohydrates than grasses, usually distributed mainly in leaf laminae (Culvenor *et al.*, 1989). Although it is usually stated that the storage carbohydrate of legumes is starch (see Jones and Wilson, 1987), there is evidence to suggest that the starch accumulation occurs only in the upper leaves of the canopy, at least in subterranean clover (Culvenor *et al.*, 1989), and then only at higher light intensities. The bulk of the NSC in the above study was WSC.

In tropical grasses, the main storage carbohydrate does appear to be starch (Norton, 1982) but in temperate grasses carbohydrate is stored as fructans (Smouter and Simpson, 1989). The total NSC content of temperate grasses is usually in the range 5–40% of the herbage DM, with the main storage organ being the stem (Ballard *et al.*, 1990; Armstrong *et al.*, 1992), especially as flowering approaches.

In general, NSC accumulates when the rate of hexose formation during photosynthesis exceeds hexose utilization in respiration and growth. The concentration usually increases with plant age and reaches a peak near anthesis (anther exertion at flowering). Storage carbohydrates are then mobilized into developing seedheads and perennating organs. The amount of NSC present thus depends not only on plant species but also on the conditions influencing plant growth. Application of N fertilizer to grass swards has often resulted in reductions in NSC content, especially in species with inherently low NSC levels (Jones and Wilson, 1987). The effect is usually attributed to accelerated plant growth and results in negative correlations between N and NSC contents (Humphreys, 1989a; Radojevic *et al.*, 1994), although these correlations should not be taken to indicate that plant selection for increased NSC content will reduce protein content (Radojevic *et al.*, 1994). Reduced light intensity due to shading or overcast weather can also reduce NSC content (Melvin and Sutherland, 1961; Birrell, 1989).

Interest in the contribution of NSC to animal performance has focused primarily on grass cultivars expressing a high-WSC phenotype (e.g. Humphreys, 1989b; Radojevic *et al.*, 1994), which is hardly surprising given the higher live weight gains reported in young animals grazing such cultivars (e.g. Davies *et al.*, 1989, 1991; Munro *et al.*, 1992). However, what is perhaps not sufficiently appreciated is the contribution of NSC to the nutritive value of grasses at maturity and during senescence. It is axiomatic that, beyond flowering, there is a rapid decline in the nutritive value of grasses, manifest in rapid falls in digestibility and CP concentration. On occasions, this decline is attributed to an increased concentration of cell wall components. However, detailed studies in annual ryegrass have demonstrated that the amount of cell wall is relatively stable after flowering (Ballard *et al.*, 1990; Fig. 8.4a). While there is a decline in the digestibility of NDF after this stage, the largest declines in NDF digestibility have already occurred by flowering (e.g. Pearce *et al.*, 1988). It is, in fact, the NDS content which falls most dramatically beyond flowering, especially in stems (Ballard *et al.*, 1990; Armstrong *et al.*, 1992), with most of the fall being attributable to WSC (Ballard *et al.*, 1990). The consequence is that the yield of digestible DM in the sward present as NDF is relatively stable beyond flowering; the reduction in digestible yield is almost entirely attributable to cell contents (Ballard *et al.*, 1990; Fig. 8.4b).

Nutrient interactions under normal grazing

In this section, I wish to examine the way in which the nutritive value of the herbage and the transactions in the rumen can influence diet composition, intake and performance in the grazing animal, with the emphasis again being on the need for synchrony between energy and protein transactions in the rumen. The examination will consist of four examples: one comparing animal performance on legumes vs. grasses, one on carbohydrate : protein interactions on green pasture, one on WSC effects on senescing pasture and, finally, one on protein : digestibility interactions on dead pasture.

Animal performance on legume vs. grass pastures

As discussed above, not only do animals consume more legume than grass when compared at the same digestibility (Freer and Jones, 1984), but there is also a higher efficiency of utilization of the products of legume digestion compared with grass digestion. The data in Table 8.2 are the estimated efficiencies of utilization of ME for maintenance and for growth, in a hypothetical 30 kg young sheep given *ad libitum* access to either 100% grass or 100% legume pastures. The values were calculated from the relationships in SCA (1990), as encapsulated in the decision-support system GrazFeed (Freer *et al.*, 1996). Over the digestibility

Table 8.2. Effects of herbage digestibility and species (grass vs. legume) on the efficiencies of utilization of metabolizable energy (ME) for maintenance (k_m) and for growth (k_g).

	k_m*		k_g*	
Digestibility	Grass	Legume	Grass	Legume
0.55	0.67	0.67	0.33	0.43
0.60	0.70	0.70	0.37	0.50
0.65	0.71	0.71	0.41	0.54
0.70	0.73	0.73	0.43	0.57
0.75	0.73	0.73	0.44	0.59
0.80	0.73	0.73	0.44	0.59

*Efficiencies were calculated using the functions in the decision-support system GrazFeed (Freer *et al.*, 1996) and assume a 30 kg young sheep with *ad libitum* access to either pure grass or pure legume pasture.

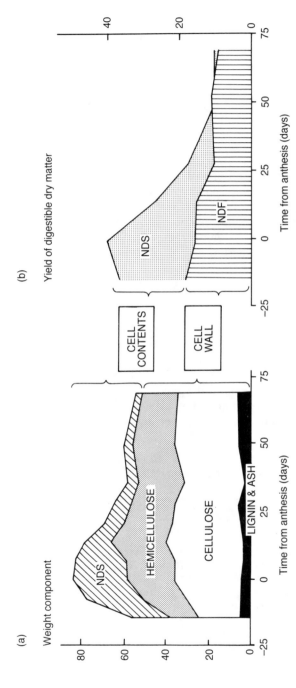

Fig. 8.4. (a) Mean weights (mg) of total dry matter (DM) as neutral detergent solubles (NDS), hemicellulose, cellulose, lignin and ash in a representative stem internode of annual ryegrass (*Lolium rigidum*), in relation to time from anthesis. (b) Mean yield (mg) of digestible DM as NDS or neutral detergent fibre (NDF) in a representative stem internode of annual ryegrass, in relation to time from anthesis. (Adapted from Ballard *et al.*, 1990).

Table 8.3. Relative liveweight gains of young sheep grazing pure swards of a range of pastures in New Zealand (Ulyatt, 1981a) or in southern Australia (Freer and Jones, 1984).

Reference	Pasture species		Relative live-weight gain*
Ulyatt (1981a)	Perennial ryegrass		100
	Short-rotation ryegrass		148
	Italian ryegrass		160
	Timothy		129
	Agrostis tenuis	spring	100
		summer	83
	White clover		186
	Lucerne		170
	Maku lotus		143
Freer and Jones (1984)	Annual ryegrass		100
	Phalaris (cv. Australian)		109
	Subterranean clover (cv. Bacchus Marsh)		124
	Lucerne (cv. Hunter River)		134

*For ease of comparison, the live-weight gain of animals grazing the ryegrass is set to 100 in each case.

range 55–75%, there is a small effect of digestibility on the efficiency of utilization of ME for maintenance (k_m), but no difference in efficiency between grass and legume. However, as digestibility increases, the efficiency of utilization of ME for growth (k_g) increases markedly, and the efficiencies expected with legumes are well above those for grasses. In effect, to achieve the same k_g as a legume, a grass must be at least 10 percentage units more digestible.

The combined effects of the higher intake and better k_g observed with legumes result in the observed superiority of legumes for growth in young stock (Table 8.3; Ulyatt, 1981a; Freer and Jones, 1984). Legumes have also been shown to support superior milk yields in dairy cows (see Thomson, 1982), but this is due principally to the effect of legumes on intake, since there is little evidence that the efficiency of utilization of ME for lactation is any higher for legumes than for grasses.

Carbohydrate : protein interactions on green pasture

High levels of WSC in pasture have frequently been reported to result in greater herbage intakes (Michell, 1973; Birrell, 1989; Jones and Roberts, 1991) and efficiency, measured either indirectly as lamb production per unit DOM intake (Davies *et al.*, 1991) or directly in calorimetry studies (Corbett *et al.*, 1966). It has also often been said that WSC content should influence the efficiency of microbial protein production, by altering the relative rates of fermentation and rumen ammonia production (e.g. Beever *et al.*, 1978; SCA, 1990). However, there have been few data to support this idea, especially with the grazing animal.

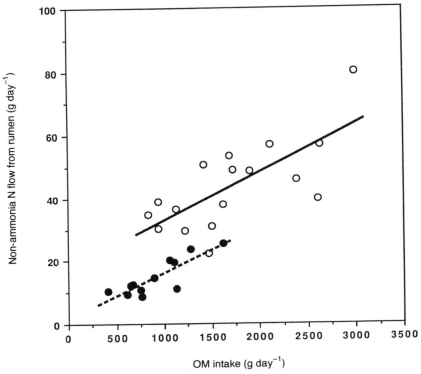

Fig. 8.5. Influence of organic matter (OM) intake on the flow of non-ammonia nitrogen from the rumen of ewes grazing perennial ryegrass in summer (○) or autumn (●). (Adapted from Dove and Milne (1994), with the permission of the *Australian Journal of Agricultural Research*.)

We recently reported an effect of herbage WSC content on microbial protein production in ewes grazing the same perennial ryegrass sward in the Scottish summer and autumn (Dove and Milne, 1994). The herbage digestibility, estimated from plant and faecal cuticular alkane levels (Dove *et al.*, 1990), was 0.80 and did not differ between seasons. Protein (non-ammonia $N \times 6.25$) flows from the rumen were positively related to herbage intake in both seasons, as would be expected (Fig. 8.5), but the relationships differed significantly between the seasons. At a given level of intake, protein flows were much lower when ewes grazed the autumn pastures. Our calculations indicated that summer herbage supported an efficiency of microbial protein synthesis of 10.4 g microbial CP MJ^{-1} ME consumed, consistent with previous studies (ARC, 1984; SCA, 1990). The equivalent value for the autumn pastures was only 5.4 g microbial CP MJ^{-1} ME.

The reduced microbial efficiency in autumn was closely related to much lower levels of rumen propionate and thus higher acetate : propionate ratios in the rumen during autumn (Dove and Milne, 1994; Fig. 8.6). Although we did not

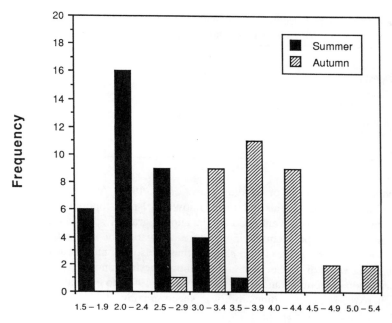

Fig. 8.6. Frequency distributions of rumen acetate : propionate ratios in ewes grazing perennial ryegrass pastures in summer or autumn. (From Dove and Milne (1994), with the permission of the *Australian Journal of Agricultural Research*.)

measure herbage WSC content directly, previous reports indicate that such a volatile fatty acid pattern in the rumen would be the likely consequence of very low herbage WSC levels (Beever *et al.*, 1978; SCA, 1990). The frequency distributions of rumen acetate : propionate ratios in our two groups of animals barely overlapped (Fig. 8.6) and nearly all the values for animals on autumn pasture were above 3.0, the suggested critical value above which WSC supply is regarded as limiting microbial protein synthesis (see SCA, 1990, for further discussion).

This example serves to make two major points. First, an increase in the WSC supply to the rumen can result in improved microbial efficiency, even with highly digestible, vegetative pastures. Second, the distinction drawn between 'protein' and 'energy' as nutrients is often questionable for the ruminant and may even be misleading. One of the main effects of increased WSC supply to the rumen can be increased protein flow from the rumen, due to the more efficient capture of N in that organ.

Carbohydrate interactions on senescing pasture

Several reports have indicated that the loss of NDS from grass pastures during senescence can be significantly delayed by spraying the pasture with the herbicide

Table 8.4. Effects of glyphosate treatment of annual grass pasture on the diet composition, digestibility and intake of grazing sheep (Adapted from Simpson et al., 1993).

	Glyphosate-treated	Untreated
OM digestibility	0.656	0.494
OM intake (g day^{-1})	1192	863
Proportion of intake derived from		
Leaf plus sheath	0.634	–
Flower spike	0.366	0.206
Stem	–	0.791

glyphosate, at the time of heading (Leys et al., 1991; Armstrong et al., 1992; Simpson et al., 1993). This treatment, known as 'spray-topping' in Australia, is done principally to prevent seed set in annual grasses so that they no longer constitute a weed problem during pasture renovation or in a subsequent crop. However, Simpson et al. (1993) showed that it also has major consequences for the feeding value of the herbage and for animal performance. These authors used natural and synthetic alkanes (Dove and Mayes, 1991) to estimate diet composition, herbage intake and digestibility in sheep grazing glyphosate-treated or untreated annual grass pastures.

In the 60 days after anthesis in the swards, animals grazing the untreated pasture only maintained live weight, whereas those grazing the treated pasture gained 40 g day^{-1}. The digestibility of the pasture consumed from the treated plots was 15 percentage units higher than that of herbage consumed from the untreated plots (Table 8.4), much of this increase being attributable to the higher WSC content in treated pasture (23.5% DM) compared with the untreated pasture (15.8% DM) (Simpson and Dove, 1994). The herbage intake of the sheep grazing the treated material was 38% higher than that of the sheep grazing the untreated material (Table 8.4).

In addition to these effects, the higher WSC of the treated pasture appeared to have a strong influence on diet selection by the sheep. Preference tests were conducted indoors, using hand-harvested pasture separated into leaf blade, leaf sheath, stem and flower spike. Housed animals showed a marked preference for treated stem over all other fractions, including treated and untreated leaf, both of which had higher digestibilities than the treated stem (Siever-Kelly et al., 1993). The patterns of cuticular wax alkanes differed between the various plant fractions and thus could be used, in conjuction with faecal alkane concentrations, to estimate the proportions of the different plant parts in the diets of the grazing animals (Dove and Mayes, 1991). This approach identified a major difference in the dietary preferences of the sheep grazing the two swards and confirmed the indoor preference tests (Dove et al., 1992). On the treated material, sheep ate no stem and mostly leaf blade and sheath, plus some flower spike (Table 8.4). This accorded with the relative digestibilies of the plant parts. However, sheep grazing the untreated pastures consumed no leaf or sheath, some flower spike, but almost 80%

stem. In both swards, stem had the lowest digestibility of all the plant fractions (untreated 0.361, treated 0.570; Simpson and Dove, 1994) but, in treated material, it had by far the highest WSC content (37.2% of DM).

These results strongly suggest a role for WSC in the process of diet choice in grazing animals, although they are not the first so to do (e.g. Bland and Dent, 1964; Michell, 1973; Arnold, 1981). In discussing the possible involvement of WSC in diet preference, Arnold (1981) commented that it was unclear what would function as a molecular signal to the grazing animal, since the term 'WSC' has no physiological significance. However, herbage levels of simple sugars are often positively correlated with total WSC levels (R.J. Simpson, Canberra, 1994, personal communication) and thus could readily fill the role of signal.

This example is presented to make two major points. First, increased WSC levels in senescing pasture can result in large increases in the digestibility of the consumed diet, in herbage intake and in animal performance. Second, diet choice on the treated, high-WSC pasture was more closely related to the WSC content of plant parts, rather than their digestibility, and strongly suggested a role for WSC in the process of diet choice in the grazing animal.

Protein : digestibility interactions in animals grazing low-quality roughage

In contrast with green herbage, the consumption of low-quality roughages, such as dead pasture or cereal residues, is usually limited by the slow rate of clearance of digesta from the rumen, as a result of the high comminution energy and low rate of digestion of such roughage. This problem is exacerbated if the roughage has a low N content (< 1%) since, at this level, RDP supply in the rumen begins to limit microbial growth and activity (see Fig. 8.1 above). The digestibility of the roughage will then be even lower than the comminution energy would suggest.

It follows that the provision of extra N should result in increased rumen microbial activity, better digestion of cell wall components and improved intakes; this is the basis of the use of urea supplements for stock grazing low-quality herbage. The beneficial effects of extra N are evident in the results of Freer *et al.* (1985, 1988). In both studies, the provision of up to 400 g day^{-1} of an oat grain : sunflower meal supplement increased the intake of low-quality roughage (Table 8.5). These increases can be attributed to the extra N provided by the oat grain (1.6% N) and especially the sunflower meal (6.6% N).

Further increments of supplement above 400 g day^{-1} reduced roughage intake, ultimately to levels below those of unsupplemented animals. This substitution between supplement and roughage probably represents the effects of both rumen fill and a change in the rumen microbial population and conditions. In a closely related study, Doyle *et al.* (1988) found that supplements of 2 : 1 oat grain : sunflower meal depressed the digestion of NDF in the rumen, and attributed the depression to pH reduction (Mould and Ørskov, 1983–1984) and

Table 8.5. Effect of supplements containing rumen-degradable protein on the roughage intake of grazing lambs.

Source	Intake of supplement (g DM day^{-1})	Intake of roughage (g DM day^{-1})
Freer et al. (1985)*	0	630
	360	692
	540	467
	720	208
Freer et al. (1988)†	0	544
	160	808
	324	859
	640	514

*Supplement 1 : 1 oat grain (1.6% N) : sunflower meal (6.6% N).
†Supplement 2 : 1 oat grain (1.3% N) : sunflower meal (6.5%N).

increased competition between cellulolytic and amylolytic organisms in the rumen (El-Shazly et al., 1961).

Although this example involves supplementary feeding as well as grazing, the results further highlight the way in which microbial growth and population dynamics, N supply, NSC supply and cell wall digestion are all interrelated in a manner which can have significant consequences for animal performance. They also illustrate that the NSC supply can have either negative or positive effects on rumen digestion and N dynamics, depending on the N supply, the cell wall content of the diet and the form of the NSC (starch vs. sugars/fructans).

CAN WE BREED PLANTS WITH IMPROVED FEEDING VALUE?

The question posed in the heading paraphrases that asked by Ulyatt (1981b). In his review, he went on to suggest that an ideal plant would have:

> high protein content, particularly increased sulphur amino acids; high levels of soluble carbohydrate; some feature such as thicker cell walls in the soft tissues, or presence of tannins, that would either slow the release of soluble protein or render it less soluble; an easily ruptured epidermis; vascular tissue that is sufficient to maintain agronomic merit but is fragile in terms of [shear] strength; concentrations of minerals sufficient to maintain animal health.

Ulyatt (1981b) went on to suggest that, 'in the near future', nutritionists and plant breeders would be able to collaborate to design and produce such plants.

Despite this optimism, the ensuing 13 years have not seen major progress. Indeed, there have been few examples of successful, deliberate breeding for improved feeding value (van Wijk et al., 1993; Stone, 1994). As van Wijk et al.

(1993) emphasized, many of the successful cultivars which have appeared were discovered *post factum*. It is illuminating to examine how this situation might have come about. Such examinations often seem to be couched in terms of the animal scientist making nutritive value demands and the plant breeder trying to meet them and, in one sense, this adversarial approach may be part of the problem. Nevertheless, it does seem that the setting of nutritive value goals has been a difficulty. The problem, according to van Wijk *et al.* (1993), was that of 'the emphasis shifting from one criterion [of nutritive value] to the other'. One is tempted to ask 'Whose emphasis?' since a striking feature of the literature on this topic is the consistency of the nutritive value goals over the last 15 years. For example, there are very similar characteristics cited for the ideal plant in Ulyatt's description above, in the Delphi survey reported by Wheeler and Corbett (1989) and in the discussions at the most recent International Grassland Congress (Poppi, 1993). A major theme of the discussion at the 1993 meeting was the need for better understanding of the interactive nature of nutrient flows in the rumen (e.g. Beever, 1993) and better definition of the nutrient limitations at particular points in the production system. From this came the suggestion that cultivars with improved feeding value should be enterprise-specific (Poppi, 1993), although the time required for and the research costs and commercial realities of cultivar development all work against such an idea.

It is not the purpose of this chapter to resolve this issue, but more attention to the following points might assist future progress.

1. There is still a need for better appreciation of the way in which nutrient interactions in the rumen can influence feeding value. For example, suggestions that we should breed cultivars with reduced rumen degradation of forage proteins demand an understanding of what individual proteins actually do in the plant and how they then behave in the rumen. Discussions couched in CP terms will not prove useful. Moreover, there is a need for a better appreciation of the fact, that even in a single animal enterprise, the feeding value goals change with the season and with the stage of growth of the animals.

2. Having set the breeding goals, there is a need for improvement in the measurement and understanding of the plant characteristics which relate to the goal. An example already mentioned is the inadequacy of *in sacco* measurements for breeding work aimed at reducing plant protein degradation in the rumen. Similarly, attempts to breed for reduced herbage lignin content as such miss the point that it is the distribution of the lignin and the nature of its bonding to cell wall polysaccharides which are the important variables from the point of view of the animal and the rumen (Stone, 1994). A further example can be found in attempts to breed for reduced leaf shear strength, which actually resulted in plants with leaves of a lower linear density and no advantage for animal production (Mackinnon *et al.*, 1988; Inoué *et al.*, 1994a, b) because the methods used did not actually measure the shear, compression and tensile properties of the leaves (Baker *et al.*, 1993; Henry *et al.*, 1993, 1996).

3. Once cultivars with improved nutritive value are developed, they must survive in sown swards and contribute nutrients in a timely way before they contribute to improved feeding value. Even then, their presence may have unexpected consequences, such as extra demands for pasture management or even overutilization due to a change in animal preferences. These issues have recently been canvassed by Clark and Wilson (1993) and require the attention of both nutritionists and breeders. One approach to the evaluation of both the breeding objectives themselves and then the outcomes of having the new cultivars in a grazing system is the use of computer models of those systems. For useful discussion of this aspect, the reader is referred to Donnelly *et al.* (1994).

In a recent discussion of the possible contributions of advances in molecular biology to improved feeding value, Ulyatt (1993) indicated that, as he had earlier predicted (Ulyatt, 1981b), nutritionists and plant breeders were now collaborating to 'design' better pasture plants. The major advances in plant molecular biology in the intervening period, coupled with 'traditional' plant breeding methods, had allowed this collaboration to evolve. To date, progress has been mainly methodological, but encouraging nevertheless. Two major challenges which remain are, first, the development of simple, routine procedures for the genetic transformation of pasture grasses and, second, as Ulyatt (1993) has put it, 'the ability of the nutritionist to define in specific biochemical terms the characters that require improvement'. If plant breeders and nutritionists can rise to these challenges, exciting progress should result.

CONCLUSIONS

Much has been written about the 'plant–animal interface' or 'interaction'. From the perspective of the nutritional or digestive ecology of the grazing animal, it is probably better to think in terms of a 'pasture–rumen–animal interaction'. This term would serve to remind us that the way in which the host animal interacts with its pasture is inextricably linked to the needs of and processes within the rumen itself. The processes we refer to as 'ingestive behaviour' and 'grazing management' are not separate from the nutritional interactions described in this chapter, but are part of the same continuum. A better appreciation of this fact will lead almost inevitably to a better understanding of the grazing system and eventually, to better advice about how pasture can best be converted into sustainable profit.

REFERENCES

Agricultural and Food Research Council (AFRC) (1992) The nutrient requirements of ruminant animals: protein. Report No. 9, AFRC Technical Committee on Responses to Nutrients. *Nutrition Abstracts and Reviews* 62, 787–835.

Agricultural Research Council (ARC) (1980) *The Nutrient Requirements of Ruminant Livestock*. Commonwealth Agricultural Bureaux, Farnham Royal, 368 pp.

Agricultural Research Council (ARC) (1984) *The Nutrient Requirements of Ruminant Livestock, Supplement 1*. Commonwealth Agricultural Bureaux, Farnham Royal, 45 pp.

Armstrong, R.D., Simpson, R.J., Pearce, G.R. and Radojevic, I. (1992) Digestibility of senescing annual ryegrass following applications of glyphosate. *Australian Journal of Agricultural Research* 43, 871–885.

Arnold, G.W. (1981) Grazing behaviour. In: Morley, F.H.W. (ed.) *Grazing Animals*. Elsevier, Amsterdam, pp. 79–104.

Baker, S.K., Klein, L., de Boer, E.S. and Purser, D.B. (1993) Genotypes of dry, mature subterranean clover differ in shear energy. *Proceedings of the 17th International Grassland Congress*, 592–593.

Ballard, R.A. (1988) The digestibility and chemical composition of dead pasture grasses. Unpublished M.Agr.Sci. thesis, University of Melbourne.

Ballard, R.A., Simpson, R.J. and Pearce, G.R. (1990) Losses of the digestible components of annual ryegrass (*Lolium rigidum* Gaudin) during senescence. *Australian Journal of Agricultural Research* 41, 719–731.

Barry, T.N. (1981) Protein metabolism in growing lambs fed on fresh ryegrass (*Lolium perenne*)–clover (*Trifolium repens*) pasture *ad lib*. 1. Protein and energy deposition in response to abomasal infusion of casein and methionine. *British Journal of Nutrition* 46, 521–532.

Beever, D.E. (1993) Ruminant animal production from forages: present position and future possibilities. *Proceedings of the 17th International Grassland Congress*, 535–541.

Beever, D.E., Terry, R.A., Cammell, S.B. and Wallace, A.S. (1978) The digestion of spring and autumn harvested perennial ryegrass by sheep. *Journal of Agricultural Science (Cambridge)* 90, 463–470.

Birrell, H.A. (1989) The influence of pasture and animal factors on the consumption of pasture by grazing sheep. *Australian Journal of Agricultural Research* 40, 1261–1275.

Bland, B.F. and Dent, J.W. (1964) Animal preference in relation to the chemical composition and digestibility of varieties of cocksfoot. *Journal of the British Grassland Society* 19, 306–315.

Brady, C.J. (1976) Plant proteins, their occurrence, quality and distribution. In: Sutherland, T.M., McWilliam, J.R. and Leng, R.A. (eds) *Reviews in Rural Science II: From Plant to Animal Protein*. University of New England, Armidale, pp. 13–16.

Clark, D. and Wilson, J.R. (1993) Implications of improvements in nutritive value on plant performance and grassland management. *Proceedings of the 17th International Grassland Congress*, 543–549.

Colburn, M.W., Evans, J.L. and Ranage, C.H. (1968) Apparent and true digestibility of forage nutrients by ruminant animals. *Journal of Dairy Science* 51, 1450–1457.

Corbett, J.L., Langlands, J.P., McDonald, I. and Pullar, J.D. (1966) Comparison by direct energy calorimetry of the net energy values of an early and a late season growth of herbage. *Animal Production* 8, 13–27.

Culvenor, R.A., Davidson, I.A. and Simpson, R.J. (1989) Regrowth of swards of subterranean clover after defoliation. 1. Growth, non-structural carbohydrate and nitrogen content. *Annals of Botany* 64, 545–556.

Davies, D.A., Fothergill, M. and Jones, D. (1989) Assessment of contrasting perennial ryegrasses, with or without white clover, under continuous sheep stocking in the uplands. 1. Animal production from the grass varieties. *Grass and Forage Science* 44, 431–439.

Davies, D.A., Fothergill, M. and Jones, D. (1991) Assessment of contrasting perennial ryegrass, with and without white clover, under continuous sheep stocking in the uplands. 3. Herbage production, quality and intake. *Grass and Forage Science* 46, 39–49.

Donnelly, J.R., Freer, M. and Moore, A.D. (1994) Evaluating pasture breeding objectives using computer models. *New Zealand Journal of Agricultural Research* 37, 269–275.

Dove, H. and McCormack, H.A. (1986) Estimation of the ruminal degradation of forage rape after incubation in nylon bags in the rumen of sheep. *Grass and Forage Science* 41, 129–136.

Dove, H. and Mayes, R.W. (1991) The use of plant wax alkanes as marker substances in studies of the nutrition of herbivores: a review. *Australian Journal of Agricultural Research* 42, 913–952.

Dove, H. and Milne, J.A. (1994) Digesta flow and rumen microbial protein production in ewes grazing perennial ryegrass. *Australian Journal of Agricultural Research* 45, 1229–1245.

Dove, H., Milne, J.A. and Mayes, R.W. (1990) Comparison of herbage intakes estimated from *in vitro* or alkane-based digestibilities. *Proceedings of the New Zealand Society of Animal Production* 50, 457–459.

Dove, H., Siever-Kelly, C., Leury, B.J., Gatford, K.L. and Simpson, R.J. (1992) Using plant wax alkanes to quantify the intake of plant parts by grazing animals. *Proceedings of the Nutrition Society of Australia* 17, 149.

Doyle, P.T., Dove, H., Freer, M., Hart, F.J., Dixon, R.M. and Egan, A.R. (1988) Effects of a concentrate supplement on the intake and digestion of a low-quality forage by lambs. *Journal of Agricultural Science (Cambridge)* 111, 503–511.

El-Shazly, K., Dehority, B.A. and Johnson, R.R. (1961) Effect of starch on the digestion of cellulose *in vitro* and *in vivo* by rumen micro-organisms. *Journal of Animal Science* 20, 268–273.

Fraser, D.L., Poppi, D.P. and Fraser, T.J. (1991) The effect of protein or amino acid supplementation on growth and body composition in lambs grazing white clover. *Proceedings of the Third International Symposium on the Nutrition of Herbivores*, 22.

Freer, M. (1981) The control of food intake by grazing animals. In: Morley, F.H.W. (ed.) *Grazing Animals*. Elsevier, Amsterdam, pp. 105–124.

Freer, M. and Jones, D.B. (1984) Feeding values of subterranean clover, lucerne, phalaris and Wimmera ryegrass for lambs. *Australian Journal of Experimental Agriculture and Animal Husbandry* 24, 156–164.

Freer, M., Dove, H., Axelsen, A., Donnelly, J.R. and McKinney, G.T. (1985) Responses to supplements by weaned lambs grazing mature pasture or eating hay in yards. *Australian Journal of Experimental Agriculture* 25, 289–297.

Freer, M., Dove, H., Axelsen, J.R. and Donnelly, J.R. (1988) Responses to supplements by weaned lambs when grazing mature pasture or eating hay cut from the same pasture. *Journal of Agricultural Science (Cambridge)* 110, 661–667.

Freer, M., Moore, A.D. and Donnelly, J.R. (1996) GRAZPLAN: decision support systems for Australian grazing enterprises. II The animal biology model for feed intake,

production and reproduction and the GrazFeed DSS. *Agricultural Systems* 46 (in press).
Goering, H.K. and Van Soest, P.J. (1970) *Forage Fiber Analysis*. Agriculture Handbook No. 379, United States Department of Agriculture. Agricultural Research Service, Washington, DC.
Harris, P.J. (1983) Cell walls. In: Hall, J.L. and Moore, A.L. (eds) *Isolation of Membranes and Organelles from Cell Walls*. Academic Press, London, pp. 25–53.
Hawke, J.C. (1973) Lipids. In: Butler, G.W. and Bailey, R.W. (eds) *Chemistry and Biochemistry of Herbage*, Vol. 1. Academic Press, London, pp. 213–264.
Henry, D.A., Macmillan, R.H., Roberts, F.M. and Simpson, R.J. (1993) Assessment of the variation in shear strength of leaves of pasture grasses. *Proceedings of the 17th International Grassland Congress*, 553–555.
Henry, D.A., Macmillan, R.H. and Simpson, R.J. (1996) Measurement of the shear and tensile fracture properties of leaves of pasture grasses. *Australian Journal of Agricultural Research* 46, 587–603.
Hogan, J.P. (1982) Digestion and utilization of proteins. In: Hacker, J.B. (ed.) *Nutritional Limits to Animal Production from Pastures*. Commonwealth Agricultural Bureaux, Farnham Royal, pp. 245–257.
Humphreys, M.O. (1989a) Water-soluble carbohydrates in perennial ryegrass breeding. I. Genetic differences among cultivars and hybrid progeny grown as spaced plants. *Grass and Forage Science* 44, 231–236.
Humphreys, M.O. (1989b) Water-soluble carbohydrates in perennial ryegrass breeding. III. Relationships with herbage production, digestibility and crude protein content. *Grass and Forage Science* 44, 423–430.
Inoué, T., Brookes, I.M., John, A., Hunt, W.F. and Barry, T.N. (1994a) Effects of leaf shear breaking load on the feeding value of perennial ryegrass (*Lolium perenne*) for sheep. I. Effects of leaf anatomy and morphology. *Journal of Agricultural Science (Cambridge)* 123, 129–136.
Inoué, T., Brookes, I.M., John, A., Kolver, E.S. and Barry, T.N. (1994b) Effects of leaf shear breaking load on the feeding value of perennial ryegrass (*Lolium perenne*) for sheep. II. Effects on feed intake, particle breakdown, rumen digesta outflow and animal performance. *Journal of Agricultural Science (Cambridge)* 123, 137–147.
Jones, D.I.H. and Wilson, A.D. (1987) Nutritive quality of forage. In: Hacker, J.B. and Ternouth, J.H. (eds) *The Nutrition of Herbivores*. Academic Press, Sydney, pp. 65–89.
Jones, E.L. and Roberts, J.E. (1991) A note on the relationship between palatability and water-soluble carbohydrates content in perennial ryegrass. *Irish Journal of Agricultural Research* 30, 163–167.
Kennedy, P.M. (1990) Constraints of rumen dynamics on lignocellulose digestion. In: Akin, D.E., Ljungdahl, L.G., Wilson, J.R. and Harris, P.R. (eds), *Microbial and Plant Opportunities to Improve Lignocellulose Utilization by Ruminants*. Elsevier, New York, pp. 3–15.
Klein, L., Baker, S.K., Purser, D.B., Zaknich, A. and Bray, A.C. (1994) Telemetry to monitor sounds of chews during eating and rumination by grazing sheep. *Proceedings of the Australian Society of Animal Production* 20, 423.
Leys, A.R., Cullis, B.R. and Plater, B. (1991) Effect of spraytopping applications of paraquat and glyphosate on the nutritive value and regeneration of vulpia (*Vulpia bromoides* (L.) S.F. Gray). *Australian Journal of Agricultural Research* 42, 1405–1415.

Lyttleton, J.W. (1973) Proteins and nucleic acids. In: Butler, G.W. and Bailey, R.W. (eds) *Chemistry and Biochemistry of Herbage*, Vol. 1. Academic Press, London, pp. 63–103.

Mackinnon, B.W., Easton, H.S., Barry, T.N. and Sedcole, J.R. (1988) The effect of reduced leaf shear strength on the nutritive value of perennial ryegrass. *Journal of Agricultural Science (Cambridge)* 111, 469–474.

McNabb, W.C., Higgins, C., Tabe, L. and Higgins, T.J.V. (1993) Transfer of genes encoding proteins with high nutritional value into pasture legumes. *Proceedings of the 17th International Grassland Congress*, 1085–1091.

Mangan, J.L. (1972) Quantitative studies on nitrogen metabolism in the bovine rumen: the rate of proteolysis of casein and ovalbumin and the release and metabolism of free amino acids. *British Journal of Nutrition* 27, 261–283.

Melvin, J.F. and Sutherland, M.A. (1961) Effect of shading during growth on the soluble sugar contents of short rotation ryegrass. *Australian Journal of Experimental Agriculture and Animal Husbandry* 1, 153–155.

Michell, P.J. (1973) Relations between fibre and water soluble carbohydrate contents of pasture species and their digestibility and voluntary intake by sheep. *Australian Journal of Experimental Agriculture and Animal Husbandry* 13, 165–170.

Minson, D.J. (1971) Influence of lignin and silicon on a summative system assessing the organic matter digestibility of *Panicum*. *Australian Journal of Agricultural Research* 22, 589–598.

Minson, D.J. (1976) Nutritional significance of protein in temperate and tropical pastures. In: Sutherland, T.M, McWilliam, J.R. and Leng, R.A. (eds) *Reviews in Rural Science II: From Plant to Animal Protein*. University of New England, Armidale, pp. 27–31.

Minson, D.J. (1982) Effects of chemical and physical composition of herbage eaten upon intake. In: Hacker, J.B. (ed.) *Nutritional Limits to Animal Production from Pastures*. Commonwealth Agricultural Bureaux, Farnham Royal, pp. 167–182.

Minson, D.J. (1984) Digestibility and voluntary intake by sheep of five *Digitaria* species. *Australian Journal of Experimental Agriculture and Animal Husbandry* 24, 494–500.

Minson, D.J., Harris, C.E., Raymond, W.F. and Milford, R. (1964) The digestibility and voluntary intake of S22 and H.1 ryegrass, S170 tall fescue, S48 timothy, S215 meadow fescue and germinal cocksfoot. *Journal of the British Grassland Society* 19, 298–305.

Moir, K.W. (1971) The *in vivo* and *in vitro* digestible fractions in forage. *Journal of the Science of Food and Agriculture* 22, 338–341.

Mould, F.L. and Ørskov, E.R. (1983–1984) Manipulation of rumen fluid pH and its influence on cellulolysis *in sacco*, dry matter degradation and the rumen microflora in sheep offered either hay or concentrates. *Animal Feed Science and Technology* 10, 1–14.

Munro, J.M.M., Davies, D.A., Evans, W.B. and Scurlock, R.V. (1992) Animal production evaluation of herbage varieties. 1. Comparison of Aurora with Frances, Talbot and Melle perennial ryegrasses when grown alone and with clover. *Grass and Forage Science* 47, 259–273.

Norton, B.W. (1982) Differences between species in forage quality. In: Hacker, J.B. (ed.) *Nutritional Limits to Animal Production from Pastures*. Commonwealth Agricultural Bureaux, Farnham Royal, pp. 88–110.

Nugent, J.H.A. and Mangan, J.L. (1981) Characteristics of the rumen proteolysis of fraction I (18S) leaf protein from lucerne (*Medicago sativa* L). *British Journal of Nutrition* 46, 39–58.

Oldham, J.D. (1993) Recent progress towards matching feed quality to the amino acid needs of ruminants. *Animal Feed Science and Technology* 45, 19–34.

Ørskov, E.R. and McDonald, I. (1979) The estimation of protein degradability in the rumen from incubation measurements weighted according to rate of passage. *Journal of Agricultural Science (Cambridge)* 92, 499–503.

Pearce, G.R., Lee, J.A., Simpson, R.J. and Doyle, P.T. (1988) Sources of variation in the nutritive value of wheat and rice straws. In: Reed, J.D., Capper, B.S. and Neate, P.J.H. (eds) *Plant Breeding and the Nutritive Value of Crop Residues*. ILCA, Addis Ababa, pp. 195–231.

Peoples, M.B., Beilharz, V.C., Waters, S.P., Simpson, R.J. and Dalling, M.J. (1980) Nitrogen redistribution in wheat (*Triticum aestivum* L.). II. Chloroplast senescence and the degradation of ribulose-1,5-biphosphate carboxylase. *Planta* 149, 241–251.

Poppi, D.P. (1993) Chairperson's summary paper Session 14: Nutritive value. *Proceedings of the 17th International Grassland Congress*, pp. 602–603.

Poppi, D.P., Norton, B.W., Minson, D.J. and Hendriksen, R.E. (1980) The validity of the critical size theory for particles leaving the rumen. *Journal of Agricultural Science (Cambridge)* 94, 275–280.

Radojevic, I., Simpson, R.J., St John, J.A. and Humphreys, M.O. (1994) Chemical composition and *in vitro* digestibility of lines of *Lolium perenne* selected for high concentrations of water-soluble carbohydrate. *Australian Journal of Agricultural Research* 45, 901–912.

Reis, P.J. and Schinkel, P.G. (1963) Some effects of sulphur-containing amino acids on the growth and composition of wool. *Australian Journal of Biological Sciences*, 16, 218–230.

Sheehan, W., Fontenot, J.P. and Blaser, R.E. (1985) *In vitro* dry matter digestibility and chemical composition of autumn-accumulated tall fescue, orchardgrass and red clover. *Grass and Forage Science* 40, 317–322.

Siever-Kelly, C., Leury, B.J., Gatford, K.L., Simpson, R.J. and Dove, H. (1993) Diet selection by sheep offered glyphosate-treated herbage and plant fractions. *Proceedings of the 15th International Congress of Nutrition*, Vol. 2, p. 646 (Abstract P153).

Simpson, R.J. and Dove, H. (1994) Plant non-structural carbohydrates, diet selection and intake. In: Dove, H. (ed.) The plant–animal interface revisited. *Proceedings of the Australian Society of Animal Production* 20, 59–61.

Simpson, R.J., Gatford, K.L., Siever-Kelly, C., Leury, B.J. and Dove, H. (1993) Improved feeding value in senescing grass pasture after application of glyphosate. *Proceedings of the 17th International Grassland Congress*, 558–559.

Smouter, H. and Simpson, R.J. (1989) Occurrence of fructans in the Gramineae (Poaceae). *New Phytologist* 111, 359–368.

Spencer, D., Higgins, T.J.V., Freer, M., Dove, H. and Coombe, J.B. (1988) Monitoring the fate of dietary proteins in rumen fluid using gel electrophoresis. *British Journal of Nutrition* 60, 241–247.

Standing Committee on Agriculture (SCA) (1990) *Feeding Standards for Australian Livestock: Ruminants*. CSIRO Australia, Melbourne.

Stone, B.A. (1994) Prospects for improving the nutritive value of temperate, perennial pasture species. *New Zealand Journal of Agricultural Research* 37, 349–363.

Thomson, D.J. (1982) The nitrogen supplied by and the supplementation of fresh or grazed forage. In: Thomson, D.J., Beever, D.E. and Gunn, R.G. (eds) *Forage Protein in Ruminant Animal Production*. BSAP Occasional Publication No. 6, BSAP, Thames Ditton, pp. 53–66.

Troelsen, J.E. and Campbell, J.B. (1969) The effect of maturity and leafiness on the intake and digestibility of alfalfas and grasses fed to sheep. *Journal of Agricultural Science (Cambridge)* 73, 145–154.

Ulyatt, M.J. (1973) The feeding value of herbage. In: Butler, G.W. and Bailey, R.W. (eds) *Chemistry and Biochemistry of Herbage*, Vol. 3. Academic Press, London, pp. 131–178.

Ulyatt, M.J. (1981a) The feeding value of temperate pastures. In: Morley, F.H.W. (ed.) *Grazing Animals*. Elsevier, Amsterdam, pp.125–141.

Ulyatt, M.J. (1981b) The feeding value of herbage: can it be improved? *New Zealand Agricultural Science* 15, 200–205.

Ulyatt, M.J. (1993) Chairperson's summary paper. Session 29: Molecular biology – forage quality. *Proceedings of the 17th International Grassland Congress*, 1105–1106.

Ulyatt, M.J. and Egan, A.R. (1979) Quantitative digestion of fresh herbage by sheep. V. The digestion of four herbages and prediction of sites of digestion. *Journal of Agricultural Science (Cambridge)* 92, 605–616.

Van Soest, P.J. (1967) Development of a comprehensive system of feed analysis and its application to forages. *Journal of Animal Science* 26, 119–128.

Van Wijk, A.J.P., Boonman, J.G. and Rumball, W. (1993) Achievements and perspectives in the breeding of forage grasses and legumes. *Proceedings of the 17th International Grassland Congress*, 379–383.

Webster, A.J.F. (1992) The metabolizable protein system for ruminants. In: Garnsworthy, P.C., Haresign, W. and Cole, D.J.A. (eds) *Recent Advances in Animal Nutrition 1992*. Butterworth–Heinemann, Oxford, pp. 93–110.

Weston, R.H. (1982) Animal factors affecting feed intake. In: Hacker, J.B. (ed.) *Nutritional Limits to Animal Production from Pastures*. Commonwealth Agricultural Bureaux, Farnham Royal, pp. 183–198.

Weston, R.H. and Davis, P. (1991) The significance of four forage characters as constraints to voluntary intake. *Proceedings of the Third International Symposium on the Nutrition of Herbivores*, 33.

Weston, R.H. and Hogan, J.P. (1986) Nutrition of herbage-fed ruminants. In: Alexander, G. and Williams, O.B. (eds) *The Pastoral Industries of Australia*. Sydney University Press, Sydney, pp. 183–210.

Wheeler, J.L. and Corbett, J.L. (1989) Criteria for breeding forages of improved feeding value: results of a Delphi survey. *Grass and Forage Science* 44, 77–83.

Wilson, J.R. (1982) Environmental and nutritional factors affecting herbage quality. In: Hacker, J.B. (ed.) *Nutritional Limits to Animal Production from Pastures*. Commonwealth Agricultural Bureaux, Farnham Royal, pp. 111–131.

Wilson, J.R. (1994) Cell wall characteristics in relation to forage digestion by ruminants. *Journal of Agricultural Science (Cambridge)* 122, 173–182.

Yemm, E.W. and Willis, A.J. (1954) The estimation of carbohydrates in plant extracts by anthrone. *Biochemistry Journal* 57, 508–514.

Multispecies Grazing in the Serengeti

M.G. Murray and A.W. Illius

Institute of Cell, Animal and Population Biology, Ashworth Laboratories, University of Edinburgh, West Mains Road, Edinburgh EH9 3JT, UK

INTRODUCTION

One of the most striking characteristics of African savanna ecosystems is the richness of their ungulate fauna: in parts of East Africa, ten or more species of grazing ungulate may be found in close proximity, with several species occurring together in mixed groups. Even these species-rich communities may be impoverished in comparison with their counterparts in the Pleistocene, when Africa supported some 65% more genera of large mammals (Martin, 1966). What allows so many species of grazing ungulate to coexist over relatively long time periods? This question has more than academic interest, as the answer will form the basis for managing the conservation of multispecies communities in the future. This chapter addresses the question in the context of the Serengeti ungulate community, which is one of the world's richest and has seen extensive field and theoretical research since the 1960s (Sinclair and Norton-Griffiths, 1979; Sinclair and Arcese, 1995).

In an elementary theory of species diversity, MacArthur (1972) shows how the number of species in a community can be increased in one of three ways: (i) by decreasing the range of resources that each species utilizes (i.e. by increasing specialization); (ii) by enlarging the overlap in resource use between neighbouring species; or (iii) by increasing the overall spectrum of resources available to the community (Fig. 9.1). MacArthur (1972) makes no assumptions about the nature of underlying processes, but his conceptual framework is used here to examine the influences on ungulate diversity in Serengeti of a range of ecological processes, including dietary specialization, interspecific resource competition, niche overlap,

©1996 CAB INTERNATIONAL.
The Ecology and Management of Grazing Systems (eds J. Hodgson and A.W. Illius)

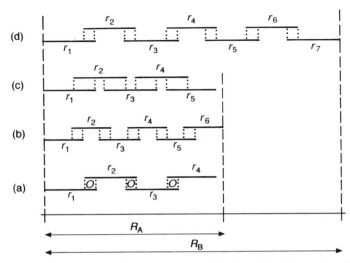

Fig. 9.1. Species diversity in relation to resource utilization. (a) Each species utilizes a range r_i of the overall spectrum R_A of resources, overlapping with neighbouring species by O. In a packed community, greater species richness arises from: (b) smaller individual ranges r_i; (c) greater overlap between neighbours in resource use; or (d) utilization of a wider spectrum R_B of resources (MacArthur, 1972).

facilitation and migration. An application of Tilman's (1982) hypothesis of resource competition in plants is investigated in the process.

Throughout the world, it is common to utilize pastures with more than one animal species or for a mixture of domestic and wild grazing animals to be present. In Africa, herbivore biomass is largely determined by rainfall and soil type, but is also positively related to the number of species present (Fritz and Duncan, 1994). Multispecies grazing systems pose a number of challenges and potential advantages over single-species systems. The foremost concern must be whether the presence of a range of species is beneficial, through increased and diversified saleable output, increased efficiency of resource use, facilitatory interactions between species, improved disease control and improved ability to manage vegetation and reduce land degradation. Potentially deleterious effects may arise through interspecific competition for resources or the transfer of disease from wild populations to domestic animals. The present chapter concentrates on the issues of resource competition and partitioning between grazing species, by reviewing the nutritional ecology of ruminants in their natural environment. The reader is referred to the many reviews of mixed grazing in temperate agricultural contexts (e.g. Nolan and Connolly, 1977, 1989, 1992; Lambert and Guerin, 1989; Wright and Connolly, 1996).

The review thus concentrates on aspects of resource use in a natural grassland ecosystem. It is probable that predation and disease also have direct impacts on ungulate community structure (Sinclair and Norton-Griffiths, 1979, 1982;

Sinclair, 1985; Borner *et al.*, 1987; Dublin *et al.*, 1990a), and further work is required to establish the importance of predation as a niche axis for small-sized ungulates, but these topics fall outside the scope of this chapter.

THE SERENGETI–MARA ECOSYSTEM

The extent of the Serengeti–Mara ecosystem is based on the annual range of migratory wildebeest (*Connochaetes taurinus*). Its main features are open grass plains to the south and southeast, where rainfall is low, with open scrub and woodlands in the remaining moister areas (map in Murray, 1995). The gradient in rainfall and in growing conditions across the system is exploited by the wildebeest, which spend the wet season on the short grasslands to the south east and the dry season in the woodlands. There is also a gradient in growing conditions within the woodlands, large areas of which are set on gently rolling hills. The period of grass growth is shorter at the top of the catena, which typically supports shorter grasses, than at the bottom, where taller grasses are found in sumps and along the lines of streams and rivers. Residential species tend to move up the catena in the wet season and to work their way down in drier conditions, mirroring in a less dramatic way the annual migrations of wildebeest (Bell, 1971).

As elsewhere in Africa, the quality and quantity of herbage in Serengeti grassland habitats is, to a large extent, defined by soil–water relations, which underlie the seasonal patterns in growth and phenology of grass swards as well as the size and structure of the grasses when they mature (Vesey-FitzGerald, 1973). Tall grass is associated with prolonged growing conditions, which are typical of the moister regions, riverine terraces and the bottom of the catena. It contains a greater ratio of structural to soluble carbohydrates and hence is typically of poor digestibility (Bell, 1970; Van Soest, 1982). Conversely, short grasses, which are comparatively digestible, are associated with abbreviated growing conditions. They are found in the more arid areas, towards the top of the catena and on soils which are shallow or of sandy texture.

Over 120 grass species have been collected in the Serengeti National Park from several distinctive grassland communities (Anderson and Talbot, 1965; Bell, 1969; Duncan, 1975; Sinclair, 1977; McNaughton, 1983). The treeless plains to the southeast are dominated by short, mat-forming grasses including *Sporobolus* and *Digitaria* species. Most of the rainfall in this region occurs from November to May, averaging 400–800 mm per annum (Norton-Griffiths *et al.*, 1975; Campbell, 1989). Herds of wildebeest, zebra (*Equus burchelli*) and gazelle (*Gazella thomsonii*) maintain a grazing lawn on the plains during the wet season, and small numbers of oryx (*Oryx beisa*) are resident in the southeastern boundary area (Walther, 1978). With the current population of over 1 million wildebeest, there is insufficient fuel remaining on the short-grass plains to support fires during the dry season.

The wooded parts of the park include the western corridor, and the central and northern woodlands. Here the annual rainfall averages from 900 to 1200 mm with scattered showers stimulating patchy growth of grass during the dry season. Dominant grasses with an erect growth form in the woodland habitats include *Bothriochloa insculpta, Digitaria macroblephara, Eriochloa fatmensis, Panicum coloratum, Panicum maximum, Pennisetum mezianum, Themeda triandra, Setaria sphacelata* and *Sporobulus fimbriatus*. Migratory ungulates are present during the dry season, along with resident populations of oribi (*Ourebia ourebi*), mountain reedbuck (*Redunca fulvorufula*), Bohor reedbuck (*Redunca redunca*), warthog (*Phacochoerus aethiopicus*), topi (*Damaliscus lunatus*), hartebeest (*Alcelaphus buselaphus*), waterbuck (*Kobus ellipsiprymnus*), roan antelope (*Hippotragus equinus*), buffalo (*Syncerus caffer*), hippopotamus (*Hippopotamus amphibius*) and elephant (*Loxodonta africana*). Wildfires may occur in any month of the dry season, with 25–90% of the area being burnt annually (Norton-Griffiths, 1979).

QUESTIONS ABOUT COEXISTENCE

Do ungulate species differ in their feeding niches?

Observations of the diet composition of African ungulates, estimated either from gut contents of shot animals or by direct observation of grazing (Gwynne and Bell, 1968; Bell, 1969; Hansen *et al*., 1975; Owaga, 1975; Jarman and Sinclair, 1979; McNaughton, 1985; Murray and Brown, 1993), lead to the following generalizations. Small-bodied species, such as Thomson's gazelle, are more selective feeders than larger animals, such as topi and buffalo. All ruminants select against stem and leaf sheath and in favour of leaf lamina, but this tendency is strongest in small and weakest in large species. By implication, the digestibility of the diet is higher in the smaller species. Small species have botanically diverse diets, including a wide range of dicotyledons (forbs) growing in the base of the sward. Large species make less use of short grass than do small species and are the first to move off swards high on the catena as they become depleted during the dry season.

Species of similar size but with different digestive systems will use different food-quality niches, but smaller hindgut fermenters, such as zebra, may compete directly with large ruminants, such as buffalo, because the former can tolerate lower food quality. Similar results have been shown in domestic sheep and cattle, and also in ponies and red deer, grazing a range of natural vegetation of temperate regions (Gordon, 1989; Hodgson *et al*., 1991). It is concluded that ungulate species do show differences in feeding niche and that differences in body size and digestive system underlie niche separation (Bell, 1970; Jarman, 1974). There remains the question of what the mechanisms are which give rise to this separation.

A series of models have predicted habitat choices of different-sized ungulates according to the relative importance of constraints in food quantity and quality (Bell, 1969; Demment and Van Soest, 1985; Owen-Smith, 1985, 1989; Illius and Gordon, 1993; Owen-Smith and Cumming, 1993). The models explain some of the differences in the distributions of ungulate species according to differences in efficiency of foraging, without invoking competition, predation or any other interspecific interaction. Illius and Gordon (1987) showed that the allometric scaling of incisor arcade morphology with body mass explained the mass-related separation of feeding niches shown by sympatric ungulate species feeding on short swards. This was demonstrated by a model of grazing processes which describes how dental morphology and vegetation structure interact to determine the weight and composition of plant material which can be removed in a single bite. Comparing animals taking 30,000 bites in a day (close to the maximum), they show how body size and variation in vegetation characteristics (such as height, biomass concentration and vertical distribution of live and dead plant parts) affect the functional response (daily energy intake rate relative to requirements). In essence, short swards impose greater constraints on the bite depth of large species, relative to the metabolic requirements, despite their having broader incisor arcades and hence larger bite areas. The result is that, on short swards, larger species are at a disadvantage *vis-à-vis* smaller ones when both graze the same areas. This suggests that there is potential for competitive exclusion of the larger species when defoliation intensity is sufficient to deplete jointly used resources.

In their far-reaching application of Kleiber's rule to wild ungulate communities, Richard Bell (1969, 1970) and Peter Jarman (1968, 1974) note that the rate of minimal metabolism of mammals tends to increase according to body mass (M) as $M^{0.75}$. Bell (1969, 1970) and Jarman (1968, 1974) infer that small ungulates with a high metabolic rate per unit of body mass will be more frequently limited by the energy content of forage, and that large ungulates with a high overall metabolic rate will be more frequently limited by the biomass of forage. In other words, small ungulates will be more limited by food quality and large ungulates by food quantity.

Demment and Van Soest (1985) and Illius and Gordon (1991, 1992) analysed the digestive constraint on forage intake, which is a central component of the Bell–Jarman principle. Both studies concluded that larger species have the advantage of larger digesta load relative to metabolic requirements, longer retention time and more extensive digestion of forages. Larger species can therefore subsist on poorer-quality food. Illius and Gordon (1991) predicted that maximum daily energy intake should scale with $M^{0.88}$ (i.e. greater than $M^{0.75}$), showing the advantage of large size for utilization of diets with slow digestion rates. It also suggests a strong selection pressure, during seasons when forage is abundant but of poor quality, for the evolution of large body size, counterbalanced by the pressure for the evolution of small size during seasons when forage is scarce but of high quality.

Do feeding adaptations limit the spectrum of habitats that can be utilized by each ungulate?

Bell and Jarman's principle can be applied to the investigation of boundaries to the distribution of grazing ungulates. To what extent do grassland habitats of the Serengeti limit the energy intake of different-sized resident ungulates? It is well established that sward characteristics play a dominant role in determining the herbage intake of grazing ruminants (Stobbs, 1973; Hodgson, 1982, 1985). The main constraints on intake arise from two distinct characteristics of the sward, namely its quality and quantity, which are often negatively correlated. On high biomass swards, the low digestibility of the herbage may impose constraints on daily intake (digestive constraints), whereas short, low-biomass swards impose limits on bite mass and hence limit daily intake. Quantification of these relationships by mathematical modelling (Illius and Gordon, 1987, 1991, 1992) allows prediction of the consequences of sward characteristics for daily energy intake and resource partitioning.

The digestibility of forage and the height of the sward in Serengeti habitats were monitored in a series of permanent plots (25 m × 25 m) in the central woodlands between Banagi and Nyaraswiga hills, which were visited at 2-monthly intervals from July 1986 to May 1987. Canopy height was measured within each plot along a transect of 30 step points, using a metre rule with sliding pointer. Samples of forage were collected by clipping all vegetation within five 25 cm × 25 cm quadrats to 2 cm. The samples were sorted, dried and stored for analysis of neutral cellulase digestibility (NCD) of the green leaf and stem fractions (Murray, 1991). Data from three of these plots, located at the bottom, middle and top of the catena, provided a broad range of sward structures for this analysis.

Using these estimates of sward characteristics as inputs to the model of Illius and Gordon (1991, 1992), the maximum daily energy intake, assuming that digestive constraints apply, was predicted for animals of the size of adult female Thomson's gazelle (16 kg), topi (109 kg) and buffalo (447 kg). The low-biomass constraint was investigated for the same sizes of animal, using the model of Illius and Gordon (1987).

Food quality

Seasonal variation in digestibility of grass leaf and stem from the central woodlands is illustrated in Fig. 9.2a. Stem is less variable and of lower digestibility than leaf, except for samples taken from the top of the catena, where a favoured forage species, *D. macroblephara*, predominates in the sward. The quality of green leaf is lower towards the bottom of the catena, and it declines during the wet season (November to May), when these pastures are only lightly grazed. In the dry season the quality of green leaf depends on grazing pressure, being high in September at the top of the catena, where heavy grazing maintains the grass in an immature growth form. Peak levels of leaf digestibility in all plots coincide with a green flush brought on by the first rains in November.

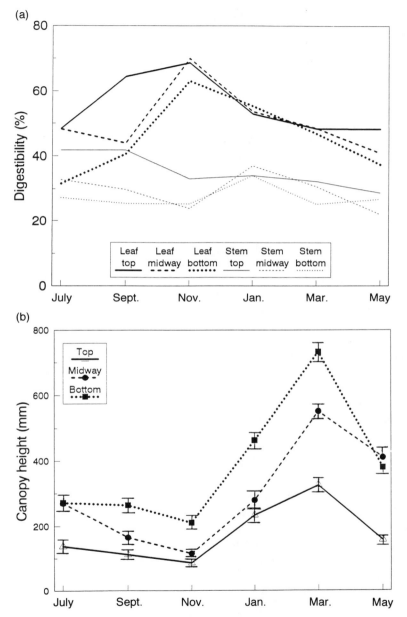

Fig. 9.2. Seasonal variation in (a) quality and (b) quantity of grass forage at different heights of the catena in the central woodlands of Serengeti.

Predicted energy intake ranges from 1.25 to 1.5 of daily maintenance requirements on the best swards (Fig. 9.3a). The energy budget of free-ranging topi utilizing the best forage species in the central woodlands was estimated by Murray (1991) to barely meet energy requirements. Free-ranging animals probably select a fraction of green leaf that is more digestible than that collected by clipping; nevertheless the low estimates of intake emphasize that the energy balance of resident ungulates foraging in the Serengeti woodlands is precarious. The figure also shows the separation of feasible diets with body mass: larger animals can tolerate poorer-quality food than small animals when high-quality food items become scarce.

With unlimited availability of green leaf, the energy intake of buffalo, topi and Thomson's gazelle would be above their respective maintenance requirements for approximately half the year only, if they were feeding at the top of the catena. At the middle and bottom of the catena, buffalo would maintain positive energy retention for approximately 4 months but the intake of gazelle would fall below maintenance requirements at all times except for a few weeks following the first rains, when a green flush is available to grazers (Fig. 9.3b, c). Remembering that no allowance has been made for energy requirements of growth and reproduction, it is not surprising to find that gazelles are nomadic grazers in Serengeti, continuously seeking out areas of new growth. It is concluded that pastures in the central woodlands are marginal as year-round habitats for resident ungulates, and that their suitability is closely dependent on production of new growth in the dry season by intensive grazing and/or fire.

The smaller-sized ungulates will have added difficulty in surviving the energy deficit of the late wet season and early dry season because of their lower fat reserves relative to metabolic weight. They have several options for improving the quality of their diet: they can maintain an energy-rich grazing lawn by intensive grazing in a restricted locality; they can expand their diet to include dicotyledons, many of which have highly digestible shoots and leaves; or they can migrate to other pastures. All three options are practised by Thomson's gazelle in Serengeti (Jarman and Sinclair, 1979).

A seasonal scarcity of energy-rich forage in mature and senescent grass swards is only part of the nutritional challenge facing grazing ungulates on marginal rangelands. In drier regions it is not uncommon for the rains to be delayed by several months or even to fail altogether. The pulse of high quality pasture at the onset of the wet season is then lost, and there is no moisture to sustain new growth following the passage of a fire or intense grazing. Such conditions will severely tax ungulates of all sizes, but particularly those of smaller size. Thus a number of factors, including poor substrate quality, prolonged growing conditions and short or sporadic wet seasons, conspire to restrict the habitats suitable for the smaller grazers. Consequently we agree with the broad generalization of Du Toit and Owen-Smith (1989) that smaller grazers have lower biomass densities and more disjunct distributions as a result of more specialized dietary requirements.

Food quantity

The height of grassland swards in Serengeti is inversely related to their digestibility (Fig. 9.2b), as would be expected from the higher ratio of cell wall to cell contents in tall grasses (Bell, 1971). The canopy is at its tallest in the middle of the wet season, and it increases in height towards the bottom of the catena, where moisture in the soil is retained for longer. During the dry season, the swards continue to decrease in height, but only slowly. Fire did not occur within the area during the study period, so the decline was largely the consequence of grazing and trampling. The lowest average height recorded was 86 mm in the plot near the top of the catena in November. Limitations on intake in ruminants of over 500 kg can be expected in sparse swards below 60 mm height (Illius and Gordon, 1987); thus it is concluded that there is usually sufficient food at all times of the year in the central woodlands for female buffalo or any of the smaller-sized ruminants.

These results suggest that forage biomass has a minor influence, at most, on the distribution of grazing ungulates in the central woodlands. But there is one part of the Serengeti where food biomass might constrain energy intake, even in the height of the wet season. The average wet-season canopy height of the short-grass plains in the southeast of the ecosystem is only 20 mm, with a maximum among a number of unfenced plots of 50 mm; grazed vegetation is reported as never taller than 30 mm (McNaughton, 1984). In order to model energy intake of ungulates on the plains, we assumed that ruminants encountered dense swards with an average height of 40 mm and that they could bite down to a depth of 20 mm. Two values of organic matter digestibility (OMD) (56% and 65%) gave either end of the range of digestibilities for green leaf, and 52% was taken for the digestibility of other material, and it was assumed that the sward contained 40% green live leaf and 60% dead (Murray, 1995). The predicted energy intakes determined in this way for a range of ruminants are plotted in Fig. 9.4. It can be seen that larger species, such as buffalo, are severely constrained by the height of grass swards found on the short-grass plains.

Thus both forage quality and quantity can constrain energy intake of ungulates in Serengeti: forage quality may constrain energy intake of smaller ruminants on medium or tall grasslands, particularly where new growth is inhibited by drought, absence of fire or light grazing pressure; forage quantity may constrain energy intake of species which are not adapted to foraging from heavily grazed lawns.

Does competition modify habitat structure and reduce the overall range of resources?

In the previous section, low forage biomass on grazing lawns is shown to limit the energy intake of larger ungulates. The possibility arises, therefore, that these same

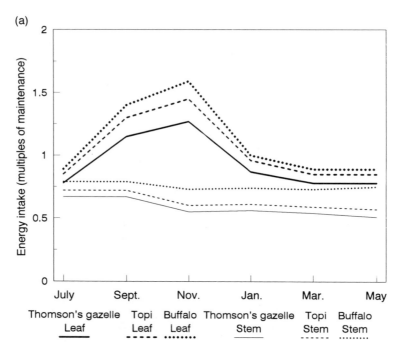

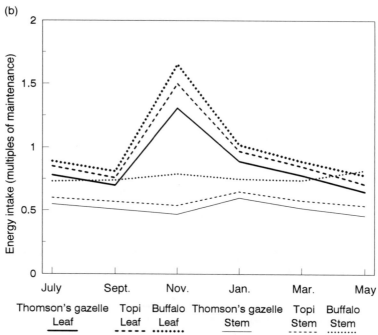

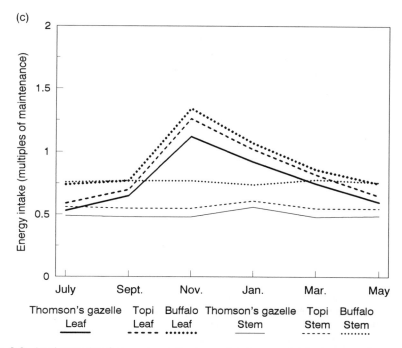

Fig. 9.3. (*and opposite*) Seasonal variation in predicted energy intake (expressed as multiples of daily maintenance requirements) of ungulates grazing at (a) the top of the catena, (b) the middle of the catena and (c) the bottom of the catena.

swards may be important loci for interspecific resource competition. Two possible mechanisms of resource competition were investigated in feeding trials with tame wildebeest and topi in the central woodlands of Serengeti (M.G. Murray, unpublished). Dry-matter intake and leaf selectivity were measured in animals feeding within the same plots, containing a range of canopy heights and biomasses. It was found that sward height declined as a direct result of grazing and that wildebeest could maintain positive energy retention on shorter swards than topi (Murray and Brown, 1993). In fact, zero energy retention was achieved on vegetative swards of height 30 mm for topi but only 20 mm for wildebeest. Somewhat surprisingly, it was found that the lower intake of topi when feeding on very short swards derived from a slower bite rate rather than a smaller bite size. The net effect is the same however: if unconstrained, topi will tend to move elsewhere, leaving wildebeest alone on the grazing lawn. Thus wildebeest, with their relatively broad mouths, are capable of controlling access to the above-ground primary production of grazing lawns.

It is not surprising to find small grazers, such as oribi and Thomson's gazelle, utilizing short grass, but it might be expected that hippopotamus, with its enormous bulk, would avoid it. Yet its broad lips enable it to crop the sward very

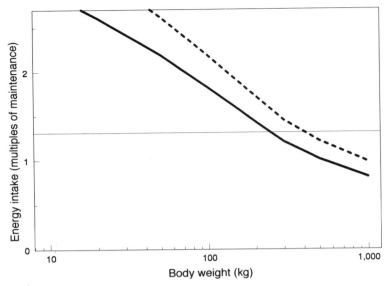

Fig. 9.4. Predicted energy intake in relation to body size of ungulates grazing on the short-grass plains in Serengeti, estimated from predicted bite mass and constrained to 30,000 bites day^{-1}. Dashed and solid lines indicate 65% and 56% OMD for green leaf respectively. The minimum energy intake to achieve a steady state under field conditions is indicated.

closely, creating the so-called 'hippo lawns', which can be too short for effective grazing by other species. Interspecific competition was revealed by the management cull of hippopotamus in Queen Elizabeth National Park, Uganda, during 1957–1958, after which buffalo and waterbuck numbers rose sharply on the pastures formerly used by hippopotamus (Eltringham, 1974). We conclude that the shape of the mouth can override the effects of body size in determining which species 'captures' the production of a grass sward by excluding others.

Narrow-mouthed ungulates may compensate for their lower intake rates by increased selectivity of high-quality components in the sward (Jarman, 1974; Hanley, 1982; Owen-Smith, 1982; Gordon and Illius, 1988; Hodgson *et al.*, 1991; Murray and Brown, 1993). In feeding trials in Serengeti, topi gathered 5–20% more green leaf from a range of mature swards than did wildebeest. Thus their narrower mouths adapt grazers such as topi, hartebeest and waterbuck to the medium-length or tall grasslands that escape fire and close grazing by wider-mouthed ungulates. The population densities of these grazers in Serengeti are much smaller than that of wildebeest, suggesting that competitive interactions may be one-way, with wildebeest (or other wide-mouthed species) dominating whatever pastures they graze, leaving narrow-mouthed species to scavenge on what is left (D. Kreulen, personal communication). However, M.G. Murray (unpublished) has found that grazing by narrow-mouthed species may leave swards in a condition that cannot be utilized profitably by wider-mouthed species. In feeding trials it was found that grazing reduced the proportion of green leaf in mature swards along with canopy height. Topi could achieve a positive energy retention provided a minimum of 12% of green leaf remained in the sward, but wildebeest grazing on this same sward would have obtained an energy intake that was 5% below maintenance. Thus the possibility arises that narrow-mouthed ungulates can exclude wider-mouthed species from grass swards by selectively removing green leaf and that competition is a two-way process.

Although the mechanisms of resource competition among grazing ungulates are now much clearer, the frequency and extent to which interspecific competition affects community structure remains uncertain. It has been argued that direct competition through food depletion has little influence on the structure of African ungulate communities in comparison with facilitatory effects on food quality, predation, disease and other historical processes (Owen-Smith, 1989; Dublin et al., 1990a). Yet our studies indicate effective exclusions of large grazers over substantial areas of the Serengeti–Mara ecosystem. One reason for this uncertainty is the difficulty in recognizing when a community is structured by competition. The difference between competitive and non-competitive distributions can be subtle, as we attempt to show by reference to the sequence of ungulates associated with grazing rotations (see Fig. 9.6 and accompanying text below).

Competition between grazing ungulates could decrease the availability of usable habitats to both generalists and selective grazers through a process of 'sward capture', which involves modification of the structure of grass swards and exclusion of ungulates lacking appropriate feeding adaptations (Gordon and Illius, 1989; M.G. Murray, unpublished). In the long term, species coexistence in African savannas may therefore depend on spatial heterogeneity in grass swards being maintained by the influence of external factors, such as variation in soil–water relations and in timing of the growing season.

In classical models of resource competition in animal communities, niche separation depends on the use of different resources or of different locations along a single resource axis (MacArthur, 1972). In a development of this theory, particularly for application to plant communities (Tilman, 1982, 1988) but subsequently also to animal communities (Abramsky and Rosenzweig, 1984; Rothhaupt, 1988), it was predicted that two or more species could share the same site by specializing on different ratios of availability of just two limiting resources. This 'resource-ratio hypothesis' has attracted much attention, particularly for its prediction that maximum diversity will occur where resources are most scarce.

The resource-ratio hypothesis hinges on the stated assumption that organisms consume more of the resource that most limits them (Tilman, 1982). Thus a plant (species A) may have evolved a specialized metabolism enabling it to use some scarce mineral (x) more efficiently. This plant can now survive in a deficient site by using very small amounts of that mineral. The same argument could apply to a second plant (species B) and a second scarce mineral (y). In one site which was deficient in both minerals, species A could survive by using more of mineral y, which is the resource that most limits it; similarly species B could survive by using more of mineral x. Thus, coexistence of the two plant species is stable at this site (Tilman, 1982).

In considering resource competition among ungulates and other herbivores, there are immediate difficulties over this critical assumption. Niche separation does not arise because one species can utilize essential resources more or less efficiently than another species, but because of different susceptibilities among herbivores to plant defences and other factors that constrain their use of a single resource – plant tissue. Consequently, herbivores will consume more of the resource in a form that least limits them. Thus, in Serengeti, the grazing community in any given location is not stable because of the possibility of sward capture by a single species and because different species could capture the same site in different years. Coexistence is possible because no single species captures the sward in all locations.

Are ungulates excluded from large areas of rangeland by interspecific resource competition?

Interspecific competition could affect ungulate distributions on a much wider geographical scale than the local catena. In order to investigate whether competition had influenced the species composition of the ungulate community in Serengeti, information was gathered on size and mouth shape of all the grazing species recorded in the park (Table 9.1), enabling average incisor breadth to be plotted against average body mass (Fig. 9.5). The most striking result is the proliferation of species between 100 kg and 300 kg body mass, all with similar breadth of incisor row. Some variation is lost in this data set, which gives measurements taken at the base of the row, rather than at the cutting surface. Even so, there is no suggestion that similarity in body size or mouth breadth has been limited by competition between neighbouring species. Nor do differences in digestive system between ruminants and non-ruminants account for the clustering of species, as removal of the latter fails to space out the remaining species.

Gaps in the assemblage of grazers appear at the upper end, between roan antelope and buffalo, between buffalo and hippopotamus, and between hippopotamus and elephant. It is possible that species which could fill these gaps formerly occurred in the ecosystem. For instance, another species of buffalo (*Pelorovis antiquus*) dominated open savannas of East Africa until quite recently (Kingdon, 1982), and a late stage of this lineage, which was much larger than the present-day buffalo, appears in several North African rock paintings (Sinclair, 1977). Larger again is the white rhinoceros (*Ceratotherium simum*), which today has a precarious northern foothold in the region of the southern Nile, although cave paintings in Tanzania and elsewhere suggest a wider distribution until very recent times (Kingdon, 1979).

Table 9.1. Size and morphology of grazing ungulates, including elephant, in the Serengeti–Mara ecosystem (data from Sachs, 1967; Kingdon, 1982; Skinner and Smithers, 1990; I.J. Gordon, personal communication; R.M. Laws, personal communication).

Species code			Abundance (animals km^{-2})	Sex	Mass (kg)	Incisor breadth (mm)
1	Oribi	Ourebia ourebi	R	M	14	18
1	Oribi	Ourebia ourebi	R	F	14.2	17
2	Thomson's gazelle	Gazella thomsoni	16	F	16	16
2	Thomson's gazelle	Gazella thomsoni	16	M	20	18
3	Mountain reedbuck	Redunca fulvorufula	R	F	29	24
3	Mountain reedbuck	Redunca fulvorufula	R	M	30	25
4	Bohor reedbuck	Redunca redunca	N	F	45	
5	Warthog	Phacochoerus aethiopicus	0.2	F	53	
4	Bohor reedbuck	Redunca redunca	N	M	55	
5	Warthog	Phacochoerus aethiopicus	0.2	M	87	
6	Topi	Damaliscus lunatus	2.8	F	109	53
7	Hartebeest	Alcelaphus buselaphus	0.8	F	126	54
6	Topi	Damaliscus lunatus	2.8	M	130	54
7	Hartebeest	Alcelaphus buselaphus	0.8	M	143	58
8	Oryx	Oryx gazella	R	F	162	48
9	Wildebeest	Connochaetes taurinus	61.2	F	163	60
10	Waterbuck	Kobus ellipsiprymnus	0.1	F	175	47
8	Oryx	Oryx gazella	R	M	176	54
9	Wildebeest	Connochaetes taurinus	61.2	M	201	66
11	Zebra	Equus burchelli	9.3	F	219	
10	Waterbuck	Kobus ellipsiprymnus	0.1	M	227	58
11	Zebra	Equus burchelli	9.3	M	248	52.6
12	Roan antelope	Hippotragus equinus	R	F	260	55
12	Roan antelope	Hippotragus equinus	R	M	280	60
13	Buffalo	Syncerus caffer	2.4	F	447	93
13	Buffalo	Syncerus caffer	2.4	M	751	107
14	Hippopotamus	Hippopotamus amphibius	N	M	1374	400[†]
14	Hippopotamus	Hippopotamus amphibius	N	F	1466	430
15	Elephant	Loxodonta africana	R	F	3000	
15	Elephant	Loxodonta africana	R	M	6000	
	Uganda kob[‡]	Adenota kob	E	F	60	32
	Uganda kob	Adenota kob	E	M	90	38

R, rare (< 1000 animals); N, not surveyed; E, locally extinct.
*Measured across base of tooth row (I.J. Gordon, personal communication).
[†]Measured in a straight line across widest part of face (R.M. Laws, personal communication).
[‡]Included for comparison only as this species does not occur in the ecosystem.

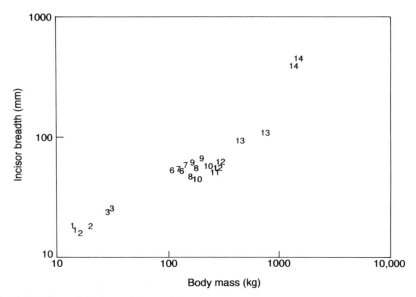

Fig. 9.5. Two niche determinants of the community of grazing ungulates in Serengeti (see Table 9.1).

There is also a possible gap in the Serengeti series at the lower end of the scale between reedbuck and topi. Warthog presently fit within this gap but they are unusual grazers in several respects. The gap would be nicely filled by the Uganda kob (*Kobus kob*), a medium-sized reduncine antelope that was formerly distributed around the northern and eastern shores of Lake Victoria, but which never extended far into the area now designated as Serengeti National Park (Kingdon, 1982). Interestingly, there is very little overlap between the former distributions of wildebeest and kob in East Africa, which is suggestive of competition between these two species, both of which specialize on short green pastures.

The evidence suggests that competition is a factor limiting some grazing ungulates from using the short grass plains in the southeast of the Serengeti–Mara ecosystem. We cannot rule out the possibility that resource competition in the past may have excluded some smaller and larger-sized herbivores from the Serengeti–Mara ecosystem. However, we found no evidence to suggest that resource competition was limiting the use of Serengeti grassland habitats by species in the 100–300 kg size range.

Is the spectrum of usable habitats in the ecosystem enlarged by ungulate grazing?

The grazing rotation in the Rukwa Valley in south western Tanzania provides an example in which grazing pressure apparently increases the overall spectrum of resources available to the community (Vesey-Fitzgerald 1965). By cropping and trampling the tall grasses, larger ungulates increase the range of sward structures, providing room for a greater variety of ungulate species. A similar grazing rotation has been described in the western corridor of Serengeti National Park (Bell, 1970). In both instances, larger-sized species are the first to leave when growth ceases at the start of the dry season, and this may be due to their poorer ability to maintain diet quality in the face of declining green leaf in the sward.

Grazing rotations could also arise from sward capture, as described above, with larger ungulates being excluded from short swards at the end of the wet season because of close grazing by smaller (or broader-mouthed) species. In much the same way, sheep may exclude red deer (*Cervus elaphus*) from high-quality 'greens' in the highlands of Scotland (Clutton-Brock and Harvey, 1983; Gordon and Illius, 1989). If grazing successions are propelled by resource competition, we predict that larger (or more selective) grazers will be absent from the ranges of the smaller (or less selective) species but, that if competition is absent, there will be partial overlap (Fig. 9.6).

Experimental studies on the southeastern grass plains of Serengeti reveal that grazing of senescent stands of *Themeda–Pennisetum* grassland by wildebeest stimulates regrowth, which subsequently attracts Thomson's gazelle (McNaughton, 1976). Here it is clear that grazing by one species has facilitated grazing by another. Facilitation of mineral nutrition may also be a feature of wet-season concentrations of resident ungulates in the woodland areas of Serengeti (McNaughton, 1988). But while the immediate effects of grazing may be nutritionally beneficial to a second species, the net effect, as measured by facilitation of the energy flux into the second ungulate population, is not easily established.

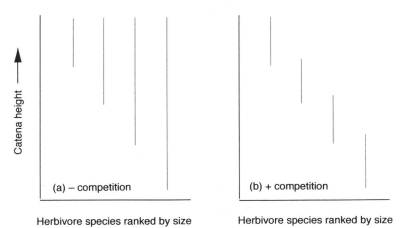

Fig. 9.6. Predicted influence of grazing competition on the overlap in habitat utilization between ungulates of different body size. It is assumed that grass swards become taller and less digestible towards the bottom of the catena.

After recording a doubling in population size of Serengeti wildebeest in the 1970s in association with a decline in Thomson's gazelle, Sinclair and Norton-Griffiths (1982) conclude that competition rather than facilitation was regulating the gazelle population. In a further study, Sinclair (1985) presents evidence that gazelle associate more closely than expected by chance with wildebeest, and argues that they thereby benefit from a reduced risk of predation. But Borner *et al.* (1987) argue that the build-up in wildebeest numbers has generated a build-up in lions (*Panthera leo*) and hyenas (*Crocuta crocuta*), which have then brought about a reduction in the gazelle population, i.e. the explanation for opposing numerical trends lies in 'apparent competition' between the wildebeest and gazelle populations. To make matters even more confusing, Dublin *et al.* (1990a) reanalysed the errors arising in the various censuses, concluding that there may have been no decline in the gazelle population after all.

It is concluded that the spectrum of usable habitats can be increased by grazing pressure, and that this has an observable effect on grazing distributions. But further work is required to establish the relative importance of facilitation and other interspecific interactions in terms of energy flow into ungulate populations. No matter what the outcome of such work, it is unlikely on present evidence that processes which involve ungulate body size, mouth shape and digestive system alone will be sufficient to account for niche separation of species in the size range of 100–300 kg.

Why do ungulates migrate?

Fryxell *et al.* (1988) investigated the benefits to wildebeest of migration, using a simulation model. They found support for the observation that migrants escape the effects of predators, which, because of the need to feed immobile young for long periods of dependence, are unable to move with the migrating herds. Consequently, the predator population, supported by the few resident prey species, is too low to have a substantial impact on migrant prey species. Of course it could be that the predator population is low because wildebeest migrate for other reasons. However, their model also suggests, on the basis of admittedly simplistic assumptions about grass growth, that utilization by migrants of an extensive plains area during the main rainy season cannot account for their great numbers. This is because population regulation occurs during the season of greatest shortage, which was predicted to be the dry season, when migrants returned to the woodland area. The model addresses an interesting question but makes no allowance for increased metabolic requirements in lactating females in the wet season, ignores the possibility of mineral or nitrogen limitations and does not allow for transfers of energy and nutrients between wildebeest ranges in the form of body stores. Additionally, it fails to provide an explanation as to why resident species of ungulate have not evolved migratory habits to avoid predation.

Mineral analyses of forage and wildebeest blood serum indicate that migratory movement on to the short-grass plains at the start of the wet season may be in response to mineral deficiencies, especially in phosphorus (Kreulen, 1975; McNaughton, 1990; Murray, 1995). Return movements to the woodlands at the end of the wet season are necessary if the animals are to find green pasture and water (Pennycuick, 1975; Maddock, 1979). Residential species may be able to avoid mineral shortages by selective foraging and flexible timing of reproduction (Murray, 1995). However, further direct evidence of electrolyte concentrations in wildebeest and resident species is required in order to test the mineral hypothesis.

Does migration allow increased coexistence of ungulates in Serengeti?

As migrants in the Serengeti–Mara ecosystem, wildebeest, zebra and Thomson's gazelle are characterized by their mobility, gregariousness, selectiveness of grazing site and square-shaped mouths. In comparison, residential species such as oribi, reedbuck, hartebeest and waterbuck are characterized by selective feeding at the level of individual bites to a greater extent than site selectivity and by solitary or group living. The contrast is greatest in the medium-sized species, which have relatively narrow mouths for selective feeding in grass swards with erect growth form. The largest residential grazers, hippopotamus and, elsewhere, white rhinoceros, have wide mouths but are adapted to low-quality diets by their large body mass.

Different foraging strategies represent distinctive solutions to the problem of surviving periods of reduced availability of food. Migrants specialize on early growth stages and move seasonally between areas with the highest nutritional quality (Murray and Brown, 1993; Murray, 1995; Albon and Langvatn, 1992). If such places are unpredictable, the foraging becomes nomadic, with herds using environmental cues to seek out areas of green flush. Thus the migrant's strategy maximizes intake of energy and nutrients when conditions on the range are poor. In contrast, sedentary species do not leave their traditional ranges when food is scarce except under unusually harsh conditions. Some cope with seasonally poor pastures by selecting higher-quality components and avoiding lower-quality components in the sward (Murray and Brown, 1993). Other residents may have a lower metabolic rate than migrants (Murray, 1993), perhaps reaching a nadir in the site-bound hippopotamus, whose well-worn trail system, leading from river or lake sanctuaries to grazing lawns, is witness to a sedentary lifestyle.

Female sexual maturity in one resident, the hartebeest, is later than in either wildebeest or topi. In addition, the timing of reproduction in residents may be less responsive to seasonal change and more susceptible to variations in female body condition (Sinclair, 1983). The significance of these differences lies in a lowered daily energy requirement. Thus the resident's strategy minimizes energy and nutrient losses when conditions are poor and intake is marginal.

Migrants and residents can also be distinguished by their predator avoidance behaviour. Migrants frequently graze in open habitats, where they must endure long-distance pursuits by coursing predators (Kruuk and Turner, 1967; Kruuk, 1972; Malcolm and Van Lawick, 1975). Pursuits of resident species usually take place in wooded and broken country, or close to cover (Reich, 1981). In a preliminary study of skeletal muscle metabolism in the two groups, aerobic (succinate dehydrogenase) capacity in topi was $2.6 \times$ that in waterbuck, with wildebeest close to that of topi (Spurway *et al.*, 1996). Endurance would be further enhanced by efficient energy storage and recovery sytems during locomotion (Alexander, 1977).

Taken together, these findings suggest the possibility of a fundamental dichotomy between migrant and resident ungulates, which finds expression in differences of behaviour, anatomy, metabolism and reproduction. Since resource competition is more likely between neighbouring species in this continuum than between species at the extremes, we suggest that this enables similar-sized species in the 100–300 kg bracket to share the same grassland habitats in Serengeti. From the perspective of MacArthur's (1972) conceptual model (see Introduction above), the spectrum of usable habitats in the ecosystem at any one season is enlarged by migration, just as it is by localized movements up and down the catena.

MANAGEMENT OF MULTISPECIES COMMUNITIES

This section examines some of the wider issues in managing multispecies communities of grazing ungulates. The main management goal in any complex natural ecosystem should be to understand the processes that give rise to habitat heterogeneity and to work with them. In grazing systems, both fire and grazing by large animals increase heterogeneity by knocking back the maturation and growth cycles of the grass swards. This creates a mosaic of growth stages, which can be utilized by many species. Fire also leaves a mineral-rich ash, which could provide a useful supplement for resident species on poor pastures, but this benefit has to be balanced against the cost of late burns. Hot fires late in the dry season consume reserve pastures, which may be crucial if rains are late, and also hold back the grassland-to-woodland cycle (Pellew, 1983; Stronach, 1988; Dublin *et al.*, 1990b). Mowing is an alternative management technique that can be used to increase habitat heterogeneity. It is more intensive to apply but has the advantage of being more easily controlled than fire.

Over large areas of Africa, soils are deficient in minerals by livestock standards, making those localities which are rich in minerals of special importance to grazing animals (Underwood, 1981; McDowell, 1985). It is necessary to protect mineral-rich grasslands in order to conserve long-distance migrations by ungulates. Thus an early step in any management plans directed at conserving or expanding a migration should be a survey of mineral concentrations across a wide spectrum of pastures (Murray 1995). The migration routes need to be protected from settlement, agriculture and construction of fences or other barriers. It is also important to ensure that access to water resources is kept free of harassment from livestock, herders or their dogs. Mass mortality and wholesale destruction of migrations have followed where one or more of these basic requirements have been denied (Child, 1972; Borner, 1985; Whyte and Joubert, 1988; Williamson and Mbano, 1988).

More and more populations of resident species are becoming fragmented and isolated by changes in land-use patterns. This leads directly to a reduction in gene flow, particularly where dispersal of young males is prevented (Murray, 1982; Nelson and Mech, 1987). Translocations may ease problems of inbreeding, but care must then be taken to avoid mixing regionally distinctive, historically isolated genotypes.

Containment of populations is also likely to exacerbate problems arising from local overabundance of ungulates (Ferrar, 1983). In many natural grazing systems in Africa, overgrazing and loss of body condition may be prevented either by predation or by management culls and hunting. If large predators are incompatible with neighbouring forms of land use and if less intrusive management practices are preferred, it may be possible to rotate the use of dry-season pastures and rangelands through voluntary movement of herds. Largely untested so far, this technique would involve careful management of water and mineral resources, ensuring that the location of alternative sources was known to animals before closing off access to any one particular source.

REFERENCES

Abramsky, Z. and Rozenzweig, M.L. (1984) Tilman's predicted productivity–diversity relationship shown by desert rodents. *Nature* 309, 150–151.
Albon, S.D. and Langvatn, R. (1992) Plant phenology and the benefits of migration in a temperate ungulate. *Oikos* 65, 502–513.
Alexander, R.McN. (1977) Terrestrial locomotion. In: Alexander, R.McN. and Goldspink, G. (eds) *Mechanics and Energetics of Animal Locomotion*. Chapman and Hall, London, pp. 168–203.
Anderson, G.D. and Talbot, L.M. (1965) Soil factors affecting the distribution of the grassland types and their utilization by wild animals on the Serengeti plains, Tanganyika. *Journal of Ecology* 53, 33–56.
Bell, R.H.V. (1969) The use of the herb layer by grazing ungulates in the Serengeti National Park, Tanzania. PhD thesis, University of Manchester.
Bell, R.H.V. (1970) The use of the herb layer by grazing ungulates in the Serengeti. In: Watson, A. (ed.) *Animal Populations in Relation to their Food Supply*. Blackwell Scientific, Oxford, pp. 111–123.
Bell, R.H.V. (1971) A grazing ecosystem in the Serengeti. *Scientific American* 224, 86–93.

Borner, M. (1985) The isolation of Tarangire National Park. *Oryx* 19, 91–96.

Borner, M., Fitzgibbon, C.D., Borner, Mo., Caro, T.M., Lindsay, W.K., Collins, D.A. and Holt, M.E. (1987) The decline of the Serengeti Thomson's gazelle population. *Oecologia* 73, 32–40.

Campbell, K.L.I. (1989) *Serengeti Ecological Monitoring Programme, Programme Report*. Serengeti Wildlife Research Centre, PO Box 3134, Arusha, Tanzania.

Child, G. (1972) Observations on a wildebeest die-off in Botswana. *Arnoldia* 31, 1–13.

Clutton-Brock, T.H. and Harvey, P.H. (1983) The functional significance of variation in body size among mammals. *Special Publications of the American Society of Mammalogists* 7, 632–663.

Demment, M.W. and Van Soest, P.J. (1985) A nutritional explanation for body-size patterns of ruminant and non-ruminant herbivores. *American Naturalist* 125, 641–672.

Dublin, H.T., Sinclair, A.R.E., Boutin, S., Anderson, E., Jago, M. and Arcese, P. (1990a) Does competition regulate ungulate populations? Further evidence from Serengeti, Tanzania. *Oecologia* 82, 283–288.

Dublin, H.T., Sinclair, A.R.E. and McGlade, J. (1990b) Elephants and fire as causes of multiple stable states in the Serengeti–Mara woodlands. *Journal of Animal Ecology* 59, 1147–1164.

Duncan, P. (1975) Topi and their food supply. PhD thesis, University of Nairobi, Kenya.

Du Toit, J.T. and Owen-Smith, N. (1989) Body size, population metabolism, and habitat specialisation among large African herbivores. *American Naturalist* 133, 736–740.

Eltringham, S.K. (1974) Changes in the large mammal community of Mweya Peninsula, Rwenzori National Park, Uganda, following removal of hippopotamus. *Journal of Applied Ecology* 11, 855–865.

Ferrar, A.A. (ed.) (1983) *Guidelines for the Management of Large Mammals in African Conservation Areas*. South African National Scientific Programmes No. 69, Council for Scientific and Industrial Research, Pretoria.

Fritz, H. and Duncan, P. (1994) On the carrying capacity for large ungulates of African savanna ecosystems. *Proceedings of the Royal Society, London, Series B* 256, 77–82.

Fryxell, J.M., Greever, J. and Sinclair, A.R.E. (1988) Why are migratory ungulates so abundant? *American Naturalist* 131, 781–789.

Gordon, I.J. (1989) Vegetation community selection by ungulates on the Isle of Rhum. II. Vegetation community selection. *Journal Applied Ecology* 26, 53–64.

Gordon, I.J. and Illius, A.W. (1988) Incisor arcade structure and diet selection in ruminants. *Functional Ecology* 2, 15–22.

Gordon, I.J. and Illius, A.W. (1989) Resource partitioning by ungulates on the Isle of Rhum. *Oecologia* 79, 383–389.

Gwynne, M.D. and Bell, R.H.V. (1968) Selection of vegetation components by grazing ungulates in the Serengeti National Park. *Nature* 220, 390–393.

Hanley, T.S. (1982) The nutritional basis for food selection by ungulates. *Journal of Range Management* 35, 146–151.

Hansen, R.M., Mugambi, M.M., Bauni, S.M. (1975) Diets and trophic ranking of ungulates of the northern Serengeti. *Journal of Wildlife Management* 49, 823–829

Hodgson, J. (1982) Influence of sward characteristics on diet selection and herbage intake by the grazing animal. In: Hacker, J.B. (ed.) *Nutritional Limits to Animal Production from Pastures*. CAB, Farnham Royal, UK, pp. 153–166.

Hodgson, J. (1985) The control of herbage intake in the grazing ruminant. *Proceedings of the Nutrition Society* 44, 339–346.

Hodgson, J., Forbes, T.D.A., Armstrong, R.H., Beattie, M.M. and Hunter, E.A. (1991) Comparative studies of the ingestive behaviour of sheep and cattle grazing indigenous hill plant communities. *Journal of Applied Ecology* 28, 205–227.

Illius, A.W. and Gordon I.J. (1987) The allometry of food intake in grazing ruminants. *Journal of Animal Ecology* 56, 989–999.

Illius, A.W. and Gordon, I.J. (1991) Prediction of intake and digestion in ruminants by a model of rumen kinetics integrating animal size and plant characteristics. *Journal of Agricultural Science* 116, 145–157.

Illius, A.W. and Gordon, I.J. (1992) Modelling the nutritional ecology of ungulate herbivores: evolution of body size and competitive interactions. *Oecologia* 89, 428–434.

Illius, A.W. and Gordon I.J. (1993) Diet selection in mammalian herbivores: constraints and tactics. In: Hughes, R.N. (ed.) *Diet Selection*. Blackwell Scientific Publications, Oxford, pp. 157–181.

Jarman, P.J. (1968) The effect of the creation of lake Kariba upon the terrestrial ecology of the Middle Zambezi Valley, with particular reference to large mammals. PhD thesis, Manchester University, UK.

Jarman, P.J. (1974) The social organization of antelope in relation to their ecology. *Behaviour* 48, 215–266.

Jarman, P.J. and Sinclair, A.R.E. (1979) Feeding strategy and the pattern of resource partitioning in ungulates. In: Sinclair, A.R.E. and Norton-Griffiths, M. (eds) *Serengeti – Dynamics of an Ecosystem*. University of Chicago Press, Chicago, pp. 130–163.

Kingdon, J. (1979) *East African Mammals Vol. 3B, Large Mammals*. Academic Press, London.

Kingdon, J. (1982) *East African Mammals Vols 3C and 3D, Bovids*. Academic Press, London.

Kreulen, D.K. (1975) Wildebeest habitat selection on the Serengeti plains, Tanzania, in relation to calcium and lactation: a preliminary report. *East African Wildlife Journal* 13, 297–304.

Kruuk, H. (1972) *The Spotted Hyena – A Study of Predation and Social Behavior*. University of Chicago Press, Chicago.

Kruuk, H. and Turner, M. (1967) Comparative notes on predation by lion, leopard, cheetah and wild dog in the Serengeti area, East Africa. *Mammalia* 31, 1–27.

Lambert, M.G. and Guerin, H. (1989). Competitive and complementary effects with different species of herbivore in their utilization of pastures. In: *Proceedings of the 16th International Grassland Congress*, Nice, 1989, pp. 1785–1789.

MacArthur, R.H. (1972) *Geographical Ecology*. Harper & Row, New York, USA.

McDowell, L.R. (1985) *Nutrition of Grazing Ruminants in Warm Climates*. Academic Press, New York.

McNaughton, S.J. (1976) Serengeti miratory wildebeest: facilitation of energy flow by grazing. *Science* 191, 92–94

McNaughton, S.J. (1983) Serengeti grassland ecology: the role of composite environmental factors and contingency in community organization. *Ecological Monographs* 53, 291–320.

McNaughton, S.J. (1984) Grazing lawns: animals in herds, plant form, and coevolution. *American Naturalist* 124, 863–886.

McNaughton, S.J. (1985) Ecology of a grazing ecosystem: the Serengeti. *Ecological Monographs* 55, 259–294.

McNaughton, S.J. (1988) Mineral nutrition and spatial concentrations of African ungulates. *Nature* 334, 343–345.
McNaughton, S.J. (1990) Mineral nutrition and seasonal movements of African migratory ungulates. *Nature* 345, 613–615.
Maddock, L. (1979) The 'migration' and grazing succession. In: Sinclair, A.R.E. and Norton-Griffiths, M. (eds) *Serengeti – Dynamics of an Ecosystem.* University of Chicago Press, Chicago, pp. 104–129.
Malcolm, J.R. and Van Lawick, H. (1975) Notes on wild dogs (*Lycaon pictus*) hunting zebras. *Mammalia* 39, 231–240.
Martin, P.S. (1966) Africa and Pleistocene overkill. *Nature* 212, 339–342.
Murray, M.G. (1982) Home range, dispersal and the clan system of impala. *African Journal of Ecology* 20, 253–269.
Murray, M.G. (1991) Maximising energy retention in grazing ruminants. *Journal of Animal Ecology* 60, 1029–1045.
Murray, M.G. (1993) Comparative nutrition of wildebeest, hartebeest and topi in the Serengeti. *African Journal of Ecology* 31, 172–177.
Murray, M.G. (1995) Specific nutrient requirements and migration of wildebeest. In: Sinclair, A.R.E. and Arcese, P. (eds) *Serengeti II: Dynamics, Management and Conservation of an Ecosystem.* University of Chicago Press, Chicago, pp. 231–256.
Murray, M.G. and Brown, D. (1993) Niche separation of grazing ungulates in the Serengeti: an experimental test. *Journal of Animal Ecology* 62, 380–389.
Nelson, M.E. and Mech, L.D. (1987) Demes within a northeastern Minnesota deer population. In: Chepko-Sade, B.D. and Halpin, Z.T. (eds) *Mammalian Dispersal Patterns.* University of Chicago Press, Chicago, pp. 27–40.
Nolan, T. and Connolly, J. (1977) Mixed stocking by sheep and steers – a review. *Herbage Abstracts* 47, 367–374.
Nolan T. and Connolly, J. (1989) Mixed versus mono grazing of steers and sheep. *Animal Production* 48, 519–533.
Nolan T. and Connolly, J. (1992) System approach to soil–plant–animal interactions. In: de Jong, R., Nolan, T. and van Bruchem, J. (eds) *Natural Resource Development and Utilization.* Wageningen Agricultural University, Wageningen, pp. 12–30.
Norton-Griffiths, M. (1979) The influence of grazing, browsing and fire on the vegetation dynamics of the Serengeti. In: Sinclair, A.R.E. and Norton-Griffiths, M. (eds) *Serengeti – Dynamics of an Ecosystem.* University of Chicago Press, Chicago, pp. 310–352.
Norton-Griffiths, M., Herlocker, D. and Pennycuik, L. (1975) The patterns of rainfall in the Serengeti Ecosystem, Tanzania. *East African Wildlife Journal* 13, 347–374.
Owaga, M.L. (1975) The feeding ecology of wildebeest and zebra in Athi-Kaputei plains. *East African Wildlife Journal* 13: 375–383
Owen-Smith, N. (1982) Factors influencing the consumption of plant products by large herbivores. In: Huntley, B.J. and Walker, B.H. (eds) *Ecological Studies, Vol. 42, Ecology of Tropical Savannas.* Springer-Verlag, Berlin, pp. 359–404.
Owen-Smith, N. (1985) Niche separation among African ungulates. In: Vrba, E.S. (ed.) *Species and Speciation. Transvaal Museum Monograph No. 4*, Transvaal Museum, Pretoria, pp. 167–171.
Owen-Smith, N. (1989) Morphological factors and their consequences for resource partitioning among African savanna ungulates: a simulation modelling approach. In: Morris, D.W., Abramsky, Z., Fox, B.J. and Willig, M.R. (eds) *Patterns in the*

Structure of Mammalian Communities. Texas Tech University Press, Lubbock, Texas, pp. 155–165.

Owen-Smith, N. and Cumming, D.H.M. (1993) Comparative foraging strategies of grazing ungulates in African savanna grasslands. *Proceedings of the 17th International Grassland Congress*, 691–697.

Pellew, R.A. (1983) The impacts of elephant, giraffe and fire upon the *Acacia tortilis* woodlands of the Serengeti. *African Journal of Ecology* 21, 41–74.

Pennycuick, L. (1975) Movements of the migrating wildebeest population in the Serengeti area between 1960 and 1973. *East African Wildlife Journal* 13, 65–87.

Reich, A. (1981) The behavior and ecology of the African wild dog (*Lycaon pictus*) in the Kruger National Park. PhD thesis, Yale University, New Haven.

Rothhaupt, K.O. (1988) Mechanistic resource competition theory applied to laboratory experiments with zooplankton. *Nature* 333, 660–662.

Sachs, R. (1967) Liveweights and body measurements of Serengeti game animals. *East African Wildlife Journal* 5, 24–36.

Sinclair, A.R.E. (1977) *The African Buffalo.* University of Chicago Press, Chicago.

Sinclair, A.R.E. (1983) The adaptations of African ungulates and their effects on community function. In: Bourlière, F. (ed.) *Tropical Savannas. Ecosystems of the World 13*, Elsevier, New York, pp. 401–426.

Sinclair, A.R.E. (1985) Does interspecific competition or predation shape the African ungulate community? *Journal of Animal Ecology* 54, 899–918.

Sinclair, A.R.E. and Arcese, P. (1995) *Serengeti II: Dynamics, Management and Conservation of an Ecosystem.* University of Chicago Press, Chicago.

Sinclair, A.R.E. and Norton-Griffiths, M. (1979) *Serengeti – Dynamics of an Ecosystem.* University of Chicago Press, Chicago.

Sinclair, A.R.E. and Norton-Griffiths, M. (1982) Does competition or facilitation regulate migrant ungulate populations in the Serengeti? A test of hypotheses. *Oecologia* 53, 364–369.

Skinner, J.D. and Smithers, R.H.N. (1990) *The Mammals of the Southern African Region.* University of Pretoria, Pretoria.

Spurway, N.C., Murray, M.G., Gilmour, W.H. and Montgomery, I. (1996) Quantitative skeletal muscle histochemistry of four East African ruminants. *Journal of Anatomy* 188, 455–472.

Stobbs, T.H. (1973) The effect of plant structure on the intake of tropical pastures. I. Variation in the bite size of grazing cattle. *Australian Journal of Agricultural Research* 24, 809–819.

Stronach, N. (1988) *Serengeti Fire Ecology Project 1985–1987.* Tanzania National Parks, Arusha, Tanzania.

Tilman, D. (1982) *Resource Competition and Community Structure.* Princetown University Press, Princetown, New Jersey.

Tilman, D. (1988) *Dynamics and Structure of Plant Communication.* Princetown University Press, Princetown, New Jersey.

Underwood, E.J. (1981) *The Mineral Nutrition of Livestock, 2nd edn.* Commonwealth Agricultural Bureaux, Farnham Royal, England.

Van Soest, P.J. (1982) *Nutritional Ecology of the Ruminant.* O&B Books, Corvallis, Oregon.

Vesey-FitzGerald, D.F. (1965) The utilisation of natural pastures by wild animals in the Rukwa Valley, Tanganyika. *East African Wildlife Journal* 3, 38–48.

Vesey-FitzGerald, D.F. (1973) *East African Grasslands*. East African Publishing House, Nairobi.
Walther, F. (1978) Behavioral observations on oryx antelope (*Oryx beisa*) invading Serengeti National Park, Tanzania. *Journal of Mammalogy* 59, 243–260.
Whyte, I.J. and Joubert, S.C.J. (1988) Blue wildebeest population trends in the Kruger National Park and the effects of fencing. *South African Journal of Wildlife Research* 18, 78–87.
Williamson, D.T. and Mbano, B. (1988) Wildebeest mortality during 1983 at Lake Xau, Botswana. *African Journal of Ecology* 26, 341–344.
Wright, I.A. and Connolly, J. (1995) Improved utilization of heterogeneous pastures by mixed species. In Journet, M., Grenet, E., Farce, M.-H., Thériez, M. and Demarquilly, C.D. (eds) *Recent Developments in the Nutrition of Herbivores*. INRA Editions, Paris, pp. 425–436.

Grazing Systems and Their Management

Complexity and Stability in Grazing Systems

10

N.M. Tainton,[1] C.D. Morris[2] and M.B. Hardy[3]

[1]*Department of Grassland Science, University of Natal, Private Bag X01, Scottsville, Pietermaritzburg 3209, South Africa;* [2]*Range and Forage Institute, Private Bag X01, Scottsville, Pietermaritzburg 3209, South Africa;* [3]*Cedara Agricultural Development Institute, Private Bag X9059, Scottsville, Pietermaritzburg 3200, South Africa*

COMPLEXITY IN ECOSYSTEMS

Ecosystems are complex because they contain a number of entities which interact with one another in a number of different ways at different levels of organization. The degree of complexity depends not only on the number of entities in the system (i.e. its diversity) or on the way in which these entities differ from each other and within themselves (heterogeneity), but also on how the system is organized and the degree of interrelationship between its components (Allen and Starr, 1982).

Biological diversity is often equated with heterogeneity and complexity. However, descriptions and measures of ecosystem diversity usually have a strong species and population basis (e.g. IUCN, 1988). Such descriptions, however, do not adequately account for variety in the functional and environmental characteristics of ecological systems and they largely ignore the contribution to complexity of other levels of organization, such as trophic levels. Functional complexity, in terms of energy and nutrient flow patterns, also needs to be considered in the description and analysis of ecosystem complexity.

Complexity in an ecosystem arises through interactions (connections) between its components, and even relatively 'simple' systems with few entities may exhibit complex behaviour as a result of such interactions (Kolasa and Rollo, 1991). Connections within the ecosystem (such as predation, competition and facilitation) may be symmetrical or asymmetrical and either strong or weak. For example, where resource levels (e.g. rainfall) remain fairly constant so that plant densities are relatively high, symmetrical competition between plants for available resources is important (McIvor, 1993). In humid environments, plants and herbivores are also strongly connected because herbivores have a strong effect on plant populations. In contrast, in arid and semiarid areas there is a weak interaction between plants as well as between grazers and their food source (Ellis and Swift, 1988; Mentis *et al.*, 1989). Here abiotic factors, such as rainfall and soil fertility, play a large role in dictating vegetation and herbivore dynamics because of the extremes which they exhibit (Noble, 1986; Scholes and Walker, 1993). The implications of the degree of 'connectivity' within a system for its long-term stability will be discussed later in this chapter.

Complexity is manifest at a number of levels or scales in ecosystems. The comprehension of complexity is, however, very much user- and scale-dependent. Observers who seek to measure and describe the complexity of a particular system will, of necessity, themselves define the scope of the system under study and the entities in the system that have relevance to their particular aims (Allen and Hoekstra, 1991). The degree of heterogeneity and complexity detected will then be a function of the observer's own perception and the degree of complexity perceived will be strictly scale-dependent (Greig-Smith, 1983). So a system may be seen to be homogeneous at one scale and heterogeneous at another. In a similar way, temporal changes in the system may be detected only at a particular spatial scale. Organisms within the ecosystem will also vary in their response to heterogeneity at various scales. What is noise to one organism at a certain scale may be pattern, containing useful information, to another (Kolasa and Rollo, 1991). Furthermore, a particular organism (a grazing animal for example) may have the ability to respond to heterogeneity in its food source at a particular scale and actively select for forage at that scale, but its ability to respond to heterogeneity at some other scale may be limited (Montague and Laca, 1993).

It has been argued that heterogeneity, as measured by an observer (i.e. measured complexity as opposed to functional complexity, the latter of which is a property of the system itself), may have limited value for the analysis of ecosystem complexity. This is so because of the restricted ability of the observer to view complexity across a number of scales and the often arbitrary nature of the entities (e.g. species) and levels of organization defined in the system (Bond, 1989; Allen and Hoekstra, 1991). However, such constructs are necessary for comprehension of a complex system (Simon, 1962).

A hierarchical model of ecosystem organization, based on discontinuities of rate processes within the system (O'Neill *et al.*, 1986), has been demonstrated to be useful for viewing ecological complexity (Allen and Wileyto, 1983; Kolasa and Pickett, 1989). Such a model allows for complexity at a particular scale to be isolated from variation at other scales. Meaningless noise from lower levels in the hierarchy can be filtered out and variation at higher levels of organization can be regarded as constant background. For example, when examining the impact of grazing on the floristic composition of vegetation at the landscape scale, daily and seasonal variations in tissue turnover within plants can be ignored, while long-term climatic conditions can conveniently be regarded as being constant.

A further advantage of a hierarchical view of ecosystems is that the multidimensional complexity in the system can be decomposed into a number of levels of organization (i.e. scales) containing relatively few interacting entities. The relationships between these entities and their connectivity with higher and lower levels of organization can then be explicitly modelled.

COMPLEXITY IN GRAZING SYSTEMS

Grazing systems vary in their complexity from that, for example, of a monospecific sown pasture grazed by a single animal species, through to a multispecies rangeland with a variety of animal species with diverse feeding behaviours. In grazing systems, complexity can arise in four major areas.

- Complexity within the food resource, i.e. the plants and the environment complex.
- Complexity within the herbivore population.
- Complexity at the plant/animal interface.
- Complexity in the management strategy.

It is important to appreciate that the manager is an integral part of the grazing system and that his or her actions contribute to complexity in the system (Naveh and Lieberman, 1984).

Plant and environmental complexity

When considered in terms of biotic diversity alone, the complexity of the vegetation utilized by herbivores will reflect the spatial and temporal heterogeneity of the environment at varying scales. This heterogeneity will affect livestock production through its influence on the amount of feed available, the acceptability of that feed and its digestibility and nutrient content, each of which will show varying degrees of spatiotemporal variation.

The floristic composition of vegetation, especially in mixed pastoral systems, is also a primary determinant of forage supply because plant species generally vary widely in their productivity (Hurd and Pond, 1958; Danckwerts *et al.*, 1985), acceptability (Clary and Pearson, 1969; Barnes *et al.*, 1984), digestibility (Van Soest, 1983) and nutrient content (Minson, 1982). A single species may also vary widely in these forage characteristics in different habitats and at different stages of maturity (Kirkman, 1988; Zacharias, 1992). The availability of nutrients also varies seasonally, depending on rainfall, temperature patterns and species mix (Tainton *et al.*, 1989).

Variability in floristics and vegetation structure between regions (gamma diversity) is dictated by historical and environmental factors. These differences, at the scale of vegetation biomes and vegetation types, determine the choice of suitable animal types, animal species mixes and livestock production systems.

The spatial diversity of vegetation across landscapes within a region (beta diversity) describes the heterogeneity that results from interactions between the spatial distribution of past and present environmental conditions (i.e. habitat diversity) and the differential responses of plant species to these conditions (i.e. habitat specificity) (Shmida and Wilson, 1985). Landscape heterogeneity along catenas is due largely to topographic variation in the physical and chemical properties of soils arising during paedogenesis, although microclimatic variation will also play a part. Major disturbances, e.g. flooding, can increase landscape heterogeneity by large-scale redistribution of soil across the landscape (Pickup, 1989).

Alpha diversity within a habitat is a function of the floristic composition of the vegetation. As discussed earlier, the composition of vegetation at a particular position in the landscape will largely determine the availability of forage and an increase in alpha diversity (in terms of species richness) will generally widen the seasonal forage flow, resulting in a lengthening of the grazing season (Tainton *et al.*, 1989). This is particularly true when constituent species differ widely in their seasonal forage production patterns (such as in a mixture of C_4 – warm season – and C_3 – cool season – grasses) (Ode *et al.*, 1980). Diversity of life-form and structure (e.g. grass/tree pattern in savanna areas) will also influence the seasonal forage flow (Scholes and Walker, 1993).

Individual plants within a species may also vary widely in their morphology, phenology and productivity. Such heterogeneity which is common in grass swards (Thorhallsdottir, 1990), has a large influence on the selectivity and intake rate of grazing animals (Illius, 1986) and thus contributes to complexity at the plant/animal interface.

Herbivore population complexity

Complexity in grazing systems associated with the nature of the herbivore population will be dictated by the number, type and combination of herbivores and their degree of interaction. There is a range in complexity from simple single-species production systems, such as commercial beef or sheep operations, to multispecies systems, such as game ranches or wildlife reserves. In the most complex systems a range of grazers and browsers coexist. In the savannas of Africa, as many as 20 different herbivore species may coexist at any one site, although it is common for most of the biomass in the herbivore population to be derived from only three or four species (Cumming, 1982). Invertebrate herbivores (e.g. insects) are also an important component of the herbivore population (Gandar, 1980).

With the exception of the differentiation between grazers and browsers and to a lesser extent between bulk and concentrate grazers, there is a considerable degree of dietary overlap between herbivores, with plant species preferred by a particular species of animal generally thought to be favoured by other species as well (Mentis, 1981). Differences in grazing habits may, however, lead to complementarity, as between cattle and sheep and between the animals which take part in the grazing succession reported from the Serengeti plains in East Africa (Bell, 1971). Such grazing complementarity may increase output per unit area. In such mixed systems, the short-grass (concentrate) grazers benefit from the modification of sward structure brought about by the long-grass (bulk) grazers, so that sheep, for example, generally perform better when grazed in mixed systems with cattle than when grazed alone (Nolan and Connolly, 1977; Hardy and Tainton, 1993). This may be so, however, only when the quality of forage on offer is low, as when large quantities of relatively mature plant material have accumulated. Cattle, particularly when young, may also benefit from the presence of sheep (Nicol and Souza, 1993), but not where only limited amounts of high-quality forage is available (Hardy and Tainton, 1993). In the Serengeti, the smaller ungulates create 'grazing lawns', following the initial grazing of long grass by the bulk grazers (McNaughton, 1984). These lawns are productive sources of high-quality forage. Therefore, herbivores themselves engender complexity in their food source through selective feeding. However, they also do this in other ways, as when areas surrounding dung pats are avoided by individuals of the species depositing the dung, but may be readily grazed by individuals of other herbivore species (Nolan and Connolly, 1977).

In spite of the tendency for different species of herbivores to select the same types of food, the impact of each on the plant community may be quite different because of differences in the way in which they physically defoliate the plant. A notable example of this is the different impact which sheep and cattle will have on a plant community (Humphries, 1987; Hardy *et al.*, 1994). Sheep have been shown, for example, to reduce the proportion of clover (Bedell, 1973) but increase the amount of *Poa trivialis* (Conway *et al.*, 1972) in the sward, whereas a high proportion of cattle results in increased amounts of clover relative to grass (Lambert and Guerin, 1989).

Complexity at the plant/animal interface

Grazing animals are selective feeders. They harvest plant species or plant parts non-randomly and so do not ingest an average of the forage on offer (Hodgson, 1990; Vallentine, 1990). The particular diet selected by a herbivore depends on a complex interaction between characteristics of the animal (e.g. dietary requirements, mouth anatomy and grazing behaviour) and characteristics of its food source (Stuth, 1991). Selection occurs across various scales because animals simultaneously perceive and interact with their feeding environment at different levels of heterogeneity in the food source (Senft *et al.*, 1987). The scales at which selection occurs are therefore not independent. Processes at large scales, e.g. the plant community, are the direct result of processes occurring at smaller scales, i.e. individual species or small patches. Patterns and processes at a fine scale are also affected by the overall characteristics of the feeding environment, such as patchiness of the food source and the relative availability of different forage components (Montague and Laca, 1993).

Although selection may occur at multiple spatial and temporal scales, it is convenient to consider selective feeding and its consequences at a number of clearly defined levels. These are outlined below.

Habitat selection

Patterning at various scales in African savannas, from microscale (alpha diversity) to regional scale (gamma diversity), results in a spatial and seasonal movement of indigenous ungulates and nomadic livestock herds (e.g. Coughenour, 1991; Homewood and Rodgers, 1991). Physical attributes of the habitat, such as the availability of water and shelter and the vegetation structure, define resource quality and are important determinants of the carrying capacity for native ungulates. Also, the spatial arrangement of resource patches within a region influences the size of an animal's home range (Carr and MacDonald, 1986; Fabricius, 1994).

Patch selection (mosaic or spot grazing)

Patch selection may be associated with spatial distribution of different forage species, with patchiness of the resource base (e.g. soil depth or nutrient content) or, most importantly, with the spatial heterogeneity or variation in sward structure resulting from prior selective grazing. The latter is often greatly influenced by urine and dung deposition on the pasture (Ring *et al.*, 1985). In seasonally productive rangelands, the abundance of forage at the commencement of grazing at the start of a new grazing season often has a major impact on the subsequent patchiness of grazing. The greater the amount of forage which accumulates prior to the commencement of grazing, the greater is the tendency for grazed patches to develop (Ring *et al.*, 1985).

Patch grazing promotes increased spatial diversity of species distribution and of the quality of the resource base (by influencing, for example, soil compaction and nutrient status) and, in so doing, promotes further patch selection. In at least some grassland types, such patch selection is seen as an important starting-point for grassland degradation in that patches, once formed, remain areas of animal concentration and extremely intense grazing (Hatch and Tainton, 1990).

Species selection

One of the important advantages in species heterogeneity is that it may broaden the seasonal availability of forage, either because different plant species exhibit different growth patterns or because of temporal differences in the preference which herbivores show for different species (e.g. Hatch and Tainton, 1993). Such complexity and the selective grazing with which it is normally associated may, however, effect long-term floristic change because of the different levels of defoliation stress which are then applied to different species. Typically, such selection would be expected to encourage the less palatable species at the expense of the more preferred species (Morris *et al.*, 1992).

Individual plant and plant-part selection

Animal selection patterns are influenced by the architectural characteristics of the grass sward (Illius, 1986). Variations in, for example, leaf table height, tensile strength and leaf percentage between individual plants of the same species (resulting, for example, from differences in their immediate past grazing history) and between different parts within the plant (e.g. lamina, pseudostem and reproductive culms) contribute to canopy heterogeneity and influence herbage intake patterns. Such variation in overall canopy characteristics may override interspecific differences in their influence on selection, even in multispecies swards (O'Reagain *et al.*, 1993).

An important aspect of individual plant selection is that, even if selection disadvantages only a small proportion of the plants of a particular species relative to their neighbours, such selection may have important long-term consequences for the community through the gradual loss of that species. Unfortunately, such effects are masked where data interpretation and field assessments are based on the mean values applying to a species, so that the likely impacts of any treatment are often not appreciated and nor are the reasons for any changes in the community when they are eventually recognized (Montague and Laca, 1993).

The effects of selective grazing at various scales

Selective grazing results in the uneven distribution of grazing pressure, both within and between plant communities and between and within plant species. While such selection is exhibited at the individual plant level or even at the plant-part level through, for example, its effect on vigour, its effects are also manifest at higher (such as population or community) levels of organization. Its impact does, however, vary in different systems.

THE CONCEPT OF STABILITY AND RESILIENCE IN GRAZING SYSTEMS

Pastures and grasslands are located near the centre of the competition–stress–disturbance (CSD) triangle of Grime (1979). All the elements are important and none is overwhelming (McIvor, 1993), although the relative balance between the three factors is likely to vary among different plant communities. This is shown in attempts to develop an understanding of complexity and stability in grazing systems and this has led to two apparently opposing views on grazing system dynamics. The mainstream, or orthodox, view adheres to the equilibrium theory, which postulates that, once disturbance has occurred in a system (i.e. the interacting components of interest – the vegetation community or pasture, herbivores and environmental impacts), the system state either returns to its former equilibrium or equilibrates within a new 'domain of attraction'. Here, grazing intensity is considered to be the main variable influencing plant–herbivore dynamics. There is, however, a substantial body of evidence which suggests that many rangeland systems follow non-equilibrium dynamics in that a steady state is never achieved. In these systems it would seem that abiotic variables (and notably rainfall distribution, amount and intensity) have an overriding influence on vegetation dynamics and hence also on herbivore populations. Here herbivores are postulated to play a secondary, and usually a relatively insignificant, role in influencing forage production potential.

We believe that both equilibrium and non-equilibrium paradigms are appropriate, given the particular set of biotic and abiotic environmental influences relating to the grazing system concerned. The equilibrium theory would seem to be appropriate where rainfall is relatively consistent and predictable and where the community is comprised totally or largely of perennial plants, as is common in humid regions. Here spatial and temporal variability of fodder is largely determined by management. In arid and semiarid environments, with unpredictable and unreliable rainfall and where the annual component of the community is large, it is likely that variable climatic conditions (and variable rainfall in particular) have a greater influence on forage availability and community dynamics than does management. The characteristics of these two different systems are shown in Table 10.1.

CONNECTIVITY AND STABILITY

The stability of an ecosystem does not depend simply on the diversity of its components (i.e. on the number of species), but rather on the nature and degree of the connections between components of the system. According to the theory of 'connective stability' (Allen and Starr, 1982), systems with an intermediate degree of 'connectivity' are the most stable, whereas both overconnected and underconnected systems may be unstable. A highly connected system may be vulnerable to disturbance because all its components interact closely and hence are subject to the effects of perturbations (May, 1973). In contrast, in systems where there are weak connections between components, there may be insufficient feedback between components to dampen fluctuations induced by perturbations and the system may be inherently unstable. In highly connected ecosystems, the removal of a single component may lead to instability. For example, sown pastures are usually

Table 10.1. Characteristics of equilibrium and non-equilibrium grazing systems.

Characteristics	Equilibrium system	Non-equilibrium system
Environment	Uniform – rainfall high and consistent	Variable – rainfall low and erratic
Floristic structure	Comprised of perennial plants	Comprised largely of annual plants
Forage flow	Relatively constant and predictable	Variable and unpredictable
Driving forces	Grazing and fire, 'management-driven' – level of management input determines response, e.g. stocking rate, fire frequency, etc.	Moisture availability, 'event-driven' – chance and contingency of non-biological (e.g. rainfall) and biological (e.g. grazing) events determine dynamics
Balance between plants and animals	Stable – negative feedback determines equilibrium position	Plant and animal populations fluctuate widely – 'non-equilibrium'
Appropriate models	Succession (range condition), stable isoclines, relatively simple dynamics	State and transition, complex dynamics
Stability	Stable and non-resilient	Unstable, resilient
Management control	Strong	Weak
Management of complexity	Manipulative to reduce heterogeneity 1. Sedentary – camping, rotation and regulation of animal nos 2. Manipulative – aim to maximize stability and uniformity 3. Control selection	Exploitation of heterogeneity 1. Migratory – transhumance to exploit resource heterogeneity 2. Opportunistic and flexible – aim to maximize production while reducing risk 3. Allow selection

highly connected to the 'manager', and the productivity and composition of the pasture will degenerate if fertilizer and water inputs are halted. Increasing the connectivity of an underconnected system may also cause the system to change to a new stable state, different from the former state. Such a change has been reported to occur in semiarid shrublands, where the erection of fences to restrict animal movement leads to a tighter coupling between animals and their food source, eventually resulting in localized degradation of the vegetation and of the soil resources (Hoffman, 1988).

THE EQUILIBRIUM PARADIGM

The basis of the equilibrium theory

That grazing systems have the potential to exhibit stability is implicit in the widely applied concept of carrying capacity for range and cultivated pastures throughout the world. Setting the carrying capacity of a system suggests that there is an equilibrium plant/herbivore space within which a defined level of sustained livestock production may be expected: hence the so-called 'economic' and 'ecological' carrying capacity concepts proposed by Caughley (1979).

The current understanding of grazing ecosystems is largely based on equilibrium theory. In its simplest form, successional pathways of pioneer, through subclimax to climax vegetation and the reversal of the pathways, due to some form of disturbance (e.g. overgrazing), are entrenched in mainstream range science (Dyksterhuis, 1949; Stoddart *et al.*, 1975; Foran *et al.*, 1978). More complex successional pathways are also recognized within which the system state may cross into one or more stable or unstable states, depending on the timing, intensity and type of disturbance and the state of the system at the time of the disturbance (Walker and Noy-Meir, 1982; Friedel, 1991). Stable and unstable equilibria are identified and the system may be defined as being resilient or non-resilient (Holling, 1973).

In this context, then, stability is defined in terms of the rate and extent of change in the system's component parts relative to their equilibrium values (Walker and Noy-Meir, 1982). A stable system is one which, when subjected to outside stress, changes little in composition and production and any change which may occur is slow (Walker, 1980; Mentis and Tainton, 1981; Walker and Noy-Meir, 1982). A resilient system may or may not be stable, and is usually unstable. A resilient community may change considerably in composition and production but will usually return quickly to its equilibrium after the disturbance has been removed. Resilience, therefore, is a measure of the degree to which a system may absorb changes without crossing a threshold into some other equilibrium position or 'domain of attraction'.

The domain of attraction within a system's space defines a system state from which it cannot change without major intervention (Noy-Meir, 1975). Many natural communities have more than one stable position and associated domain of attraction. The state toward which the system will tend depends upon which side of the boundary it finds itself on at the time of disturbance and the magnitude and nature of the disturbance (Walker and Noy-Meir, 1982). A system which has only one combination of its component parts toward which it tends over time is said to be globally stable, i.e. it has a single stable equilibrium position.

Examples of the application of equilibrium theory

Hardy and Mentis (1986) applied the above concept of equilibrium theory to model beef production in a humid, sourveld, grassland in Natal, South Africa. (Sourveld is defined as grassland (veld) in which the forage plants become unacceptable and less nutritious on reaching maturity, thus allowing the grassland to be utilized for only a portion of the year in the absence of licks (Trollope *et al.*, 1990).) The model was based on four main issues emerging from current pastoral theory. First, grazing intensity affects the floristic composition of the range because of the variable tolerance of plant species to grazing and because of the different levels of stress applied by the grazing animals to different plants. Second, animal performance depends partially on the floristic composition of the range because plant species vary in palatability and nutritive value. Third, although heavy grazing will induce the dominance of grazing-tolerant plants, with light or no grazing these plants will be replaced by grazing-intolerant species. Fourth, a single, bidirectional pathway of change in proportional species composition develops in response to the intensity of grazing. The model was developed as follows.

Plant response

Plant response was measured in terms of changes in proportional species composition (S) in relation to grazing intensity (given by the stocking rate (H) of herbivores, measured in AU ha^{-1}, where 1 AU is the equivalent of the metabolic live mass of an average 450 kg steer (Meissner *et al.*, 1983). Given empirical data derived from long-term grazing trials, a zero isocline of S to grazing intensity was developed in S–H space (Fig. 10.1a). The zero isocline, $\Delta S = 0$, is defined as the line joining all points which are stable with respect to grazing intensity.

Animal response

A zero isocline for herbivores ($\Delta H = 0$) was established in relation to S. This zero isocline joins all points in the S–H plane at which the herbivore population is constant (i.e. at maintenance) for given values of S (Fig. 10.1b).

Stability and dynamics

To reveal the stability and dynamics of the plant–herbivore system, the two zero isoclines were superimposed in the manner shown in Fig. 10.2. The shapes and relative positions of the lines might be expected to vary with both vegetation and animal type and with management. Two plausible cases are presented in Fig. 10.2. In Fig. 10.2a, point *b* represents an unstable equilibrium between overgrazed and undergrazed domains of attraction, with stable equilibria at *c* and *a* respectively.

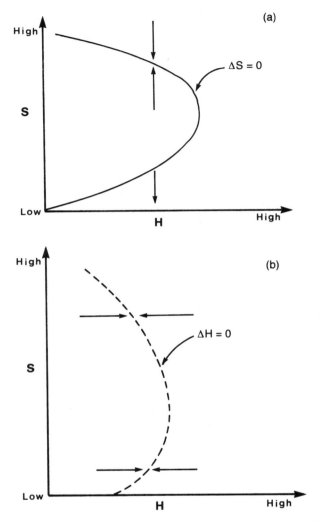

Fig. 10.1. Plant–herbivore relationships (after Hardy and Mentis, 1986). (a) The vegetation compositional zero isocline ($\Delta S = 0$) in the plane of plant species composition (S) and stocking rate (H). (b) The herbivore zero isocline ($\Delta H = 0$) drawn in the S–H plane. Arrows indicate the direction of change.

Given, at the start of grazing, that the system is located in the undergrazed domain, the state will shift to *a*, assuming no intervention. State *a* has a stable S with a high proportion of grazing-intolerant plants, and the herbivores are stabilized at a low H. In contrast, if the initial state is in the overgrazed domain, then without intervention the system will shift to either *a* or *c*. State *c* is associated with a high proportion of grazing-tolerant plants. The herbivore population again stabilizes at

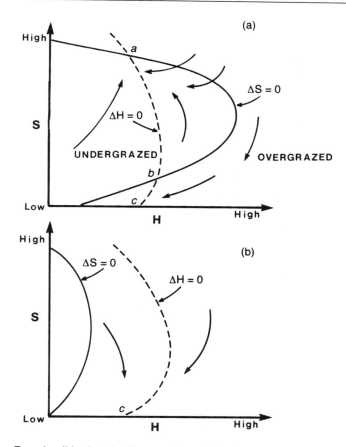

Fig. 10.2. Two plausible plant–herbivore systems. Unbroken line is $\Delta S = 0$. Broken line is $\Delta H = 0$. (a) Metastable with undergrazed and overgrazed domains with stable equilibria at *a* and *c* respectively. (b) Globally stable with only an overgrazed domain at *c*. S, plant species composition; H, herbivore stocking rate.

modest H. With more than one stable state the system is metastable. In Fig. 10.2b, there is only an overgrazed domain with a single stable equilibrium at *c*. The position of this equilibrium depends on the grazing tolerance of the species involved. If H is held low so that the S–H state is to the left of $\Delta S = 0$, the equilibrium shifts to high S. However, without controlling H, equilibrium at *c* (i.e. the overgrazing state) is inevitable. This system is globally stable.

The plant–herbivore model developed by Hardy and Mentis (1986) suggests that grazing systems in humid sourveld grasslands, where seasonal rainfall is reasonably reliable and predictable, are of the form depicted in Fig. 10.2b, i.e. they are globally stable in the overgrazed domain. If commercial graziers were to practise light stocking (low grazing intensity) and remove the annual increase in live mass, in which case the system state would remain to the left of $\Delta S = 0$, in Fig. 10.2a, a high S could be maintained. In subsistence pastoral systems where numbers of livestock, rather than individual animal performance, are the primary concern, H will invariably be high (to the right of $\Delta S = 0$). Subsistence grazing systems are therefore likely to be maintained in the overgrazed domain represented by Fig. 10.2b, with S generally at a low level.

Equilibrium theory therefore appears to be useful in understanding the dynamics of grazing systems in humid perennial rangelands and where environmental variability is low. Such theory also underpins the carrying-capacity norms for cultivated pastures and overseeded rangelands, where interventions such as fertilizer application, pasture rejuvenation, overseeding, mowing, burning and irrigation assist in maintaining a uniformity of inputs, thus sustaining a 'stable' livestock output. This has allowed the development of analytical models which relate the performance of grazing animals to the stocking rate on pasture (e.g. Jones and Sandland, 1974).

The role of competition in equilibrium systems

In equilibrium rangeland and pasture systems, it is assumed that, given that the plants which constitute a mixed sward at any particular site are sufficiently well adapted to the reliable and predictable environment to survive in competition with their neighbours, it is the defoliation management to which the community is subjected which largely determines both the relative abundance of the individual species and their productivity. The defoliation impact is brought about primarily through the influence of defoliation frequency, intensity and timing on the growth and competitiveness of different plants. These influences are magnified by differences in the defoliation treatment applied to different plants when grazing is selective.

Such differences in the defoliation regime to which different plants are subjected may produce one of two possible results. At one end of the scale are those more palatable plants which are so intensively and frequently grazed that they are unable to survive in competition with their less intensively defoliated and therefore faster-growing neighbours (Morris and Tainton, 1993). At the other end of the scale are those plants which are avoided by herbivores. Since the majority of forage plants, and most certainly those which evolved in systems which were exposed to either or both grazing and fire, are poorly adapted to lenient grazing unless fire is used at intervals to remove the accumulated top growth, such plants are also unable to survive.

Therefore, in equilibrium systems which typify areas in which abiotic conditions are such as to provide very favourable conditions for plant growth, it is defoliation management which largely determines the survival of the individual plants, through its influence on the competitive relationships between each plant and its neighbour. Competition is therefore seen as playing a crucial role in these systems.

NON-EQUILIBRIUM SYSTEMS

The basis of the non-equilibrium theory

For the most part, the world's rangelands are located in arid and semiarid environments with highly variable rainfall. Annual plants normally form an important component of the plant community. Here, it is argued, the grazing systems are at disequilibrium, since environmental variability seldom allows the system state to equilibrate (Behnke and Scoones, 1993). Fodder production follows rainfall distribution and amount, and animal production tracks fodder production patterns. When abundant moisture is available the annual plants may produce large quantities of forage. For most of the time, however, only that forage produced by the relatively more stable perennial component is available. Fodder availability therefore varies widely, both seasonally and annually.

The main characteristics of non-equilibrium grazing systems are presented in Table 10.1. The ecosystem/community is spatially extensive, with externalities critical to system dynamics (Ellis and Swift, 1988). Single- and multiyear periods of drought have major influences on forage availability and therefore on herbivore populations. Ellis and Swift (1988) illustrate the effects of such external impacts on the dynamics of a plant–herbivore system in an arid region of northwestern Kenya (Fig. 10.3) subjected to communal pastoralism.

In Fig. 10.3 the x axis (H) relates to the number of herbivores and the y axis (P) relates to forage production. While the axes are similar to those of Fig. 10.2, the understanding of plant–herbivore dynamics is based on different assumptions. In this example, which could also apply to many wildlife systems, the pastoral activities revolve around maintaining livestock numbers rather than around the marketing of livestock and hence maximizing animal mass.

- Offtake is in the form of products such as milk, draught power and dung rather than the physical offtake of animals. Livestock numbers are therefore regulated mainly by fecundity rates and the availability of forage.
- Forage production (P) responds sensitively to the variable rainfall, with an often dramatic response of the vegetation during years of good rainfall.
- Herbivore populations are never high enough during high-rainfall seasons to materially affect the amount of plant biomass (Ellis and Swift, 1988).
- Herbivore population size (i.e. the numbers of animals) remains constant during short, year-long periods of drought (although animals will lose condition) but declines during multiyear (2 years or longer) droughts.
- Multiyear droughts are frequent enough and herd recovery is slow enough to ensure that livestock numbers never approach the theoretical ecological carrying capacity.
- Droughts never last long enough to completely eliminate the herbivore population.

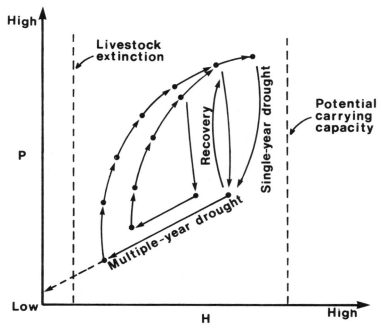

Fig. 10.3. A plant–herbivore system at disequilibrium (from Ellis and Swift, 1988), in this case influenced by frequent single- and multiyear droughts. P, forage production; H, number of herbivores.

Under these conditions, populations of plants and herbivores are unstable over time but are persistent.

According to disequilibrium theory, therefore, the grazing system has no points in plant–herbivore space which could be considered in terms of equilibrium theory. Plant–herbivore dynamics track external (abiotic) variability rather than plant and herbivore density-dependent interactions. This is illustrated in the model shown in Fig. 10.3. The upper right-hand points indicate high herbivore population and high forage availability. In a single-year drought, forage production will decline but animal numbers will remain relatively constant. Individual animals will lose condition but will survive the relatively short period of low forage availability. Given a relatively high rainfall in the next season, the animals will tend to regain lost condition and a moderate increase in population size can be expected. However, should the period of low rainfall persist for a second year (or longer), then livestock numbers will decline.

The above example assumes little or no intervention from outside the system, but not all arid and semiarid rangelands are free of some form of intervention. Not uncommonly, conserved feeds or animals are introduced from outside the system, so that animals may be held at artificially high levels (in relation to available forage) at certain times. This practice, by promoting soil loss, has been shown to lead to a progressive damping of postdrought recovery of perennial species in semiarid grasslands (Hatch, 1994). If allowed to continue, this process may lead to the loss of the perennial (and more stable) component of the community, so that forage production is limited to shallow-rooted ephemerals, and then only for short periods following sufficient rain to promote their germination and growth. Forage availability is therefore sporadic and temporally extremely unstable. Alternatively, such practice may promote the development of stable communities dominated by unpalatable perennial species, which provide little useful forage, so that the community loses its ability to sustain all but a low (but in this instance stable) level of animal production (Roux and Vorster, 1983).

The role of competition in non-equilibrium systems

As reported by McIvor (1993), the relative importance of competition declines as the level of stress increases. By inference, therefore, the influence of herbivore use on interplant competition is likely to have little impact on the plant community in non-equilibrium systems. Such impact is indeed likely to be minimal during periods of abundant moisture availability, when forage production might be expected to outstrip herbivore demand, but we find it difficult to envisage such low levels of competitive interactions between plants during periods of low moisture availability. However, in many, but not all, arid and semiarid systems, palatability differences between species or between plants at different levels of maturity are often small, so that grazing is not normally as selective as it commonly is in humid regions. There are, however, exceptions to this, as, for example, in some semiarid shrublands, where palatability differences between species may be large (Vorster, 1982). It is difficult to believe, therefore, that herbivore grazing does not exert some considerable influence on such communities, although it is true that rainfall is still likely to exert the dominant influence on the plant community and may be expected to mask the influence of grazing on such systems, particularly during extended periods of drought.

Another major potential impact of grazing in arid and semiarid environments is the negative impact on the seed production of annual species and of perennials which are obligate seed reproducers (O'Connor, 1991).

THE MANAGEMENT OF COMPLEXITY

As previously indicated, grazing systems vary widely in their complexity, depending *inter alia* on the diversity of the herbivore population, the heterogeneity of the grazing resource, the complexity of the interactions between the herbivores and their food source and the intricacy of the management system.

To discuss how management strategies in grazing systems vary according to the complexity of the system and to highlight general principles governing the relationship between the management of complexity and system stability, a broad classification of grazing systems is needed. Of relevance here is how complexity, stability and the management of complexity differs between equilibrium and non-equilibrium grazing systems (Ellis and Swift, 1988; Westoby et al., 1989; Behnke and Scoones, 1993) (Table 10.1). Most grazing systems will lie between these two extremes or comprise a combination of the two. For example, in Zimbabwe, during dry years when rainfall is unpredictable, vegetation biomass and cattle numbers fluctuate widely and do not equilibrate at a single 'stable' position. In contrast, during higher than average rainfall years, the cattle population displays the characteristics of an equilibrium system (Scoones, 1993).

Management of equilibrium systems

In equilibrium systems the aim of management, in many cases, is to reduce the heterogeneity of the system in order to facilitate management. In such systems, the perception is that the more diverse the system, the more complex the management required to optimize production and maintain system stability. Management actions aimed at dampening fluctuations in the system are effective because the manager has relatively strong control over the dynamics of the system via manipulation of management inputs, such as fertilizer levels, stocking rates and burning schedules (Table 10.1).

In both rangelands and sown pastures, considerable attention has been paid to grazing systems which either reduce the extent to which animals selectively graze (e.g. non-selective grazing (NSG) (Acocks, 1966)) or reduce the impact of selective grazing on the forage resource (e.g. controlled selective grazing (CSG) (Pienaar, 1968), high-performance grazing (HPG) (Booysen, 1969), short-duration grazing (SDG) (Savory, 1978) and deferred-rotation systems (Stoddart et al., 1975)). Such systems usually involve land subdivision (paddocking) in order to reduce the variability of the forage resource available to the animals at any one time. Burning is also employed to remove accumulated low-quality residual material and induce a uniformly palatable flush of new growth. Regular burning will also reduce the degree of heterogeneity caused by patch grazing (Mott, 1986).

In areas where rainfall is relatively high and reliable, multispecies rangeland is often replaced with monospecific sown pastures to ensure a steady and reliable forage flow. Variability in the seasonality of available forage from rangelands can be reduced by planting pasture species with different seasonal growth patterns or by conserving material produced in one season for feeding at some other season, either as foggage (standing hay) or by harvesting it mechanically for longer-term storage (as hay or silage).

Management of non-equilibrium systems

In non-equilibrium systems, diversity and heterogeneity are an integral part of the system and the manager should not attempt to dampen such complexity. Attempts to reduce such variability often have negative consequences for the stability and sustainability of the system. For example, in regions where forage resources are widely dispersed, both seasonally and spatially, prevention of free migration of herbivores by paddocking may lead to local soil and vegetation degradation (Hoffman and Cowling, 1990). Perturbations, such as fire and herbivory, also play a role in maintaining diversity and resilience in non-equilibrium systems, and the elimination of disturbance may alter landscape pattern and diversity, as well as reducing the ability of the ecosystem to survive similar perturbations in the future (Scholes and Walker, 1993).

In disequilibrium grazing systems, because the grazier has little influence on system dynamics, the appropriate management strategy is to exploit the spatio-temporal variation in forage availability in a flexible, opportunistic manner, rather than attempting to manipulate the system to suit the grazier's requirements. This is the basis of traditional nomadic livestock systems, such as those in East Africa. Nomadic pastoralists employ an adaptive resource-use strategy involving migration of animals in small groups over large areas to 'track' pulses in forage production, manipulation of herd structure and herd composition to meet subsistence requirements, while minimizing risk, and the use of 'fall-back' areas for periods of severe forage scarcity (Ellis and Swift, 1988). In such systems, animals select nutrient-rich patches in the landscape to meet their forage requirements (Scholes and Walker, 1993).

In many regions of Africa today, however, such nomadic pastoralism is no longer possible, due to various sociopolitical changes, and a compromise solution which is gaining credence is to encourage diversity in the herbivore populations on rangelands in order to exploit the vegetation heterogeneity and to broaden the consumptive (culling, hunting) and non-consumptive (tourism) options for wildlife (Owen-Smith and Cumming, 1993). Multispecies game systems and combined cattle/game enterprises are replacing monocultures of domestic livestock (e.g. campfire programme in Zimbabwe (Cumming, 1993)) and, even in domestic livestock systems, there is a strong move towards including goats as browsers in traditional cattle production systems in savanna areas (Aucamp, 1976).

Although non-equilibrium systems are generally unpredictable in their response to management, opportunities to manipulate the system in a certain direction may occasionally occur under a particular set of conditions. Dynamics in disequilibrium systems are characterized by periods of rapid change resulting from the coincidence of various factors (e.g. intense grazing following a drought (Danckwerts and Stuart-Hill, 1987)), followed by periods when the system is relatively insensitive to management manipulation. Managers can make use of these 'windows of opportunity' to induce or prevent significant changes in the system (Westoby et al., 1989). For example, in grasslands where the timing of the onset of spring rains determines the relative performance of different grass species, the timing of grazing in spring may alter their relative abundance in the sward (Walker et al., 1986; Westoby et al., 1989).

CONCLUSION

There can be no single approach to the management of complexity in grazing systems. It may, in circumstances where the system state can be relatively rigorously controlled by management (as, for example, in equilibrium systems typical of humid rangelands), be advisable to attempt to dampen the spatiotemporal heterogeneity of the system in order to simplify management and maximize production. Usually, however, such control can be achieved only at some cost to the land manager, but the required intervention is none the less often financially worthwhile. In the majority of rangelands, however, variability cannot be effectively dampened even with extensive management intervention and here it would, in any event, seem unwise to attempt to do so. The preferred option in such so-called non-equilibrium systems would be for the manager to exploit rather than attempt to dampen such heterogeneity. It must none the less be appreciated that the extent to which the natural heterogeneity of any rangeland can be effectively exploited is often severely restricted by, for example, settlement patterns and/or population pressures, both of which will often restrict animal movement and so deny access to the natural heterogeneity within any particular landscape.

REFERENCES

Acocks, J.P.H. (1966) Non-selective grazing as a means of veld reclamation. *Proceedings of the Grassland Society of Southern Africa* 1, 33–39.
Allen, T.F.H. and Hoekstra, T.W. (1991) Role of heterogeneity in scaling of ecological systems under analysis. In: Kolasa, J. and Pickett, S.T.A. (eds) *Ecological Heterogeneity*. Springer-Verlag, New York, pp. 47–68.
Allen, T.F.H. and Starr, B. (1982) *Hierarchy: Perspectives for Ecological Complexity*. University of Chicago Press, Chicago.
Allen, T.F.H. and Wileyto, E.P. (1983) A hierarchical model for the complexity of plant communities. *Journal of Theoretical Biology* 101, 529–540.
Aucamp, A.J. (1976) The role of the browser in the bushveld of the Eastern Cape. *Proceedings of the Grassland Society of Southern Africa* 11, 135–138.
Barnes, D.L., Rethman, N.F.G., Beukes, B.H. and Kotze, G.D. (1984) Veld composition in relation to grazing capacity. *Journal of the Grassland Society of Southern Africa* 1, 16–19.
Bedell, T.E. (1973) Botanical composition of a subclover–grass pasture as affected by single and dual grazing by cattle and sheep. *Agronomy Journal* 65, 502–504.

Behnke, R.H. and Scoones, I. (1993) Rethinking range ecology: implications for rangeland management in Africa. In: Behnke, R.H., Scoones, I. and Kerven, C. (eds) *Range Ecology at Disequilibrium.* Overseas Development Institute, London, pp. 1–30.

Bell, R.H.V. (1971) A grazing ecosystem in the Serengeti. *Scientific American* 225, 86–93.

Bond, W.J. (1989) Describing and conserving biotic diversity. In: Huntley, B.J. (ed.) *Biotic Diversity in Southern Africa: Concepts and Conservation.* Oxford University Press, Cape Town, pp. 2–18.

Booysen, P. de V. (1969) An evaluation of the fundamentals of grazing systems. *Proceedings of the Grassland Society of Southern Africa* 4, 84–91.

Carr, C.M. and MacDonald, D.W. (1986) The sociality of solitary forages: a model based on resource dispersion. *Animal Behaviour* 34, 1540–1549.

Caughley, G. (1979) What is this thing called carrying capacity? In: Boyce, M.S. and Hayden-Wing, L.D. (eds) *North American Elk: Ecology, Behaviour and Management.* University of Wyoming, Laramie, Wyoming.

Clary, W.P. and Pearson, H.A. (1969) Cattle preferences for forage species in Northern Arizona. *Journal of Range Management* 22, 114–116.

Conway, A., McLoughlin, A. and Murphy, W.E. (1972) *Development of a Cattle and Sheep Farm.* Animal Management Series No. 2, An Foras Taluntais, Dublin.

Coughenour, M.B. (1991) Spatial components of plant–herbivore interactions in pastoral, ranching and native ungulate ecosystems. *Journal of Range Management* 44, 530–542.

Cumming, D.H.M. (1982) The influence of large herbivores on savanna structure in Africa. In: Huntley, B.J. and Walker, B.H. (eds) *Ecology of Tropical Savannas.* Springer-Verlag, Berlin, pp. 217–246.

Cumming, D.H.M. (1993) Multi-species systems: progress, prospects and challenges in sustaining range animal production and biodiversity in east and southern Africa. *Proceedings of the 7th World Conference on Animal Production,* Edmonton, Alberta, pp. 145–159.

Danckwerts, J.E. and Stuart-Hill, G.C. (1987) Adaption of an increaser and decreaser grass species to defoliation in a semi-arid grassveld. *Journal of the Grassland Society of Southern Africa* 5, 218–222.

Danckwerts, J.E., Aucamp, A.J. and Barnard, H.J. (1985) Forage production of some grass species in the False Thornveld of the Eastern Cape. *Journal of the Grassland Society of Southern Africa* 2, 31–34.

Dyksterhuis, E.J. (1949) Condition and management of rangeland based on quantitative ecology. *Journal of Range Management* 2, 104–114.

Ellis, J.E. and Swift, D.M. (1988) Stability of African pastoral systems: alternate paradigms and implications for development. *Journal of Range Management* 41, 450–459.

Fabricius, C. (1994) The relation between herbivore density and relative resource density at the landscape level: kudu in semi-arid savanna. *African Journal of Range and Forage Science* 11, 7–10.

Foran, B.D., Tainton, N.M. and Booysen P. de V. (1978) The development of a method for assessing veld condition in three grassveld types in Natal. *Proceedings of the Grassland Society of Southern Africa* 13, 27–33.

Friedel, M.H. (1991) Range condition assessment and the concept of thresholds: a viewpoint. *Journal of Range Management* 44, 422–426.

Gandar, M.V. (1980) Short term effects of the exclusion of large mammals and insects in broad leaf savanna. *South African Journal of Science* 76, 29–31.

Greig-Smith, P. (1983) *Quantitative Plant Ecology*. University of California Press, Berkeley, pp. 19–53.
Grime, J.P. (1979) *Plant Strategies and Vegetation Processes*. John Wiley & Sons, New York.
Hardy, M.B. and Mentis, M.T. (1986) Grazing dynamics in sour grassveld. *South African Journal of Science* 82, 566–572.
Hardy, M.B. and Tainton, N.M. (1993) Mixed-species grazing in the Highland Sourveld of South Africa: an evaluation of animal production potential. *Proceedings of the 17th International Grassland Congress*, 2026–2028.
Hardy, M.B., Tainton, N.M. and Morris, C.D. (1994) Defoliation patterns in three grass species in sourveld grazed by cattle and sheep. *African Journal of Range and Forage Science* 11, 37–42.
Hatch, G.P. (1994) The bioeconomic implications of various stocking strategies in the semi-arid savanna of Natal. Unpublished PhD thesis, University of Natal, Pietermaritzburg.
Hatch, G.P. and Tainton, N.M. (1990) A preliminary investigation of area-selective grazing in the Southern Tall Grassveld of Natal. *Journal of the Grassland Society of Southern Africa* 7, 238–247.
Hatch, G.P. and Tainton, N.M. (1993) Effect of veld condition and stocking intensity on species selection patterns by cattle in the Southern Tall Grassveld of Natal. *African Journal of Range and Forage Science* 10, 11–20.
Hodgson, J. (1990) *Grazing Management: Science into Practice*. Longman Handbooks in Agriculture, Longman Scientific and Technical, co-published with John Wiley, New York, 203 pp.
Hoffman, M.T. (1988) Rationale for Karoo grazing systems: criticisms and research implications. *South African Journal of Science* 84, 556–559.
Hoffman, M.T. and Cowling, R.M. (1990) Vegetation change in the semi-arid eastern Karoo over the last 200 years: an expanding karoo – fact or fiction? *South African Journal of Science* 86, 286–294.
Holling, C.S. (1973) Resilience and stability of ecological systems. *Annual Review of Ecology and Systematics* 4, 1–23.
Homewood, K. and Rodgers, W. (1991) *Masailand Ecology. Pastoralist Development and Wildlife Conservation in Ngorongoro, Tanzania*. Cambridge University Press, Cambridge.
Humphries, L.R. (1987) *Tropical Pastures and Fodder Crops*, 2nd edn. Intermediate Tropical Agriculture Series, Longman Scientific and Technical, co-published with John Wiley, New York, 155 pp.
Hurd, R.M. and Pond, F.W. (1958) Relative preference and productivity of species on summer cattle ranges, Big Horn Mountains, Wyoming. *Journal of Range Management* 11, 109–114.
Illius, A.W. (1986) Foraging behaviour and diet selection. In: Gudmundsson, O. (ed.) *Grazing Research at Northern Latitudes*. Plenum Press, New York, pp. 227–236.
IUCN (1988) *General Assembly of the International Union for the Conservation of Nature and Natural Resources*, Costa Rica.
Jones, R.J. and Sandland, R.L. (1974) The relation between animal gain and stocking rate: derivation of the relation from the results of grazing trials. *Journal of Agricultural Science (Cambridge)* 83, 335–342.

Kirkman, K.P. (1988) Factors affecting seasonal variation of forage quality in summer rainfall areas of South Africa. Unpublished MSc (Agriculture) thesis, University of Natal, Pietermaritzburg.

Kolasa, J. and Pickett, S.T.A. (1989) Ecological systems and the concept of biological organisation. *Proceedings of the National Academy of Science (USA)* 86, 8837–8841.

Kolasa, J. and Rollo, C.D. (1991) The heterogeneity of heterogeneity: a glossary. In: Kolasa, J. and Pickett, S.T.A. (eds) *Ecological Heterogeneity*. Springer-Verlag, New York, pp. 1–23.

Lambert, M.G. and Guerin, H. (1989) Competitive and complementary effects with different species of herbivore in their utilisation of pastures. *Proceedings of the 16th International Grassland Congress*, Nice, pp. 1785–1790.

McIvor, J.G. (1993) Distribution and abundance of plant species in pastures and rangeland. *Proceedings of the 17th International Grassland Congress*, 285–290.

McNaughton, S.J. (1984) Grazing lawns: animals in herds, plant form and co-evolution. *American Naturalist* 124, 863–886.

May, R.M. (1973) *Stability and Complexity in Model Ecosystems*. Princeton University Press, New Jersey.

Meissner, H.H., Hofmeyer, H.S., van Rensburg, W.J.J. and Pienaar, J.P. (1983) *Classification of Livestock for Realistic Prediction of Substitution Values in Terms of a Biologically Defined Large Stock Unit*. Technical Communication, No. 175, Department of Agriculture, Pretoria.

Mentis, M.T. (1981) The animal factor. In: Tainton, N.M. (ed.) *Veld and Pasture Management in South Africa*. Shuter & Shooter, Pietermaritzburg, pp. 287–312.

Mentis, M.T. and Tainton, N.M. (1981) Stability, resilience and animal production in continuously grazed sour grassveld. *Proceedings of the Grassland Society of Southern Africa* 16, 37–43.

Mentis, M.T., Grossman, D., Hardy, M.B., O'Connor, T.G. and O'Reagain, P.J. (1989) Paradigm shifts in South African range science, management and administration. *South African Journal of Science* 85, 684–687.

Minson, D.J. (1982) Effects of chemical and physical composition of herbage upon intake. In: Hacker, J.B. (ed.) *Nutritional Limits to Animal Production from Pastures*. Commonwealth Agricultural Bureaux, Farnham Royal, pp. 167–182.

Montague, W.D. and Laca, E.A. (1993) The grazing ruminant: models and experimental techniques to relate sward structure and intake. *Proceedings of the 7th World Conference on Animal Production*, Edmonton, Alberta, pp. 439–460.

Morris, C.D. and Tainton, N.M. (1993) The effect of defoliation and competition on the regrowth of *Themeda triandra* and *Aristida junciformis* subsp. *junciformis*. *African Journal of Range and Forage Science* 10, 124–128.

Morris, C.D., Tainton, N.M. and Hardy, M.B. (1992) Plant species dynamics in the Southern Tall Grassveld under grazing, resting and fire. *Journal of the Grassland Society of Southern Africa* 9, 90–95.

Mott, J.J. (1986) Patch grazing and degradation in natural pastures of the tropical savannas in northern Australia. In: Horne, F., Hodgson, J., Mott, J.J. and Brougham, R. (eds) *The Plant–Animal Interface*. Winrock International, Morrilton, Arkansas, pp. 153–161.

Naveh, Z. and Lieberman, A.S. (1984) *Landscape Ecology: Theory and Application*. Springer-Verlag, New York.

Nicol, A.M. and Souza, S.N. (1993) The effect of co-grazing with sheep on the grazing intake of young and adult cattle. *Proceedings of the 7th World Conference on Animal Production*, Edmonton, Alberta, pp. 334–335.

Noble, I.R. (1986) The dynamics of range ecosystems. In: Joss P.J., Lynch P.W. and Williams O.B. (eds) *Rangelands: a Resource under Siege*. Cambridge University Press, Cambridge, pp. 3–5.

Nolan, T. and Connolly, J. (1977) Mixed stocking by sheep and steers – a review. *Herbage Abstracts* 47, 367–379.

Noy-Meir, I. (1975) Stability of grazing systems: an application of predator–prey graphs. *Journal of Ecology* 63, 459–481.

O'Connor, T.G. (1991) Local extinction in perennial grasslands: a life-history approach. *American Naturalist* 137, 753–773.

Ode, D.J., Tieszen, L.L. and Lerman, J.C. (1980) The seasonal contribution of C_3 and C_4 plant species to primary production in a mixed prairie. *Ecology* 61, 1304–1311.

O'Neill, R.V., DeAngelis, D.L., Waide, J.B. and Allen, T.F.H. (1986) *A Hierarchical Concept of Ecosystems*. Princeton University Press, Princeton.

O'Reagain, P.J., Haller, M. and Zacharias, P.J.K. (1993) Relationship between sward structure and dietary quality and intake, in cattle grazing humid sour grassveld in southern Africa. *Proceedings of the 17th International Grassland Congress*, 730–731.

Owen-Smith, N. and Cumming, D.H.M. (1993) Comparative foraging strategies of grazing ungulates in African savanna grasslands. *Proceedings of the 17th International Grassland Congress*, 691–698.

Pickup, G. (1989) New land degradation survey techniques for arid Australia: problems and prospects. *Australian Rangeland Journal* 1, 74–82.

Pienaar, A.J. (1968) Beheerde selektiewe beweiding (Controlled selective grazing). *Landbouweekblad* Junie 11, 40–41.

Ring, C.B., Nicholson, R.A. and Launchbaugh, J.L. (1985) Vegetational traits of patch grazed rangelands in west-central Kansas. *Journal of Range Management* 38, 51–55.

Roux, P. and Vorster, M. (1983) Vegetation change in the karoo. *Proceedings of the Grassland Society of Southern Africa* 18, 25–29.

Savory, A. (1978) A holistic approach to ranch management using short duration grazing. *Proceedings of the 1st International Rangeland Congress*, Denver, Colorado, pp. 555–557.

Scholes, R.J. and Walker, B.H. (1993) *An African Savanna: Synthesis of the Nylsvlei Study*. Cambridge University Press, Cambridge.

Scoones, I. (1993) Why are there so many animals? Cattle population dynamics in the communal areas of Zimbabwe. In: Behnke, R.H., Scoones, I. and Kerven, C. (eds) *Range Ecology at Disequilibrium*. Overseas Development Institute, London, pp. 62–76.

Senft, R.L., Coughenour, M.B., Bailey, D.W., Rittenhouse, L.R., Sala, O.E. and Swift, D.M. (1987) Large herbivore foraging and ecological hierarchies. *Bioscience* 37, 789–799.

Shmida, A. and Wilson, M.V. (1985) Biological determinants of species diversity. *Journal of Biogeography* 12, 1–20.

Simon, H.A. (1962) The architecture of complexity. *Proceedings of the American Philosophy Society* 106, 467–482.

Stoddart, L.A., Smith, A.D. and Box, T.W. (1975) *Range Management*. McGraw Hill, New York.

Stuth, J.W. (1991) Foraging behaviour. In: Heitschmidt, R.K. and Stuth, J.K. (eds) *Grazing Management: an Ecological Perspective.* Timber Press, Portland, pp. 65–83.

Tainton, N.M., Zacharias, P.J.K. and Hardy, M.B. (1989) The contribution of veld diversity to the agricultural economy. In: Huntley B.J. (ed.) *Biotic Diversity in Southern Africa: Concepts and Conservation.* Oxford University Press, Cape Town, pp. 107–120.

Thorhallsdottir, T.E. (1990) The dynamics of a grassland community: a simultaneous investigation of spatial and temporal heterogeneity at various scales. *Journal of Ecology* 78, 884–908.

Trollope W.S.W., Trollope, L. and Bosch, O.J.H. (1990) Veld and pasture management terminology in southern Africa. *Journal of the Grassland Society of Southern Africa* 7(1), 52–61.

Vallentine, J.F. (1990) *Grazing Management.* Academic Press, San Diego.

Van Soest, P.J. (1983) *Nutritional Ecology of the Ruminant.* O & B Books, Corvallis, Oregon.

Vorster, M. (1982) The development of an ecological index method for assessing veld condition in the karoo. *Proceedings of the Grassland Society of Southern Africa* 17, 84–89.

Walker, B.H. (1980) Stable production versus resilience: a grazing management conflict?. *Proceedings of the Grassland Society of Southern Africa* 15, 79–83.

Walker, B.H. and Noy-Meir, I. (1982) Aspects of the stability and resilience of savanna ecosystems. In: Huntley, B.H. and Walker, B.H. (eds) *Ecology of Tropical Savannas.* Springer Verlag, Berlin, pp. 556–590.

Walker, B.H., Mathews, D.A. and Dye, P.J. (1986) Management of grazing systems – existing versus event-orientated approach. *South African Journal of Science* 82, 172.

Westoby, M., Walker, B.H. and Noy-Meir, I. (1989) Opportunistic management for rangelands not at equilibrium. *Journal of Range Management* 42, 266–274.

Zacharias, P.J.K. (1992) Factors affecting the seasonal change in plant quality in various ecological zones in Natal. *Proceedings of the 4th International Rangeland Congress,* Montpellier, pp. 438–441.

Management of Grazing Systems: Temperate Pastures

G.W. Sheath[1] and D.A. Clark[2]

[1]AgResearch, Whatawhata Research Centre, Private Bag 3089, Hamilton, New Zealand; [2]Dairying Research Corporation Ltd, Private Bag 3123, Hamilton, New Zealand

INTRODUCTION

Terrestrial ecosystems are structured from a hierarchy of interacting levels. Grassland farming systems are no exception (Fig. 11.1). Physical resources of the environment form the foundation upon which these production systems depend. Worldwide, there is an increasing determination to sustain the level and quality of these resources, i.e. soil, water and air. From a grasslands perspective, the productive use of these resources is very much dependent on the condition and availability of forage plant and grazing animal resources. Genetic merit and health status of these biota are most significant. The three resource levels of our proposed hierarchy are interactive and the degree of this interaction is strongly influenced by system management.

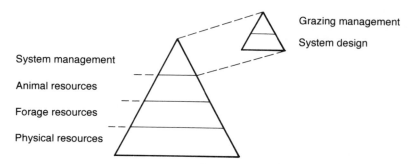

Fig. 11.1. A hierarchical structure of grassland systems.

©1996 CAB INTERNATIONAL.
The Ecology and Management of Grazing Systems (eds J. Hodgson and A.W. Illius)

Strategies for managing grassland systems have been refined to a greater degree in temperate zones than in other climatic areas. This chapter will focus on systems management within the context of temperate grasslands. In this case, we propose that systems management has two parts, namely systems design and grazing management. For both parts, general principles and objectives will be discussed. Specific decision rules and action plans will then be detailed for grazing management. Finally, case-studies with simulated scenarios will be developed for a beef and sheep farm and a dairy farm to illustrate the process of setting and achieving production targets, particularly where resource conditions (e.g. climate) are variable.

SYSTEM DESIGN

The design of any farming system must fit within the context of defined business goals. It must satisfy the product specifications of the target market and operate within the constraints of the resources that are economically and socially available. Farm systems that produce a continuous supply of products (e.g. fresh milk, chilled lamb) are most commonly associated with predictable and uniform feed supply. As forage growth becomes more variable and/or unpredictable, then seasonal supply of product becomes more dominant (e.g. processed dairy products, frozen meat). In this latter situation breeding animals are more common. Seasonal flexibility in their live weights and feed requirements can be used more readily to buffer the impacts of variable feed supply (Nicol and Nicoll, 1987; Rattray et al., 1987).

Pasture, grazed *in situ*, is the common forage base of temperate grassland systems. Seasonal variation in pasture growth depends on climatic extremes, but generally 60–70% of annual production occurs in 4 months. It is difficult to design profitable systems which can directly consume current growth during this peak period. Therefore, surpluses are either conserved (e.g. silage, hay) or transferred into summer as standing, low quality feed. Conservation of forage has been a long-standing practice for smoothing out feed surpluses and deficits. Where price : cost ratios have allowed, conserved forages and high energy/protein supplements have become integral feed sources within intensive dairy and meat production systems.

The degree of dependence on conservation and supplements most readily distinguishes between the temperate grassland systems currently practised in the northern and southern hemispheres. With greater control of input variation, the 'ideal' pasture management is more easily implemented in supplemented systems (Maxwell, 1985). In contrast, where *in situ* grazing dominates, both pasture plants and grazing animals have to adjust to the effects of any imbalance in forage supply and demand. In this situation, grazing management decisions must be informed and flexible.

New Zealand's intensive grassland systems are almost entirely reliant on *in situ* grazing. These systems have important constraints on animal production. Most significant is the need to compromise individual animal performance to achieve high levels of pasture utilization and still maintain an annual cycle of animal production. An example in dairying systems is where full feeding of cows in early lactation is compromised in order to allow high stocking rates that are capable of converting most spring pasture growth into milk (Bryant, 1981). It is because stocking rate has such a dominant effect on animal demand and pasture use that it is seen as a major factor governing high animal output per hectare from grasslands (McMeekan, 1956).

Because of the interrelationship between stocking rate and individual animal performance, it is important that high system efficiency is not over-emphasized when product quality is important or where year to year variation in forage supply is considerable. Conservative stocking rates of capital stock should be chosen in environments where summer pasture growth is unreliable (Cacho and Bywater, 1994). This reduces negative carry over effects into subsequent years. If high levels of animal performance are to be achieved under variable pasture supply, maintenance of high stocking rates requires either very flexible stock policies or a willingness to feed supplements.

Flexibility in stock policy and feed demand can be achieved through altering the purchase and disposal dates of finishing animals and by adjustments to the commencement and duration of lactation of breeding animals (Fig. 11.2). Unlike

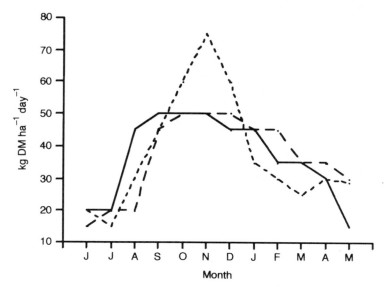

Fig. 11.2. Pasture growth (kg DM ha^{-1} day^{-1}) (- - - - -) and daily feed demand for cows calving in July or (————) August (- - - -), both stocked at 3.3 cows per hectare (from Holmes and Wilson, 1987). DM, dry matter.

changes in capital stocking rates (e.g. ewes, cows), these policies are the main instrument for varying animal demand within a year and coping with seasonal and between-year variation in forage supply. Appropriate stock policies that better align feed demand with forage supply are the hallmark of profitable animal production in low input systems (Bryant and Sheath, 1987).

Choice of parturition time is also a critical decision because of the threefold increase in feed demand during lactation. The optimal time in year-round grazing systems will depend on the relative contribution of stocking rate and performance per head to profit per hectare. In dairying systems, earlier concentrated calving and moderate feeding in early lactation ensure that all cows are at peak lactation during peak spring pasture growth. This gives maximum conversion of high-quality spring pasture into milk and maximizes production per hectare (Bryant, 1981). Flexible decisions on the timing of drying off dairy cows are also important in unlocking the effects of one year's production from the next. Delays in drying off can result in reduced winter feed reserves and reductions in cow condition at a time when conversion of pasture to milk is inefficient. New Zealand experiments have consistently shown that drying off in autumn is a more cost-effective option than feeding supplements in late lactation (Bryant and McDonald, 1987).

Within sheep production systems, earlier lambing increases the time available for lambs to achieve acceptable slaughter weights before the onset of low pasture quality and drought. However, there is a balance. Excessively early lambing can exacerbate feed deficits during early spring lactation and have negative consequences on ewe fleece weight, next year's lambing and the performance of other stock on the farm (Sheath *et al.*, 1990). This highlights the importance of avoiding focus on a single component when designing grassland systems.

The annual pattern of feed demand in sheep and beef production systems is also influenced by the ratio of lactating to dry stock. Figure 11.3 illustrates these patterns for sheep and beef systems that have similar May–September daily feed demands. Finishing steers within the system reduces the spring peak in feed demand and the ultimate consequence is poor summer–autumn feed quality. Inclusion of bull finishing increases summer feed demand and exposes the system to greater risk where summer feed supply is variable. Feeding imbalances in finishing systems can be corrected through supplementation, but these practices are generally costly.

The eventual design of a profitable system will be a compromise between the feed demands of the chosen stock policy, the level and pattern of pasture production and the affordable level of supplementation. There will be periods of feed deficit and oversupply. It is within these terms of reference that grazing management decisions have to be made and subsequent actions have to be implemented.

The curvilinear relationship between live-weight gain per hectare and stocking rate is well known (Jones and Sandland, 1974). For any grazing system, an optimum stocking rate can be defined. Above this optimum rate, the extra animals being carried do not fully compensate for the decline in individual live-weight gain. Eventually a point is reached where individual live-weight gain is zero and

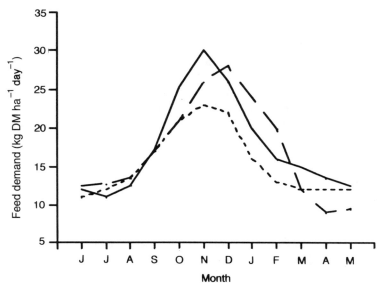

Fig. 11.3. Daily feed demand (kg DM ha^{-1} day^{-1}) for stock policies A (———), B (- - - -) and C (– – – –) when average daily demand is similar for May–September. A, breeding ewe 70% of demand, breeding cow 30%; B, breeding ewe 30%, breeding cow 30%, 2½-year steer finishing 40%; C, breeding ewe 30%, 20-month bull finishing 70%; DM, dry matter.

ingested pasture is used only for maintenance. Often, this short-term situation is necessary where seasonal feed supply is highly variable. It allows annual animal output to be optimized. Where markets require specified carcass weights and penalties are incurred for failing to meet these targets, biological optima in stocking rate have to be compromised.

GRAZING MANAGEMENT OBJECTIVES

Grazing management should be seen as the second tier of integration within systems management. It is a fine-tuning mechanism and can be simply summed up as 'where and when to move the grazing animals'. The direct and indirect effects of these decisions are less simple, for they can influence the condition and performance of the animal, plant and physical resources of the system. This section will outline the general objectives of some key grazing management decisions.

Animal resource

The most obvious effect of grazing management is on the nutrition and subsequent performance of grazing animals. Feed intake is mainly controlled by the daily allowance of pasture that is provided on a per-animal basis (Nicol and Nicoll, 1987; Rattray et al., 1987). The main modifying effect is that of accessibility. There are minimum pasture mass/height limits below which animals are unable to achieve full intake, irrespective of total pasture offered. The decision rules and actions that centre on nutrition levels will be covered in later sections of this chapter.

Grazing systems vary widely in the maintenance feed demands that must be satisfied. Milk production systems have a relatively constant maintenance demand throughout the year. When matched with an efficient buffering system based on fat reserves, this means that highly stocked systems with relatively low production per cow can utilize a high proportion of the pasture grown in a profitable manner. Only when cheap, high-quality concentrate feed is available is it more profitable to minimize the maintenance fraction of intake by maximizing milk yield per cow.

In contrast, the maintenance component of a finishing beef system is extremely important. This is because each unit of live-weight gain adds to the maintenance cost until the end of the finishing phase. This fact highlights the need for beef finishing systems to achieve maximum gains throughout the finishing phase in order to reduce maintenance costs. This requires offering high allowances of high-quality pasture. Except in the short term, this is often difficult to achieve, because of the accumulation of dead herbage at low levels of pasture utilization.

A less obvious effect of grazing management on feeding animals is the use of rotational grazing to ration existing pasture and transfer it *in situ* to another time period. This is most commonly practised during winter in New Zealand. The purpose of winter rotational grazing has not been to grow more pasture during that season. More importantly, the transfer of pasture into early spring moves it to a period of greater animal response (Rattray et al., 1987). It also helps maintain critical early spring pasture cover, which benefits both pasture accessibility and growth rate (Parker and McCutcheon, 1992). Rapid rotations or continuous grazing during winter can lead to lower spring covers and subsequently poorer animal performance, especially at high stocking rates (Smeaton and Rattray, 1984; Clark et al., 1986).

Planned grazing management can take advantage of the differences in seasonal feed requirements that exist between different stock classes (Milligan et al., 1987). By concentrating animals into flocks/herds and controlling their grazing and feeding levels, there is the opportunity to prioritize feed and/or integrate different stock classes. Advantages occur through better feed conversion efficiencies or greater pasture utilization. In times of competition, where increased utilization is necessary, pasture can be channelled to the stock class generating the highest marginal revenue. For example, there is little financial return from overfeeding a pregnant ewe in winter when greater financial returns would come from improved feeding of finishing cattle. In this situation, priority can be given to the cattle by either grazing the priority class ahead in a leader/follower rotation or by adjusting the areas allocated to separate classes. Another example of leader/follower management occurs during summer in New Zealand, when lambs are followed by animals of lower nutritional priority, such as ewes and cows (Smith et al., 1976). This arrangement achieves high levels of performance by one stock class, and yet improves overall utilization during a period of low pasture quality.

The planned integration of different animal species in a grazing sequence can also benefit the health status of the animal resources. Lamb performance benefits from integrated grazing with cattle (Brunsdon, 1989), principally because pasture contamination of internal sheep parasites is reduced by cattle grazing. Avoiding the challenge of animal diseases can be a most valuable outcome of wise grazing management and is extremely relevant for those farming systems that seek to reduce chemical use. A further example is that of the mycotoxic disease facial eczema, where its impact can be reduced by identifying and avoiding paddocks with high spore loadings and by ensuring that animals do not graze into the pasture base (Sheath et al., 1987).

Plant resource

While the status of physical resources (e.g. rainfall, soil fertility) and pasture sowing practices dominate the genetic make-up of pastures, grazing management can influence pasture species balance. The timing and severity of grazing has the greatest effect on annual species and vegetatively spreading plants that seek to fill gaps between fixed plant units (Sheath and Boom, 1985). Grazings which increase the level of reseeding and/or the amount of bare ground will encourage these 'filler' species (e.g. white clover). Conversely, where species have the ability to dominate other plants, then they are most advantaged when grazing pressure is low during their most active growth period. The dominance of *Agrostis capillaris* where summer feed surpluses are common is evidence of this process.

Grazing preference–rejection, whether it is active or passive, can also influence pasture composition. For this reason the species of grazing animal can alter composition. The integrated grazing of sheep in cattle-dominant systems is a means of reducing the weed ragwort, *Senecio jacobaea* (Betteridge et al., 1994). Another example is the inclusion of goat grazing to reduce the content of weed species such as gorse, *Ulex europaeus* (Rolston et al., 1982). In addition to the effects of grazing, animals may also influence pasture composition through their hoof action. Cattle cause greater levels of hoof damage to the pasture/soil surface, which can lead to a greater content of filler species, such as white clover (Lambert et al., 1983).

The effects of defoliation on pasture growth have been covered in detail in previous chapters in this volume (Lemaire and Chapman, Chapter 1; Briske, Chapter 2). Suffice it to say here that only at the extremes of over- and undergrazing is the net herbage growth or accumulation of pastures significantly affected (Bircham and Hodgson, 1983). Within temperate grassland systems that are locked into year-round grazing, the role of grazing management in avoiding overgrazing is most important in early to mid-spring. It is during this period that animal production is most responsive to current pasture supply (Smeaton and Rattray, 1984). Grazing management strategies during drought or in recovery from drought do little to influence pasture and system performance (Vartha and Hoglund, 1983). More importantly, reduced stock numbers or the injection of supplements or nitrogen (N) allows more effective recovery of the production system.

When pastures are undergrazed, death and decay of herbage tissue is the dominant process. While these losses balance out the contribution of new growth, this process is least important from the perspective of feed quantity, as it generally operates within the context of pasture surpluses. More important is the effect on pasture quality, as dead herbage in the diet causes a large reduction in digestibility (Rattray and Clark, 1984). During times of pasture surplus, the key role of grazing management is to create a balance between high animal intakes and the maintenance of pasture quality. While animal performance differs little between various forms of grazing management when pasture supply is adequate (Clark *et al.*, 1986), rotational grazing has some advantages during periods of surplus. Higher stock densities can help reduce preferential grazing and surpluses can be more readily identified (Sheath, 1983). Mixed-species grazing is also an important strategy for reducing patch grazing and maintaining pasture quality. It is this process that helps underpin the production advantages of mixed sheep and cattle grazing during times of surplus feed (Nolan and Connolly, 1977; McCall *et al.*, 1986).

Maximum herbage intake per animal on grazed areas and the concentration of surpluses for conservation are key ingredients of Scottish grassland systems (Maxwell, 1985). Similar principles have been developed for New Zealand hill systems with surpluses being concentrated on designated, easier contoured paddocks (Sheath *et al.*, 1984). Where conservation of these surpluses is not possible because of land terrain, animals of low nutritional priority (e.g. beef cows) can be used to graze these paddocks in autumn. Some dairy farmers are also using this deferred grazing strategy as an alternative to conservation (McCallum *et al.*, 1991).

Pasture conservation is a key component of many animal production systems in regions with long periods of low or zero pasture growth. Its value is not contested in these situations. However, there is less compelling evidence on the profitability of conservation where all-year grazing is possible (Thomson *et al.*, 1989). This occurs because of the 'paradox of conservation'. At low stocking rates large amounts of conserved fodder can be made, but it is not needed because winter demand can be met by current pasture growth. Conversely, at high stocking rates there is little surplus available in spring to make more than token amounts of conserved fodder. Low product prices, coupled with high dry matter (DM) losses during conservation and high conservation costs, may make options such as deferred grazing attractive.

Where conservation is justified and a high-quality silage is required, early closure and harvesting no later than 40 days after closure is recommended. However, closure date should not be so early that current milk production is decreased, because the production response from even high-quality silage is likely to be less than half that directly obtainable from spring pasture (Rogers, 1985).

Physical resource

Under the umbrella of sustainability, there is an increasing interest and awareness in the impact of farm practices on environmental conditions. While it is becoming apparent that these impacts can be quite diverse, the effects of grazing management decisions are primarily confined to the soil–water interface. Grazing animals can modify soil physical conditions through their hoof actions. Rotational grazing systems during winter which involve high stock densities (200–500 sheep ha^{-1}) have caused soil compaction in some areas of New Zealand (Greenwood and McNamara, 1992). Anaerobic conditions resulting from this compaction have led to lower pasture production. A further example is that of cattle-treading damage, which not only displaces surface soil but also reduces water infiltration rates. These processes increase the potential for increased water runoff and sediment contamination of waterways (Lambert *et al.*, 1985).

Nitrogen is a key component in the productivity of intensive grazing systems, but there are important implications in N losses to the environment. These losses mainly occur because of the aggregation of N into dung and urine by the grazing animal. Aggregation will always be a problem where 'efficient' grazing systems harvest most of the pasture growth.

An important challenge for the future will be the design of optimal landscape arrangements and managements which combine the objectives of production and resource care within temperate grassland systems. Grazing management will undoubtedly be an important tool in implementing these arrangements.

DECISION RULES AND ACTIONS

Planning and management decisions are made in relation to a defined production objective and modified as a result of monitoring system performance. We identify a time-based decision structure that should be followed in making management decisions. Unfortunately, pastoral system managers often give incorrect weighting to the different levels of the decision structure (Fig. 11.4).

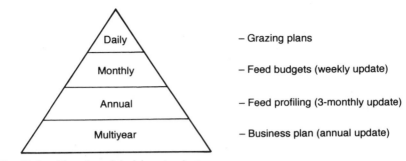

Fig. 11.4. Time-based decision structure.

The concept of sustainability underpins these planning steps and involves gathering information on past climatic events and any future predictions, long-term market and financial forecasts and likely environmental legislation. Personal factors, such as age, lifestyle requirements, managerial and technical strengths and weaknesses, are also important. Long-term farm development plans should be laid only once decisions involving these factors have been made. These plans will include subdivision of pasture, water-supply, fertilizer input, weed and pest control, pasture establishment and alternative uses for fragile pasture land. The above indicates that decision rules and actions on grazing management will be futile unless firmly set in a long-term, overall farm plan. The key to sustainable pastoral farming is the setting and annual updating of appropriate long-term business plans. The benefits of sophisticated computer programs to generate short-term grazing plans are greatest when they operate within focused, but flexible, business plans.

Decision rules

The implementation of grazing systems requires information on the following variables:

- *Feed demand*, which is a function of the feed required to meet a specified production target, and the number of animals, integrated for each class of animal.
- *Feed supply*, which is a function of pasture growth rate, pasture cover and the availability of supplementary feed.

For convenience these are usually expressed in kg DM, but feed quality (megajoules metabolizable energy (MJME) kg^{-1} DM) may be such that production targets cannot be met. Where feed quality varies significantly, demand and supply are better expressed as MJME. There are various ways of linking feed supply and demand, i.e. feed budgeting and grazing plans (Lucas and Thompson, 1990).

- *Feed profiling* is the setting of long-term policy, such as stocking rate, timing of parturition and general stock buying and selling policy. A key indicator of success is the profile of average farm pasture cover (kg DM ha^{-2}) throughout the year. It should show no marked peaks or troughs that could lead to sustained periods of reduced pasture quality or availability. Condition score and live-weight of stock should show planned seasonal variation, but should not be significantly different at the beginning and end of each annual cycle.
- *Feed budgeting* determines the optimum ways of using a feed surplus or alleviating a feed deficit. Key indicators are similar to those for feed profiling and, in addition, predicted pasture growth rates over weeks or months are used in conjunction with production targets.
- *Grazing plans* seek to achieve immediate production targets. They include decisions on rotation length, daily supplement fed and when to move stock. Key indicators are daily lactation yield, short-term live-weight gain and pasture residual mass. At this level, tactical decisions on access to water, avoidance of pugging and proximity to handling facilities become important.

Pasture and livestock monitoring

The large volume of scientific literature relating pasture parameters to livestock performance is of little value if they cannot be, or are not, monitored on individual farms. An important recent development is the concept of group farm monitoring, a team approach to technology transfer involving farmers, consultants and researchers. Biological and financial components of farm production are monitored and shared among the group (Webby and Sheath, 1991). Key pasture parameters are pregrazing and postgrazing pasture mass, pasture height and botanical composition. Key animal parameters are live weight, condition score, milk yield, wool production and reproductive indices.

Pasture height is more easily measured than pasture mass and has been adopted widely in the UK (Le Du *et al.*, 1979). In New Zealand, visually estimated pasture mass has been widely used, especially in the dairy sector (Milligan *et al.*, 1987). The method used is unimportant provided sound relationships between pasture and animal parameters are available. Nicol (1987) provides an excellent summary of these relationships for sheep, dairy and beef cattle, deer and goats in New Zealand production systems. Condition scoring of livestock is most easily learnt from an experienced practitioner, but charts and videos make the skill widely accessible. Scales provide an objective estimate of stock performance and regular weighing eliminates much of the guesswork in management decisions. These simple measures are grossly underutilized in too many farm systems.

Work by Bircham and Hodgson (1983) and Parsons *et al.* (1983) has shown that it is not possible to maximize gross photosynthesis and net pasture production in the same system (Fig. 11.5). On the demand side it is not possible to maximize intake per animal and animal production per hectare in the same system (Bryant

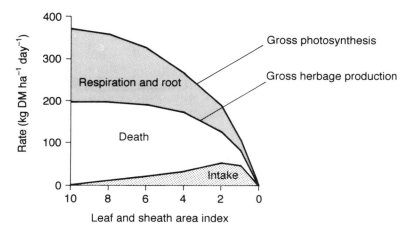

Fig. 11.5. The effects of the intensity of continuous stocking on the balance between photosynthesis, gross tissue production, herbage intake and death (from Parsons *et al.*, 1983). DM, dry matter.

and Holmes, 1985). Therefore grazing management should aim to keep pasture cover in the optimum range for net pasture growth. This range is 4 to 6 cm for continuously stocked sheep pastures in the UK and between 2000 to 3000 kg DM ha^{-1} for rotationally grazed dairy pastures in New Zealand. Importantly, as pasture cover approaches the lower end of this optimum, pasture quality is likely to increase, because of the greater proportion of green leaf in the sward. However, continued severe grazing will push the pasture into a zone of lower productivity. A consequence of this management will be some restriction of intake per animal. The required relation between performance per animal and per hectare will depend largely on the change in value of a product unit as stocking rate increases. For example, a fine-wool wether flock can be carried at high stocking rates without affecting price kg^{-1} wool. However, a prolific prime lamb flock producing contract lambs to specified carcass weights, at specified times, will be highly sensitive to inaccurate feed planning.

In addition, seasonal variation in pasture growth (see Fig. 11.2) means that stocking rates and parturition times that are designed to optimize the direct intake of high-quality pasture will inevitably lead to shortages of pasture supply at other times. In New Zealand where little supplementation occurs, targets for pasture cover on dairy farms at calving are 1800–2200 kg DM ha^{-1} and for sheep farms at lambing they are 900–1100 kg DM ha^{-1}. To achieve these targets requires informed decisions on drying-off date for cows, culling proportion and times for sheep, sale of store/prime stock, the purchase of supplementary feed, grazing off and N fertilizer use.

This brief summary of decision rules highlights the complexity of managing grassland systems, particularly when it involves a range of stock classes. Further complexity is added by climatic, topographic and market variation. However, this variation can provide opportunities for profitable short-term decisions when the principles of feed planning and performance monitoring are understood and applied.

CASE-STUDIES

Previous sections have highlighted key system and grazing management principles. Production goals are achieved through a combination of time-based planning processes, monitoring indicators of condition and performance, and implementing grazing decisions. Correctly used, these tools should provide flexibility in decision-making and implementation in order that the farm system can quickly respond to changing economic and climatic conditions.

In this section two case-studies based on New Zealand sheep and dairy farming systems are presented to illustrate management responses to predictable and unpredictable variation in feed supply. The format is the same for both sections. Multiyear, annual, monthly and daily planning are discussed and key decision rules outlined. The different scenarios used a computer simulation and their outcomes are then presented and discussed.

Sheep and beef breeding and finishing farm

The sheep/cattle enterprise discussed here is a New Zealand hill-country farm carrying Romney ewes, rearing replacements and selling surplus lambs in prime or store condition, depending on pasture supply. It also has a herd of Angus breeding cows, with surplus weaners sold at 18 months of age. This case-study focuses on the planning and actions that are necessary to cope with variable summer-dry conditions. It is an important scenario because year-to-year variation in pasture growth is generally greatest in this season within temperate environments. Recent developments in decision support software (e.g. Stockpol; McCall *et al.*, 1991) can assist, particularly at the multiyear and feed profiling level. Key principles and action plans upon which the farm system was based are as follows.

Multiyear business plan

Long-term profitability was sought by maximizing profits in the average year and avoiding disaster in drought years. The stock policy that was used provided a balance between market requirements and available economic and farm resources, i.e. ewes 40%, hoggets 10%, cows 20% and 20-month finishing cattle 30% of May–August daily feed demand (av. 12 kg DM ha^{-1}). Pasture composition had been improved by fertilizer and lime applications, judicious subdivision and the sowing of improved pasture cultivars. These developments decrease the seasonal variation in forage supply and increase the range of feasible stock policies (Lambert et al., 1983).

Annual feed profiling plan

Peak lambing and calving date was set for early September, i.e. 4 weeks before current pasture growth rate equals system demand. The consequences of highly variable December–March pasture growth rate (av. 15 kg DM ha^{-1} day^{-1}, SD 7 kg DM ha^{-1} day^{-1}) were managed through flexible disposal of surplus stock, the conservation and feeding of silage and the strategic use of N fertilizer to correct deficits in early spring pasture cover. Key condition indicators were 460 kg cow at mating, 55 kg ewe in January and average August pasture covers of 1000 and 1500 kg DM ha^{-1} on sheep and cattle paddocks respectively.

Monthly feed budgets

Autumn

Pasture required for flushing ewes was prepared in January by grazing to 1000–1500 kg DM ha^{-1} residual, which gave a pasture mass of 3500 kg DM ha^{-1}, containing 2500 kg green DM ha^{-1} in March. Summer pasture growth rates may often make this target unachievable; hence the need to reach target live weights in January or supplement ewes with pasture silage. At the start of winter grazing an average farm pasture cover of 1600 kg DM ha^{-1} was required. Sales of surplus and cull stock in the summer and autumn were managed to achieve this target.

Winter

Winter feed budgets and grazing plans were set to achieve an average farm pasture cover at lambing of 900–1100 kg DM ha^{-1}. Pregnant ewes require 0.9 kg DM day^{-1} for maintenance, which was achieved by grazing to a 500 kg DM ha^{-1} residual. In contrast, replacement hoggets required 1.2 kg DM day^{-1} to grow at 50 g day^{-1} over winter, and this required a residual of at least 800 kg DM ha^{-1} (Rattray et al., 1987). Similar residuals allowed heifers calving at 2 years of age to gain 0.5 kg live weight (LW) day^{-1} net of fetal growth. Mature breeding cows experienced a weight loss of up to 15% during early winter. Provided pregrazing pasture mass was > 2000 kg DM ha^{-1}, cows were able to graze to 500 kg DM ha^{-1} residual during this period. In late pregnancy a residual grazing pasture mass of 700 kg DM ha^{-1} led to a small weight loss, which did not compromise subsequent rebreeding. There was little opportunity to adjust stocking rate at this time of year without foregoing substantial future revenue. Therefore, where feed budgets showed a potential deficit, N fertilizer was one of the few options used to generate extra pasture. Pasture silage was also used to supplement cattle in late winter.

Spring

To achieve high milk production and recovery of ewe live weight, pasture allowances of 6–8 kg DM ewe^{-1} day^{-1} and postgrazing residuals of 1500 kg DM ha^{-1} were required. For beef cows a condition score at mating was set at 5 (scale 1–9). This provided a postcalving anoestrum period of < 80 days to allow for an annual breeding cycle and a pregnancy rate of > 90% within a restricted 2-month mating period. Cow live-weight gain will vary from zero at 2 kg DM 100 kg LW^{-1} day^{-1} allowance to 1.8 kg day^{-1} at 5 kg DM 100 kg LW^{-1} day^{-1} on New Zealand hill country (Nicol and Nicoll, 1987).

Summer

During summer, lambs for sale and replacement ewe lambs were the highest-priority stock class. At this stage live-weight gain and postgrazing pasture mass relationships are the most convenient to use (Rattray *et al.*, 1987). Target lamb weight gain was 100 g day^{-1}, which required a postgrazing mass of 1200 kg DM ha^{-1} for pastures containing low amounts of dead material. These pastures were prepared by grazing with cows. Note that gains of 200–250 g day^{-1} are only possible on clover-dominant swards.

Daily grazing plans

Maintenance feeding of breeding stock during winter was strictly rationed through the use of an 80-day grazing rotation and minimum pasture residuals of 700–800 kg DM ha^{-1}. Maximum feeding opportunities were provided for all lactating animals and finishing cattle in the spring. During summer, weaned lambs were provided with paddocks previously grazed by cattle. Once pregraze pasture levels exceed 3000 kg DM ha^{-1} during the late spring–summer, the grazing of ewes and cows was integrated.

Simulated scenarios

Using the decision-support model Stockpol (Marshall *et al.*, 1991), the performance of this base system was compared for three scenarios.

- *Scenario A*. Average December–March pasture growth rates (15 kg DM ha^{-1} day^{-1}) occurred in both year 1 and year 2. Lamb disposal sought to have minimum 13.5 kg carcasses and all surplus lambs gone by April. Silage was conserved (5% land) and fed to cattle in June–August.
- *Scenario B*. Low December–March pasture growth rates (7 kg DM ha^{-1} day^{-1}) occurred in year 1 with a return to average conditions in year 2. Flexible lamb disposal sought to have all surplus lambs gone by February. Silage was first

Table 11.1. The condition and performance of three management scenarios where summer pasture growth is variable: sheep/cattle system.

	Scenario A	Scenario B	Scenario C
Year 1			
August cover*	1250	1250	1250
Cattle (wt)†	419	419	419
Lambing (%)	102	102	102
Lamb (wt ha^{-1})‡	117	102	110
Nitrogen (kg ha^{-1})	3	17	0
Gross margin ($ ha^{-1})	383	342	365
Year 2			
August cover	1250	1240	970
Cattle (wt)	419	395	334
Lambing (%)	102	102	92
Lamb (wt ha^{-1})	117	117	93
Nitrogen (kg ha^{-1})	3	3	0
Gross margin ($ ha^{-1})	383	358	296
July cover	1310	1310	1400

*Pasture cover, kg DM ha^{-1}.
†Live weight of 20-month cattle sold in December.
‡Aggregate live weight (kg) of lambs sold ha^{-1}

used to maintain ewe body weights for mating and then cattle performance. N fertilizer was applied in July to ensure target spring covers were achieved.
- *Scenario C.* As for B but no silage was conserved, lamb disposal was inflexible and sought minimum 13.5 kg carcasses, and no N fertilizer was applied.

The outcome of these scenarios is given in Table 11.1 from the perspective of direct (year 1) and carry-over (year 2) effects of drought conditions. Live weights of breeding stock were similar at the start and end of the 2-year period. For scenario B, earlier disposal reduced lamb revenue, silage feeding to ewes maintained their body weight and subsequent lambing and any remaining feed deficits were reflected in the weight and value of finishing cattle. For scenario C, ewe live weight at mating was 4 kg lower and lambing percentage declined, and cattle performance was most severely affected by inadequate pasture covers. This situation is typical of unbuffered systems where annual revenue, animal condition and pasture covers see-saw. Two important points are highlighted by these comparisons: the impact of system management needs to be considered for more than 1 year; and the competitive effect between cattle and sheep is most strongly reflected in poor cattle performance when winter–spring pasture covers are inadequate.

Dairy farm case study

This case-study focuses on the planning and actions needed to cope with variable early-spring feed supply on a New Zealand seasonal-supply dairy farm. It is an important scenario because mistakes in feed allocation at this time can influence milk solids production for the whole season. Key principles and action plans upon which the farm system was based are as follows.

Multiyear business plan

Long-term profitability was sought by maximizing profits in the average years and avoiding disaster in years of low spring growth. To achieve this objective, key decisions involved stocking rate, calving date, culling policy and feed allocation.

Annual feed profiling plan

Start of calving was set at 20 July, with a planned 6-week calving spread. The consequences of variable early-spring feed supply were managed through flexible feed allocation procedures and the use of supplementary feed. Target cow condition scores at drying off are given in Table 11.2 and an average farm pasture cover in August of 1700 kg DM ha^{-1} was set.

Monthly feed budgets

Autumn

Drying off was a critical decision because it not only ends the cash flow for the year, but it also reduces feed demand by approximately 50%. Resolution of this conflict required information on cow condition score, current production level, current pasture cover and predicted winter feed supply and demand.

Table 11.2. Condition score criteria for drying off heifers and cows in a New Zealand spring-calving herd (A.M. Bryant, personal communication).

	Heifers	Cows
Early March	3.0	2.5
Early April	3.5	3.5
Late April	4.5	4.0

Winter

When cows were dried off close to the condition score required at calving, then winter feeding was at maintenance level. To increase condition score by one unit required 135–235 kg DM cow^{-1}, depending on breed and pasture quality (Holmes and Wilson, 1987). The risk of lower than normal winter pasture growth was covered by either drying off earlier at a higher condition score or increasing the level of supplementary feed available. Grazing plans and feed budgets were most useful from late autumn to early spring. A feed budget that was frequently updated ensured that cow condition score and farm cover were kept on target for calving. Decisions on supplementary feeding were able to be made before pastures fell into the zone of lower productivity. Drier soil conditions in late autumn enabled hay or silage to be fed with very limited pasture, ensuring high utilization of both feeds.

Spring

Average farm cover at calving is a critical determinant of annual milk solids yield ha^{-1} (Bryant, 1990). At high stocking rates, pasture must be rationed for 6–8 weeks after calving to avoid pasture entering the zone of lowered productivity (Fig. 11.5). This was achieved by the continued maintenance feeding of uncalved cows and an allocation of pasture to lactating cows that ensured average farm cover was no less than 1700 kg DM ha^{-1} by the time pasture growth exceeded pasture demand. As pasture growth rates continued to rise during spring, it was difficult to maintain these optimum growth rates and intakes with a fixed stocking rate. For example, with a pasture growth rate of 100 kg DM ha^{-1} day^{-1} and a residual of 2000 kg DM ha^{-1}, pastures moved from the optimal zone in 10 days, and after 15 days gave a pregrazing mass of 3500 kg DM ha^{-1}. At this stage decision rules for conservation had to be applied.

Note that there was no standard time for closing conservation areas. This was determined by herd feed requirements, actual pasture growth rates and existing farm cover. Of course, the calculations will then predict either the decrease in intake if a large area is closed or the amount of N-boosted pasture or supplements required to maintain maximum intakes. Similar calculations can be used to determine cutting date, but other factors such as silage/hay quality requirements affect this decision.

Summer

Variation in rotation lengths was insufficient to overcome any mismatch between supply and demand in drought conditions. Therefore, in late summer a decision rule to remove culls from the herd was applied. When current or expected average herd intakes fell below 10 kg DM cow^{-1} day^{-1}, cull cows were removed. This gave up to a 20% reduction in stocking rate, which allowed condition score and milk solid yield of remaining cows to be maintained or reduced at a slower rate.

Daily grazing plans

In winter, strip-grazing of autumn-grown pasture achieved high utilization, as only maintenance intake was required. Grazing residuals were low (< 1000 kg DM ha^{-1}) without affecting future regrowth (Clark, 1978). Use of drier ground and allowing pasture access for only 3–4 h day^{-1} minimized damage after heavy rain. This attention to detail concentrated pasture damage on relatively small areas and was important in reducing potential soil compaction and weed ingress.

In spring, cows were fully fed by offering allowances of 60 kg DM cow^{-1} day^{-1}. With pre- and postgrazing pasture masses of 2700 and 2000 kg DM ha^{-1}, intakes and net pasture growth rates were maximized without compromising pasture quality at future grazings. During summer, grazing plans involved longer recovery intervals between grazings in order to ration out pastures at a time of decreasing growth rate.

Simulated scenarios

The potential importance of grazing management is demonstrated for an average New Zealand dairy farm using a dairy farm simulation model, UDDER (Larcombe, 1989). Annual pasture growth of 13.7 t DM ha^{-1} year^{-1} is assumed with stocking rate optimized at 3.1 cows ha^{-1} and calving starting on 20 July for maximum gross margin with 990 kg milk solids ha^{-1} produced. Pasture allocation is optimized with rotation lengths of 60–100 days for dry cows, 40 days in early lactation, decreasing to 25 days by mid-September and 17 days by early November.

Two scenarios are compared, both being based on a 50% reduction in pasture growth rate for the August–September period.

- *Scenario A.* Rotation lengths that have proved optimal in an 'average' year are maintained.
- *Scenario B.* Rotation lengths and hence feed allocation are altered in response to changes in a key indicator such as average pasture cover (1700 kg DM ha^{-1} in early August).

Table 11.3 outlines the consequences of these management strategies on whole-system performance.

In this extreme, but perfectly feasible example, the use of a standard grazing management plan that optimizes profit in an 'average' year (scenario A) resulted in lower production compared with a flexible management system (scenario B). Only 70% of possible milk solids production and 51% of possible gross margin was achieved. It is important to note that major shortfalls in pasture production at critical points of the production cycle require major changes in management. In this case, rotation length had to increase from 28 to 120 days, i.e. only 50% of the

Table 11.3. The effect of feed allocation decisions on dairy farm performance parameters.

	Scenario A: grazing management assuming 'good' spring	Scenario B: grazing management assuming 'poor' spring
Aug–Sept. pasture growth rate (kg DM ha^{-1} day^{-1})	20	24
Aug–Sept. rotation length (days)	28	120
Aug–Sept. cow intake (kg DM day^{-1})	9.9	6.5
Av. pasture cover on 1 Oct. (kg DM ha^{-1} day^{-1})	1220	2170
Pasture consumed (t DM ha^{-1} year^{-1})	10.5	12.8
Milk solids (kg ha^{-1})	590	840
Gross margin ($ ha^{-1})	890	1740

DM, dry matter.

farm should be grazed in August and September. This drastic step decreased herbage intake, and milk production continued with input from body reserves.

Scenario A resulted in a pasture cover on 1 October that was too low for potential spring growth rates to be fully realized. Combined with the increased intake drive of cows in low body condition and the high stocking rate, this would have kept pastures in a zone of lowered productivity for the remainder of the lactation. As a result, 2.3 t DM ha^{-1} year^{-1} less would have been grown and consumed in scenario A.

In reality, managers would be unlikely to perceive and immediately react to low pasture growth rates. A pragmatic approach can be to use supplements (hay, silage or meal) to avoid very low pasture intakes and to gamble on a quick return to more normal conditions. This approach often works because normality does return quickly and responses to supplement are high. It also has psychological benefits for the farmer because positive action is taken and 'expected' milk levels are maintained.

CONCLUSION

Grassland farms consist of a hierarchy of interacting levels, namely physical resources, biota and systems management. This latter level is an integrating process. The design of any system must take into account the farm business goals (e.g. market requirements) and economically available resources (e.g. level and variability of feed supply). Policy decisions on stocking rate, parturition time, conservation and supplementary feeding play a major part in determining the economic and biological optima of a grazed grassland system.

Grazing decisions form the second tier of systems management and operate as a fine-tuning mechanism. These decisions seek to further improve the alignment of feed supply and demand and to achieve a balance between acceptable individual animal performance and efficient forage utilization. From both a planning and a monitoring perspective, pasture cover is a key integrator and indicator of this balance. Where supplementation is minimal, low pasture cover in early spring is restrictive to annual performance of most grassland systems. Conservation practices and/or the integration of stock and land class are important in maintaining pasture quality during periods of feed surplus.

Forward planning through annual feed profiling, short-term feed budgets and daily feed allocation is an important aid to achieving animal performance targets. Equally important is ongoing monitoring of animal and pasture indicators. Rapid response to changing circumstances is the hallmark of robust and productive grazing systems.

REFERENCES

Betteridge, K., Costall, D.A. and Hutching, S.M. (1994) Ragwort (*Senecio jacobaea*) controlled by sheep in a hill country bull beef systems. *Proceedings 47th New Zealand Plant Protection Conference*, 53–57.

Bircham, J.S. and Hodgson, J. (1983) The influence of sward conditions on rates of herbage growth and senescence in mixed swards under continuous stocking management. *Grassland Forage Science* 38, 323–331.

Brunsdon, R.V. (1989) Principles of helminth control. *Veterinary Parasitology* 6, 185–215.

Bryant, A.M. (1981) Maximising milk production from pasture. *Proceedings of the New Zealand Grassland Association* 42, 82–91.

Bryant, A.M. (1990) Present and future grazing systems. *Proceedings of the New Zealand Society of Animal Production* 50, 35–38.

Bryant, A.M. and Holmes, C.W. (1985) Utilisation of pasture on dairy farms. In: Phillips, T.I. (ed.) *The Challenge: Efficient Dairy Production*. Australian Society of Animal Production, Warragul, pp. 48–63.

Bryant, A.M. and Macdonald, K.A. (1987) Management of the dairy herd during a dry summer. *Proceedings of the Ruakura Farmers' Conference* 39, 53–56.

Bryant, A.M. and Sheath, G.W. (1987) The importance of grazing management to animal production in New Zealand. *Proceedings of the 4th AAAP Animal Science Congress*, 13–17.

Cacho, O.J. and Bywater, A.C. (1994) Use of a grazing model to study management and risk. *Proceedings of the New Zealand Society of Animal Production*, Lincoln University, Canterbury, New Zealand, pp. 377–381.

Clark, D.A. (1978) Effect of pasture reserves and stocking rate on ewe and lamb performance from mid-pregnancy to weaning. *Proceedings of the New Zealand Grassland Association* 40, 81–88.

Clark, D.A., Lambert, M.G. and Grant, D.A. (1986) Influence of fertiliser and grazing management on North Island 5. Animal production. *New Zealand Journal of Agricultural Research* 29, 407–420.

Greenwood, P.B. and McNamara, R.M. (1992) An analysis of the physical conditions of two intensively grazed Southland soils. *Proceedings of the New Zealand Grassland Association* 54, 71–75.

Holmes, C.W. and Wilson, G.F. (1987) Feeding the herd: growth and utilization of feed. In: Holmes C.W. and Wilson, G.F. (eds) *Milk Production from Pasture.* Butterworths of New Zealand Limited, Wellington, pp. 12–24.

Jones, R.J. and Sandland, R.L. (1974) The relation between animal gain and stocking rate. Derivation of the relation from the results of grazing trials. *Journal of Agricultural Science (Cambridge)* 83, 335–342.

Lambert, M.G., Clark, D.A., Grant, D.A. and Costall, D.A. (1983) Influence of fertiliser and grazing management on North Island moist hill country 1. Herbage accumulation. *New Zealand Journal of Agricultural Research* 26, 95–108.

Lambert, M.G., Devantier, B.P., Nes, P. and Penny, P.E. (1985) Losses of nitrogen, phosphorus and sedimentation run-off from hill country under different fertiliser and grazing management regimes. *New Zealand Journal of Agricultural Research* 28, 371–379.

Larcombe, M.T. (1989) The effect of manipulating reproduction on the productivity and profitability of dairy herds which graze pasture. PhD thesis, Melbourne.

Le Du, Y.L.P., Combellas, J., Hodgson, J. and Baker R.D. (1979) Herbage intake and milk production by grazing dairy cows. 2. The effects of level of winter feeding and daily herbage allowance. *Grass and Forage Science* 34, 249–260.

Lucas, R.J. and Thompson, K.F. (1990) Pasture assessment for livestock managers. In: Langer, R.H.M. (ed.) *Pastures – Their Ecology and Management.* Oxford University Press, Auckland, New Zealand, pp. 241–262.

McCall, D.G., Smeaton, D.C., Gibbison, M.L., McKay, F.J. and Hockey, H.-U.P. (1986) The influence of sheep to cattle ratios on liveweight gain on pastures grazed to different levels in late spring-summer. *Proceedings of the New Zealand Society of Animal Production* 46, 121–124.

McCall, D.G., Marshall, P.R. and Johns, K.L. (1991) An introduction to Stockpol: a decision support model for livestock farms. In: *Proceedings of the International Conference on Decision Support Systems for Resource Management*, April 1991, Texas A & M University, pp. 27–30.

McCallum, D.A., Thomson, N.A. and Judd, T.G. (1991) Experiences with deferred grazing at the Taranaki Agricultural Research Station. *Proceedings of the New Zealand Grassland Association* 53, 79–83.

McMeekan, C.P. (1956) Grazing management and animal production. *Proceedings of the 7th International Grassland Congress*, 146–156.

Marshall, P.R., McCall, D.G. and Johns, K.L. (1991) An introduction to Stockpol: a decision support model for livestock farms. *Proceedings of New Zealand Grassland Association* 53, 137–140.

Maxwell, T.J. (1985) Systems studies in upland sheep production: some implications for management and research. *HFRO Biennial Report 1984–85*, 155–163.

Milligan, K.E., Brookes, I.M. and Thompson, K.F. (1987) Feed planning on pasture. In: Nicol, A.M. (ed.) *Livestock Feeding on Pasture.* Occasional Publication No. 10, New Zealand Society of Animal Production, pp. 75–88.

Nicol, A.M. (1987) *Livestock Feeding on Pasture.* New Zealand Society of Animal Production, Occasional Publication No. 10, 145 pp.

Nicol, A.M. and Nicoll, G.B. (1987) Pastures for beef cattle. In: Nicol, A.M. (ed.) *Livestock Feeding on Pasture.* Occasional Publication No. 10, New Zealand Society of Animal Production, pp. 119–132.

Nolan, T. and Connolly, J. (1977) Mixed stocking by sheep and steers – a review. *Herbage Abstracts* 47, 367–374.

Parker, W.J. and McCutcheon, S.N. (1992) Effect of sward height on herbage intake and production of ewes of different rearing rank during lactation. *Journal of Agricultural Science (Cambridge)* 118, 383–395.

Parsons, A.J., Leafe, E.L., Collett B., Penning, P.D. and Lewis, J. (1983) Physiology of grass production under grazing. II Photosynthesis, crop growth, and animal intake of continuously grazed swards. *Journal of Applied Ecology* 20, 127–139.

Rattray, P.V. and Clark, D.A. (1984) Factors affecting intake of pasture. *New Zealand Agricultural Science* 18(3), 141–146.

Rattray, P.V., Thompson, K.F., Hawker, H. and Sumner, R.M.W. (1987) Pastures for Sheep Production. In: Nicol, A.M. (ed.) *Livestock Feeding on Pasture.* Occasional Publication No. 10, New Zealand Society of Animal Production, pp. 89–103.

Rogers, G.L. (1985) Pasture and supplements in the temperate zone. In: Phillips, T.I. (ed.) *The Challenge: Efficient Dairy Production.* Australian Society of Animal Production, Warragul, pp. 85–108.

Rolston, P., Lambert, M.G. and Clark, D.A. (1982) Weed control options in hill country. *Proceedings of the New Zealand Grassland Association* 43, 196–205.

Sheath, G.W. (1983) Pasture utilisation in hill country 1. Influence of grazing duration and land contour. *New Zealand Journal of Experimental Agricuture* 11, 309–319.

Sheath, G.W. and Boom, R.C. (1985) Effects of November–April grazing pressure on hill country pastures 2. Pasture species composition. *New Zealand Journal of Experimental Agriculture* 13, 329–340.

Sheath, G.W., Webby, R.W. and Pengelly, W.J. (1984) Management of late spring–early summer pasture surpluses in hill country. *Proceedings of the New Zealand Grassland Association* 45, 199–206.

Sheath, G.W., Webby, R.W. and Boom, R.C. (1987) Facial eczema in hill country – potential toxicity and effects on ewe performance. *Proceedings of the New Zealand Society of Animal Production* 47, 45–48.

Sheath, G.W., Webby, R.W., Pengelly, W.J. and Boom, C.J. (1990) Finishing lambs on hill country. *Proceedings of the New Zealand Grassland Association* 51, 181–186.

Smeaton, D.C. and Rattray, P.V. (1984) Winter–spring nutrition and management effects on ewe and lamb performance. *Proceedings of the New Zealand Grassland Association* 45, 190–198.

Smith, M.E., Dawson, A.D. and Short, W.D. (1976) Increasing hill country profit without raising costs. *Proceedings of the Ruakura Farmers' Conference* 28, 22–26.

Thomson, N.A., McCallum, D.A. and Prestidge, R.W. (1989) Is making hay or silage worth the effort? *Proceedings of the Ruakura Farmers' Conference* 41, 50–56.

Ulyatt, M.J. and Waghorn, G.C. (1993) Limitations to high levels of dairy production from New Zealand pastures. In: Edwards N.J. and Parker W.J. (eds) *Improving the Quality and Intake of Pasture Based Diets for Lactating Dairy Cows.* Occasional Publication No. 1, Department of Agricultural and Horticultural Systems Management, Massey University, pp. 11–32.

Vartha, E.W. and Hoglund, J.H. (1983) What is the make up of a dryland pasture? *Proceedings of the New Zealand Grassland Association* 44, 204–210.

Webby, R.W. and Sheath, G.W. (1991) Group monitoring, and basis for decision making and technology transfer on sheep and beef farms. *Proceedings of the New Zealand Grassland Association* 53, 13–16.

Management of Rangelands: Paradigms at Their Limits

M. Stafford Smith

National Rangelands Program, CSIRO Division of Wildlife and Ecology, PO Box 2111, Alice Springs, NT 0871, Australia

INTRODUCTION

The rangelands of the world are those areas which are too dry, too unreliable, too infertile or too remote to warrant intensive management inputs. These lands overlap with savannas and grade into the more intensive pastures, which are considered elsewhere in this volume. This chapter focuses on the semiarid to arid grazed rangelands of the world, where management units are large commercial fenced paddocks, regions grazed by nomadic herders or other extensively managed areas. Nearly 30% of the world's land surface falls into this definition.

The thesis of this chapter is that the characteristics of the rangelands stretch the management paradigms obtained from other, more mesic systems, often past their limits. Issues such as temporal and spatial heterogeneity, which may be sidestepped in many systems, become driving forces which cannot be ignored in the rangelands, because of their implications at the scale of management. As a result, approaches to vegetation change, optimal animal production and even human decision-making and decision support must be reassessed and redesigned. Yet, by definition, rangelands have low productivity, so that research effort per unit area has been much less than in more intensively used systems. Conceptual thinking has thus taken a long time to break free from the paradigms imposed from other regions. It is timely to review the changes of the past decade, with their advances and limitations.

The chapter identifies some special features of rangelands and then examines current paradigms of vegetation change, animal production and decision support in the specific context of their use for management, but from the point of view of research. I then turn to how paradigms impinge directly on management, both on the ground and at a policy level, from the point of view of the managers. I conclude by discussing how rangelands should be classified to avoid problems of inappropriate generalization in the future.

WHAT IS SPECIAL ABOUT RANGELANDS?

Rangelands usually fall at one end of at least four continua, reviewed many times (see Stafford Smith and Pickup, 1993, for sources) and summarized in Table 12.1. Although rangelands are loosely unified by these characteristics, the term still encompasses a huge variety of ecosystems. Important axes of variation include fertility, moisture means and variances, seasonality, temperature, soils, history, and so on. I return to discuss this diversity later.

Since management priorities and options are determined by land use, it is essential to be cognizant of the purpose of land use when discussing research and research paradigms. The uses of the rangelands are dominated by grazing, but with a variety of social goals, summarized in Table 12.2. There are many other

Table 12.1. Some continua which tend to critically characterize rangelands, with consequences compared with other grazing systems.

Factor	Consequences	
Low productivity (water ± nutrient limited)	Large management units (paddocks, parks, nomadic grazing areas, home ranges of wild animals)	Units encompass (rather than segregate) spatial heterogeneity
Variable rainfall*	Extreme interannual variability	Difficult to separate the effects of management from the effects of temporal variability
Mainly natural vegetation	> 100 interacting plant species, and multiple life forms (tree, shrubs, grasses, herbs, etc.)	Ecologically complex (compared with improved pastures dominated by two or three species)
Limited scientific attention	Less information (despite greater complexity)	Cannot know much about all individual localities ('adaptive management' approaches become vital for managers)

*Globally, rainfall variability increases with decreasing latitude, decreasing annual rainfall and in regions subject to the El Niño–southern oscillation phenomenon (Nicholls and Wong, 1990). Most rangelands satisfy two or even three of these criteria, although there are cold rangelands in northern Eurasia and in southern South America.

Table 12.2. The principal forms and goals of grazing use in rangelands.

Grazing use	Intensity of inputs	Management goals
Commercial grazing (ranching)	Low inputs of labour, materials and capital per unit area (e.g. fences, nutrients, etc.)	Singular goals – efficient production of meat or fibre for markets Quality of production is increasing in importance over quantity
Subsistence pastoralism	High inputs of labour, but very low inputs of capital	Mixed goals – animals for traction, milk, skins, meat, fibre, as well as integration with other agriculture for subsistence – with some market links for surpluses Numbers of animals may be more important than condition
Wildlife grazing	Low inputs	For tourism, recreation or conservation, not always compatibly Visitors need large numbers of animals to ensure sightings, but conservation may emphasize smaller numbers of large herbivores

non-grazing uses (mining, tourism, etc.; see Fisher *et al.*, 1996), which are not considered in detail here.

Because of the low productivity of rangelands, the main biological options open to managers are:

- to manipulate vegetation, using cheap, extensive tools like fire and grazing;
- to manage production primarily by adjusting the timing, amount and location of grazing.

Research must therefore focus on the use and implications of these simple approaches, at an appropriate scale in space and time. Although capital-intensive works (e.g. for rehabilitation) are occasionally justifiable, these are usually beyond the resources of private users, since they do not provide sufficient return on investment. The following sections examine how such constraints mean that paradigms drawn from the more intensively used grazed systems are pressed beyond their limits in rangelands.

THE GRAZED RESOURCE: VEGETATION PRODUCTION AND CHANGE

In rangelands as elsewhere, grazing production depends on vegetation production, which in turn is driven by climate, soil conditions, the feedback effects of grazing itself and management interventions, such as the use of fire (Fig. 12.1). Unlike most systems elsewhere, rainfall is usually variable, the vegetation composition is complex and the management feedbacks are spatially diverse. Forage production is usually based on grasses and herbs and is a fast process (Stafford Smith and Pickup, 1993), varying from month to month with the weather. The stability of response of fast processes is governed by slow processes, such as changes in vegetation composition. Thus information needs for management can be loosely classified under the two components, plant production and vegetation composition.

Plant production

Models of plant production vary from simple regressions of annual net primary productivity (ANPP) on rainfall-related measures (reviewed by Wisiol, 1984), through growth relationships based on a water-use efficiency and soil-moisture budget (e.g. Hobbs *et al.*, 1994), to detailed process models (Noble, 1975; Redetzke and van Dyne, 1979; Hanson *et al.*, 1988; Seligman and van Keulen, 1989; McKeon *et al.*, 1990; Moore *et al.*, 1991; Richardson *et al.*, 1991), which allow for nutrient pools, litter and perhaps several plant pools on a weekly or daily time step, to mechanistic grass-sward models (e.g. Thornley *et al.*, 1994), which may incorporate leaf geometry, internal plant shading and the physiology of photosynthesis. Most models for which parameterization has been attempted in rangelands treat forage as a homogeneous sward, effectively like a crop. Tree competition may be included (Scanlan, 1992) and a litter pool is important, but very few rangeland models have more than one forage pool with an account of competition between pools. Noble (1975) and Hanson *et al.* (1988) both allow multiple pools, but this makes the models exponentially more difficult to parameterize and validate. The simplifications are quite adequate in higher-productivity pastures. Although all such pastures have more than one species in them, a few species of broadly comparable quality are dominant; consequently, the changes in quality of the sward over time dominate over the differences between species.

For the drier rangelands, the main failings of forage production models derived from high productivity pastures are as follows.

- They usually fail to account for plant establishment thresholds (most such models assume that growth will occur once soil moisture is above wilting point, which is more realistic in mesic environments).
- They do not usually model plant competition other than that between tree and forage pools (some exceptions are noted above).
- They mostly treat the pasture as an aggregated point without spatial variation (Goodall (1967), Noble (1975), Schlesinger *et al.* (1989), Coughenour *et al.* (1990), and Pickup (1995) are examples of exceptions using a variety of approaches, but none has been widely applied to management).

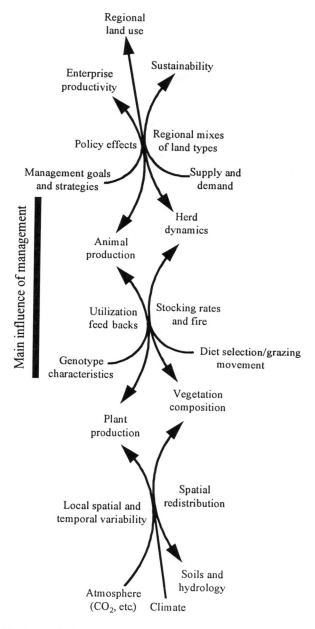

Fig. 12.1. Key factors in the rangeland environment, building from the driving forces of climate acting on the resource base, through the production system, up to management and regional land use. CO_2, carbon dioxide.

Despite these limitations, the simplified models probably predict forage production adequately for most rangelands dominated by grass or grass and trees, where slow and fast processes are not greatly confounded. Trees have a long turnover time relative to the ground layer, so that establishment issues can be ignored for the former. Problems arise, however, in rangelands containing long-lived palatable shrubs (e.g. shrub steppes, such as Australia's chenopod shrublands). These plants represent a significant part of the forage supply, especially in dry years when forage is at a premium, and yet their establishment requirements rarely occur. Their turnover times fall between those of trees and grasses and thus they blur the boundary between compositional change and annual forage production. They remain the least well-handled type of rangeland production system.

Vegetation composition and succession

The importance of vegetation composition in rangelands has long been recognized. It has been held up as the index by which vegetation condition and productivity can be judged (Dyksterhuis, 1949; see reviews in Lauenroth and Laycock, 1989), originally based on the equilibrium dynamics of Clementsian succession towards a climax vegetation state (Clements, 1916). Changes in paradigms relating to succession in recent years have often been reviewed (see S. Archer, Chapter 4, this volume), but these are particularly significant in the rangelands. In brief, the elegant concept of Clementsian succession has been found wanting in most rangelands, for numerous reasons (reviewed in Lauenroth and Laycock, 1989; Westoby *et al.*, 1989; Friedel, 1991; Stafford Smith and Pickup, 1993; Walker, 1993b). After an interregnum of uncertainty, range scientists have now enthusiastically embraced a new non-equilibrium paradigm of states and transitions (S&T), most clearly defined by Westoby *et al.* (1989). Both Clements (1916) and Westoby *et al.* (1989) were clear about the limitations and applicability of their conceptual models; in both cases, however, subsequent users have found the concepts easy to grasp and have tended to apply them without heed for the original product warnings.

Stafford Smith and Pickup (1993) note two limitations of the S&T approach. First, it does not readily account for spatial links across the landscape (Ash *et al.*, 1995); grass loss and erosion in one area may lead to deposition and shrub increase in another, depending on the spatial layout of the landscape. Walker (1993b) counters that the S&T concept must be applied at an appropriate scale of vegetation unit, in order to ensure that the pressures which trigger a transition are locally homogeneous. However, the scale of landscape function does not necessarily accord with the scale of management. Second, the emphasis of S&T on the event of a transition seems to be taking attention away from chronic precursors of change and the different rates of transitions. In fact, some changes happen gradually, others rapidly, relative to management time-frames. This was definitely not intended by the proponents, who note the importance of describing precisely these aspects of transitions.

The reality of vegetation change is that, on a fine enough temporal scale, it is a continuous process. The problem in rangelands is that different vegetations, and indeed different parts of one complex vegetation patch, show different characteristic rates of change relative to the tempo of management activity (which itself varies between systems). Thus trees, shrubs, grasses and herbs all respond differently, but within each category there is also a wide range of life-history strategies in response to the many opportunities provided by climatic uncertainty (Shmida *et al.*, 1986; Stafford Smith and Morton, 1990). Again, the spatial scale of management means that these may all be encompassed by a single management unit.

From the point of view of modelling, there have been attempts to comprehend this continuous system of change. In the simpler tree/grass subtropical systems, the competition models of Scanlan (1992) have worked well (McKeon *et al.*, 1990). Additional vegetation layers complicate the process, especially where some shrubs are palatable. As mentioned, Hanson *et al.* (1988) and Noble (1975) handle multiple vegetation pools, but without attempting to model establishment of the long-lived pools. IMAGES (Hacker *et al.*, 1991) addresses establishment in the problematic shrub steppes, but with a coarse time step and too few other pools (Hacker, 1992). None of these models is truly spatial. Archer *et al.* (1988) consider spatial dynamics and, at a genuine landscape scale, Pickup (1995) has begun to model cover responses from multitemporal satellite data.

Ultimately, models must capture the processes of establishment, persistence and competition in the context of fire and grazing, as expressed by different plant functional types and life histories. We still lack a comprehensive theory of how these different plant characteristics respond to various water and nutrient limitations, with less emphasis on the radiation limitations which prevail in mesic systems. However, all the elements are present for a more comprehensive treatment of this problem. In my opinion, this will arise by drawing together four strands, as Reynolds *et al.* (1995) have begun to do. An improved understanding of (i) the role of establishment characteristics (building on work reviewed in Hughes *et al.*, 1994) will be coupled with that of (ii) the persistence of established plants under varying resource regimes (Tilman, 1990; Campbell and Grime, 1993; Reynolds and Pacala, 1993; Chesson, 1994; etc.), in order (iii) to parameterize transition matrix approaches (Moore, 1990; Moore and Noble, 1990; Law and Morton, 1993) in terms of plant functional types (Walker, 1993a). This understanding can then be coupled (iv) with traditional production models (Hanson *et al.*, 1988; McKeon *et al.*, 1990) to assess how the probability of alternative vegetation changes varies under different management regimes, given particular patterns of weather. Results from analyses such as this will allow S&T management models to be formulated with greater objectivity and to be linked mechanistically to economic outcomes (Walker, 1993a).

This approach will still not encompass spatial connectivity. This can be approached, on the one hand, from remotely sensed data, as employed by Pickup

(1995), and, on the other, from detailed models of spatial processes (e.g. Schlesinger *et al.*, 1989; Tongway and Ludwig, 1994), which will eventually need to be constructed on digital elevation models. The latter are hopelessly data-intensive for most of the rangelands at present, but small area analyses will provide insights which help to parameterize the remotely sensed models. The advent of cheap radar-satellite technology may provide new topographic information in a decade or so.

Even if all these strands can be drawn together to provide comprehensive, continuous models of vegetation change, conceptual simplifications such as the S&T model will continue to be essential for management. However, it is important that the latter does not move from paradigm to dogma (I. Noy-Meir, personal communication). In my opinion (and in that of Walker, 1993a), the S&T model is an exceedingly valuable management and classification paradigm when used in the appropriate systems (those where rates of transition are fast relative to management activities), but is an inadequate scientific paradigm. The latter requires a more generalized concept, recognizing the complexity of change at different temporal scales, from which one should then simplify in an appropriate way for management rules of thumb in different systems. Thus S&T simplifications will be most effective for arid systems with threshold changes, whereas a greater emphasis on equilibrium concepts may remain useful for management in subtropical rangelands (Danckwerts and Tainton, 1993; see Table 12.4 below).

THE GRAZING PRODUCT: ANIMAL PRODUCTION AND POPULATION DYNAMICS

There are two reasons why we need to understand how animals use and affect their forage resource – first, to predict the short-term economic outputs of different management strategies and, second, to predict the long-term impacts on the sustainability of production. Animal production is understood in fine detail under well-controlled conditions, with models such as AUSPIG (Black *et al.*, 1993) even able to respond to the amino acid balance of intake proteins. However, such models are not useful in rangelands, due to the uncertainty of what is actually going into the animal, which is complicated by diet selection in complex swards, both in space and time. The feedbacks on vegetation production, as indicated in the previous section, are even more complex.

Defining stocking rates: beyond Jones–Sandland relationships

Concepts of stocking rates (and 'carrying capacity' – see Caughley (1979) for confusion about this term) have probably generated more rangeland literature than any other single element of management. The definition of a 'correct' stocking rate is seen as the basis for optimizing production and avoiding damage to the resource. Animal scientists (as well as agricultural science and economics courses) still widely base their conceptual thinking about stocking rates on the Jones–Sandland relationships between stocking rate and productivity (Jones and Sandland, 1974). A good review of this approach in the context of assessing land degradation is given by Wilson and MacLeod (1991) and summarized in Fig. 12.2. Wilson and MacLeod (1991) argue that land degradation ('overgrazing' in their terms) can only be said to have occurred when the linearity of the basic relation-

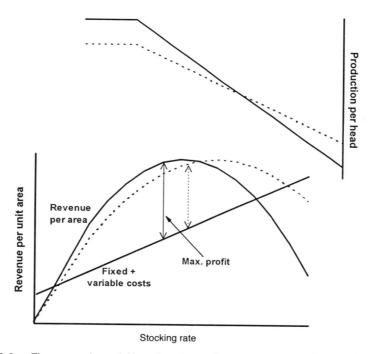

Fig. 12.2. The economic model based on Jones–Sandland relationships (after Wilson and MacLeod, 1991). The upper solid line shows a plateau of production per animal at low stocking rates, which then linearly decreases with increasing stocking rate; this relationship is usually obtained from grazing trials. The lower solid line re-expresses productivity in terms of economic return per unit which has a quadratic relationship with stocking rate. Given some fixed costs of production, and others that depend on the numbers of animals, the straight line then shows the total costs of production per unit area, so that the maximum profit occurs where these lines are furthest separated. This point putatively represents the optimum stocking rate from the economic point of view (ignoring risk). In contrast, the biological optimum is at the peak of the parabola. The corresponding dotted lines show how a different element of production (e.g. wool instead of meat growth) may be less sensitive to stocking rate in the same system, and consequently will have a higher economic optimum stocking rate.

ship breaks down at higher stocking rates, and conclude that this happens less often than has been assumed. Thus the paradigm is being used not only to establish stocking-rate recommendations for management, but also to identify when damage may be caused.

There is no doubt that the linear Jones–Sandland relationship is generally valid in small enclosures, having been measured many times, especially in high-production grasslands (e.g. references in Wilson and MacLeod, 1991). It is based on an underlying functional response between forage intake and availability, reviewed, for example, by Short (1987), who shows that, on a metabolic weight basis, kangaroos, sheep and rabbits have a similar maximum intake rate. However, in rangelands there are many practical problems with the use of the Jones–Sandland paradigm.

- The relationship varies from year to year with rainfall in ways which can differ between vegetation types and climates. Small amounts of interannual variation can be averaged, but this is not realistic in many rangeland climates. The economic paradigm of a single optimum stocking level is valid if interannual variability is small, but is not conducive to considering stock numbers which vary from year to year with temporal autocorrelation.
- Large paddocks are not used evenly by herbivores (Coughenour, 1991; Pickup and Stafford Smith, 1993) – the spatial use of the paddock might trigger severe thresholds in productivity when one vegetation type is exhausted, or it may enable animals to buffer their production to some extent. Precisely this form of spatial buffering was also found by Scoones (1993) in a communal cattle area of Zimbabwe. Although it is theoretically possible to perform a grazing trial at the whole-paddock scale, it is prohibitively expensive and it would be difficult to find true replicates anyway. Spatial problems should matter least for reasonably homogeneous pastures types.
- There are practical measurement problems. The relationship must be determined for a particular herbivore/pasture combination by a stocking rate trial over several years. Not only do small enclosures fail to mimic spatial aspects of large paddocks, but also vegetation change often occurs at high stocking levels, so that some of the enclosures are really measuring the relationship in a different vegetation type. Thus the relationships are of doubtful practical value.

What are the solutions to these problems?

First, developments from optimal foraging theory and detailed mechanistic models (Belovsky, 1986; Ungar and Noy-Meir, 1988; Baker *et al.*, 1992; Spalinger and Hobbs, 1992; Gross *et al.*, 1993; Parsons *et al.*, 1994; Thornley *et al.*, 1994) have a great deal to offer in the near future by improving our understanding of why different types of plants are preferred by herbivores with different body constructions. However, these models have yet to be parameterized for complex vegetation types. It also remains to be seen whether both herbivores and plants can be classified into functional types which adequately capture the complexity of natural rangelands vegetation, but, if so, these types could be used to parameterize management effects in the vegetation models described earlier.

Second is the question of spatial selection of diets and distribution of grazing impacts. The assumption of even use of paddocks breaks down as paddocks become large, a fact that is well documented in rangelands (Low *et al.*, 1981; Senft *et al.*, 1983; Pickup and Chewings, 1988; Stafford Smith, 1988). The assumption is also incorrect under management by herders (e.g. Scoones, 1993; review in Coughenour, 1991), albeit for different functional reasons. The spatial distribution of grazing impact can be studied at both a broad landscape scale (Low *et al.*, 1981; Senft *et al.*, 1983; Stafford Smith, 1988) and on the basis of local decision-making by animals (e.g. Stafford Smith *et al.*, 1985; Senft *et al.*, 1987; Bailey *et al.*, 1989; El Aich *et al.*, 1989). Ultimately the former depends on the latter but, from the point of view of application to paddock design (Cridland and Stafford Smith, 1993), the problem may be more efficiently approached directly at the landscape scale (Pickup and Chewings, 1988). Thus these spatial models need greater incorporation into the diet selection and grazing impact models. Of course, a much greater intensity of fencing can obviate the need for this understanding – this is precisely a benefit identified for intensive grazing systems (Hart *et al.*, 1993) – but this is often not economic in rangelands.

Third is handling the variation through time effectively. This is already achieved by dynamic animal production models. These are based on Jones–Sandlands relationships, but driven by realistic forage availabilities and re-expressed in terms of utilization levels (McKeon *et al.*, 1990). Instead of the static average view of a single curve, therefore, these models permit a full range of strategies with varying stocking rates over time to be analysed, thus capturing the temporal dynamics of stock numbers. We will return to the use of these models. However, they retain one general limitation, which is that they predict only animal growth rates. In rangelands, reproduction and mortality rates are equally important in the derivation of economic outputs.

Beyond growth – reproduction and mortality

The paradigm for productivity in intensive grazing systems is usually growth rates – body weights and wool or milk production. This is reasonable, since animal reproduction and mortality are closely monitored and controlled in small paddocks, and enterprises may even purchase their young stock. Concern about the population parameters often revolves around the competence of different breeds (e.g. twinning in sheep for fat lamb production) or rare extreme environmental events (e.g. cold deaths (Donnelly, 1984)).

In rangelands, however, most enterprises breed their own stock. Birth and death rates become as important as growth for comparing alternative management strategies (Foran and Stafford Smith, 1991). In Australia, for example, mortalities of breeding cattle may vary from 4 to 25% in different years (Foran *et al.*, 1990), and sheep losses can be as high as 65%. Reproduction rates for cattle, even in good years, vary from 50 to 90% due to management alone (i.e. this range can be obtained on adjoining properties that have similar resources but differ in stocking rates and other factors such as breed (Foran *et al.*, 1990)), and the mean lambing rate for whole shires in Queensland varied between 22 to 79% during 1967–1985 (O'Dempsey, 1987). Comparable figures can be found in subsistence grazing systems in Africa (e.g. Scoones, 1993). Reproduction is usually more sensitive to stocking rate (or pasture utilization) than is wool growth (O'Reagain and Turner, 1992), so an assessment of different management strategies under climatic variability which does not account for herd dynamics will greatly underestimate the effects of management (Stafford Smith and Foran, 1992).

In wildlife populations, it is normal to estimate population parameters in the form of numerical response functions (e.g. Bayliss, 1987; Choquenot, 1991). However, such analyses assume that the populations are unconstrained, always tending towards their carrying capacity for the current conditions. This is reasonable for unmanaged wildlife populations, but is insufficiently sensitive for domestic animals and perhaps even managed wildlife. In these cases management strategies usually keep population numbers below the environmental carrying capacity to increase the efficiency of production. The effectiveness of other management practices, such as predator control, weaning and disease management, also has significant impacts on survival. To assess the economic significance of such strategies, response models need to be more sensitive than is possible with the wildlife approach.

In the absence of predators and extreme environmental conditions, mortality ought to be dependent on body condition, or the rate of change of this. Likewise, in the absence of special management, conception and reproduction is mainly dependent on body condition and physiological status, providing an interaction with a male animal is assured (not always the case in large paddocks). There are models of these factors (e.g. White *et al.*, 1983), although these have usually been restricted to the low end of mortality and high end of reproduction. The need for more comprehensive relationships has been recognized before (e.g. Gillard and Monypenny, 1988), but these are only now being developed for the full range of conditions in the Australian rangelands (Moore *et al.*, 1995). For the first time, this will enable a true economic reckoning of the implications of different pasture utilization levels to be made, by modelling the entire system represented in Fig. 12.1. In the meantime, we must shed the presumption that growth alone matters in the rangelands.

THE GRAZING MANAGER: SUSTAINABLE MANAGEMENT DECISIONS

Problems with scientific paradigms do not end at the boundary between the biophysical and cultural system. If you are managing a spatially diverse,

Table 12.3. Some key differences of emphasis in rangelands compared with more intensively used pastures and their consequences for assisting management decisions with models and decisions support systems.

Issue	Improved pastures	Rangelands
Spatial heterogeneity	Between management units ⇒ Point/aggregated models often adequate	Within management units ⇒ Spatial models often needed
Temporal variability	Reasonable annual replicability Management cycle mainly within a year ⇒ Static annual models adequate	Large year-to-year variability Management cycle over several years ⇒ Dynamic models required
Information availability/functional complexity	Relatively good, both on ecological processes and resource base ⇒ Optimization feasible	Poor, leading to flat optimizing surfaces ⇒ Scenario/'what if' analyses more useful than optimal solutions
Commercial management goals	Reasonably concise, incorporating economics ⇒ Can develop optimal answers for specific strategies; must put in economic context	Often mixed or muddled, but still incorporating economics ⇒ Need manager to balance 'what if' scenarios; must put in economic context
Adaptive feedback	Easy and quick ⇒ Value of decision-support tools soon obvious	Difficult and slow ⇒ Scepticism about decision-support tools
Research/extension	Managers seek information ⇒ Linear transfer more often adequate	Managers rely on experience ⇒ Participatory research more often needed

temporally variable resource with limited information, your decision-support needs are philosophically rather different to someone using intensively managed, improved pastures. The differences, outlined in Table 12.3, are exemplified by the 'three Rs' of the agronomic experimental paradigm described as the Rothamsted model – replication in space, repeatability in time and reductionism in studying processes. All three are reasonable in intensive pasture management, where management units are small enough to break down spatial heterogeneity, where rainfall is sufficiently regular from year to year to allow repeated experiments and where biological processes are sufficiently simplified and well known for it to be possible to progress by reductionism. All three break down at real management scales in extreme rangeland environments. Spatial heterogeneity unavoidably occurs within management units, 'no two years are ever alike' and biological complexity means that many processes are best studied for their emergent properties at a more holistic level.

In all agricultural research, there has recently been an increased recognition of the importance of understanding the goals and socioeconomic aspirations of managers, and hence of the need to express biological results in an economic context. This is true in the rangelands as elsewhere. However, whereas enterprises with short management cycles (crops, fat lambs, etc.) see rapid and well-defined feedback between interventions and results, many rangeland management activities take years to pay off. Even then, the effects of management are hard to separate from the background noise caused by rainfall variability. Understandably, managers tend to be sceptical of research recommendations, especially if they are contrary to current practice. This scepticism may in fact be well justified if the results are based on reductionist research which has not been verified in the context of the whole complex system.

Just as ecological and production paradigms must be reassessed carefully in rangelands, so too must paradigms for the process of research, extension and decision support (Table 12.3). Some points arise directly out of the ecological paradigms: the need for spatial models when assessing individual management units; the importance of dynamic, not static, models of the system; and a genuine difficulty with optimization approaches, which, coupled with unclear objectives, limited information and difficulty in obtaining convincing adaptive feedback, means that a 'what if' scenario approach to decision support is often more appropriate. This is not to deny all value of optimizing approaches – some recent developments which couple economics with complex ecology for the first time are exciting (e.g. Walker, 1993a; Beard, 1994).

Other problems are more socially orientated, however. Various surveys in Australia (O'Keefe, 1992) and the USA (Rowan *et al.*, 1994) show that rangeland producers differ from the wider farming population. In Australia, for example, a distinctive mode of learning is recognized to dominate in rangelands (O'Keefe, 1992); this mode emphasizes learning by experience over learning by observation, the latter being more common in intensive farming operations. Such differences have serious consequences for the appropriate 'extension' paradigm for management. Jiggins (1993) discusses six approaches which are in use, and begins to identify circumstances when different paradigms are more or less appropriate. The historical researcher–extension–manager linear transfer paradigm works adequately for information which is well defined, has a small variance and fast feedback, and which is often sought by the managers themselves (e.g. insecticide doses, crop planting dates or supplement mixes). In contrast, many rangeland problems are ill defined, recommendations have a large variance in that they differ greatly for different circumstances, there is poor and slow feedback on success and the learning mode is not usually one of seeking information from others. In these circumstances, participatory paradigms are more appropriate. There are a number of these (see Ison, 1993; Jiggins, 1993; Rural Extension Centre *et al.*,

1993), but the participatory problem-solving approach adopted recently in Queensland seems to be working well (Clark and Filet, 1994).

These are differences between rangelands and more intensive systems. There are also differences between rangelands in different parts of the world. It is essential to recognize where control occurs in different societies, so as to identify the appropriate path by which societal interests may operate. Balent and Stafford Smith (1993) consider the triangle of interactions between the resource, the livestock and human beings, and suggest that four tiers of influence on land management can be identified: individual land managers or herders, local communities or peer groups, regional government and national government. With this framework, it is useful to look at how strong the links are around the triangle and how strong they are at the different tiers (Fig. 3 in Balent and Stafford Smith, 1993). In the past, the community level of African systems has often been stronger in allocating land-use rights than either individuals or regional government, although privatization of tenure rights in recent years has reduced this strength. In contrast, Australia has never had a strong level of management at this level, with most power being at the individual level. It is arguable that the recent Land Care movement (Campbell, 1994) is beginning to empower a new (for Australia) peer-group level. The significance of this analysis is that it identifies the weak points in the management hierarchy, and helps to highlight whether paradigmatic solutions in one social system are transferable to another.

PARADIGMS FOR THE MANAGER

So far, I have been emphasizing problems for management in rangelands from the research point of view. It is now time to turn more positively to the ways in which paradigm changes may affect different types of rangeland managers.

Management by commercial ranchers

At a biological level, an individual manager can manage the vegetation and the level and spatial distribution of grazing. This must be done within a realistic economic context which recognizes variability through time, and must go hand in hand with effective feedback from monitoring.

Managing the resource

There seem to be two important new messages for the management of vegetation in rangelands (and to some extent the soil resource directly). One is the use of a properly implemented S&T concept. 'Properly implemented' means observing all the steps suggested in the original Westoby *et al.* (1989) paper and emphasizing management for a transition rather than just at a transition. The second is the adoption of adaptive management (*sensu* Walters, 1986; Danckwerts *et al.*, 1993). But, while the S&T idea has rapidly captured the imagination of researchers, funding agencies and managers alike, the adaptive management philosophy of critical experiments (e.g. Chapters 9 and 10 in Walters, 1986), yielding maximum information for minimum risk to the resource, has been poorly researched in rangelands and little presented to managers. The scale of the rangelands is such that experimental data will never be gathered by scientists from all land types and combinations. Managers themselves must therefore indulge in some well-directed experimentation if they wish to achieve the best results possible from their own land (Burnside and Chamala, 1994). In many ways, the philosophy of adaptive management is thus more important than a particular paradigm of vegetation change, although the latter will influence the experiments undertaken through the former.

Stocking rate strategies

The critical paradigm shift in relation to managing grazing pressure has been to move away from static notions of optimal stocking rates and to recognize that these inevitably vary through time. The ways in which stocking rates are varied through time, deliberately or otherwise, we may term 'stocking rate strategies'. The paradigm shift has been greater for management researchers than for managers themselves, who have been practising such variations throughout history, but it does impinge on the latter as a result of the expectations of extension and legislation.

Recommendations on fixed stocking rates have gradually evolved to incorporate more sophisticated risk-related criteria, such as a stocking rate which would not have exceeded (say) 30% utilization in 70% of years when simulated through past decades of weather data (McKeon *et al.*, 1990). In reality, however, producers can vary their stocking rates over time in ways which lie on the generalized continuum from constant to highly variable (termed opportunistic (Sandford, 1983), unlimited (Riechers *et al.*, 1989), reactive (McKeon *et al.*, 1993) or trading (Stafford Smith *et al.*, 1994)). The trade-off is that, the more constant a producer attempts to be, the lower the target stocking rate must be (Fig. 12.3). There are considerable benefits to be gained from stocking rates that are lower than those regarded on short term criteria as economically optimal (Purvis, 1986), but in the extreme case one has so few stock that avoiding ecological risk is unprofitable. The point of balance depends on the resilience of the vegetation to short-term overgrazing, the variability of weather over time, the economic position (e.g. indebtedness) of producers and their attitude towards risk (e.g. Rodriguez and Jameson, 1988; Bernardo and Engle, 1990; Stafford Smith and Foran, 1992). As discussed earlier, only dynamic models can simulate this range of behaviour; these are now becoming available.

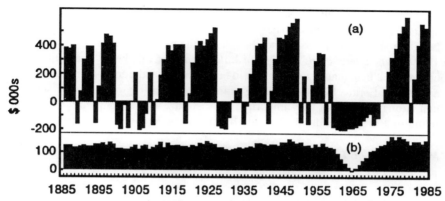

Fig. 12.3. An economic comparison over 100 years of real climatic data, showing the annual cash surplus of (a) a relatively high-stocking-rate trading-management strategy that tracks forage availability to have a close to optimal utilisation rate all the time compared with (b) a very low-stocking-rate strategy (after Foran and Stafford Smith, 1991). (a) averages a higher cash return (with much greater variance) than (b), although different taxation policies and debt levels can reverse this order. (a) must reduce animal numbers very rapidly in a dry year, thus being subject to the vagaries of the market, whereas (b) is largely 'drought-proof' (e.g. the 1960s were the driest decade on record, the 1970s the wettest); more implications are discussed in Stafford Smith et al. (1994). This example was modelled using a herd dynamics and economics model which flexibly mimics manager tactics in detail, but does not have true biological feedbacks at the level of pasture growth and animal biological rates; models incorporating the latter are now becoming available.

Spatial distributions

The other major paradigm rediscovery is that of the spatial element of grazing. This has been aided by ecologists' discovery of patch dynamics, but retarded by the preconception of even use in small intensive pastures. Models of grazing distribution have already been mentioned. One major issue for management is that, although distance to water is an important controlling variable in paddocks exceeding about 500 ha, the animals' preference for different types of vegetation and the spatial heterogeneity of the latter determines whether management of water-points alone can affect the patchiness of impact (Cridland and Stafford Smith, 1993). A second concern is that land degradation does not normally occur uniformly over a landscape, but develops through patch processes, which are consequent on very uneven usage (Andrew, 1988; Fuls, 1992; Pickup and Stafford Smith, 1993). Grazing pressure can be adjusted for the areas of greatest impact, but this may be financially inefficient (Pickup and Stafford Smith, 1993). Curiously, early commercial producers, who herded their animals much more closely than today, probably recognized the spatial variability of grazing more than their modern counterparts. In this case the paradigm reassessment has been among managers as well as researchers.

At a broader regional scale, Danckwerts *et al.* (1993) note that the use of spatial heterogeneity, through nomadism, can buffer temporal variability. This has been the approach of subsistence herders in the past (Sandford, 1983; Scoones, 1993). Danckwerts *et al.* (1993) see little chance of this being compatible with commercial land tenure systems, but in fact 'mechanized transhumance' (Stafford Smith and Foran, 1993) – trucking animals between regions to track seasonal conditions – is increasingly important in Australia, where companies own leases in multiple areas or individual pastoralists make agreements with others.

Monitoring and feedback

Good management critically depends on feedback which assesses its success (Danckwerts *et al.*, 1993; Burnside and Chamala, 1994). Although there are genuine problems with achieving such feedback in the variable conditions of rangelands, it is also true that rangeland managers are particularly poor at maintaining records. In Australia, it remains common for managers to keep no more than their sale records and the limited financial details which are required by law; they have little animal production information, no vegetation records and no records of differences in paddock management and response other than those in the hazy memory of the practitioner. As management becomes more intensive, records improve and managers rapidly become better at appraising situations realistically. Adaptive management cannot function in the absence of such records; a record-keeping paradigm shift is waiting to take place, perhaps with an incoming younger generation.

Meanwhile, government agencies do see that monitoring is important, and have recognized that monitoring on-farm and for regional land management are quite different activities (Pickup and Stafford Smith, 1993). The former requires easily measured, acceptable *aides-mémoire*, while the latter may need to stand the test of courts on a much more comprehensive, more regional, but perhaps less detailed basis. Considerable work is therefore progressing on indicators of sustainability (Hamblin, 1992) and on manager-derived performance indicators (Clark and Filet, 1994).

Management by subsistence pastoralists

While subsistence pastoralists use most of the same resources as commercial ranchers, they have different goals. This means that numbers of animals are often more important than efficient growth, and mixtures of animals and of land uses are common. Labour inputs are usually more intensive, so that animals may be herded continuously; this opens many more opportunities for control of the spatial distribution of grazing, although there are still foci of activity, such as water-points and villages. Historically, many subsistence users were nomadic and used very large areas, but new borders, wars, increased populations and political changes mean that such movements are now constrained in most areas of the world. The majority of subsistence herding is thus semisedentary.

Hence all the concepts described for commercial ranching are important here too. However, important paradigm shifts have come from a different point of view.

The demise of 'carrying capacity'

Colonial administrations in the past tended not to recognize the implications of the way in which management goals of subsistence pastoralists differ from those of commercial ranchers. High stocking rates were perceived to have devastated the land, regulations were established (but rarely enforced) to restrict stock numbers and undesirable land tenure changes were implemented. This history has been documented by numerous commentators, such as Sandford (1983) and Behnke (1994), and has led to a significant shift in thinking about ecological processes and their implications for subsistence pastoralists.

The concept of 'carrying capacity' in wildlife populations has always meant the maximum stocking rate that the system could support, towards which an uncontrolled population will tend. In Africa this usage became confounded (Caughley, 1979) with an optimum stocking rate (economic optimum, at that), based on the Jones–Sandland analysis for commercial ranches. Even the original usage has little meaning in highly variable systems – Shepherd and Caughley (1987) suggest that a threshold is crossed when the coefficient of variation of rainfall exceeds about 30%, above which the equilibrium concept of carrying capacity becomes no more than theoretical. Furthermore, even if one could define an optimum stocking rate, this will usually be higher for objectives such as milk, traction or animal numbers than for meat production (see Fig. 12.2). All of these factors have led to the demise of the confused concept of 'carrying capacity' and a major shift to the paradigm of 'non-equilibrium dynamics', reviewed well in Behnke *et al.* (1993).

In this context, the demise of the carrying capacity paradigm is a good thing. Not so good is a new confusion arising in the way that the implications of its demise are applied. Commentators are inferring that the land is somehow far more resilient to subsistence uses than to commercial ranching, and that there is therefore some sort of inbuilt resistance to land degradation in subsistence systems. The fact that subsistence managers can meet their production goals at considerably higher stocking rates than are optimal for commercial producers (Abel, 1993, p.194) does not mean that the land is more resilient; it just means that some factors of production are far less sensitive to stocking rate than others (see Fig. 12.2). Some ecosystems are more sensitive than others, as has long been known (e.g. Walker *et al.*, 1981), and results are being extrapolated wrongly from more to less resilient systems. Furthermore, under true nomadism with lower populations, systems may have had longer recovery times than today, making them appear more resilient.

In short, the pendulum is swinging too far. On the one hand there is no doubt that past colonial and postcolonial policies in Africa overstated the supposed degradation caused by subsistence pastoralists, who have, despite this, somehow persisted for hundreds of years (Behnke, 1994). On the other hand, there is no doubt that some areas have suffered severe damage, and some regions are now more susceptible to the impacts of drought than they used to be, as a result of lost productivity as well as population increases and climatic trends. Even if some regions can still support large numbers of cattle, future options may be being closed off, given that subsistence pastoralists are becoming increasingly linked to markets (Behnke, 1987; Kerven, 1994) and other land uses are becoming more important. Ironically, it is the mixed-agriculture subsistence systems where the importance of mixed land-use goals was most apparent in the past; yet this is where the impacts of one type of use (maintaining stock numbers) on others is being poorly considered. The paradigms of Africa still have some development ahead of them (e.g. de Wit, 1990), not least in reconciling local interests with those of society at large, to which we now turn.

Management for policy and societal interests

Grazing management is also carried out indirectly by local, regional and national levels of policy-makers. Although this is more at arm's length, it is vital that policy-makers understand grazing ecology, since different policies can have radical impacts on management. This level of management is usually aimed at (i) provision of advisory services to the commercial or subsistence managers themselves; we have already discussed changes in the appreciation of management goals, and other participatory aspects of extension paradigms. It may also help (ii) to buffer variability through activities such as drought assistance; subsidies interact greatly with ecology, to an extent beyond my present scope. Finally, regional managers must (iii) monitor the impacts of grazing on societal interests in one way or another.

Sustainable use, not exploitation

It is the issue of protecting societal interests – for example, implementing soil conservation requirements, water quality controls and other elements of sustainability at a regional level – that is probably most complicated for policy-makers. As Behnke (1994, pp. 21–23) remarks, 'Foremost among the problems of administration is that of cost. Administration should pay for itself. This unexceptionable principle . . . places unusually severe strictures on routine administrative activity in pastoral settings' because of the large areas involved relative to their productivity, the complexity induced by seminatural systems driven by climatic variability and the 'sometimes truculent nature of pastoral populations'.

There are various ways in which society's goals may be met; these include establishing the appropriate land tenure, enacting soil conservation and other legislative controls and maintaining broad-scale monitoring systems from which to assess (or enforce) the success of other measures. There is a large literature on land tenure and associated financial instruments such as rents or rates, albeit with limited recognition of ecological realities until recently. Holmes and Knight (1994) review the approaches in some major commercial rangelands systems (the USA, Australia and New Zealand), while there is a continuing debate on the 'tragedy of the commons' (see Behnke, 1994; Hardin 1994) and other aspects in subsistence systems. Paradigms remain confused and I see no clear resolution of all the issues in these debates as yet. However, they all imply the need for some monitoring of outcomes, in which there has been progress.

The major problem with monitoring rangelands is to obtain sufficient areal coverage while collecting enough detail, so that the effects of human activities can be separated from the natural background 'noise' caused by climate variability at various temporal scales. In the past, systems have had to rely on ground measurements; these can provide an indication of regional status but, by themselves, are unable to provide the accurate and repeatable measures needed to detect trends at any realistic level of resources (Pickup, 1989). Recent work with remote sensing, based on an understanding of spatial erosion processes, grazing distributions and key elements of vegetation change, has allowed the development of a more satisfactory index of land productivity which can show trends on a decadal time-scale (Bastin *et al.*, 1993; Pickup *et al.*, 1994). The best approaches unite this remote-sensing technology with some well-targeted elements of ground assessment, being then cheaper and far more efficacious that traditional methods (Bastin *et al.*, 1993). Although these techniques have been demonstrated in some Australian systems, the new paradigm has yet to gain full political acceptance and requires proving for different vegetation, climate and management systems. Furthermore, implementing a cost-effective monitoring scheme is one thing; developing mechanisms and the political will to act on the results is another (Stafford Smith and Pickup, 1993).

Sustainable habitation, not sustainable use

Another major paradigm shift has been occurring in rangelands, especially commercial rangelands in developed countries, over the past decade. This has been a move away from exploitation of rangelands (albeit hopefully sustainable) towards a more benign concept of 'sustainable habitation' or occupation (B.H. Walker, personal communication), where a complex of mixed land uses downplays the sole significance of grazing. This has come about for two reasons. First, the assumption that grazing products from commercial rangelands are themselves of great importance has been questioned, especially in the developed nations of the world (Stafford Smith and Foran, 1993), given their small total productivity and the risks of degradation. A reassessment of the full subsidy being paid into these regions has led people to believe that management should be targeted more towards public goals of land management rather than private goals of economic production. Second, a broader appreciation of the different meanings of 'sustainable' has begun to contribute to the philosophical basis for carrying this out. In short, the difference between the enterprise goal of sustainable land use and the regional goal of ecologically sustainable land management has been recognized (Morton et al., 1995). The latter incorporates the maintenance of ecological function – water quality, biodiversity, aesthetic values and so on – at a regional scale and in the context of mixed land uses.

Exactly how the land-use conflicts involved in implementing this regional-scale paradigm should be resolved remains to be seen. However, there are many implications for policy-makers. Among them is the problem that the institutional structures of government in most countries are monolithic with respect to different land uses; departments competitively represent agriculture, mining, conservation, tourism and native lands. This leads to conflicts between land-use groups just when an integrated approach is required. A major rethink of these institutional structures governing land use in rangelands is therefore required.

Another implication is that these public bodies must facilitate decisions about whether some uses or combinations of uses are actually sustainable, especially where government policies and societal expectations in the past have led pastoralists into non-viable enterprises (e.g. Passmore and Brown, 1992). Pickup and Stafford Smith (1993) identify three criteria for determining whether a land use is locally sustainable: whether it is economically viable (usually determined by market forces, but often distorted by government intervention), whether it causes land degradation (an ecological issue, sometimes complex) and whether land management strategies which satisfy the first two criteria are socially feasible (that is, it may be possible to be viable and non-degrading but only by implementing management strategies or lifestyles which it is unreasonable to expect of the majority of managers). The latter two issues may be very hard to assess, but must be addressed. Such assessments must not be top–down; government must act as a facilitator rather than an enforcer, developing community ownership rather than community antagonism.

Thus paradigm shifts in ecology and extension flow through to changes at the level of societal expectations and political management. Given the significance of such changes, it is worth a final reflection on whether some of our past paradigm failures could have been avoided.

MAKING SENSE OF DIVERSITY

Some paradigms have changed in rangelands as a result of the general evolution of society and science. However, a significant number of paradigm failures have arisen from an inappropriate transfer of concepts from well-studied mesic environments into the rangelands' extremes. Still others have resulted from the unchallenged generalization of findings from one rangeland system to those elsewhere, which has inevitably tended to happen because of the limited work carried out in rangelands. For example, the debate on overcompensation by grazed grasses and its implications for managers has been well inflamed by such problems (Noy-Meir, 1993). What general lessons can be drawn from these changes?

As a result of repeated calls for the recognition of the differences between rangeland systems, a typology of rangelands is beginning to emerge. This typology cannot be simple, because the classification is needed at different hierarchical levels for different purposes (Allen and Starr, 1982; Pickett et al., 1987), both in terms of processes and physical scale. The importance of different elements of a typology depends on the purpose and level of management and the scale at which it is carried out. Not only do these differ between national cultures, but they also differ greatly between ranchers, subsistence pastoralists and regional managers, and even between individuals within one category.

Drawing mainly on the discussion earlier in this chapter, we can make the following statements.

1. Expectations and needs differ greatly, depending on whether research is carried out from the point of view of private individuals or society.
2. The form of production that pastoralists are seeking to create will affect conclusions about stocking levels, strategies and sustainability.
3. As the level of primary productivity increases, the options open to managers in terms of investment (and consequently in terms of possible subsidies, such as supplementary feeding and pasture management) increase.
4. As the level of rainfall (or other climatic) reliability decreases, mean productivity decreases and management decision-making becomes harder.
5. As management units become larger (through decreased productivity and increased temporal variability, as well as for cultural reasons), spatial heterogeneity becomes more important, providing both problems and opportunities for management.
6. The previous statements are 'all other things being equal'. However, systems differ significantly in terms of their mean levels of fertility, their absolute levels of temporal and spatial heterogeneity, their evolutionary history and their recent history of management; furthermore, the human cultures that occupy them differ in their expectations of land use at both individual and collective levels.

In a recent *magnum opus*, Milchunas and Lauenroth (1993) tested the concepts of Milchunas et al. (1988). They found that there are differences in ecosystem responses to grazing which relate to levels of ANPP (acting as a surrogate for rainfall, edaphic factors and fertility) and grazing history at an evolutionary time-scale. From the point of view of arid rangelands, which of course sit at the lower end of the ANPP scale, the relationships seem to be different for shrublands and grasslands. Both productivity and species composition responses change along these gradients in broadly predictable ways. At regional and local levels, where these global factors are reasonably constant, other levels of variability become important and modulate the effects of the global factors. At this level there has been no single classification of all rangelands. For example, the plant-available moisture/available nutrients (PAM/AN) plane concept (Frost *et al.*, 1986; Walker, 1993b) has been used to separate savanna environments with different levels of PAM and AN. However, this classification does not catch other factors, such as seasonality, rainfall reliability in time and space and levels of spatial redistribution, nor does it address the issue of different land users.

These elements of diversity can be considered as the different expressions of spatial, temporal and historical heterogeneity at various scales. The expressions include mean, variance, kurtosis (e.g. extreme events) and autocorrelation (or system memory). In reality, they all have a continuous distribution across scales, but it is useful to break this continuum up according to human perceptions in management. Thus the global scale sets a general regional context, a regional scale, which is larger than individual management areas, sets the context for the latter, and a local scale, within the management unit, incorporates the arrangement of landscapes and patches within landscapes.

Table 12.4 outlines what seem to me to be dominant axes of variation at the different scales and their consequences. No generalization about rangelands should be made without at least considering its validity in terms of these variables, and there is no excuse for research to confuse managers or confound policy by inappropriately conflating systems which differ with regard to these factors. With a new global research effort evolving in rangelands, as a result of international conventions relating to global change and desertification (Stafford Smith *et al.*, 1995), it is increasingly important to place results correctly on the various axes of rangelands variation.

Ultimately the mechanism for exploring the differences between all these axes of variation will be appropriate process models. However, it would be foolish and inefficient to suggest that all scales and processes should be incorporated in one 'megamodel'. Hence conceptual simplifications will still be needed for managers in particular regions. It is important that these management simplifications are developed from a consideration of the whole framework of variability, with a clear indication of their applicable systems, rather than being promulgated *de novo* as generalizable across all conditions. The paradigms discussed earlier in this chapter will then have their proper place in management.

CONCLUSION

There have been some exciting developments in the rangelands over the past decade, with more to come in the near future. These developments have drawn on ecology outside rangelands, but have been driven by the nature of rangelands, in particular by the way in which the scale of management in space and time interacts with the rangeland environment. This scale is simply different from that of management in most mesic systems. A review of why past paradigms have failed

Table 12.4. Some criteria which would help to interpret rangelands research, with some implications for management; a focus on any single criterion may be misleading. Other factors (e.g. fossorial animals (S. Archer, Chapter 4, this volume), chance taxonomic differences (Westoby, 1988), etc.) may also be important locally.

Scale	Criterion	General management implication	Some sources
General context	ANPP	Defines category of rangeland; also determines options open to management in terms of viable investments	Milchunas and Lauenroth, 1993
	Coevolutionary grazing history	General form of grazing impact on species change	Milchunas and Lauenroth, 1993
	Type of management targeted – societal or individual (ranching or subsistence)	Affects scale of interpretation below and objectives of land use – a single sustainable land use or mixed uses locally, compared with a regional balance of land uses	Morton et al., 1995
Regional	Average PAM/AN balance	Determines whether system is water- or nutrient-limited, and hence whether forage is limited by quantity or quality	Frost et al., 1986; O'Reagain and Mentis, 1990
	Interannual variability of rainfall + predictability of season of rainfall	Determines temporal variability and hence appropriateness of equilibrium (where less variable, more predictable) or non-equilibrium concepts in simplified models for management. Seasonality/temperature of production also affects life-form responses (herbs/grasses); extreme events affect establishment of deep-rooted vegetation forms	Shepherd and Caughley, 1987; Danckwerts and Tainton, 1993; Ellis et al., 1993
	Broad-scale* patchiness of fertility	Determines regional spatial heterogeneity, and hence whether land-use conflict is focused, and the options for transhumance	Morton et al., 1995; Scoones, 1993
Local	Soil texture and fertility	Determines patch resilience and constraints to average productivity	Walker, 1993a
	Fine-scale* heterogeneity of landscape (run-on, fertility, vegetation types)	Determines local spatial heterogeneity, and hence: productivity in smaller rains; landscape-scale resilience (susceptibility to degradation by local overgrazing); importance of distance to water relative to vegetation preferences in grazing distributions; with temporal variability, the applicability of Jones–Sandlands models	Shmida et al., 1986; Pickup and Stafford Smith, 1993; Cridland and Stafford Smith, 1993; This Chapter; Ash et al., 1995
	Recent (last 50–100 years) management history (fire, grazing)	Current state of landscape has an important bearing on its responses, but probably more so in regions with limited coevolutionary grazing history	Ash et al., 1995; Archer, Chapter 4, this volume
Human	Management goals for production	For a given landscape/climate, production per unit area is optimal at increasing stocking rates for, in approximate order, production objectives of meat, fibre, milk, traction and numbers	Wilson and MacLeod, 1991; Abel, 1993

*Broad-scale here means larger than typical management units; fine-scale means within one management unit – thus these definitions depend on management system.

points to a major underlying problem: ideas from one rangeland (or non-rangeland) environment have repeatedly been imposed across others, without due regard for the differences among systems. There is sufficient variety among rangelands for a good typology of the differences and their functional causes and consequences to be vital. It is no longer good enough to recognize the importance of spatial and temporal heterogeneity and then seek to ignore it.

REFERENCES

Abel, N.O.J. (1993) Reducing cattle numbers on Southern African communal range: is it worth it? In: Behnke, R.H., Scoones, I. and Kerven, C. (eds) *Range at Disequilibrium*. Overseas Development Institute/International Institute for Environment and Development/Commonwealth Secretariat, London, pp. 173–195.

Allen, T.F.H. and Starr, T.B. (1982) *Hierarchy: Perspectives for Ecological Complexity*. University of Chicago Press, Chicago.

Andrew, M.H. (1988) Grazing impact in relation to livestock watering points. *Trends in Ecology and Evolution* 3, 336–339.

Archer, S., Scifres, C., Bassham, C.R. and Maggio, R. (1988) Autogenic succession in subtropical savanna: conversion of grassland to thorn woodland. *Ecological Monographs* 58, 111–127.

Ash, A.J., McIvor, J.G., Corfield, J.P. and Winter, W.H. (1995) How land condition alters plant–animal relationships in Australia's tropical rangelands. *Agriculture, Ecosystems and Environment* (in press).

Bailey, D.W., Rittenhouse, L.R., Hart, R.H. and Richards, R.W. (1989) Characteristics of spatial memory in cattle. *Applied Animal Behaviour Science* 23, 331–340.

Baker, B.B., Bourdon, R.M. and Hanson, J.D. (1992) FORAGE: a model of forage intake in beef cattle. *Ecological Modelling* 60, 257–279.

Balent, G. and Stafford Smith, D.M. (1993) Conceptual model for evaluating the consequences of management practices on the use of pastoral resources. *Proceedings 4th International Rangelands Congress*, April 1991, pp. 1158–1164.

Bastin, G.N., Pickup, G., Chewings, V.H. and Pearce, G. (1993) Land degradation assessment in central Australia using a grazing gradient method. *Rangeland Journal* 15, 190–216.

Bayliss, P. (1987) Kangaroo dynamics. In: Caughley, C., Shepherd, N. and Short, J. (eds) *Kangaroos: Their Ecology and Management in the Sheep Rangelands of Australia*. Cambridge University Press, Cambridge, UK, pp. 119–134.

Beard, R. (1994) *Optimal Stochastic Control of Renewable but Irreversible Resources with Interspecific Competition*. Agricultural Economics Discussion Paper Series 2/94, Department of Agriculure, University of Queensland, Australia, 15 pp.

Behnke, R.H. (1987) Cattle accumulation and the commercialisation of the traditional livestock industry in Botswana. *Agricultural Systems* 24, 1–29.

Behnke, R.H. (1994) Natural resource management in pastoral Africa. *Development Policy Review* 12, 5–27.

Behnke, R.H., Scoones, I. and Kerven, C. (eds) (1993) *Range Ecology at Disequilibrium*. Overseas Development Institute/International Institute for Environment and Development/Commonwealth Secretariat, London.

Belovsky, G.E. (1986) Optimal foraging and community structure: implications for a guild of generalist grassland herbivores. *Oecologia* 70, 35–52.

Bernardo, D.J. and Engle, D.M. (1990) The effect of manager risk attitudes on range improvement decisions. *Journal of Range Management* 43, 242–249.

Black, J.L., Davies, G.T. and Fleming, J.F. (1993) Role of computer simulation in the application of knowledge to animal industries. *Australian Journal of Agricultural Research* 44, 541–555.

Burnside, D.G. and Chamala, S. (1994) Ground-based monitoring: a process of learning by doing. *Rangeland Journal* 16, 221–237.

Campbell, A. (1994) *Landcare: Communities Shaping the Land and the Future.* Allen and Unwin, St Leonards, Australia, 344 pp.

Campbell, B.D. and Grime, J.P. (1993) Prediction of grassland plant responses to global change. *Proceedings of the 17th International Grasslands Congress,* Palmerston North, New Zealand, pp. 1109–1119.

Caughley, G. (1979) What is this thing called carrying capacity? In: Boyce, M.S. and Hayden-Wing, L.D. (eds) *North American Elk: Ecology, Behavior and Management.* University of Wyoming, Laramie, pp. 2–8.

Chesson, P. (1994) Multispecies competition in variable environments. *Theoretical Population Biology* 45, 227–276.

Choquenot, D. (1991) Density-dependent growth, body condition, and demography in feral donkeys: testing the food hypothesis. *Ecology* 72, 805–813.

Clark, R. and Filet, P. (1994) Local best practice, participatory problem solving and benchmarking to improve rangeland management. *Proceedings of 8th Biennial Conference,* Australian Rangelands Society, Katherine, 1994, pp. 70–75.

Clements, F.E. (1916) *Plant Succession: An Analysis of the Development of Vegetation.* Publication 242, Carnegie Institute, Washington.

Coughenour, M.B. (1991) Spatial components of plant–herbivore interactions in pastoral, ranching, and native ungulate ecosystems. *Journal of Range Management* 44, 530–542.

Coughenour, M.B., Coppock, D.L. and Ellis, J.E. (1990) Herbaceous forage variability in an arid pastoral region of Kenya: importance of topographic and rainfall gradients. *Journal of Arid Environments* 19, 147–159.

Cridland, S. and Stafford Smith, D.M. (1993) *Development and Dissemination of Design Methods for Rangeland Paddocks which Maximise Animal Production and Minimise Land Degradation.* Miscellaneous Publication 42/93, Western Australian Department of Agriculture, Perth, Western Australia, 79 pp.

Danckwerts, J.E. and Tainton, N.M. (1993) Range management: optimising forage production and quality. *Proceedings of the 17th International Grasslands Congress,* Palmerston North, New Zealand, pp. 843–850.

Danckwerts, J.E., O'Reagain, P.J. and O'Connor, T.G. (1993) Range management in a changing environment: a southern African perspective. *Rangeland Journal* 15, 133–144.

de Wit, C.T. (1990) International agricultural research for developing countries. In: Rabbinge, R., Goudriaan, J., van Keulen, H., Penning de Vries, F.W.T. and van Laar, H.H. (eds) *Theoretical Production Ecology: Reflections and Prospects.* Pudoc, Wageningen, pp. 295–301.

Donnelly, J.R. (1984) The productivity of breeding ewes grazing on lucerne or grass and clover pastures on the tablelands of southern Australia. III. Lamb mortality and weaning percentage. *Australian Journal of Agricultural Research* 35, 709–721.

Dyksterhuis, E.J. (1949) Condition and management of rangeland based on quantitative ecology. *Journal of Range Management* 2, 104–115.

El Aich, A., Moukadem, A. and Rittenhouse, L.R. (1989) Feeding station behavior of free grazing sheep. *Applied Animal Behaviour Science* 24, 259–265.

Ellis, J.E., Coughenour, M.B. and Swift, D.M. (1993) Climate variability, ecosystem stability, and the implications for range and livestock development. In: Behnke, R.J., Scoones, I. and Kerven, C. (eds) *Range Ecology at Disequilibrium*. Overseas Development Institute/International Institute for Environment and Development/Commonwealth Secretariat, London, pp. 31–41.

Fisher, J., Stafford Smith, M., Cavazos, R., Manzanilla, H., Ffolliott, P., Saltz, D., Irwin, M., Sammis, T., Swietlik, D., Moshe, I. and Sachs, M. (1996) Land use and management: research implications from three arid and semi-arid regions of the world. In: Irwin, M.E., Hoekstra, T.W., Shachak, M. and Hewett, K. (eds) *Arid Lands Management – Towards Ecological Sustainability*. University of Illinois Press, Champaign, Illinois, (in press).

Foran, B.D. and Stafford Smith, D.M. (1991) Risk, biology and drought management strategies for cattle stations in Central Australia. *Journal of Environmental Management* 33, 17–33.

Foran, B.D., Stafford Smith, D.M., Neithe, G., Stockwell, T. and Michell, V. (1990) A comparison of development options on a Northern Australian beef property. *Agricultural Systems* 34, 77–102.

Friedel, M.H. (1991) Range condition and the concept of thresholds: a viewpoint. *Journal of Range Management* 44, 422–426.

Frost, P., Medina, E., Menaut, J.-C., Solbrig, O., Swift, M. and Walker, B. (1986) Responses of savannas to stress and disturbance: a proposal for a collaborative programme of research. *Biology International*, Special Issue 10.

Fuls, E.R. (1992) Ecosystem modifications created by patch-overgrazing in semi-arid grasslands. *Journal of Arid Environments* 23, 59–69.

Gillard, P. and Monypenny, R. (1988) A decision support approach for the beef cattle industry of tropical Australia. *Agricultural Systems* 26, 179–190.

Goodall, D.W. (1967) Computer simulation of changes in vegetation subject to grazing. *Journal of the Indian Botanical Society* 46, 356–362.

Gross, J.E., Shipley, L.A., Hobbs, N.T., Spalinger, D.E. and Wunder, B.A. (1993) Functional response of herbivores in food-concentrated patches: tests of a mechanistic model. *Ecology* 74, 778–791.

Hacker, R.B. (1992) IMAGES: an integrated model of an arid grazing ecological system for improving rangeland management. *Agricultural Systems and Information Technology* 4(2), 17–19.

Hacker, R.B., Wang, K.-M., Richmond, G.S. and Lindner, R.K. (1991) IMAGES: an integrated model of an arid grazing ecological system. *Agricultural Systems* 37, 119–163.

Hamblin, A. (1992) *Environmental Indicators for Sustainable Agriculture*. Bureau of Rural Resources, Canberra, Australia.

Hanson, J.D., Skiles, J.W. and Parton, W.J. (1988) A multi-species model for rangeland plant communities. *Ecological Modelling* 44, 89–123.

Hardin, G. (1994) The tragedy of the unmanaged commons. *Trends in Ecology and Evolution* 9, 119.

Hart, R.H., Bissio, J., Samuel, M.J. and Waggoner Jr, J.W. (1993) Grazing systems, pasture size, and cattle grazing behavior, distribution and gains. *Journal of Range Management* 46, 81–87.

Hobbs, T.J., Sparrow, A.D. and Landsberg, J.J. (1994) A model of soil moisture balance and herbage growth in the arid rangelands of Central Australia. *Journal of Arid Environments* 28, 281–298.

Holmes, J.H. and Knight, L.R. (1994) Pastoral lease tenure in Australia: historical relic or useful contemporary tool? *Rangeland Journal* 16, 106–121.

Hughes, L., Dunlop, M., French, K., Leishman, M.R., Rice, B., Rodgerson, L. and Westoby, M. (1994) Predicting dispersal spectra: a minimal set of hypotheses based on plant attributes. *Journal of Ecology* 82, 933–950.

Ison, R.L. (1993) Changing community attitudes. *Rangeland Journal* 15, 154–166.

Jiggins, J. (1993) From technology transfer to resource management. *Proceedings of 17th International Grasslands Congress,* Palmerston North, New Zealand, pp. 615–622.

Jones, R.J. and Sandland, R.L. (1974) The relation between animal gain and stocking rate: derivation of the relation from the results of grazing trials. *Journal of Agricultural Science* 83, 335–342.

Kerven, C. (1994) *Customary Commerce.* ODI Publications, London.

Lauenroth, W.K. and Laycock, W.A. (1989) *Secondary Succession and the Evaluation of Rangeland Condition.* Westview Press, Boulder, Colorado.

Law, R. and Morton, R.D. (1993) Alternative permanent states of ecological communities. *Ecology* 74, 1347–1361.

Low, W.A., Dudzinski, M.L. and Müller, W.J. (1981) The influence of forage and climatic conditions on range community preference of Shorthorn cattle in central Australia. *Journal of Applied Ecology* 18, 11–26.

McKeon, G.M., Day, K.A., Howden, S.M., Mott, J.J., Orr, D.M., Scattini, W.J. and Weston, E.J. (1990) Northern Australian savannas: management for pastoral production. *Journal of Biogeography* 17, 355–372.

McKeon, G.M., Howden, S.M., Abel, N.O.J. and King, J.M. (1993) Climate change: adapting tropical and sub-tropical grasslands. *Proceedings of the 17th International Grasslands Congress,* Palmerston North, New Zealand, pp. 1181–1190.

Milchunas, D.G. and Lauenroth, W.K. (1993) Quantitative effects of grazing on vegetation and soils over a global range of environments. *Ecological Monographs* 63, 327–366.

Milchunas, D.G., Sala, O.E. and Lauenroth, W.K. (1988) A generalized model of the effects of grazing by large herbivores on grassland community structure. *American Naturalist* 132, 87–106.

Moore, A.D. (1990) The semi-Markov process: a useful tool in the analysis of vegetation dynamics for management. *Journal of Environmental Management* 30, 111–130.

Moore, A.D. and Noble, I.R. (1990) An individualistic model of vegetation stand dynamics. *Journal of Environmental Management* 31, 61–81.

Moore, A.D., Donnelly, J.R. and Freer, M. (1991) GrazPlan: an Australian DSS for enterprises based on grazed pastures. In: Stuth, J.W. and Lyons, B.G. (eds) *Proceedings of International Conference on Decision Support Systems for Resource Management.* College Station, Texas, USA, April 1991, pp. 23–26.

Moore, A.D., Mayer, D., Pepper, P., Freer, M. (1995) *Modelling of Research Data on Reproduction and Mortality of Cattle and Sheep*. DroughtPlan Working Paper No. 5, CSIRO, Alice Springs, Australia.

Morton, S.R., Stafford Smith, D.M., Friedel, M.H., Griffin, G.F. and Pickup, G. (1995) The stewardship of arid Australia: ecology and landscape management. *Journal of Environmental Management* 43, 195–217.

Nicholls, N. and Wong, K.K. (1990) Dependence of rainfall variability on mean rainfall, latitude, and the southern oscillation. *Journal of Climate* 3, 163–170.

Noble, I.R. (1975) Computer simulations of sheep grazing in the arid zone. Unpublished PhD thesis, University of Adelaide, Australia, 308 pp.

Noy-Meir, I. (1993) Compensating growth of grazed plants and its relevance to the use of rangelands. *Ecological Applications* 3, 32–34.

O'Dempsey, N. (1987) What are extra lambs worth to you? *The Grazier (Queensland)* 8(87), 2–3.

O'Keefe, M. (1992) *A Qualitative Project into the Adoption of Pasture Research and the Potential for GRAZFEED*. Report to CSIRO Institute of Plant Production and Processing, Canberra, Australia, 64 pp.

O'Reagain, P.J. and Mentis, M.T. (1990) The effect of veld condition on the quality of diet selected by cattle grazing the Natal Sour Sandveld. *Journal of the Grassland Society of South Africa* 7, 190–195.

O'Reagain, P.J. and Turner, J.R. (1992) An evaluation of the empirical basis for grazing management recommendations for rangeland in southern Africa. *Journal of the Grassland Society of South Africa* 9, 38–49.

Parsons, A.J., Thornley, J.H.M., Newman, J. and Penning, P.D. (1994) A mechanistic model of some physical determinants of intake rate and diet selection in a two-species temperate grassland sward. *Functional Ecology* 8, 187–204.

Passmore, J.G.I. and Brown, C.G. (1992) Property size and rangeland degradation in the Queensland mulga rangelands. *Rangeland Journal* 14, 9–25.

Pickett, S.T.A., Collins, S.L. and Armesto, J.J. (1987) A hierarchical consideration of causes and mechanisms of succession. *Vegetatio* 69, 109–114.

Pickup, G. (1989) New land degradation survey techniques for arid Australia – problems and prospects. *Australian Rangeland Journal* 11, 74–82.

Pickup, G. (1995) A simple model for predicting herbage production from rainfall in rangelands and its calibration using remotely-sensed data. *Journal of Arid Environments* 30, 227–245.

Pickup, G. and Chewings, V.H. (1988) Estimating the distribution of grazing and patterns of cattle movement in a large arid zone paddock: an approach using animal distribution models and Landsat imagery. *International Journal of Remote Sensing* 9, 1469–1490.

Pickup, G. and Stafford Smith, D.M. (1993) Problems, prospects and procedures for assessing sustainability of pastoral land management in arid Australia. *Journal of Biogeography* 20, 471–487.

Pickup, G., Bastin, G.N. and Chewings, V.H. (1994) Remote-sensing-based condition assessment for nonequilibrium rangelands under large-scale commercial grazing. *Ecological Applications* 4, 497–517.

Purvis, J.R. (1986) Nurture the land: my philosophies of pastoral management in central Australia. *Australian Rangeland Journal* 8, 110–117.

Redetzke, K.A. and van Dyne, G.M. (1979) Data-based, empirical, dynamic matrix-modeling of rangeland grazing systems. In: French, N.R. (ed.) *Perspectives in Grassland Ecology – Results and Applications of the US/IBP Grassland Biome Study.* Springer-Verlag, New York, pp. 157–173.

Reynolds, H.L. and Pacala, S.W. (1993) An analytical treatment of root-to-shoot ratio and plant competition for soil nutrient and light. *American Naturalist* 141, 51–70.

Reynolds, J.F., Virginia, R.A. and Schlesinger, W.H. (1995) Defining functional types for models of desertification. In: Smith, T.M., Shugart, H.H. and Woodward, F.I. (eds) *Plant Functional Types.* Cambridge University Press, Cambridge, UK (in press).

Richardson, F.D., Hahn, B.D. and Wilke, P.I. (1991) A model for the evaluation of different production strategies for animal production from rangeland in developing areas: an overview. *Journal of the Grassland Society of South Africa* 8, 153–159.

Riechers, R.K., Conner, J.R. and Heitschmidt, R.K. (1989) Economic consequences of alternative stocking rate adjustment tactics: a simulation approach. *Journal of Range Management* 42, 165–171.

Rodriguez, A. and Jameson, D.A. (1988) Rainfall risk in grazing management. *Ecological Modelling* 41, 85–100.

Rowan, R.C., White, L.D. and Conner, J.R. (1994) Understanding cause/effect relationships in stocking rate change over time. *Journal of Range Management* 47, 349–354.

Rural Extension Centre, Woods, E., Moll, G., Coutts, J., Clark, R. and Ivin, C. (1993) *INFORMATION EXCHANGE. Report Commissioned by Australia's Rural Research and Development Corporations.* Land and Water Resources Research and Development Corporation, Canberra, Australia.

Sandford, S. (1983) *Management of Pastoral Development in the Third World.* Wiley, Chichester, UK.

Scanlan, J.C. (1992) A model of woody–herbaceous biomass relationships in eucalypt and mesquite communities. *Journal of Range Management* 45, 75–80.

Schlesinger, W.H., Fonteyn, P.J. and Reiners, W.A. (1989) Effects of overland flow on plant water relations, erosion, and soil water percolation on a Mojave Desert landscape. *Soil Science Society of America Journal* 53, 1567–1572.

Scoones, I. (1993) Why are there so many animals? Cattle population dynamics in the communal areas of Zimbabwe. In: Behnke, R.H., Scoones, I. and Kerven, C. (eds) *Range Ecology at Disequilibrium.* Overseas Development Institute/International Institute for Environment and Development/Commonwealth Secretariat, London, pp. 62–76.

Seligman, N.G. and van Keulen, H. (1989) Herbage production of a Mediterranean grassland in relation to soil depth, rainfall and nitrogen nutrition: a simulation study. *Ecological Modelling* 47, 303–311.

Senft, R.L., Rittenhouse, L.R. and Woodmansee, R.G. (1983) The use of regression models to predict spatial patterns of cattle behavior. *Journal of Range Management* 36, 553–557.

Senft, R.L., Coughenour, M.B., Bailey, D.W., Rittenhouse, L.R., Sala, O.E. and Swift, D.M. (1987) Large herbivore foraging and ecological hierarchies: landscape ecology can enhance traditional foraging theory. *BioScience* 37, 789–799.

Shepherd, N. and Caughley, G. (1987) Options for management of kangaroos. In: Caughley, C., Shepherd, N. and Short, J. (eds) *Kangaroos: Their Ecology and Management in the Sheep Rangelands of Australia.* Cambridge University Press, Cambridge, UK, pp. 188–219.

Shmida, A., Evenari, M. and Noy-Meir, I. (1986) Hot desert ecosystems: an integrated view. In: Evenari, M., Noy-Meir, I. and Goodall, D.W. (eds) *Hot Deserts and Arid Shrublands, B*. Elsevier, Amsterdam, pp. 379–387.

Short, J. (1987) Factors affecting food intake of rangelands herbivores. In: Caughley, C., Shepherd, N. and Short, J. (eds) *Kangaroos: Their Ecology and Management in the Sheep Rangelands of Australia*. Cambridge University Press, Cambridge, UK, pp. 84–99.

Spalinger, D.E. and Hobbs, N.T. (1992) Mechanisms of foraging in mammalian herbivores: new models of functional response. *American Naturalist* 140, 325–348.

Stafford Smith, D.M. (1988) *Modelling: Three Approaches to Predicting how Herbivore Impact is Distributed in Rangelands*. Regional Research Report 628, New Mexico State University Agriculture Experiment Station, Las Cruces, New Mexico, 56 pp.

Stafford Smith, D.M. and Foran, B.D. (1992) An approach to assessing the economic risk of different drought management tactics on a South Australian pastoral sheep station. *Agricultural Systems* 39, 83–105.

Stafford Smith, D.M. and Foran, B.D. (1993) Problems and opportunities for commercial animal production in the arid and semi-arid rangelands. *Proceedings of the 17th International Grasslands Congress*, Palmerston North, New Zealand, pp. 41–48.

Stafford Smith, D.M. and Morton, S.R. (1990) A framework for the ecology of arid Australia. *Journal of Arid Environments* 18, 255–278.

Stafford Smith. D.M. and Pickup, G. (1993) Out of Africa, looking in: understanding vegetation change and its implications for management in Australian rangelands. In: Behnke, R.H., Scoones, I. and Kerven, C. (eds) *Range Ecology at Disequilibrium*. Overseas Development Institute/International Institute for Environment and Development/Commonwealth Secretariat, London, pp. 196–226.

Stafford Smith, D.M., Noble, I.R. and Jones, G.K. (1985) A heat balance model for sheep and its use to predict shade-seeking behaviour in hot conditions. *Journal of Applied Ecology* 22, 753–774.

Stafford Smith, D.M., McNee, A., Rose, B., Snowdon, G. and Carter, C. (1994) Goals and strategies for Aboriginal cattle enterprises. *Rangeland Journal* 16, 77–93.

Stafford Smith, D.M., Campbell, B., Archer, S., Ojima, D. and Steffen, W. (1995) *GCTE Focus 3 – Pastures and Rangelands Task: An Implementation Plan*. Report No. 3, Global Change and Terrestrial Ecosystems, Canberra.

Thornley, J.H.M., Parsons, A.J., Newman, J. and Penning, P.D. (1994) A cost–benefit model of grazing intake and diet selection in a two-species sward. *Functional Ecology* 8, 5–16.

Tilman, D. (1990) Constraints and tradeoffs: toward a predictive theory of competition and succession. *Oikos* 58, 3–15.

Tongway, D.J. and Ludwig, J.A. (1994) Small-scale resource heterogeneity in semi-arid landscapes. *Pacific Conservation Biology* 1, 201–208.

Ungar, E.D. and Noy-Meir, I. (1988) Herbage intake in relation to availability and sward structure: grazing processes and optimal foraging. *Journal of Applied Ecology* 35, 1045–1062.

Walker, B.H. (1993a) Stability in rangelands – ecology and economics. *Proceedings of the 17th International Grasslands Congress*, Palmerston North, New Zealand, pp. 1885–1890.

Walker, B.H. (1993b) Rangeland ecology: understanding and managing change. *Ambio* 22, 80–87.

Walker, B.H., Ludwig, D., Holling, C.S. and Peterman, R.M. (1981) Stability of semi-arid savanna grazing systems. *Journal of Ecology* 69, 473–498.

Walters, C. (1986) *Adaptive Management of Renewable Resources.* Macmillan Press, New York.

Westoby, M. (1988) Comparing Australian ecosystems to those elsewhere. *BioScience* 38, 549–556.

Westoby, M., Walker, B.H. and Noy-Meir, I. (1989) Opportunistic management for rangelands not at equilibrium. *Journal of Range Management* 42, 266–274.

White, D.H., Bowman, P.J., Morley, F.H.W., McManus, W.R. and Filan, S.J. (1983) A simulation model of a breeding ewe flock. *Agricultural Systems* 19, 149–189.

Wilson, A.D. and MacLeod, N.D. (1991) Overgrazing: present or absent? *Journal of Range Management* 44, 475–482.

Wisiol, K. (1984) Estimating grazingland yield from commonly available data. *Journal of Range Management* 37, 471–475.

Management of Mediterranean Grasslands

N.G. Seligman

Agricultural Research Organization, Volcani Center, Bet Dagan, 50250, Israel

INTRODUCTION

'Mediterranean grasslands' – an oxymoron?

Mediterranean grasslands are a rich and complex ecological resource of considerable economic importance. But, in a sense, 'mediterranean grasslands' is a contradiction in phytosociological terms because broadleaved, sclerophyllous, evergreen trees and shrubs characterize regions with a mediterranean-type* climate (Raven, 1973) and annual grasslands are an extreme degradation stage (Table 13.1) or an artefact maintained by cultivation, fire and grazing against the continuing pressures of invading and regenerating woody species. However, the term does suggest the polarity that underlies much of its vegetation dynamics. Vestiges of the woody vegetation from which the grasslands were derived continue to play a role, which can vary from threat to benefit. In Spain the *dehesa* oak savanna supports complex livestock husbandry enterprises. On red earth in Greece and Italy, grasslands occur at lower elevations on sites cleared from scrub oak forest (Zinke, 1973). Annual Eurasian species from the Mediterranean basin have invaded the other mediterranean regions of the world and have replaced much of

---

* In this chapter, 'Mediterranean', with upper-case 'M' refers to countries or zones around the Mediterranean basin in North Africa, southern Europe and the Near East, where the climate is 'typically mediterranean'; 'mediterranean', with a lower-case 'm', refers to mediterranean zones or climates generally.

Table 13.1. Regression stages and bio-indicators of Iberoatlantic mesomediterranean live-oak woodlands (derived from Table 28 in Rivas Martinez, 1987.)

Regression stage	Characteristic species
I. Forest, woodland (*bosque*)	*Quercus rotundifolia* *Pyrus bourgaeana* *Paeonia broteroi* *Doronicum plantagineum*
II. Dense shrubland (*matorral denso*)	*Phillyrea angustifolia* *Quercus coccifera* *Cytisus multiflorus* *Retama sphaerocarpa*
III. Scrub (*matorral degradado*)	*Cistus ladanifer* *Genista hirsuta* *Lavandula sampaiana* *Halimium viscosum*
IV. Grassland (*pastizales*)	*Agrostis castellana* *Psilurus incurvus* *Poa bulbosa*

the original savannas and woodlands (Axelrod, 1973), especially in California and Australia (Specht, 1973; Baker, 1988).

The Old World and the New World: 10,000 vs. 400 years

The Mediterranean zone of the Old World, lying astride southern Europe, North Africa and the Near East, is where agriculture and domestication of crops and livestock began some 10,000 years ago. Land use, and with it the status and role of grasslands, have undergone change during millennia as political and economic fortunes have shifted with the rise and fall of empires (Naveh and Dan, 1973; Noy-Meir and Seligman, 1979; Edelstein and Milevski, 1994). More than 2000 years ago Plato complained that the hills around Athens were 'like the skeleton of an old man, all the fat and soft earth wasted away' (cited by Attenborough, 1987). Yet, throughout the rise and fall of civilizations, the land has continued to support agriculture (Aschmann, 1973) and the natural vegetation has continued to supply materials for human use, as well as forage for the livestock that has been so integral a part of the economy of the region. Today, both soils and vegetation are still remarkably resilient under disturbance (Seginer *et al.*,1963; Morin *et al.*, 1979; Noy-Meir and Walker, 1986; Malanson and Trabaud, 1987). This robustness of Mediterranean ecosystems is both an intrinsic and an acquired characteristic: intrinsic, because the soils on the ubiquitous limestone, chalk and basalt substrate are normally well structured and because all the woody species evolved millions of years ago, before the Miocene (Axelrod, 1973); acquired, because many of the annual species and most of the numerous genotypes have evolved and are still evolving under conditions of extensive cultivation and intensive grazing use (Raven, 1973; Woodward and Morley, 1974; Cocks, 1992b).

In the mediterranean regions of the New World, where introduction of domestic ruminants and intensive cultivation is much more recent, Eurasian annual species have invaded favourable habitats and have replaced many native species (Baker, 1988), especially on better soils (Jackson, 1985) and sites where soil fertility has been enhanced (Specht, 1973). In some cases, there has been severe landscape dislocation, involving loss of indigenous species, soil erosion, salinization of bottom lands, and soil acidification (Specht, 1973; Aschmann, 1973; Ovalle *et al.*, 1993). While this could well have occurred in prehistoric times in the Mediterranean basin, today the robustness of its landscape, the resilience of its sclerophyllous species, the richness of its annual vegetation (Shmida, 1981) and the millennia-long association with intensive human occupation set it apart from the other mediterranean regions in matters of grassland and landscape management.

MEDITERRANEAN GRASSLANDS

Natural grassland and savanna

The cool, wet mediterranean winter and hot desiccating summer impose a strong seasonal growth cycle that favours annual species and drought-resistant perennials. The longer and drier the summer, the stronger the seasonality and the greater the predominance of annual species. Where summer conditions are milder, perennial grasses and forbs are more prominent (Jackson, 1985). Normally, grazing and depletion of soil water by the herbaceous vegetation prevent seedling establishment, and periodic fires keep the woody species from dominating (Biswell, 1956). Established woody species survive, often on protected sites, through access to water that seeps beyond the main root zone of the herbaceous species (Walker and Noy-Meir, 1982). For the competition to be effective against shrub encroachment, soil fertility should be high enough to allow for vigorous seasonal growth of the herbaceous species. Such circumstances exist, *inter alia*, on deeper soils in California (Biswell, 1956) and on some basaltic soils in the Mediterranean region (Seligman, 1973; Zohary, 1973) where the herbaceous vegetation includes many perennial species (Fig. 13.1) with an annual growth cycle very similar to that of the annual species. Perennial grass species are rare on productive soils of Californian annual rangeland (Wester, 1981) but are more common on poorer sites with shallow soils, where competition with annuals is less severe (Edwards, 1992).

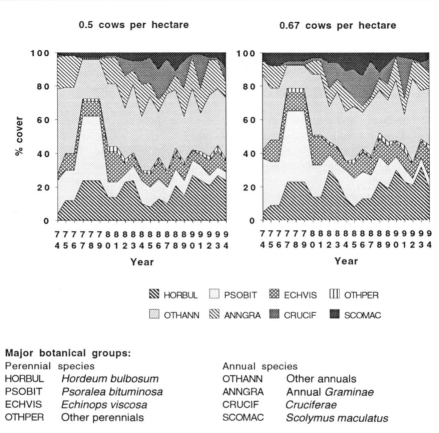

Fig. 13.1. Botanical composition of Mediterranean grassland on basaltic grumosol grazed by beef cattle at two stocking rates, Karei Deshe Experimental Range, Israel, 1974–1994. Note episodic changes in cover of *Psoralea bituminosa* and crucifers. (Derived from M. Gutman, ARO, Israel, 1994, unpublished data.)

'Vestigial' grassland, savanna and shrubland

In all mediterranean regions large tracts of land have been cleared of woody vegetation for cereal cultivation and for plantation of olives, vines, almonds, figs, carobs and other fruit trees. In countries around the Mediterranean basin, deep, fertile valley soils have been cultivated since the dawn of agriculture and have supported one of the more stable, sustainable systems of land use in the world. Olive trees hundreds of years old are common. But, as these lands are finite, population pressures forced people to clear and cultivate shallower, rockier and more fragmented land areas (Naveh and Dan, 1973; Attenborough, 1987). During more prosperous times, terraces were built on the hills, especially in the vicinity of

Table 13.2. Effect of fertilizer application and of seeded annual legumes on the production potential and on the variablity of the herbage production on annual grasslands in Spain, Israel and Syria.

	Peak biomass (Mg ha^{-1}, DM)			Standard deviation		
Site	Native pasture	Fertilized pasture	Sown and fertilized	Native pasture	Fertilized pasture	Sown and fertilized
SPAIN*						
Real de la Jara	2.38	2.92	3.98	0.82	1.10	1.76
El Gaitan	1.92	2.66	3.22	0.43	0.80	1.33
Navalmoral de la Mata	0.92	1.42	2.02	0.44	0.69	0.72
Esparragalejo	1.46	2.80	3.12	0.79	1.93	1.77
Navalvillar de Pela	1.36	1.90	2.96	0.15	0.64	0.27
Cheles	1.20	1.80	2.32	0.76	0.95	1.24
Valencia de las Torres	1.36	1.24	1.18	0.36	0.34	0.48
Mean	1.51	2.11	2.69	0.32	0.53	0.67
ISRAEL						
Migda†	2.63	5.87		1.11	3.20	
Matat‡	1.56	4.28		0.12	1.57	
Ein Yaakov‡	1.09	4.04		0.32	1.14	
Hatal, terraces‡	1.09	3.04		0.25	1.00	
Hatal, slopes‡	1.00	2.08		0.22	0.66	
SYRIA						
Tel Hadya§	1.24	2.15		0.51	1.17	

*Olea Márquez de Prado (1988); P fertilizer; sown species, mixed annual pasture legumes; sandy loams on schists and granites; climate, mesomediterranean; trials on grazed swards; duration, 5 years.
†van Keulen and Seligman (1992); N fertilizer on deep loessial soil; harvested ungrazed swards; climate, xeromediterranean; trial duration, 11 years, simulated for 9 additional years.
‡Henkin (1944); P fertilizer on shallow terra rossa or brown mediterranean soils; swards lightly grazed; climate, subhumid mediterranean; trial duration, 5/6 years.
§Osman et al. (1991) and Osman and Cocks (1993); P fertilizer on semiarid, shallow stony soil; swards heavily grazed; climate, xeromediterranean; trial duration, 7 years.
DM, dry matter; P, phosphate; N, nitrogen.

larger towns and trade centres (Edelstein and Milevski, 1994). With changing historical fortunes, many of these marginal sites were abandoned. Terraces fell to ruin and erosion. Land reverted to the ubiquitous sclerophyllous shrubby woodland, or became open grazing lands composed of dwarf shrubs interspersed with grassland species, mainly annuals but also many hemicryptophytes, including gramineous perennials (Naveh and Dan, 1973; Noy-Meir and Seligman, 1979; Shmida, 1981). For centuries these open, devastated vestiges of previous cultivation have been the fodder base for livestock husbandry in most countries around the Mediterranean basin. They are still quite productive and, when mineral deficiencies are made good, can attain impressively high levels of production (Ofer and Seligman, 1969; Table 13.2; Figs 13.2 and 13.3), albeit with considerable spatial and interannual variability (Fig. 13.4). These 'vestigial grasslands' are maintained by heavy grazing, wood cutting for fuel or building material, and fire.

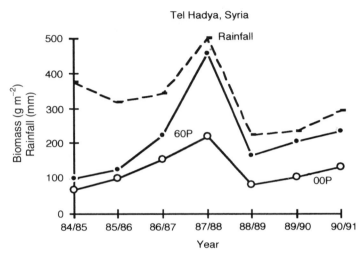

Fig. 13.2. Response of grazed native pasture to annual applications of superphosphate (60 kg ha^{-1}, P), Tel Hadya, Syria, 1984–1991. (Data from Osman *et al.*, 1991 and Osman and Cocks, 1993.)

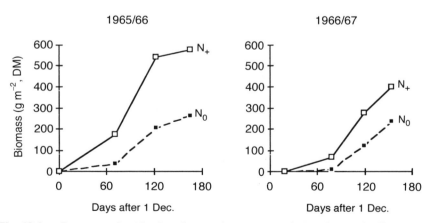

Fig. 13.3. Response of predominantly annual range vegetation to nitrogen fertilizer (15 g m^{-2}, N) on a shallow (10–30 cm) dark rendzina soil on soft limestone. Hills of Menashe, Israel. Annual precipitation 1965/66, 540 mm; 1966/67, 760 mm. DM, dry matter. (Data from Seligman and Meron, 1968.)

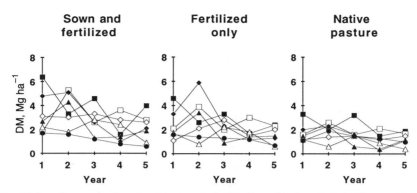

Fig. 13.4. Response of pasture biomass to fertilization with phosphate fertilizer and to seeding with selected annual pasture legume species on seven sites throughout Extremadura, southwest Spain. DM, dry matter. (Data from Olea Márquez de Prado, 1988.)

Some California woodland ranges, kept open mainly by fire, also fall into this category (Biswell, 1956).

Cultivated grassland

'Weed fields', fallows or ruderal grasslands

Where traction (by animal or by machine) is feasible, grasslands have been maintained by periodic cultivation. Such grasslands, like the extensive grassland savannas of Spain, the *dehesa*, are mostly on soils with fertility constraints that preclude the continuous cultivation practised in the more fertile, and usually deeper and heavier, valley soils. Cultivation of cereals for grain and straw and, in recent years, oat and vetch mixtures for hay provide supplementary feed for periods when pasture is inadequate for animal requirements. In these 'weed fields', the sward is composed overwhelmingly of annual species, but, as time between cultivations increases, perennial species become established, mainly grasses (*Poa bulbosa*, *Dactylis glomerata*) and hemicryptophytic thistles. Sometimes shrubs (*Retama sphaerocarpa*, *Cistus landifera*, *Rosmarinus stoechas*, etc.) occur and without intervention would increase and dominate the vegetation (Rivas Goday, 1964). The time needed for domination by the shrubby species and conversion of the vegetation into *mattoral* (Table 13.1) depends on the composition of the seed bank and on the soils and climatic conditions. In some cases, the elimination of perennial species by cultivation prevents rapid re-establishment and such ruderal 'weed-field' vegetation formations are maintained for many decades by grazing only (Pineda *et al.*, 1984).

Much of the California annual rangeland on more fertile soils can be classified as 'ruderal grassland' because it is a secondary vegetation following cultivation during the early settlement period (Baker, 1988; Edwards, 1992); it is now maintained mainly by grazing and fire (Biswell, 1956) and sometimes by periodic cultivation of cereals and even legume pasture (Menke, 1988). In Chile, also, many of the marginal agricultural lands in the foothill regions have been abandoned over the years and today support a mixed-species grassland that serves as pasture for domestic livestock (Ovalle *et al.*, 1993).

'Grassland fallows' are grazed in rotation with cultivated crops throughout the mediterranean regions (Cocks and Gintzburger, 1993) and the system can vary from very long to very short rotations, sometimes no more than an occasional fallow when cultivation is withheld for logistic and economic reasons or because of reduced fertility. In the Jewish tradition, fallow every 7 years, the *shmittah*, is obligatory and anchored in religious law.

Sown leys

While forages, especially lucerne (*Medicago sativa*), have been grown for centuries, sown pastures are relatively recent newcomers to the agricultural scene, particularly in the mediterranean regions (Davies, 1952). In Australia, the introduction of sown annual pasture legumes and later also annual grasses (mainly Wimmera ryegrass, *Lolium rigidum*) in rotation with wheat during the 1930s was led by subclover (*Trifolium subterraneum*), closely followed by the medicks (*Medicago truncatula, M. littoralis* and others) on drier and more basic soils. The practice became widespread and by 1970 about three-quarters of the area in the wheat–sheep belt was under sown self-seeding legume and mixed annual-species pasture. Productivity was maintained by routine application of superphosphate fertilizer and persistence was promoted by use of hard-seeded cultivars (Puckridge and French, 1983).

The 'sub and super' approach was emulated in other mediterranean regions: first in California (Williams *et al.*, 1956; Murphy *et al.*, 1973; Menke, 1988), then in South Africa and later in Chile (Ovalle *et al.*, 1993). In the Mediterranean basin itself, pasture legume trials were established in Israel in the early 1950s (Arnon and Dovrat, 1956) and were followed up by a Food and Agriculture Organization (FAO) project to promote the introduction of the legume ley (Anonymous, 1967). Similar internationally funded and directed projects were launched throughout the Mediterranean region, usually with expatriate expertise from Australia (Puckridge and French 1983; Chatterton and Chatterton, 1984). In 1969 a major project, funded by the World Bank, was launched in Spain by the United Nations Development Program (UNDP) (Anonymous, 1975). It initiated a long-term research programme to improve Mediterranean grasslands and livestock production and was conducted in collaboration with Spanish agricultural research organizations, in parallel with an ambitious pasture and animal development project, the Agencia de Desarollo Ganadero. The efforts to introduce annual pasture legumes in the mediterranean regions of the world are still continuing (van Heerden and Tainton, 1987; Olea Márquez de Prado, 1988; Ovalle *et al.*, 1993).

Australian pasture technology, even though actively promoted and shown to be technically feasible, did not revolutionize dryland agriculture in other mediterranean regions as it did in Australia. In very few cases did the technology spread beyond the limits of a funded project. It has not 'caught on' (Puckridge and French, 1983) even after heavy subsidization in some countries, especially Portugal and Spain. The causes were traced to socioeconomic conditions, often related to land tenure patterns, where cropping and animal husbandry were conducted by different interest groups in a non-integrated context (Springborg, 1990; Cocks and Gintzburger, 1993), or to subsidization of feed grains and concentrates (Crespo, 1993). The importance of the wider economic scene for the role of the legume ley in relation to cropping has become increasingly evident in Australia too. In southern Australia, home of the annual legume ley rotation, the pasture : crop ratio of land has fallen from 3 : 1 to less than 1 : 1, mainly as a consequence of low wool prices in relation to grain crops (especially legume grains and oil-seeds) and the availability of relatively cheap nitrogen fertilizer and also because of increasing problems with soil acidification, weeds, pests and diseases in the pasture crop itself. The change has been so dramatic that at the last International Grassland Congress, the leading paper on legume leys was entitled 'Is ley farming in mediterranean zones just a passing phase?' (Reeves and Ewing, 1993).

Beyond the economics and sociology, there are also important ecological factors that diminish the feasibility of a legume ley in the various mediterranean regions of the world. The Australian agroecological context is characterized by severe phosphorus deficiency in many soils. Consequently, vegetative growth, especially of annual N-fixing leguminous species, is very responsive to phosphorus fertilizer applications. The legumes provide high-quality forage plus a modest (Cocks, 1980) but significant and cheap source of N, which boosts the yield of wheat in the rotation and the growth of other nitrophilous annuals. The extensive nature of the Australian agricultural scene, with labour the main limiting factor and land available on a generous scale, made it possible to remain competitive even with relatively low wheat yields (mean grain yield in South Australia by 1980 being only 1.24 Mg ha^{-1} (Puckridge and French, 1983)). In addition, wool, especially when prices are relatively high, is a commodity ideally suited to extensive farming in areas where products must be transported long distances to market.

The agroecological context in other mediterranean regions is quite different. In the Mediterranean basin itself most soils, even after millennia of cultivation, are quite fertile in comparison with Australian soils (Wild, 1958) and the annual ruderal vegetation associated with cultivation and grazing has evolved a vast array of species, subspecies and 'ecotypes', among which the *Leguminosae* are prominent (e.g. Feinbrun-Dotan, 1978). As a rule, selected local ecotypes outperform the improved Australian varieties (e.g. Olea *et al.*, 1989; Cocks, 1992a; Falcinelli *et al.*, 1993). Consequently, Australian varieties that were introduced into the

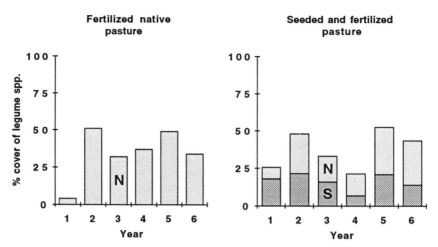

Fig. 13.5. Cover of spontaneous legume species (N) and selected *Trifolium* varieties (S) in native and seeded pasture after annual applications of phosphate fertilizer, Naval de la Mata, Extremadura, Spain. (Derived from L. Olea, SIA, Badajoz, Spain, 1944, unpublished data.)

highly heterogeneous grasslands of the region seldom persisted as dominants in the sward (e.g. Figs 13.5 and 13.7). Furthermore, wool has been a secondary product in most mediterranean regions for many years. Today, even in heartland merino country in Spain, fine wool has become secondary to lamb production (M. Granda Losada, Badajoz, Spain, 1994, personal communication). Milk and meat for subsistence and processing, and lamb and kid for both culinary and ceremonial functions, are perishable high-value products. Management of such systems tends to shift more attention to the animals than to the pasture.

In California, most rangeland is used for beef production, supplemented mainly with concentrates, urea and molasses, and alfalfa hay to overcome forage supply inadequacies. Introduction of legume leys peaked during the 1960s but has since subsided, partly because of expensive phosphate fertilizer (Menke, 1988). Still, a roughly estimated 20,000–40,000 ha year^{-1} are sown to annual pasture legumes in California (Melvin George, University of California; Davis, 1994, personal communication). In the wheat–sheep belt of the mediterranean climate zone in South Africa, efforts to introduce legumes over the past 50 years have not succeeded in establishing the technology as routine practice. In Chile there are still ongoing research efforts devoted to developing a legume ley for the semiarid zone (Ovalle *et al.*, 1993), but to date there is very little application of the technology. In the humid zone (> 1300 mm annual precipitation) there are sown legume pastures (Acuña *et al.*, 1993).

It is hazardous to judge the future of the annual legume ley in mediterranean-type dryland agriculture, despite the experience of the past 50 years. Changes in the prices of wool, wheat and other grains could conceivably swing the emphasis back to more pasture in the South Australian crop rotation. On the other hand, lower fodder grain prices, together with higher prices for animal products derived from milk and meat, tend to decrease dependence on pasture (Crespo, 1993).

GRASSLAND MANAGEMENT

Management goals

Traditionally, the central goal of grassland management has been to maximize utilization of the forage resource for animal production. The introduction of modern agricultural technology made higher production and conservation of forage another goal. Improved pasture legumes and grasses, soil nutrient amelioration with fertilizers and trace elements, weed, pest, disease and acidity control (Reeves and Ewing, 1993) and even computerized system analysis (Morrison *et al.*, 1986; Ungar, 1990) have created new opportunities for grassland management. More recently, an urban society with a post-technological ethos is putting new values on the non-forage benefits and costs of grassland improvement and use (Campos Palacin and Pearce, 1992). Management goals have become more complex. Landscape values, resource conservation, biodiversity, recreation, ecotourism and wildlife husbandry are challenging the forage utilization and production goals of grassland management. Balancing and resolving conflicting goals are becoming unavoidable management roles.

Riding a tiger

Current approaches to grassland management need to be continually reassessed because of the shrinking number of people employed in agriculture, the continuing innovations in technology and the expansion of communication facilities, which has removed logistic barriers to worldwide transport of feed grains and animal products. In less developed economies in North Africa and the Middle East, growing population has led to higher animal numbers on the land and stretched the flexibility of the systems, sometimes at a high, often unavoidable, cost in hardship and land degradation. In such cases, technological interventions in grassland management without changes in the demographic and macroeconomic context have made little headway, despite continuing efforts of local and international research and development organizations (Chatterton and Chatterton, 1984). Unfortunately, such interventions are relevant only where the rural sector has the means to choose between investment options (Milton *et al.*, 1994).

On the whole, 'management' is a strong word when applied to mediterranean grasslands, because of the great interannual variability in forage production (Table

13.2 and Fig. 13.4), and because of the resilience and 'independence' of these grasslands as they go their own way even when subjected to drastic management manipulation (Heady, 1958; Fig. 13.1). The limited possibilities for direct control over the productivity, quality and botanical composition of the grassland vegetation are compounded by the complexity of animal nutrient requirements and ingestive behaviour, which often frustrates efforts to achieve management goals (Willoughby, 1959).

Forage utilization

Forage utilization systems depend on the type and seasonality of the forage resource, scale of the enterprise, class of livestock, stocking rate, grazing system, infrastructure, land tenure and socioeconomic context.

Resources

In most mediterranean regions, grassland is not the only feed resource and any one animal production unit may depend not only on grasslands, shrublands, cereal stubble and other crop residues, but also on conserved forages (straw, hay, silage), concentrates and grain (mainly barley and oats) and unconventional feeds (like urea, molasses and poultry litter). Supplementary fodder resources have traditionally been part of the feed supply for livestock in the Mediterranean region ('we have plenty of straw and fodder' Genesis 24:25). Today, fodder grains are a widely accessible commodity. In addition, the by-products of the large-scale processing industries in many parts of the region (cottonseed, gin-trash, citrus peel, olive, grape and sugar-beet pulp, etc.) are locally available for feeding livestock (Benjamin *et al.*, 1981; Elena *et al.*, 1985). They provide alternatives to many conventional grassland management practices designed to overcome periodic forage inadequacy. These resources can support a considerable increase in animal numbers (Wagner, 1988) and must be considered in any discussion of grassland management practices in the mediterranean zone (Ungar, 1990; Crespo, 1993).

A reliable inventory of the heterogeneous mediterranean grasslands and supplementary feed resources is necessary for efficient management. As a rule, producers have a fair estimate of the value of their forage resources. Today, where rapidly changing economic and demographic circumstances are forcing innovation, imposing new constraints and raising unfamiliar management dilemmas, a more formal assessment of forage resources can be useful. This aspect of resource assessment has been studied in recent years, especially in France (Hubert, 1994), and in future may well command research and extension attention in other mediterranean countries.

Scale of operation

Grasslands can vary from a few hectares with a few animals in parts of the Middle East and North Africa to large estates with thousands of hectares and animals, especially in Australia, but also in parts of California, Spain (Elena *et al.*, 1985) and South Africa. Economies of scale are sometimes countered by improvements in production efficiency. Small producers who are dependent on limited facilities may employ more labour-intensive practices and manage their land and animals more efficiently. Lamb mortality, for instance, is often lower in small flocks.

Seasonality

The annual cycle of change that is of greatest significance for animal production in mediterranean-type pastoral systems is the cycle of forage availability and quality. 'In summer the country is of a monotonous dismal character; whilst in the spring [it is] overgrown with grass to the height of a man' (Schumacher, 1888). The moderate winters allow for year-long grazing and, if there is any confinement of livestock, it is limited to relatively short periods. Livestock nutrition must cope with a forage cycle that swings from low availability of very high-quality forage early in the growing season to surplus feed of medium to high quality during the spring flush of growth, followed by medium- to low-quality dry pasture, decreasing in availability as summer grazing progresses, until even dry pasture can become scarce in the early autumn. The reproductive cycle of ruminant livestock can be adjusted to the forage seasonality only to a limited extent, and animal production is severely impeded by extended periods of inadequate or low-quality pasture (Willoughby, 1959).

Livestock performance is dependent on the duration of the high quality pasture phase. It can last from less than 3 to more than 6 months of the year, depending on region and annual rainfall distribution (e.g. Fig. 13.6). The traditional management tactic to extend the productive phase was to exploit the heterogeneity of the vegetation by grazing early-season growth on warmer slopes and sites, late-season growth on cooler and wetter sites, dry-season utilization of deeper-rooted perennial species, including hemicryptophytic thistles and evergreen shrubs and other woody species, which despite relatively low quality are sometimes more nutritious than the dry annual herbage. Stubble fields and other crop residues, when available, are commonly used to make up shortfalls in forage availability and quality. In some areas, the solution has been transhumance, with animals being moved to summer pastures, usually in mountainous areas (e.g. Abercrombie, 1991).

Active intervention to lengthen the green season and to improve the quality of dry-season pasture has been an ongoing research objective, involving annual legume and grass species, perennial grasses (Menke, 1988), fertilizer applications to provide more early growth (Fig. 13.3) and forage shrubs to provide some green feed in the summer and during drought (Le Houérou, 1980). The improvement of

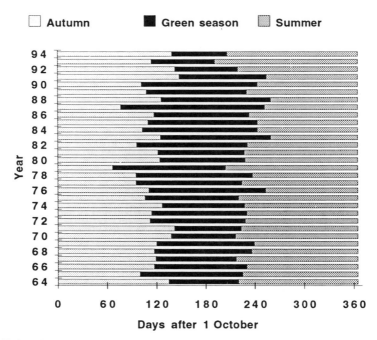

Fig. 13.6. Duration of the effective winter–spring green season at the Karei Deshe Experimental Range between 1964 and 1994. Beginning of autumn set arbitrarily at 1 October, the month when 'first rains' can be expected. Germination of annuals and regeneration of perennials can occur between October and December, but 'range readiness', the beginning of the 'green season', normally occurs during December and January. (M. Gutman, ARO, Israel, 1994, unpublished data.)

forage quality in the summer by promoting legume growth has been shown to be feasible (Rossiter, 1966; Puckridge and French, 1983; Menke, 1988; Olea *et al.*, 1989).

Livestock

The characteristically short green season is a factor that determines the preponderance of small ruminants in countries around the Mediterranean basin and in most other dry mediterranean-type regions. The proportion of goats to sheep generally increases with greater cover of shrubby species. Under moister conditions where the green season is longer, there are more cattle, kept for both beef and extensive milk production (Wagner, 1988). In some mediterranean areas, livestock have been replaced by wildlife, mainly for hunting. In others, attempts have been made to adapt larger-scale cattle ranching with low labour requirements to rangelands that are dominated by shrubby vegetation and were traditionally grazed by mixed herds, including goats (Gutman *et al.*, 1990). Changes in livestock husbandry and considerations of recreation, fire prevention, landscape enhancement and ecotourism have made it necessary to develop new systems for the management of the livestock/wildlife/vegetation complex (Etienne, 1989; Perevolotsky *et al.*, 1993).

Stocking rate and grazing pressure

Animal performance and pasture productivity are closely related to stocking rate (Jones and Sandland, 1974; Noy-Meir, 1976, 1978). Because it can be manipulated by the producer, this is one of the main tools of grassland management. On mediterranean grasslands, stocking rate is also an important determinant of shrub cover (Biswell, 1956), biodiversity (Naveh and Whittaker, 1979) and botanical composition of the herbaceous sward (Heady, 1958; Noy-Meir *et al.*, 1989). In the context of the strongly seasonal mediterranean grassland, stocking rate is meaningful mainly as an expression of grazing pressure during the cool-season growing period. During the dry period, removal of dead litter, even total removal by fire, has a very small effect on subsequent germination and growth of the annual species (Heady, 1960). Seed consumption by domestic ruminants is limited by many mechanisms, including dispersal, seed burial, spiny protection, hard coats, small seeds and so on. However, granivory, mainly by ants, can remove much of the annual seed crop (Loria, 1982; Beattie, 1988).

Animal performance on mediterranean grassland is sensitive to heavy grazing but, as a rule, total animal production increases with stocking rate until quite high levels (Eyal *et al.*, 1975; Table 13.3). Heavy stocking, ten sheep to the hectare and higher, is the rule on legume swards in good condition (Carter and Day, 1970). Herbage production is severely reduced by excessively heavy grazing (Puckridge and French, 1983) but mediterranean grasslands persist under high stocking rates, because reproductive processes that determine seed production of annual species operate mainly during the spring flush of growth, when growth rate of the herbaceous vegetation is usually much higher than the defoliation rate imposed by the

Table 13.3. Mean animal production and supplementary feed requirement of a herd of beef cattle grazing a Mediterranean grassland for 5 years at the Karei Deshe Exprimental Range, Galilee, Israel (M. Gutman, ARO, Israel, 1992, unpublished data).

Grazing regimen	Stocking rate (cows ha^{-1})	Calf crop (weaned live weight)		Net energy allocation (10^3 Mcal cow^{-1})	
		(kg cow^{-1})	(kg ha^{-1})	Annual requirement	Supplementary feed
Deferred	0.83	150	121	4.49	1.42
Deferred	0.67	160	108	4.53	1.60
Continuous	0.67	141	94	4.51	1.12
Continuous	0.50	158	79	4.68	1.24
LSD		21.8	14.2	0.22	0.14

LSD, least significant difference.

Table 13.4. Consumption rate of pasture and growth rate of the vegetation (over and above consumption) during the winter and spring on a mediterranean grassland stocked at 0.67 beef cows per hectare at the Karei Deshe Experimental Range, Galilee, Israel (M. Gutman, ARO, Israel, 1992, unpublished data).

	Winter		Spring	
Year	Ingestion (kg ha^{-1} day^{-1}, DM)	Growth (kg ha^{-1} day^{-1}, DM)	Ingestion (kg ha^{-1} day^{-1}, DM)	Growth (kg ha^{-1} day^{-1}, DM)
1986	2.6	7.2	9.0	16.6
1987	2.5	16.6	9.0	21.3
1988	2.6	11.1	8.0	24.1
1989	2.7	15.6	11.0	24.5
1990	3.6	8.9	9.0	7.7
Mean	2.8	11.9	9.2	18.8
SD	0.45	4.11	1.10	6.98

DM, dry matter; SD, standard deviation.

livestock (Table 13.4). Excess growth over consumption is such that year after year the land is 'clothed in verdure ... overgrown with grass' (Schumacher, 1888). Prolific seed production is the rule, normally well above the requirements for re-establishment of the pasture in the following season (de Ridder *et al.*, 1981; Loria, 1982). The amount of seed produced can vary widely according to the degree of defoliation during the early vegetative phase and current weather conditions. Low seed stocks can limit growth, especially during the early part of the season, but it is rare that no seed at all is produced or that the seed stocks in the soil are severely depleted. Consequently, annual grasslands are exceptionally resilient under grazing (Noy-Meir and Walker, 1986; Naveh, 1989).

Resilience of perennial species under grazing in Mediterranean grasslands depends on different mechanisms. *P. bulbosa*, which often dominates on shallow, poor soils, yields only modest amounts of forage for grazing, because most of the foliage is situated too low for prehension. *Hordeum bulbosum*, a much more productive species, which occurs widely on more favourable sites (Fig. 13.1), assumes a more prostrate habit under grazing, much like *Danthonia californicum* (Edwards, 1992). A perennial leguminous species, *Psoralea bituminosa*, like many other hemicryptophytes, is not readily grazed during the green-season flush in spring, apparently because of high coumarin content, but is heavily grazed early in the green season and is preferred to the dry pasture at the beginning of summer (Seligman and Gutman, 1975). Finally, most Mediterranean woody species persist under heavy stocking by assuming dense shrubby forms. On release from grazing, or even when grazing pressure is relaxed, most woody species resume upward growth, often forming dense thickets.

Grazing pressure influences the species composition of the grasslands. Heavy grazing tends to suppress taller species and favours the more prostrate and often less productive species (Noy-Meir *et al.*, 1989). However, the annual fluctuations in the botanical composition of the sward are poorly correlated with stocking rate (Heady, 1958; George *et al.*, 1992; Fig. 13.1). Periodic vole population explosions have a more drastic effect (Lidicker, 1988; Noy-Meir, 1988). Heavy grazing during the flowering period in early spring may lower persistence of sown legumes by depressing seed production (Rossiter, 1966; Olea *et al.*, 1989).

Many of the ills of Mediterranean landscapes have been blamed on overgrazing (Attenborough, 1987; Foran *et al.*, 1989; Cocks and Gintzburger, 1993; Tükel *et al.*, 1993, among many others). In fact, grazing is a relatively benign factor when compared with cultivation, fire, timber and fuel exploitation (Naveh, 1989; Seligman and Perevolotsky, 1994). Run-off and erosion from Mediterranean grasslands are low, with soil loss seldom more than 0.02 to 0.06 mm year^{-1} (Morin *et al.*, 1979). The well-structured, well-drained soils, derived from the calcareous and basaltic substrate, the rocks and the stone mulch, together with the residual plant litter, all contribute to a very stable habitat, which cannot be described as fragile. Even on the light sandy soils on granites and schists in the southwestern Iberian peninsula, infiltration rates are high and old stone walls indicate that the landscape has been stable for centuries. Among all the factors that contribute to landscape degradation in the Mediterranean basin, high stocking rates must be placed low on the list.

Grazing systems

Experiments on specialized grazing systems for annual pastures in California were reviewed by Heady (1960), who concluded that animal production was not improved by various systems of rotational and deferred grazing. Australian experience (Puckridge and French, 1983) and preliminary studies in Spain (Olea *et al.*, 1989) have led to similar conclusions. In Israel, beef cattle performance was slightly but not significantly lower under rotational than under continuous grazing, while the proportion of gramineous species in the sward and pasture production were higher under rotational grazing (Gutman and Seligman, 1979). The rotational system tended to stress the animals because grazing pressure was high in the grazed paddocks, but the intermediate rest periods favoured the taller annual grass species. Benefits from short-duration grazing have yet to be demonstrated on annual grassland (Menke, 1988).

Deferment of grazing at the beginning of the growing season allows the pasture biomass to exceed a threshold beyond which it can sustain heavy grazing. Theoretical and experimental studies (Smith and Williams, 1973; Noy-Meir, 1978; Ungar, 1990) suggest that it can help maintain production on mediterranean-type pastures when stocking rates are high. Otherwise benefits to animal production are doubtful. Over a 5-year period at a moderately high stocking rate, animal production was similar in deferred and continuously grazed treatments, but the amount of supplementary feed consumed in the deferred treatment was greater (Table 13.3).

The accumulated evidence confirms Heady's (1960) conclusion that, as a rule, specialized grazing systems can improve animal production and pasture condition in annual-species mediterranean grasslands only where rest is necessary to allow a severely degraded sward to recover (Noy-Meir, 1976). In practice, animals are moved from one paddock to another mainly to facilitate animal husbandry requirements or to achieve more uniform pasture utilization or special effects like reducing inflammable biomass before the summer in fire-prone areas.

Water, fencing and access

During the dry mediterranean summer, lack of accessible water can limit use of grassland in remote areas and on large ranches with inadequate water-supply. In the cool, green season, water is not an important constraint to grassland management because dry-matter content of the live annual vegetation is seldom more than 25% and the water requirement for sheep is about 3 litres kg^{-1} of consumed dry matter (Benjamin *et al.*, 1975).

In many mediterranean regions, the animals, especially the small ruminants, are herded and corralled at night. However, modern fencing technology has become common. Barbed-wire fencing is still the standard but Australian wire-netting fence, mainly for sheep, is widely used and in recent years movable, solar-powered electric fencing has become available. Flexible fencing systems have facilitated more continuous, year-long, untended grazing with much less intervention by the herder.

Forage production

Soil nutrients

Most mediterranean grasslands are on soils that are deficient in one or more plant nutrients. The poor, leached skeletal soils in Australia are severely deficient in phosphorus and N, and some also have trace mineral deficiencies (Rossiter, 1966; Puckridge and French, 1983). In the Mediterranean region, soils deficient in phosphorus include the sandy loams on granite or on Cambrian and Precambrian schists, as well as some red soils on limestone formations. Phosphorus fertilizer alone can increase herbage growth, mainly by inducing dominance of annual legumes (Ofer and Seligman, 1969; Osman and Cocks, 1993; Henkin, 1994; Fig. 13.4). Amelioration of phosphorus and sulphur deficiencies on serpentine soils in California boosts pasture growth (Williams *et al.*, 1956; Jones, 1964). In addition, generally low organic-matter content provides less available N than the requirement for potential growth of the vegetation, given the seasonal moisture conditions. Applications of N fertilizer can stimulate early growth and shorten the period of early-season pasture scarcity (Fig. 13.3) even in the semiarid margins of the mediterranean zone (van Keulen and Seligman, 1992). Grass leys based on relatively cheap (and renewable) N fertilizer (Benjamin, 1992) are sometimes more viable than legume pastures, which are dependent on the finite resources of phosphorus.

Only in Australia has fertilizer been applied to annual mediterranean-type pasture as a routine practice on a commercial scale, possibly because of the severe phosphorus deficiency in many Australian soils (Wild, 1958). Where deficiency is not so severe, fertilizer application increases production but also the amplitude of variability from year to year (Table 13.2; Fig. 13.4). Response is usually greatest in 'good' years when there is a surplus of forage and poorest in 'bad' years when pasture is scarce (van Keulen and Seligman, 1992; Fig. 13.2). If the benefits of fertilizer application are to be realized, stocking rate should be increased (Spharim and Seligman, 1990). But then the risks of heavy losses in a 'bad' year are greater.

Species composition

Mediterranean grasslands are dominated by annual species, and include not only highly productive and palatable grasses, legumes, forbs and, in some regions, perennial grasses but also many unpalatable species – crucifers, annual and perennial thistles and short-lived grasses and legumes. Shifting the balance in favour of the more desirable species on mediterranean grasslands (as opposed to cultivated leys in a crop rotation) has been an aim of research and management worldwide (Menke, 1988). In Israel, promising species have been identified and performance trials conducted since the early 1950s (Arnon and Dovrat, 1956). An intensive long-term programme designed to identify, introduce and disseminate selected annual pasture legume species into permanent grasslands has been conducted in Spain and in Portugal since the early 1970s (Anonymous, 1975). However, performance of these introduced desirable species has been disappointing. Persistence of the reseeded varieties over the years could seldom be maintained (Fig. 13.7). Often, phosphorus amendment alone led to an equivalent increase of resident legume species (e.g. Fig. 13.5). Generally, with N enhancement of the soil through fixation by the legumes, legume dominance is followed by dominance of the taller, more aggressive grasses and other nitrophilous species.

Major species changes have occurred after clearing shrublands, originally for timber, charcoal or cultivation. In Australia, widespread bush clearing for cultivation and pasture establishment has caused hydrological imbalances that have led to salinization of large tracts of bottom land. In California, 'brush conversion' by fire, selective herbicides for controlling brush regeneration, and reseeding with annual and perennial pasture species became a well-developed technology during the 1950s (Love and Jones, 1952; Love, 1961). Today, brush conversion is restricted to seeding annual ryegrass and native perennial grasses over wildfire burns (Menke, 1988; Melvin George, University of California at Davis, 1994, personal communication). In Israel, this technology was applied on relatively large areas in the Galilee, to establish grassland for beef ranching based on

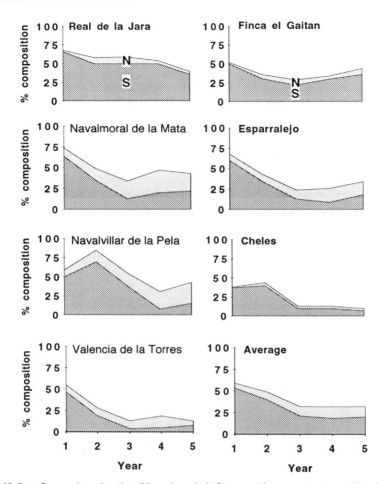

Fig. 13.7. Proportion of native (N) and seeded (S) annual legumes on seven sites in Extremadura, Spain. (Data from Olea Marquez de Prado, 1988.)

perennial gramineous species (*Oryzopsis miliacea, O. holciformis, Phalaris tuberosa*). There were technical successes (Naveh, 1989) but the practice was discontinued as soon as relief labour became unavailable. Without heavy grazing or periodic control of shrub regrowth, most of the brush conversion projects reverted to woodland or to thorny shrubland within 10 years.

There is a widespread need to maintain an open park-like savanna for grazing, recreation areas and control of major conflagrations. 'Biological control' of shrub regrowth by imposition of heavy grazing with beef cattle on woodland that had been lightly cleared was studied in Israel. This was reasonably successful but was accompanied by the spread of spiny dwarf shrubs (*Calicotome villosa*), which required periodic control with herbicide (Gutman *et al.*, 1990). Heavy supplementation was necessary during the summer and autumn. Management of such 'vestigial grasslands' under present-day conditions in most of the region is an ongoing challenge (Biswell, 1956; Etienne, 1989; Perevolotsky et al., 1993).

Forage shrubs and trees

During the long, dry mediterranean summer any green forage is a welcome addition to the generally low-quality diet of hayed-off grass, even if it is from shrubs that are usually inferior forage. Some, however, play an important role in animal production. The oaks of the Spanish *dehesa* (*Quercus ilex* ssp. *rotundifolia* and *Q. suber*) are pruned periodically (every 8 to 15 years) to increase yield and availability of acorns, mainly for fattening the Iberian pigs that are grown for the highly prized (and priced) cured ham, *jamon serrano* (Martin Bellido, 1989). Other woody species of the *dehesa* can improve pasture conditions. *R. sphaerocarpa* is not highly palatable, but it is an N-fixing leguminous species that provides some green feed, including pods and seeds, when all around is dry. It is valued by the shepherds, possibly for the protection it provides at lambing. There is a saying: *'Debajo cada retama se cría un borrego'* (Under every retama bush there grows a lamb) (Rivas Martinez, 1987). Some shrubs provide highly palatable forage. Outstanding among these are tree medick (*Medicago arborea*) and tagasaste (*Chamaecytisus palmensis*).

It is therefore no wonder that the propagation of selected forage shrubs has been an attractive research and development objective (Le Houérou, 1980), especially in the more arid margins of the Mediterranean basin in north Africa, where large areas of forage shrubs were planted as strategic reserves. Among the most commonly planted species were *Atriplex* species, mainly *A. nummularia* and *A. halimus,* and spineless cactus (*Opuntia ficus-indica*). Plantations of *Atriplex* spp. were established on many sites in Israel during the 1950s when relief labour was plentiful.

However, where controlled studies were conducted in a grazing system context, the benefits of shrubs for animal production could not be demonstrated (Leigh et al., 1970). In Israel, under equivalent stocking rates on ley systems in a semiarid mediterranean region, there was no significant difference in animal production between systems with and systems without forage shrubs (Benjamin et al., 1987). Benefits have been demonstrated under experimental conditions in Spain, but with reservations about persistence of palatable shrubs under grazing (Correal, 1993).

The protein content of forage shrub foliage is often regarded as an important nutritional supplement to low-quality dry-summer pasture. However, the role of shrubs in the nutrition of ruminants is complicated by their antinutritional characteristics. Condensed tannins, non-protein N and high salt concentrations all make standard feed quality analyses irrelevant (Nastis, 1993). Interpretations of crude protein calculated from N content as true, available protein are not borne out in feeding trials (Benjamin et al., 1992). Digestibility estimates, inflated by high salt content, especially in *Atriplex* species, seldom take into account the substantial energy cost of salt disposal to the animal (Arieli *et al.*, 1989).

Commercial plantations have not been established even in the developed countries. In the USA, a company that mass produced shrub seedlings for land reclamation projects found no market for the seedlings for pasture improvement (C.M. McKell, Logan, Utah, 1989, personal communication). The lack of proved benefits for livestock production from forage shrubs has restricted plantations to research and to subsidized development projects.

Risk and economic constraints

Finally there is consideration of how to spend limited available money most efficiently to tide over periods of pasture inadequacy: on fertilizer, reseeding, forage shrubs or on supplementary feed. The very wide use of supplementary feed and the very limited use of improvements to increase production of mediterranean pasture suggests that livestock managers trust feed supplementation more than investment in the land. According to a Spanish adage, 'With feed in the store there are no bad years.'

As indicated earlier, the dependence of the pastoral industry in the mediterranean zones of Australia on routine investments in pasture establishment and improvement is a consequence of a specific agroecological environment. Supplementary feeding is subject to both cost and labour constraints, especially on large extensive enterprises. In the Mediterranean region, where most animal husbandry systems are comparatively labour-intensive and where pasture improvement on rough hilly grasslands is expensive and often of dubious benefit, the supplementary feed option has definite attractions. However, economic conditions can change and so there is a need to determine the technological coefficients that define pasture response to improvements in different ecological conditions, for use whenever and wherever the conditions are right.

Grassland as environment

Non-forage dimensions

Grasslands, especially native rangelands and permanent pastures, have characteristics that allow for activities additional to agricultural production. In parts of Spain, many estates are maintained as hunting preserves (*coto de caza*), where habitat for wild boar, deer, hare and partridge is as important as pasture for livestock (Campos Palacin and Pearce, 1992). Other forms of recreation that are influencing grassland management include hiking, camping, recreational vehicles and agrotourism. In many countries on the northern rim of the Mediterranean basin recreational and tourist aspects of landscape use have become more important than pastoral use. Environmental benefits (hedgerows, windbreaks, erosion control, sand-dune stabilization, wildlife refuge, landscaping, fruit (e.g. *O. ficus-indica*), fuel, honey) have been recruited to justify forage shrub plantations on Mediterranean grasslands (Le Houérou, 1993). From another angle, wider ecological issues of biodiversity, endangered species, water quality, riparian habitats and landscape aesthetics have surfaced as lively public issues. In Mediterranean countries, the conservation of traditional agropastoral landscapes commands considerable European Community support. In the words of Wagner (1994): 'We are ending an era when natural resources are viewed largely as a commodity ... an increasingly urban public has broadened its values ... esthetic and non-consumptive uses, preservation and animal welfare attitudes ... now play an increasingly important role.'

Multiple use

The need to accommodate a number of different and, in certain respects, conflicting goals on the same landscape unit has been recognized for many years. 'Multiple use' is a standard management concept on the public lands in the USA and a reality in much of Mediterranean Europe. However, the growing interest in non-economic goals tends to polarize attitudes to resource use, so making the application of multiple use increasingly difficult. Livestock grazing on the mediterranean grasslands of California has become a thorny issue (Melvin George, University of California at Davis, 1994, personal communication). Dealing with conflicting goals may well become a major task of managers and administrators in agriculture, with special problems in grasslands where the number of stakeholders (and steak producers!) can be quite large. Explicit definition of goals and identification of the costs and trade-offs associated with implementation or non-implementation of any one goal is becoming an important research and administrative objective.

Assessing non-economic benefits

Cost accounting for non-economic goal attainment on mediterranean rangelands presents difficulties that have not yet been resolved. Indeed, similar difficulties have given rise to 'ecological economics', a new discipline that is concerned with concepts and methodologies for resource accounting at all levels, from the site and farm to the global. It reflects the changing values that are shaping the future use and role of grasslands.

CONCLUSION

Successional anticlimax

If sclerophyllous woodland is the 'climax vegetation' in mediterranean regions, then mediterranean grasslands, the ultimate regression stage (e.g. Table 13.1), are a 'successional anticlimax'. Despite their lowly phytosociological status, most Mediterranean grasslands are productive, robust, resilient formations, rich in biodiversity. They have proved to be sustainable through the ages – as long as they are intensively used. The long association of humans and their livestock with cultivated land has accelerated nutrient cycling (Mooney and Hobbs, 1994) and so created conditions favourable for productive, but aggressive, species. Among these, the annual legumes have played a synergistic role, increasing soil fertility by fixing atmospheric N. Grasslands in the other mediterranean regions are recent derivations that have been accommodated relatively successfully, but not without environmental dislocation.

One-way genetic flow

Over the past 400 years, European colonialism, with its domesticated crops and ruminants, has engendered a flow of annual species from the Mediterranean basin to other mediterranean regions, where, on extensive areas, they have invaded and dominated the native vegetation. The large number of species and genotypes that have become naturalized 'on foreign shores' and constitute their 'mediterranean grasslands' (Baker, 1988) are a small part of the gene bank in the highly heterogeneous Eurasian region of origin (Shmida, 1981). It is therefore not surprising that, to date, the genetic flow has been overwhelmingly one-way and that the introduction of newly selected pasture varieties, mainly from Australia, back to the Mediterranean basin is much less successful.

Essential redundancy

The strong and characteristic seasonality of mediterranean grasslands is typically 'boom-and-bust'. Growth rate of the vegetation during the spring flush is much higher than the forage consumption rate of the animals, even at high stocking rates. This 'redundant' growth imposes a low level of forage use efficiency and presents difficult nutrition management problems during the rest of the year. On the other hand, it does guarantee that seed will be produced each year and so ensures the sustainability of the annual-type pasture and of the livestock production system, even under intensive use.

'A farewell to farms'?

The technological revolution, which has converted agriculture in the developed regions of the world from a labour-intensive, low-yield activity into a capital-intensive, highly productive economic sector, has inevitably changed land and product values, land-use practices and demography. These developments have affected the mediterranean regions very unevenly, but none are immune to its consequences. Concentration of production on better land and in regions that are better endowed with agricultural resources has led to the marginalization of poorly endowed regions, abandonment or consolidation of marginal farm units and rural depopulation. This polarization of land resource use has been accompanied by a growing public and private interest in the non-productive aspects of the marginal areas and, consequently, an increase in their value for rural living, recreation, environmental quality and conservation. The consequences for grassland management have been complex. In countries on the northern rim of the Mediterranean basin and in other developed mediterranean regions, livestock husbandry has retreated from some of the more difficult or more remote areas. Grasslands have been abandoned and woody vegetation has taken over. The greater fire hazard in the fire-prone mediterranean environment threatens life, property, watershed and recreational values. Undergrazing on these grasslands is often a greater problem than overgrazing. In areas where grassland husbandry is still viable, macro-economics is complicating conditions of animal production: easily available, relatively cheap inputs encourage intensification, while market surplus, low livestock product prices, excessive subsidization costs and environmental concerns discourage intensification of grassland management. In countries on the southern rim of the Mediterranean basin, demographic pressures and supplementary feed are forcing ever-increasing livestock numbers (Seligman and Perevolotsky, 1994). In more populated areas, the already excessive pressure on the grassland resources is causing concern about sustainability (Chatterton and Chatterton, 1984; Milton *et al.*, 1994).

Where do we go from here?

For grassland managers in the numerous and varied situations in the different mediterranean regions, the answer, as always, depends on the goals and the capabilities of the people involved and on the current situation. For most grassland scientists the answer will be determined by the policy adopted by their funding organizations. But for grassland science there are issues of a general nature that have become more urgent than in the past, when agrotechnical innovation was the central objective of applied agricultural research. The confidence that rested on a belief that science would find the answers has been humbled by the socio-economic realities of recent years.

Beyond the basics of input/output price ratios, economic sustainability of many grassland enterprises depends more than ever on careful consideration of the ever-changing interconnections and synergisms between the different enterprise activities. Recognition of the nature of the change and its implications for grassland husbandry is necessary for a clearer definition of the goals of all who have a stake in the grassland resource. The continuity of change, which makes analysis of technological innovation for survival at the farm level and resolution of conflicting goals at all levels a difficult exercise, is giving rise to novel approaches (e.g. de Wit *et al.*, 1988), which, however, have not yet had much effect on grassland science. A recent survey concludes that the grassland profession today is 'rather detached from current social debate. It almost looks like a profession talking to itself' (Nores and Vera, 1993). For grassland research to widen its horizons to include ecological and social issues more explicitly, it will need to improve its capability in both these disciplines. The need is universal but the Mediterranean basin, in the words of David Attenborough (1987), 'has a special claim to our interest . . . It is the place where mankind's exploitation of the land began and where it has run its full cycle. What happened here during past millennia is, elsewhere on earth, just beginning.'

REFERENCES

Abercrombie, T.J. (1991) Extremadura, cradle of conquerors. *National Geographic* 177, 117–134.
Acuña, H., Soto, P. and Klee, G. (1993) Subterranean clover pasture improvement and utililization in the Andes foothills of the Mediterranean subhumid zone of Chile. *Proceedings, 17th International Grassland Congress*, Palmerston North, New Zealand, pp. 263–264.
Anonymous (1967) *Pilot Project in Watershed Management on the Nahal Shikma Watershed*. Final Report, UNDP/FAO/SF: 6/ISR, Rome.
Anonymous (1975) *Improvement and Utilization of Grazing Resources in Mediterranean-type Climates*. Proceedings of an International Seminar, INIA/CRIDA, Badajoz, Spain, 290 pp.
Arieli, A., Naim, E., Benjamin, R.W. and Pasternak, D. (1989) The effect of feeding saltbush and sodium chloride on energy metabolism in sheep. *Animal Production* 49, 451–457.
Arnon, I. and Dovrat, A. (1956) *Progress Report, 2nd Meeting of Working Party on Mediterranean Pasture and Fodder Development, Algiers*. Agricultural Research Station Report, Rehovot.
Aschmann, H. (1973) Man's impact on the several regions with Mediterranean climate. In: Di Castri, F. and Mooney, H.A. (eds) *Mediterranean Type Ecosystems: Origin and Structure*. Springer-Verlag, Heidelberg, pp. 363–371.
Attenborough, D. (1987) *The First Eden: The Mediterranean World and Man*. Collins/BBC Books, London.
Axelrod, D.I. (1973) History of the Mediterranean ecosystem in California. In: Di Castri, F. and Mooney, H.A. (eds) *Mediterranean Type Ecosystems: Origin and Structure*. Springer-Verlag, Heidelberg, pp. 225–277.
Baker, H.G. (1988) Sources of the naturalized grasses and herbs in California. In: Huenneke, L.F. and Mooney, H.A. (eds) *Grassland Structure and Function – California Annual Grassland*. Kluwer Academic Publishers, Dordrecht, The Netherlands, pp. 29–38.

Beattie, A.J. (1988) The effects of ants on grasslands. In: Huenneke, L.F. and Mooney, H.A. (eds) *Grassland Structure and Function – California Annual Grassland.* Kluwer Academic Publishers, Dordrecht, The Netherlands, pp. 105–116.

Benjamin, R.W. (1992) Sheep husbandry for lamb production in a semi-arid Mediterranean environment. In: Alberda, Th., van Keulen, H., Seligman, N.G. and de Wit, C.T. (eds) *Foods from Dry Lands.* Kluwer Academic Publishers, Dordrecht, The Netherlands, pp. 83–100.

Benjamin, R.W., Degen, A.A., Brieghet, A., Chen, M. and Tadmor, N.H. (1975) Estimation of food intake of sheep grazing green pasture when no free water is available. *Journal of Agricultural Science (Cambridge)* 85, 403–407.

Benjamin R.W., Barkai, D., Lavie Y. and Forti, M. (1987) Effect of adding a fodder shrub component to sheep production systems in summer grazing. In: Dovrat, A. (compiler), *Fodder Production and its Utilization by Small Ruminants in Arid Regions* (FOPAR). Fifth Annual Report, No. BGUN-ARI-48-87, Ben-Gurion University of the Negev, Beer-Sheva, pp. 21–39.

Benjamin, R.W., Oren, E., Katz, E. and Becker, K. (1992).The apparent digestibility of *Atriplex barclayana* and its effect on nitrogen in sheep. *Animal Production* 54, 259–264.

Benjamin, Y., Kali, J., Barkai, D., Benjamin, R.W. and Eyal, E. (1981) Wheat straw, poultry litter and corn grain in nutrition of Awassi mutton sheep. *Hassadeh* 61, 815–819. (Hebrew with English summary.)

Biswell, H.H. (1956) Ecology of California grasslands. *Journal of Range Management* 9, 19–24.

Campos Palacin, P. and Pearce, D.W. (1992) Asessment of the economic and environmental benefits of the cork tree dehesa system. CSIC/CSERGE, Madrid, unpublished document.

Carter, E.D. and Day, H.R. (1970) Inter-relationships of stocking rate and superphosphate rate on pasture as determinants of animal production. *Australian Journal of Agricultural Research* 21, 473–491.

Chatterton, B. and Chatterton, L. (1984) Alleviating land degradation and increasing cereal and livestock production in north Africa and the Middle East using annual *Medicago* pasture. *Agriculture, Ecosystems and Environment* 11, 117–129.

Cocks, P.S. (1980) Limitations imposed by N-deficiency on the productivity of subterranean clover-based annual pasture in South Australia. *Australian Journal of Agricultural Research* 31, 95–107.

Cocks, P.S. (1992a) Changes in the size and composition of the seed bank of medic pasture grown in rotation with wheat in north Syria. *Australian Journal of Agricultural Research* 43, 1571–1581.

Cocks, P.S. (1992b) Evolution of sown populations of subterranean clover (*Trifolium subterraneum*) in South Australia. *Australian Journal of Agricultural Research* 43, 1583–1595.

Cocks, P.S. and Gintzburger, G. (1993) Long-term sustainability of livestock producing farming systems in contrasting regions within mediterranean-type climates. *Proceedings, 17th International Grassland Congress.* Palmerston North, New Zealand, pp. 247–251.

Correal, E. (1993) Grazing use of fodder shrub plantations. In: Papanastasis, V. (ed.) *Fodder Trees and Shrubs in the Mediterranean Production Systems: Objectives and*

Expected Results of the EC Research Contract. Commission of the European Communities, Luxemburg, pp. 99–118.

Crespo, D. (1993) Opening speech. *Proceedings, 17th International Grassland Congress*, Palmerston North, New Zealand, pp. 2–3.

Davies, W. (1952) *The Grass Crop: Its Development, Use and Maintenance.* E. & F.N. Spon, London.

de Ridder, N., Seligman, N.G. and van Keulen, H. (1981) Analysis of environmental and species effects on the magnitude of biomass investment in the reproductive effort of annual pasture plants. *Oecologia (Berlin)* 49, 253–271.

de Wit, C.T., van Keulen, H., Seligman, N.G. and Spharim, I. (1988) Application of interactive multiple-goal linear programming techniques for analysis and planning of regional agricultural development. *Agricultural Systems* 26, 211–230.

Edelstein, G. and Milevski, I. (1994) The rural settlement of Jerusalem re-evaluated. *Palestine Exploration Quarterly* 126, 1–23.

Edwards, S.W. (1992) Observations on the prehistory and ecology of grazing in California. *Fremontia* 20(1), 3–11.

Elena, M., Cornut, E. and Lopez, J.A. (1985) *Estructura del sistema productiva del ecosistema de dehesa.* Servicio de Extension y Capacitacion Agraria, Badajoz, Spain.

Etienne, M. (1989) Protection of Mediterranean forests against fire: an ecological approach for redevelopment. *5th European Ecological Symposium*, Sienna, Italy.

Eyal E., Benjamin, R.W. and Tadmor, N.H. (1975) Sheep production on seeded legumes, planted shrubs and dryland grain in a semiarid region of Israel. *Journal of Range Management* 28, 100–107.

Falcinelli, M., Veronesi, F., Russi, L. and Pollodori, P. (1993) Persistence and productivity of some forage varieties and landraces of different origin grown in central Italy. *Proceedings, 17th International Grassland Congress*, Palmerston North, New Zealand, pp. 266–268.

Feinbrun-Dotan, N. (1978) *Flora Palaestina*, Vol. 3. Israel Academy of Sciences and Humanities, Jerusalem.

Foran, B.D., Friedel, M.H., MacLead, N.D., Stafford-Smith, D.M. and Wilson, A.D. (1989) *Policy Proposals for the Future of Australia's Rangelands.* CSIRO National Rangelands Program, CSIRO, Lyneham, ACT, Australia, 27 pp.

George, M.L., Brown, J.R. and Clawson, W.J. (1992) Application of non-equilibrium ecology to management of Mediterranean grasslands. *Journal of Range Management* 45, 436–440.

Gutman, M. and Seligman, N.G. (1979) Grazing management of herbaceous Mediterranean foothill range in the Upper Jordan Valley. *Journal of Range Management* 32, 86–92.

Gutman, M., Henkin, Z., Holzer, Z., Noy-Meir, I., and Seligman, N.G. (1990). Plant and animal responses to beef cattle grazing in a Mediterranean oak scrub forest. *Proceedings, 6th Meeting, FAO–European Cooperative Network on Pasture and Fodder Crop Production*, 191–196.

Heady, H.F. (1958) Vegetation changes in the California annual type. *Ecology* 39, 402–415.

Heady, H.F. (1960) Continuous vs. specialized grazing systems: a review of applications to the California annual type. *Journal of Range Management* 14, 182–193.

Henkin, Z. (1994) The effect of phosphorus nutrition, shrub control and fire on the dynamics of Mediterranean Batha vegetation in the Galilee. PhD thesis, Hebrew University of Jerusalem, Jerusalem.

Hubert, B. (1994) Pastoralisme et territoire: modélisation de pratiques d'utilisation. *Cahiers Agricultures* 3, 9–22.
Jackson, L.E. (1985) Ecological origins of California's Mediterranean grasses. *Journal of Biogeography* 12, 349–361.
Jones, M.B. (1964) Effect of applied sulfur on yield and sulfur uptake of various California dryland pasture species. *Agronomy Journal* 56, 235–237.
Jones, R.J. and Sandland, R.L. (1974) The relation between animal gain and stocking rate in grazing trials: derivation of a model from experimental results. *Journal of Agricultural Science (Cambridge)* 83, 355–342.
Le Houérou, H.N. (1980) Browse in North Africa. In: Le Houérou, H.N. (ed.) *Browse in Africa, the Current State of Knowledge*. ILCA, Addis Ababa, Ethiopia, pp. 55–82.
Le Houérou, H.N. (1993) Environmental aspects of fodder tree and shrub plantations in the mediterranean basin. In: Papanastasis, V. (ed.) *Fodder Trees and Shrubs in the Mediterranean Production Systems: Objectives and Expected Results of the EC Research Contract.* Commission of the European Communities, Luxemburg, pp. 11–34.
Leigh, J.H., Wilson, A.D. and Williams, O.B. (1970) An assessment of the value of three perennial chenopodiaceous shrubs for wool production of sheep grazing semiarid pastures. *Proceedings, 11th International Grassland Congress*, 55–59.
Lidicker, W.Z., Jr (1988) Impacts of non-domesticated vertebrates on California Grasslands. In: Huenneke, L.F. and Mooney, H.A. (eds) *Grassland Structure and Function – California Annual Grassland.* Kluwer Academic Publishers, Dordrecht, The Netherlands, pp. 135–150.
Loria, M. (1982) Granivory by harvester ants (*Messor* spp.) and seed dynamics in a semiarid mediterranean-type annual species sward. (Unpublished manuscript.)
Love, R.M. (1961) The range – natural plant communities or modified ecosystems? *Journal of the British Grassland Society* 16, 89–99.
Love, R.M. and Jones, B.J. (1952) *Improving California Brush Ranges.* California Agricultural Extension Service Circular 371, Berkeley, 38 pp.
Malanson, G.P., and Trabaud, L. (1987) Ordination analysis of components of resilience of *Quercus coccifera* garrigue. *Ecology* 68, 463–472.
Martin Bellido, M. (1989) Animal production in the south-west of Spain. *II Reunion Iberica de Pastos y Forrajes*, Badajoz-Elvas, 10–14 April 1989, pp. 309–333.
Menke, J.W. (1988) Management controls on productivity. In: Huenneke, L.F. and Mooney, H.A. (eds) *Grassland Structure and Function – California Annual Grassland.* Kluwer Academic Publishers, Dordrecht, The Netherlands, pp. 173–200.
Milton, S.J., Dean, W.R., du Plessis, M.A. and Siegfried, W.R. (1994) A conceptual model of arid rangeland degradation. *BioScience* 44, 70–76.
Mooney, H.A. and Hobbs, R.J. (1994) Resource webs in Mediterranean-type climates. In: Arianoutsou, M. and Groves, R.H. (eds) *Plant–Animal Interactions in Mediterranean-type Ecosystems.* Kluwer Academic Publishers, Dordrecht, The Netherlands, pp. 73–81.
Morin, J., Michaeli, A., Agassi, M., Atzmon, B. and Rosenzweig, D. (1979) Rainfall–run-off–erosion relationships in the Kinneret (Lake of Galilee) catchment. Research Report R42, Soil Erosion Research Station, Soil Conservation and Drainage Division, Ministry of Agriculture, Tel Aviv, Israel, 148 pp.
Morrison, D.A., Kingwell, R.S., Pannell, D.J. and Ewing, M.A. (1986) A mathematical programming model of a crop–livestock farm system. *Agricultural Systems* 20l, 243–268.

Murphy, A.H., Jones, M.B., Clawson, J.W. and Street, J.E. (1973) *Management of Clovers on California Annual Grasslands*. California Agricultural Extension Service Circular 564, Davis, 19 pp.

Nastis, A. (1993) Nutritive value of fodder shrubs. In: Papanastasis, V. (ed.) *Fodder Trees and Shrubs in the Mediterranean Production Systems: Objectives and Expected Results of the EC Research Contract.* Commission of the European Communities, Luxemburg, pp. 75–83.

Naveh, Z. (1989) Mediterranean Europe and east Mediterranean shrublands. In: McKell, C.M. (ed.) *The Biology and Utilization of Shrubs.* Academic Press, London, pp. 93–117.

Naveh, Z., and Dan, J. (1973) The human degradation of Mediterranean landscapes in Israel. In: Di Castri, F. and Mooney, H.A. (eds) *Mediterranean-type Ecosystems: Origin and Structure.* Springer-Verlag, Berlin, pp. 373–390.

Naveh, Z., and Whittaker, R.H. (1979) Measurements and relationships of plant species diversity in Mediterrranean shrublands and woodlands. In: Grassle, J.F., Patil, G.P., Smith, W.K. and Taille, C. (eds) *Ecological Diversity in Theory and Practice.* Statistical Ecology Series 6, International Cooperative Publishing House, Maryland, USA, pp. 219–239.

Nores, G.A. and Vera, R.R. (1993) Science and information for our grasslands. *Proceedings, 17th International Grassland Congress*, Palmerston North, New Zealand, pp. 33–37.

Noy-Meir, I. (1976) Rotational grazing in a continuously growing pasture: a simple model. *Agricultural Systems* 1, 87–112.

Noy-Meir, I. (1978) Grazing and production in seasonal pastures: Analysis of a simple model. *Journal of Applied Ecology* 15, 809–835.

Noy-Meir, I. (1988) Dominant grasses replaced by ruderal forbs in a vole year in ungrazed Mediterranean grasslands. *Journal of Biogeography* 15, 579–587.

Noy-Meir, I. (1992) Structure and dynamics of grazing systems on seasonal pastures. In: Alberda, Th., van Keulen, H., Seligman, N.G. and de Wit, C.T. (eds) *Foods from Dry Lands.* Kluwer Academic Publishers, Dordrecht, The Netherlands, pp. 7–24.

Noy-Meir, I. and Seligman, N.G. (1979) Management of semi-arid ecosystems in Israel. In: Walker, B.H. (ed.) *Management of Arid Ecosystems.* Elsevier Scientific Publishing, Amsterdam, pp. 113–160.

Noy-Meir, I., and Walker, B.H. (1986) Stability and resilience in rangelands. In: Joss, P.J., Lynch, P.W. and Williams, O.B. (eds) *Rangelands Under Siege.* Australian, Academy of Sciences, Canberra, pp. 21–25.

Noy-Meir, I., Gutman, M. and Kaplan, Y. (1989) Response of Mediterranean grassland plants to grazing and protection. *Journal of Ecology* 77, 290–310.

Ofer, Y. and Seligman, N. (1969) Fertilization of annual range in northern Israel. *Journal of Range Management* 22, 337–341.

Olea, L., Paredes, J., and Verdasco, P. (1989) Caracteristicas productivas de los pastos de la dehesa del S.O. de la peninsula Iberica. *II Reunion Iberica de Pastos y Forrajes*, Badajoz-Elvas, 10–14 April 1989, pp. 147–172.

Olea Márquez de Prado, L. (1988) Persistencia y produccion de pastos en el S.O. de España: Introducción de Trébol subterráneo. Colleccion Tesis Doctorales INIA Núm. 74, Instituto Nacional de Investigaciones Agrarias, Madrid, 201 pp.

Osman, A.E. and Cocks, P.S. (1993) Effects of phosphate and stocking rate on mediterranean grasslands in northern Syria. *Proceedings, 17th International Grassland Congress*, Palmerston North, New Zealand, pp. 259–260.

Osman, A.E., Cocks, P.S., Russi, L. and Pagnotta, M.A. (1991) Response of mediterranean grassland to phosphate and stocking rates: biomass and botanical composition. *Journal of Agricultural Science (Cambridge)* 116, 37–46.

Ovalle, C., Avendaño, J., Del Pozo, A. and Crespo, D. (1993). Germplasm collection, evaluation and selection of naturalized *Medicago polymorpha* in the mediterranean zone of Chile. *Proceedings, 17th International Grassland Congress*, Palmerston North, New Zealand, pp. 222–223.

Perevolotsky, A., Ettinger, E., Yonatan, R. and Gutman, M. (1993). The management of fire breaks for decreasing fire damage: problems and prospects. *Proceedings, 7th Meeting of the FAO–European Network on Mediterranean Pastures and Fodder Crop Production*, 196–200.

Pineda, F.D., Peco, B., Levassor, C., Casada, M.A. and Galiano, E.F. (1984) Some regularities in the organization and phenology of Mediterranean pastures along ecological succession. In: Riley, H. and Skjelvåg, A.O. (eds) *The Impact of Climate on Grass Production and Quality*. European Grassland Federation, Norwegian State Agricultural Research Stations, Ås, Norway, pp. 236–240

Puckridge, D.W. and French, R.J. (1983) The annual legume pasture in cereal–ley farming systems of southern Australia. *Agriculture, Ecosystems and Environment* 9, 229–267.

Raven, P.H. (1973) The evolution of Mediterranean floras. In: Di Castri, F. and Mooney, H.A. (eds) *Mediterranean Type Ecosystems: Origin and Structure*. Springer-Verlag, Heidelberg, pp. 213–224.

Reeves, T.G. and Ewing, M.A. (1993) Is ley farming in mediterranean zones just a passing phase? *Proceedings, 17th International Grassland Congress*, Palmerston North, New Zealand, pp. 2169–2177.

Rivas Goday, S. (1964) *Vegetacion y Florula de la Cuenca Extremeña del Guadiana*. Publicaciones de la Excma, Diputation Provincial de Badajoz, Madrid, 777 pp.

Rivas Martinez, S. (1987) *Memoria del Mapa de Series de Vegetation de España*. ICONA, Publicaciones del Ministerio de Agricultura, Pesca y Alimentacion, Madrid, 268 pp.

Rossiter, R.C. (1966) Ecology of the mediterranean annual type pasture. *Advances in Agronomy* 18, 1–56.

Schumacher, G. (1888) *The Jaulan*. Richard Bentley and Son, London, 302 pp.

Seginer, I., Morin, Y. and Shachori, A. (1963) *Experiments on Runoff and Erosion from the Western Slopes of Mount Carmel*. Research report no. 8, Soil Conservation Service, Ministry of Agriculture, Tel Aviv, Israel, 23 pp. (in Hebrew).

Seligman, N.G. (1973) A quantitative geobotanical analysis of the vegetation of the Golan. PhD thesis, Hebrew University of Jerusalem, 136 pp. (Hebrew with English summary).

Seligman, N.G. and Gutman, M. (1975) Utilization of *Psoralea bituminosa* in herbaceous Mediterranean pasture. *Hassadeh* 55, 2083–2086 (Hebrew with English summary).

Seligman, N.G. and Meron, Y. (1968) *Response of Annual Range Herbage in the Hills of Ephraim to Applications of Nitrogen Fertilizer*. Research Report R22, Soil Erosion Research Station, Soil Conservation and Drainage Division, Ministry of Agriculture, Tel Aviv, Israel. 148 pp.

Seligman, N.G. and Perevolotsky, A. (1994) Has intensive grazing by domestic livestock degraded Mediterranean Basin rangeland? In: Arianoutsou, M. and Groves, R.H. (eds)

Plant–Animal Interactions in Mediterranean-type Ecosystems. Kluwer Academic Publications, Dordrecht, The Netherlands, pp. 93–103.
Shmida, A. (1981) Mediterranean vegetation in California and Israel: similarities and differences. *Israel Journal of Botany* 30, 105–123.
Smith, R.C.G. and Williams, W.A. (1973) Model development for a deferred-grazing system. *Journal of Range Management* 26, 454–460.
Specht, R.L. (1973) Structure and functional response of ecosystems in the Mediterranean climate of Australia. In: Di Castri, F. and Mooney, H.A. (eds) *Mediterranean Type Ecosystems: Origin and Structure*. Springer-Verlag, Heidelberg, pp. 113–120.
Spharim, I. and Seligman, N.G. (1990) Evaluation of technological improvements in beef herd management in the light of local and national goals. *Israel Agresearch* 4, 21–44 (Hebrew with English summary).
Springborg, R. (1990). Human constraints in extending the use of forage legumes in mediterranean areas. In: Osman, A.E., Ibrahim, M.H. and Jones, M.A. (eds) *The Role of Legumes in the Farming Systems of the Mediterranean Areas*. Kluwer Academic Publishers, Dordrecht, The Netherlands, pp. 283–294.
Tükel, T., Saglamtimur, T., Gülcan, H., Tansi, V. and Anlarsal, A.E. (1993) Constraints and opportunities for Turkish grasslands use patterns and the expected development of forage crops, with the south-eastern Anatolian Project (GAP) in Turkey. *Proceedings, 17th International Grassland Congress*, Palmerston North, New Zealand, pp. 261–263.
Ungar, E.D. (1990) *Management of Agropastoral Systems in a Semiarid Region*. Simulation Monographs, Pudoc, Wageningen, The Netherlands.
van Heerden, J.M. and Tainton, N.M. (1987) Potential of medic and lucerne pastures in the Rûens area of the southern Cape. *Journal of the Grassland Society of South Africa* 4, 95–99.
van Keulen, H. and Seligman, N.G. (1992) Moisture, nutrient availability and plant production in the semi-arid region. In: Alberda, Th., van Keulen, H., Seligman, N.G., and de Wit, C.T. (eds) *Foods from Dry Lands*. Kluwer Academic Publishers, Dordrecht, The Netherlands, pp. 25–81.
Wagner, F.H. (1988) Grazers, past and present. In: Huenneke, L.F. and Mooney, H.A. (eds) *Grassland Structure and Function – California Annual Grassland*. Kluwer Academic Publishers, Dordrecht, The Netherlands. pp. 151–162.
Wagner, F.H. (1994) Changing institutional arrangements for setting natural resources policy. In: Vavra, M., Laycock, W.A. and Pieper, R.D. (eds) *Ecological Implications of Livestock Herbivory in the West*. Society of Range Management, Denver, Colorado, USA, pp. 281–288.
Walker, B.H. and Noy-Meir, I. (1982) Aspects of stability and resilience of savanna ecosystems. In: Huntley, B.J. and Walker, B.H. (eds) *Ecology of Tropical Savanna*. Ecological Studies, Volume 42, Springer-Verlag, Berlin, pp. 556–590.
Wester, L. (1981) Composition of native grasslands in the San Joaquin Valley, California. *Madroño* 28, 231–241.
Wild, A. (1958) The phosphate content of Australian soils. *Australian Journal of Agricultural Research* 9, 193–204.
Williams, W.A., Love, R.M. and Conrad, J.P. (1956) Range improvement in California by seeding clovers, fertilization and grazing management. *Journal of Range Management* 9, 28–33.

Willoughby, W.M. (1959) Limitations to animal production imposed by seasonal fluctuations in pasture and management procedures. *Australian Journal of Agricultural Research* 10, 248–268

Woodward, R.G. and Morley, F.H.W. (1974) Variation in Australian and European collections of *Trifolium glomeratum* L. and the provisional distribution of the species in southern Australia. *Australian Journal of Agricultural Research* 25, 73–78.

Zinke, P.J. (1973) Analogies beween the soil and vegetation types of Italy, Greece, and California. In: Di Castri, F. and Mooney, H.A. (eds) *Mediterranean Type Ecosystems: Origin and Structure*. Springer-Verlag, Heidelberg, pp. 61–82.

Zohary, M. (1973) *Geobotanical Foundations of the Middle East*. Gustav Fischer Verlag, Stuttgart, and Swets & Zeitlinger, Amsterdam.

Grasslands in the Well-watered Tropical Lowlands

14

M.J. Fisher, I.M. Rao, R.J. Thomas and C.E. Lascano

Centro Internacional de Agricultura Tropical, Apartado Aéreo 6713, Cali, Colombia

INTRODUCTION

The theme of this chapter is strategies to deal with limitations of forage utilization and nutritive value for pastoral livestock systems in the well-watered tropics. Because our own experiences are in the tropics of America and Australia, we have emphasized these regions. Readers are referred to more comprehensive reviews (e.g. Crowder and Chheda, 1982; Humphreys, 1991) for a fuller coverage of Africa and Asia.

The primary limitation to production by domestic grazing ruminants in the lowland tropics is low quality of the native forage. Usually the forage is deficient in one or more of protein, dry matter digestibility and/or in one or more mineral nutrients, so that utilization of the forage is low. Unless the forage resource is changed to overcome these limitations, the management options are restricted to increasing utilization and providing supplements to correct mineral and protein deficiencies. Very little can be done to increase digestibility, apart from burning the forage to take advantage of the higher digestibility of young regrowth.

These strategies are similar to those employed in other extensive range systems, and are discussed by M. Stafford Smith (Chapter 12, this volume). They can bring about improvements in animal performance, but performance remains low, in contrast to intensive systems based on more nutritious species. They may be profitable on an individual farm basis, but the need to feed increasing populations brings considerable pressure to bear to intensify production.

Intensive systems are still evolving in much of the lowland tropics. We shall discuss the limitations of the native pasture species, leading to a discussion of their replacement with introduced pastures and of the problems that arise.

To most people, management means the manipulation of stocking rate (or grazing pressure or forage allowance) and grazing system to optimize animal performance while maintaining the pasture in an acceptable condition, however defined. In our experience, the interactions between soil, plant and animal are complex and we believe that management in this sense is species-specific and probably site-specific as well.

The literature on temperate pastures shows that there have been many grazing experiments with a limited suite of species. Despite this effort, it is only recently that Hodgson and his colleagues (Hodgson, 1990) have provided the understanding necessary to give a simple management prescription for many of the places where perennial ryegrass is grown (discussed later in this chapter). Understanding of how tropical pastures function is much more limited. We shall try to analyse some of the interactions between soil, plants and animals in the well-watered tropical lowlands. There will be less emphasis on stocking rates and grazing systems in different pasture associations and more on attempting to elucidate what are the common and specific factors that control their behaviour.

LIMITATIONS OF FORAGE QUALITY AND UTILIZATION

Definition of the lowland humid tropics

We include the seasonally dry semiarid tropics with the subhumid and humid tropics in the target region. Thornthwaite (1948) defined semiarid climates as those in which a crop could be grown reliably each year, as opposed to arid climates, where it could not. In practice, this sets a lower limit of rainfall of about 1000 mm in South and Central America and probably 750 mm in Australia and Africa. These limits are somewhat arbitrary, and they depend as much on soil and rainfall distribution as on total annual rainfall.

As well as the native tropical savanna and grassland pasture communities, tropical grasslands also include the areas cleared from forest in the humid tropics that are sown to pastures, typified by those of Pará, Acre and Rondônia states in Brazil, Ucayali and Yurimaguas in Peru, Caquetá in Colombia and the lands of Indonesia, Thailand and Malaysia. The highlands of eastern Africa (Kenya, Ethiopia) and of the South American Andean countries are excluded because temperatures are low enough for temperate or subtropical species to be grown. The altitude threshold depends on latitude, ranging from about 1800–2000 m close to the equator to about 1000 m at the tropics. Within the tropical lowlands, plant growth is not restricted by low temperatures during the growing season.

Close to the equator, mean monthly maximum and minimum temperatures vary little while, closer to the tropics, the dry season is cooler than the wet season, although frosts are not common except at higher altitudes.

Much of Central America and tropical South America (the neotropics), has abundant rainfall. For example, Carimagua on the Colombian eastern plains (llanos) has a mean rainfall of 2240 mm with a range as high as 3300 mm and a growing season for pastures from April to December (Table 14.1). Excess water is more frequently a problem, rather than drought (Fisher and Thomas, 1989). In contrast, much of tropical Australia and Africa has lower annual rainfall than the neotropics, and moreover they are both subject to recurrent droughts, which may last for several years.

Brief characteristics by continent

Savanna communities have a dominant grass layer and a discontinuous tree layer, which may make up 10–50% of the total plant cover (Johnson and Tothill, 1985),

Table 14.1. Principal environmental characteristics of sites on the Colombian eastern plains, central Brazil, Peruvian Amazonia, west Africa and northern Australia.

Characteristic	Carimagua, Colombia*	Planaltina, Brazil*	Pucalpa, Peru*	Kano, Nigeria*	Katherine, Australia[†]
Ecosystem	Isohyperthermic savanna	Isothermic savanna	Tropical forest	Semiarid savanna	Semiarid savanna
Latitude, longitude	4° 30′ N 71° 19′ W	15° 44′ N 47° 59′ W	8° 24′ S 74° 36′ W	12° 00′ N 8° 31′ E	14° 3′ S 132° 3′ E
Altitude (m)	167	1050	147	472	110
Temperature (°C)					
Mean	26.5	24.0	26.4	26.2	27.2
Max.	31.0	29.4	27.1	33.3	34.6
Min.	22.1	17.6	25.6	19.2	19.8
Precipitation (mm)	2343	1271	1607	922	925
Number of dry months	5	6	2	7	7
Soil	Oxisol	Oxisol	Ultisol	Red-brown earth	Red-brown earth

*Source: CIAT database
[†]Source: Slatyer (1960).

as in the savannas of Australia and Africa and 52% of the 205 million ha (Mha) of the Cerrados of Brazil (Macedo, 1994). In contrast, the grasslands of the Colombian and Venezuelan llanos and 24% (about 50 Mha) of the Brazilian Cerrados with less fertile soils (*campo limpo* and *campo sujo* (Eiten, 1972)) have no important tree component. In the absence of any intervention, these lands in the neotropics and Australia are usually managed in extensive holdings of low productivity, while nomadism and smallholdings are common in Africa and Asia. In both cases, management inputs are minimal. Where the natural vegetation is replaced, in part or wholly, by introduced species, more intensive management is required. To put this in its context, however, the major pasture resource in the well-watered tropical lowlands is still, and will remain for the foreseeable future, the native savannas and grasslands (Fisher *et al.*, 1992). When cleared, the more fertile soils of the Brazilian Cerrados, which were previously either open savanna woodland or forest, give way to arable cropping, largely because cattle production on pasture cannot pay for the high costs of mechanical clearing. In contrast, in the humid forests, colonists have cleared lands by hand. After usually no more than 3 years of cropping, which takes advantage of the fertility derived from burning the woody vegetation, the farmers normally plant pasture. This is grazed for a number of years before it reverts to either secondary forest, bush fallow or a low-productivity 'tururco' pasture (Toledo, 1987). In other cases, impelled largely by favourable economic incentives, large landholders consolidate smaller cleared areas for cattle ranching, as in Pará state and in the recently colonized areas in Acre and Rondônia in the Brazilian Amazon. In the absence of fertilizer application, a legume component or proper management, these pastures degrade and become weedy and unproductive. This need not occur, as demonstrated in the Peruvian Amazon, where productive pastures have been maintained for 18 years at Yurimaguas with appropriate inputs of fertilizer.

In most descriptions of African tropical savannas, emphasis is placed on their use by nomadic or seminomadic herdsmen and large herds of wild animals (Crowder and Chheda, 1982). Recently there has been some collectivization of nomads into ranches and discussion of the problems of overgrazing, but there is little emphasis on management in the sense of rational utilization of the grazing resource. The problem is complicated by the substantial economic returns from tourists, who come to see the wild animals. But the needs of wild animals in large herds conflict with sedentary pastoralism, particularly if the pastoralists fence their land.

The objectives of production and conservation may be in conflict, particularly in the developed world, where capital and food supplies are secure. However, the majority of the well-watered tropical lowlands lie in developing countries, where the social value of pristine landscapes is less important than food security.

ENVIRONMENTAL EFFECTS ON PLANT GROWTH AND NUTRITIVE VALUE

Grassland productivity

In the extensive Australian systems on native savanna, commercial stocking rates are 50–60 ha head^{-1}, although on well managed research-station pastures stocking rates of 12 ha head^{-1} were achieved without any apparent botanical changes in the pastures (Norman, 1966). In the neotropics, savanna commonly supports stocking rates of 5–10 ha head^{-1}, with maximum live-weight gains of 20 kg ha^{-1} under research station conditions, although the figure in commercial herds is commonly 7–12 kg ha^{-1} (Fisher et al., 1992).

It is common for most pasture scientists to contrast pastures in terms of maximum standing biomass. In areas of limiting rainfall and short growing seasons, losses due to senescence are probably not great, but, in areas with longer growing seasons, losses due to senescence become increasingly important and net primary productivity (including below-ground root turnover) may be underestimated by as much as fivefold (Long et al., 1989, 1992). This point is important in considering carbon and nutrient flows in grazed pastures (Fisher et al., 1994).

Soil limitations

The common feature of the soils of the lowland tropics is that they are low in nutrients, particularly nitrogen (N) and phosphorus (P), and in many areas they are acid to very acid, with high levels of aluminium saturation. In nutrient-poor soils, plants can adapt to use nutrients more efficiently and produce more dry matter per unit of nutrient acquired, or not increase their efficiency and produce less. The ecological advantages for the alternative strategies are not clear but they have important implications.

The African grasses appear to have evolved to be efficient in terms of dry matter produced per unit of N and P absorbed (Rao et al., 1992). In contrast, the native grasses of the neotropical grasslands generally have similar nutrient concentrations in their tissues, but lower growth rates, leading to less efficiency in acquisition and utilization of nutrients (Table 14.2). Their lower growth rates are associated with substantially higher contents of structural material, leading to lower digestibility. Why do the introduced grasses of the neotropics, which almost invariably are native species from the African savannas, perform so much better than the native savanna species of Australia and the neotropics? It could be because the growing seasons in much of the African tropics are shorter and the soils somewhat more fertile, so that higher growth rates could confer an ecological advantage, but there is no clear evidence to support this notion. The African grasses are also more responsive to fertilizer application and their higher shoot and

Table 14.2. Comparison of plant growth and nutrient characteristics of native savanna and the introduced African grass *Brachiaria dictyoneura* on a clay–loam Oxisol at Carimagua, Colombia, during the rainy season.

Plant characteristics	Vegetation	
	Native savanna*	Introduced grass[†]
Shoot growth rate (kg ha^{-1} day^{-1})	4–10	20–120
Above-ground biomass (Mg ha^{-1})	3.26	3.62
Digestibility (%)	23.6–43.0[‡]	54–70[§]
Below-ground biomass (Mg ha^{-1})	2.10	5.10
Nutrient composition (% of dry matter)		
Shoot N	0.67	0.49
P	0.06	0.06
Ca	0.14	0.14
Root N	0.34	0.29
P	0.03	0.02
Ca	0.06	0.04
Litter N	0.28	0.50
P	0.14	0.04
Ca	0.75	0.27

*Except where noted, G. Rippstein, unpublished data.
[†]Except where noted, I.M. Rao, unpublished data.
[‡]Fisher *et al.*, 1992.
[§]Lascano *et al.*, 1991.
N, nitrogen; P, phosphorus; Ca, calcium.

root growth rates may help them to acquire greater amounts of nutrients in less time, compared with neotropical native savanna species.

There are two options in considering the replacement of native savannas with more nutritious species: modify the soil by the application of fertilizer and/or other soil amendments (lime, dolomite or gypsum), so that more demanding and nutritious species can be grown, or seek for higher-quality plants that are adapted to the soil limitations. By 'adapted' in this context we mean plants that are able to produce as well, or almost as well, as those that require large amounts of soil amendment. The economics of cattle production are such that the only real option in many areas is the use of adapted plants. A possible exception is where pastures are integrated with crops in ley systems, where the pasture takes advantage of the higher levels of fertilizer and other soil amendments applied to the crop, especially during the pioneer phase (Thomas *et al.*, 1995). This integration of pastures and crops in ley systems is a clear trend in the more mature areas (as opposed to the frontiers) of the Brazilian Cerrados (Vera *et al.*, 1992).

Climatic limitations

The main limitation to pasture quality in the seasonally dry areas of the tropics is the dry season, which may last for 6 months or more, but which in the neotropics is frequently shorter. Apart from the cleared areas of the humid forests, where there is no rainless month, there are at least one and sometimes two dry seasons, depending on whether rainfall has a unimodal or a bimodal distribution. Even in areas with a unimodal distribution, it is common to have a period of lower rainfall in mid-wet season, the so-called *veranicos* (little summers) of the neotropics. The effect on pasture growth depends very much on the region. For example in the Colombian llanos (eastern plains), where they usually occur in August, *veranicos* can be a welcome break from the heavy rains (up to 800 mm) of the preceding month. They do not appear to affect animal performance.

As soil water is exhausted in the dry season, pastures cease growth, the rate of senescence of existing leaf tissue increases and ultimately much of it dies. As far as plant survival is concerned, the remaining young leaf tissue is usually buried deeply in a protecting mass of senescing and dead leaves and culms. The apices are further protected by successive layers of dead tissue, which insulates them from the lethal effects of high temperatures if and when the pasture is burned. As the plants senesce, the senescing tissue is much reduced in nutrients, especially N and P, by the processes of remobilization. The animals have no choice but to change their diet and consume the senescent material, which also has a higher proportion of structural tissue and is therefore less digestible.

Effects of forage quality on intake

The extreme case of low nutritional quality of native pastures during the dry season allows the definition of some effects of quality on intake. Fisher *et al.* (1992) reported on the contrast between the savannas of Australia and of the neotropics:

> In northern Australia, Norman (1966) showed that savanna grasses were protein deficient, especially during the dry season [Fig. 14.1]. When cattle grazing native pasture were supplied with a protein supplement, their liveweight gain was proportional to the digestible crude protein of the supplement. In contrast, cattle grazing savanna pasture at Carimagua showed no clear relation between protein content [of the forage on offer] and animal liveweight gain [Fig. 14.2a]. More specifically, even when crude protein was in excess of 9%, the animals still lost weight, which is compelling evidence that protein content of the South American savannas does not control liveweight. On the other hand, there was a clear relation between in vitro dry matter digestibility (IVDMD) and liveweight gain [Fig. 14.2b], and the regression also indicates that the critical digestibility to meet the animals' energy needs is about 35%, somewhat less than the commonly accepted 40% (Alvarez and Lascano, 1987).

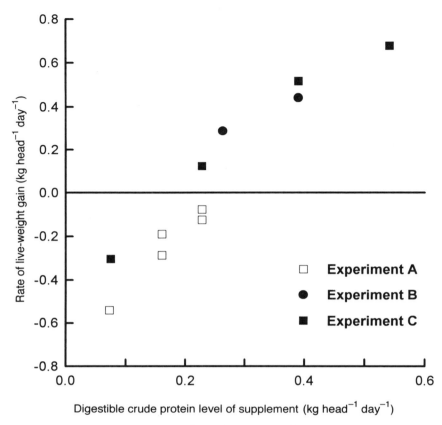

Fig. 14.1. Relation between live-weight changes on native pasture during the dry season and digestible crude protein in the supplement at Katherine, Australia (from Norman, 1966).

Mean digestibility of neo-tropical savanna grasses is rarely more than 40% [Fig. 14.3], so that animals grazing them are unable to satisfy their energy needs. In contrast, mean crude protein content during the rainy season is 8%–9%, and 6% during the dry season [Fig. 14.3] (Alvarez and Lascano, 1987). In northern Australia, savanna grasses have higher digestibility but lower protein contents, so that there the animals' energy needs are met, and they respond to supplementary protein.

It is of interest that in both cases the species have presumably evolved under low levels of available soil N and appear to have developed a tight control over N cycling. In the more seasonal environment of northern Australia, as the grasses mature, much of the N in the aerial parts is remobilized to the roots to support root growth in order to acquire water during the dry season, and the remaining mature

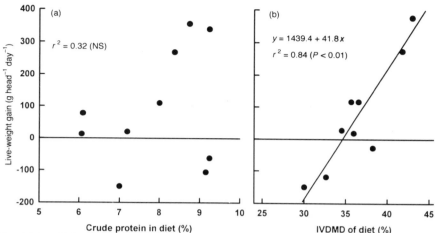

Fig. 14.2. Relationship of live-weight gain with crude protein (a) and digestibility (b) of the diet selected by cattle grazing savanna at Carimagua, Colombia (from Alvarez and Lascano, 1987). NS, not significant; IVDMD, *in vitro* dry-matter digestibility.

forage has levels less than that required by domesticated ruminants (Norman, 1966).

Influence of grass anatomy on forage quality

Most tropical grasses have the C_4 pathway of photosynthesis. A pivotal feature of C_4 plants is their ability to modify their internal carbon dioxide (CO_2) microenvironments such that CO_2 is not rate-limiting for photosynthesis. C_4 grasses have leaves with Kranz anatomy, resulting in closely spaced vascular bundles and a distinct thick-walled parenchyma bundle sheath surrounding each bundle. Leaves of C_3 species have a high proportion of loosely arranged mesophyll cells, which are usually the first to be digested. The bundle-sheath cells are more highly lignified and resist digestion in the rumen, so that C_4 grasses have lower digestibilities than C_3 plants (Wilson and Minson, 1980).

Nevertheless, there are some tropical grasses that are highly digestible by any standards, notably the Mott cultivar of *Pennisetum purpureum*, a dwarf form of common elephant grass. Live-weight gains of cattle grazing this cultivar in Florida have been recorded at 1 kg head^{-1} day^{-1} during the summer growing period (Sollenberger *et al.*, 1987, quoted by Williams and Hanna, 1995). Because of difficulties in obtaining true seed, the species has not been grown widely, but its

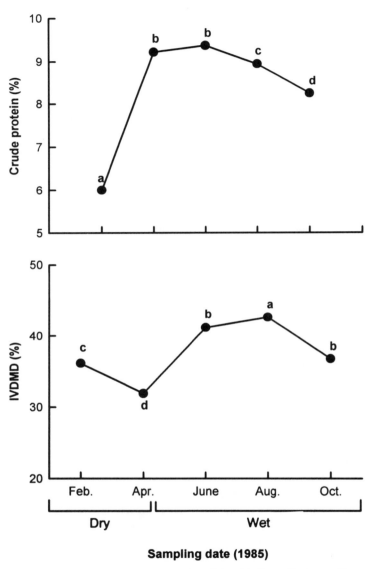

Fig. 14.3. Diet quality of savanna at Carimagua, Colombia (from Alvarez and Lascano, 1987). Points with different letters differ significantly ($P < 0.05$). IVDMD, *in vitro* dry-matter digestibility.

agronomy warrants further study to overcome this limitation. The greater digestibility of the leaf and stem tissue may be due to faster digestion of the more easily digested cells and not to greater digestion of the rigid, thick-walled, lignified structural tissues (Akin *et al.*, 1991).

There have not been widespread concerted efforts to produce transgenic tropical grasses, in contrast with many temperate grasses. It is perhaps too early in the process of introduction and selection of tropical grasses to expect that the limitations of many of them have been adequately evaluated to allow the clear definition of breeding objectives in terms of nutritional characteristics, such as digestibility, and to seek genetic variation for the desirable traits.

Many tropical species in common use, typified by *Andropogon gayanus,* have digestibilities of about 50%, which constrains the performance of cattle grazing them. Efforts to increase digestibility by breeding and selection have been successful in coastal bermudagrass in Georgia, USA (Burton *et al.*, 1967), and the criteria have been incorporated in the breeding programme in the Tropical Forages Program at the Centro Internacional de Agricultura Tropical (CIAT) in Colombia. Although the *Brachiaria* species have moderate to high digestibilities (*c.* 65%), there is a wide spread of digestibilities in the CIAT *Brachiaria* collection of some 200 accessions (Fig. 14.4). While the main criterion of breeding activities with *B. decumbens* is to increase its resistance to damage caused by attacks of the xylem-feeding insect spittlebug (Circopidae), a secondary criterion is to maintain, or enhance, the levels of digestibility (Lapointe and Miles, 1992). However, many tropical grasses are apomicts, so breeding only becomes possible when a source of true sexuality is found (Valle and Miles, 1994).

Grassland degradation and N supply

With the exception of the vertisols, the soils in tropical regions are almost universally low in available N. To discuss this further, the forms in which N exists in the soil organic matter (SOM) pools must be considered. It is commonly accepted that there are three distinct SOM pools (Parton *et al.*, 1988). The first is an active fraction, consisting of live microbes and microbial products (2- to 4-year turnover time). The second (called the passive pool) is a protected fraction, which is more resistant to decomposition (20- to 50-year turnover time) as a result of physical or chemical protection. The third is a fraction that is physically protected or chemically resistant and has a long turnover time of 800 to 1200 years. Current plant growth probably derives most of its N from the active pool and a little from the passive pool.

Cultivation for pasture establishment releases N from the passive to the active pool by disrupting the soil particles that formerly protected it. This N enters into the plant growth–decay cycle, but it is gradually lost in the utilization of herbage

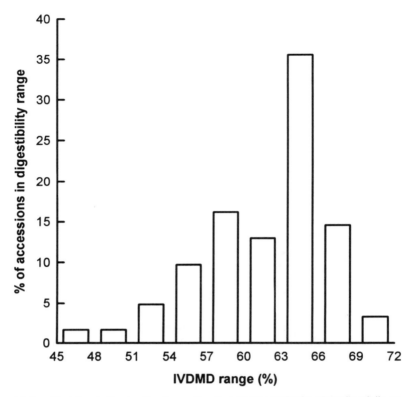

Fig. 14.4. Variation in the *in vitro* dry-matter digestibility (IVDMD) of the first fully expanded leaf of 66 *Brachiaria* accessions 4 weeks after a uniformity cut on an Oxisol during the rainy season at Quilichao, Colombia (M.J. Fisher, unpublished data).

by grazers, principally via excreta (Thomas, 1992). Inevitably the amount of N available in the system declines to its former level unless there are external inputs. If there are none, the production of a grass sward will be limited by its N use efficiency. It is clear that dry-matter production, even of grasses with high use efficiency, must inevitably decline and degradation will result. Myers and Robbins (1991) described this not as degradation but as a return to the base level of production. They described the initial flush of production in contrast as run-up (as opposed to the normal thinking of degradation as run-down). They proposed that stocking rates must be reduced as pasture productivity declines, in order to avoid damage to the pasture composition. Clearly, also, mechanical pasture renovation without an external source of N is simply releasing more of the protected N from the passive pool for a further run-up cycle. It follows that there is no alternative to a return to the base level of production in the absence of an external source of N.

N inputs from external sources are associative N fixation (by free-living organisms other than rhizobia), the background supply by rainfall, fertilizer and symbiotic N fixation. Although associative fixation undoubtedly occurs, the amount, and therefore its role, is unclear in many regions. Quantifying associative N fixation is difficult, although there are data for sugar cane that put the amount as high as 170 kg N ha^{-1} year^{-1} (Uriquiaga et al., 1992). In grass-alone pastures the rates of fixation may be 5–35 kg N ha^{-1} year^{-1}, although the monthly rate may be as high as 10 kg ha^{-1}, depending heavily on environmental conditions (Boddey and Victoria, 1986; Miranda et al., 1990). There are accession differences in the rates of associative fixation in *Panicum maximum* and differences between species of *Brachiaria*. It is not yet clear how this could be manipulated to enhance N inputs in grass-alone pastures. We identify this as a clear priority area for further research, especially its implications for sustainability of grass-alone pastures.

N inputs from rainfall through atmospheric conversion depend on the meteorology of the area in question, but the high level of lightning phenomena in most tropical areas, coupled with actual measurements of N in rainfall, suggest that it could contribute about 8–10 kg ha^{-1} year^{-1} (Szott et al., 1991). Although the amounts coming from this source may be important, it can never support the full productive potential of a vigorous tropical grass.

Many tropical grasses respond well to N fertilizer in terms of dry-matter production as long as water or other essential nutrients are not limiting (Caro-Costas and Vicente-Chandler, 1972). However, the economics of beef production in the tropics are such that widespread use in the absence of subsidies is unlikely. Nevertheless, application of relatively small amounts of N fertilizer could delay or even prevent degradation of pure grass pastures and could be cheaper than periodic renovation (J. Smith, personal communication). On the other hand, the recent reduction in fertilizer subsidies in Venezuela appears to have reduced the amount of N fertilizer applied to pastures. Farmers in Costa Rica frequently apply N fertilizer to forages for milk production, and continue to do so even when they have planted a legume. They seem not to realize that the legume contributes N to the system (P.C. Kerridge, personal communication), or that N fertilizer is detrimental to the legume (Fisher and Cruz, 1994).

We conclude that there are only two alternatives: either to accept that pure grass pastures will inevitably return from their run-up level of productivity to a lower level, or to provide N as fertilizer or symbiotically through legumes (Boddey et al., 1995; Thomas, 1995). Lower production may be acceptable if a pure grass pasture does not degrade, but degradation is likely to result in an unsustainable system.

This analysis suggests that, apart from the unusual circumstance of sustainable pure grass pastures at low stocking rates (or the possibility of manipulating associative fixation), the only route to sustainability is by incorporating a legume into the system (Thomas *et al.*, 1995). This was understood many decades ago for temperate pastures, and has been the goal of tropical pasture agronomists for almost 50 years. Target levels of legume contents in pastures have recently been defined in terms of maintaining the balance of soil N. They range from 20 to 40% of pasture dry matter, depending, among other factors, on rates of pasture utilization (Thomas, 1992, 1995).

SENSITIVITY OF TROPICAL FORAGE SPECIES TO GRAZING

Studies of the management of introduced grasslands in the tropical lowlands have largely been concentrated in Australia and the neotropics as part of the process of the introduction and selection of more productive exotic grasses and legumes. Although we are trying to arrive at principles of management that apply to all pastures in the tropics, each grass–legume association appears to have its own critical set of constraints. Even the behaviour of pure grass pastures appears to be site-specific.

Life-forms and case-studies

In a briefly summarized list of case-studies of pastures in the tropics, there are legumes and grasses with very contrasting life-forms and growth habits. The legumes vary from an annual (*Stylosanthes humilis*) to short-lived perennials (*S. hamata* and *S. capitata*), all three of which are free-seeding herbs. *S. scabra* is a shrubby perennial. *Arachis pintoi* is a strongly stoloniferous prostrate perennial, in many of its characteristics similar to white clover, while *Macroptilium atropurpureum* is a twining, weakly stoloniferous perennial. *Centrosema acutifolium* and *C. macrocarpum* are twining perennials but are not stoloniferous. The most common grasses in the lowland tropics are the *Brachiaria* species, ranging from weakly tussock-forming (*B. brizantha* and *B. ruziziensis*) to sward formers (*B. humidicola*, *B. dictyoneura* and *B. decumbens*). One *B. decumbens* cultivar, Basilisk, probably accounts for more area than all the others together, being the main species sown on the 35 Mha of the Brazilian Cerrados sown to pasture. *A. gayanus* and most forms of *P. maximum* are tall-growing tussock grasses.

S. humilis was relatively successful in northern Australia for many years until the introduction of the fungal pathogen *Coletotrichum* spp., which causes anthracnose. As an annual, it was not suitable for growth with vigorous perennial grasses (Fisher, 1970) and, when sown in pure swards, it was rapidly invaded by annual grasses (Torssell, 1966; Ive and Fisher, 1974). The short-lived perennial *S. hamata* was more compatible with perennial grasses (Torssell *et al.*, 1976), but it too was susceptible to anthracnose. Selections were made within *S. capitata* for resistance to anthracnose, and the resulting selection was released in Colombia as cv. Capica. In mixtures with *B. decumbens* cv. Basilisk on sandy soils it persists for many years, although not in the longer term. It suffers severe competition for potassium (K) from the grass *A. gayanus* (Valencia, 1983), and there is evidence that there is competition for K with *B. dictyoneura* (M.J. Fisher, unpublished data). *S. scabra* cv. Seca has been successful when oversown in the savannas of northern Queensland and is currently being used extensively by commercial producers. Loss of the savanna grass component under the increased stocking rates that the mixture can support appears to be a problem (Miller and Stockwell, 1991).

M. atropurpureum was sown on as much as 100,000 ha in the tropics and subtropics of Queensland in the 1970s (Shaw and Whiteman, 1977), but a combination of drought and low beef prices led to heavier grazing pressures and loss of the legume component during the early 1980s. Twining legumes are in general unable to tolerate heavy grazing, largely, it seems, because of the consumption of growing points by the grazing animals (Clements, 1989). There has been similar experience under grazing with other twining legumes, such as *C. acutifolium* and *C. macrocarpum* (Lascano *et al.*, 1989). In contrast, the proportion of *A. pintoi* in mixtures with both *B. humidicola* and *B. dictyoneura* increases as the stocking rate increases (Fisher and Cruz, 1994; Lascano, 1994).

Different plant species have different vulnerabilities to defoliation at different stages in their growth, although the mechanisms are usually not well understood. For example, the tussock grasses *A. gayanus* and *P. maximum* appear to be more sensitive to overgrazing at the end of the dry season than the sward-forming *Brachiaria* spp.

Regrowth

Humphreys (1991) postulated that 'a certain minimum level of energy residual is necessary for plant persistence, but the influence of reserves on *the rate of regrowth* after severe defoliation appears to be limited to the first three days of regrowth' (our emphasis). The vulnerability of many species of tropical pasture plants to overgrazing at the start of the wet season (Arndt and Norman, 1959) is not inconsistent with this notion. Presumably, after a long dry season the plants, usually tropical tallgrasses, mobilize their resources and, before they are able to replenish them, they fall below the critical minimum level of energy residual. However, the evidence to support or reject this hypothesis is limited.

Nevertheless, some species are especially vulnerable when they have resumed growth at the start of the wet season. It is easy to understand why annual grasses might be so vulnerable, because the newly established plants have scant reserves, and if the growing points are damaged then the plant must die. Perennials have presumably mobilized most of their resources to recommence growth. The relative roles of destruction and survival of growing points is clearly important.

Selection criteria

Many tropical grasses have been selected on simple criteria, namely the ability to grow and survive in the new environment, that is to withstand its biotic and abiotic constraints, and then to produce acceptable live-weight performance in grazing animals. Frequently the main criterion for selection in the early stages of the process is superior dry-matter production. However, as far as the grazing animal is concerned, the product of digestibility and dry-matter production could be a more satisfactory goal and akin to the use of harvest index (the ratio of harvested yield to total yield) in crop plants. A species with a relatively small advantage in digestibility might need to produce substantially less dry matter to produce the same animal performance as one with lower digestibility. Moreover, the lower-yielding but more digestible grass may be less competitive with any associated legume species. It may also partition a greater proportion of nutrients to above-ground biomass and, in doing so, more closely match plant growth with soil nutrient availability.

Although there has been some attempt at CIAT to use this as a selection criterion, there have been no definitive experiments to test the hypothesis. The success of breeding more highly digestible lines of coastal bermudagrass and the striking effects that relatively small differences in digestibility have on the performance of the grazing animal (Burton *et al.*, 1967) suggest that this should be more widely used as a primary selection criterion.

ANIMAL SPECIES, AND CONSEQUENCES TO PRODUCTION SYSTEMS

Species differences

The use of sheep and cattle in mixed grazing systems is not a widespread practice in the American tropical lowlands. However, farmers with an Andean background who have settled in the Amazon basin are keen to raise sheep for home consumption. As population density increases in forest margins, one would expect an increase in sheep numbers, which in turn could result in mixed grazing systems in the future.

There are few cases where improved grass–legume pastures in the tropics have been evaluated with both sheep and cattle. At Quilichao, Colombia, pastures of *B. dictyoneura* with *Desmodium ovalifolium* and *A. gayanus* with *Centrosema* spp. were grazed separately and simultaneously with sheep and cattle (Cárdenas and Lascano, 1988). Results showed that pastures grazed only by sheep had less forage on offer than those grazed by cattle, regardless of level of forage allowance. This was related to greater selection of leaves from the grass by sheep than by cattle. Another striking difference between animal species was related to legume selectivity. In pastures grazed by sheep the legume component was practically eliminated after 1 year of grazing, which was not the case in pastures grazed by cattle. When sheep and cattle together grazed the same grass–legume pastures, selection indices for legume were 0.7 and 1.5 for cattle and sheep respectively – that is, sheep selected for the legumes while cattle selected against them.

It seems, therefore, that these large differences in selectivity between sheep and cattle grazing improved grass–legume pastures could make mixed grazing systems in tropical lowlands difficult to manage. However, there are certain cases where sheep could be used to control unpalatable legumes, such as *D. ovalifolium*, which tend to dominate the grass component in pastures grazed by cattle, regardless of the grazing system used, but which appear to be more acceptable to sheep (Toro, 1990).

Heat limitations

High ambient temperatures constrain the productivity of high producing European beef and dairy breeds in the tropics. Animals are able to control excessive metabolic heat load to a certain extent by reducing movement, seeking for shade or standing in water, but these behavioural changes reduce grazing time and forage intake (Leng *et al.*, 1993). In addition, metabolic heat load is further increased when cattle consume low-digestibility forages common in tropical lowlands. The heat burden is one of the main reasons why milk production is so low in the tropics compared with temperate regions, even with animals of high genetic potential.

As a result of this negative effect of climate on specialized breeds, together with poor-quality forages, farmers in the lowland tropics are forced to work with cross-bred cows of low genetic potential to produce milk. However, pasture quality can be improved with legumes, and can probably stimulate farmers to upgrade the genetic potential of their cows, which in turn can have a large impact on milk production. Experimental results have shown that milk yield response to legume is about 10% with cows of low genetic potential, but increases to 20% with cows of medium genetic potential (Lascano and Avila, 1993). These results have even more significance if one considers that in certain tropical American countries milk production is rapidly changing from specialized intensive systems in the highlands to more extensive dual-purpose systems in the lowlands, mainly due to an increase in land value in the highlands.

Nevertheless, in much of the tropical lowlands, the major form of livestock production is from beef cattle grazing pasture. This is true even of subsistence settlers in the Amazon region, where cattle represent a form of saving of the wealth generated by the labour of clearing the land of trees. The roads are so bad, especially during the rainy season, that conveying other forms of produce to market is very difficult.

UTILIZATION AND MANAGEMENT OF GRASS/LEGUME PASTURES

Temperate pastures

Having developed the rationale for the use of legume-based pastures, we shall discuss aspects of their management. Before doing so, it is instructive to consider the uncertainty surrounding the accurate definition of management of temperate pastures. Despite the very large amount of research carried out in these systems, there are few really good guiding principles, and it is obvious that the systems are still imperfectly understood.

Briske and Silvertown (1993) reviewed 80 years' data of the Park Grass Experiment at Rothamsted in the United Kingdom. Harris (1993), in discussing their paper, concluded that 'our understanding of physiological and ecological processes [of temperate pastures] is limited'. This is for an experiment that started in 1856. In contrast, in the lowland tropics a handful of tropical pasture scientists are working with species that were unknown, except as herbarium specimens, as little as 10 years ago. Moreover, there is a whole portfolio of possibilities, in the neotropics alone for as much as 250 Mha spread over 30 degrees of latitude (Miles and Lapointe, 1992).

Grass vs. legume

Most selection among accessions of newly introduced species is carried out on the basis of high dry-matter yields, which means, other things being equal, that the outcome of the process is grasses with high potential growth rates (as discussed above). The legumes not only have the less efficient C_3 pathway, but have the added burden of providing energy to the symbiont organisms in the nodules. So it is little wonder that it is difficult to find legumes that coexist in association with grasses in tropical pastures.

In addition to the fundamentally differing photosynthetic pathways of tropical grasses and legumes, there are more subtle characteristics of contrasting life-forms, as already discussed. The major unifying concept is the ability of the components of a pasture community to withstand all the components of utilization by grazing animals (consumption, trampling and excretion) and to maintain their presence in a resilient dynamic equilibrium. It is easy to describe animal management that leads to a loss of the equilibrium as mismanagement, but it is much harder to define the converse, good management.

Limits of resiliency

Unless grazing demand matches forage production within quite narrow limits, it is inevitable that one or other component of the pasture, or the pasture as a whole, will be overgrazed. If the ability of the component(s) involved to recover is exceeded, permanent damage will result. In discussing this concept, Williams and Chartres (1991) proposed a conceptual model of the changes in system properties with time in sustainable and unsustainable systems (Fig. 14.5). While this is a useful concept, for any given system the question remains, what are the properties and what are the mechanisms? The property will obviously be the criterion or criteria by which sustainability is judged (pasture composition, amount of bare ground or weed invasion, for example), but the mechanisms will undoubtedly vary with the species and cultivar involved. Some of the mechanisms are discussed below, but in the main they are poorly understood for tropical species.

'Tropical pastures are grown in environments [that] periodically impose considerable stresses on plants and the systems of utilization devised need to exert a singular resilience' (Humphreys, 1991). The main conclusions from Humphreys' (1991) exhaustive review of the physiology, ecology, agronomy and management of tropical pastures in the tropics were that there is frequently a conflict between the long-term goals of sustainability in the sense of the landscape and the short-term needs of pasture managers. Pasture managers will usually place the welfare of their livestock ahead of the condition of their pasture. The conflict between the two is often more acute in subsistence systems than in extensive systems, because subsistence producers have the immediate food needs of their families to consider. The pressures of the need to service the sometimes crushing debt burdens to financial institutions who have provided credit cannot be ignored, however, and may lead even the large producer to making decisions that seem to be short-term.

Obviously, the more resilient a pasture community is, the easier it will be to manage, and the more resistant it will be to catastrophes such as injudicious overgrazing or accidental burning. However, the ideal of an infinitely robust system has not yet been demonstrated in the tropics, and there are few examples in temperate regions. Most tropical pasture workers regard as their goal a tropical equivalent of the white clover–perennial ryegrass community at its best, thinking that it is a universally stable system in temperate regions. However, this is not

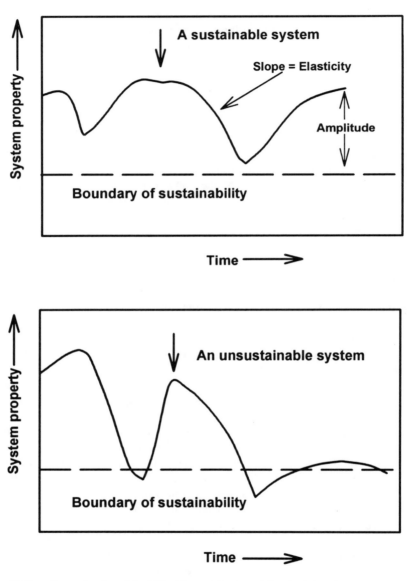

Fig. 14.5. Concepts of sustainability. The amplitude and the elasticity of the response of the system property to applied stress for both a sustainable and an unsustainable system (from Williams and Chartres, 1991).

supported by the evidence: the association is not stable on the hills of Scotland, for example (Frame and Newbould, 1986).

Demographics

Pasture plants establish from seed or vegetatively and after a finite period they die. The concepts of half-life, recruitment rates and death rates, which quantify these processes, describe the dynamics of the population. The influence on them of pasture management in its broadest sense is crucial in understanding the survival or disappearance of the particular component. However, if the processes are not understood on a mechanistic basis, demographics do not lead to a predictive ability on the part of either pasture scientist or manager.

Although demographic studies are usually descriptive, they can be useful to generate hypotheses about the cause of variations in either recruitment rates or death rates. Unfortunately, they are frequently site-specific and related to a particular set of favourable conditions (recruitment rates) or calamities (death rates). Moreover, while these may be easy enough to identify, it is not always so easy to do much about them. Nevertheless, plants that have large reserves of new individuals, as long as they can successfully establish in the face of whatever constraints are imposed by the manager or the environment, are robust from a demographic point of view.

Dynamics of growing points

Clements (1989) compared the dynamics of growing points in pasture associations of four legumes representing prostrate, twining and upright plant types (Table 14.3). He concluded that the exposure of the growing points to consumption and trampling was an important factor in the relation between plant survival and

Table 14.3. Loss of growing points to consumption and trampling in four legumes of contrasting growth habit (Clements, 1989).

	Mean number of nodes eaten per grazed runner or branch	Growing points destroyed by trampling in 3 weeks	
		Mean	Maximum
Macroptilium atropurpureum	6.1	4.5	17
Centrosenna virginianum	6.4	4.8	12
Chamaecrista rotundifolia*	9.9	7.0	13
Trifolium repens	–	2.2	5

*Formerly *Cassia rotundifolia*.

stocking rate. Twining legumes have their growing points vulnerable to consumption, so that, in the conditions of his experiment, at a stocking rate of about four animals per hectare all the growing points of *M. atropurpureum* were eaten in each grazing cycle, with obvious implications for subsequent plant growth and survival (Fig. 14.6). In contrast, with *Trifolium repens* no growing points were eaten, regardless of stocking rate, and only a small number were lost to trampling.

THE INFLUENCE OF PLANT CHARACTERISTICS ON SPECIES BALANCE

This topic has been reviewed several times in the last few years, either in its totality (Humphreys, 1991) or in parts (Coleman, 1992; Matches, 1992; McIvor, 1993). We shall emphasize here those plant characteristics that are important in determining species balance in tropical pastures.

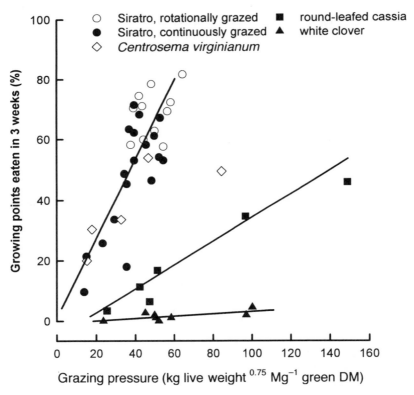

Fig. 14.6. Percentage of growing points grazed during 3 weeks as a function of metabolic grazing pressure (from Clements, 1989). DM, dry matter.

Legume persistence

The advantage of grasses associated with legumes compared with pure grass pastures was demonstrated in a review by Lascano (1994), who summarized the literature on the *Arachis* forage legumes *A. pintoi* and *A. glabrata*. Despite this, there are few commercial pastures sown to grass/legume mixtures in the tropics. In seeking a reason, Lascano concluded that lack of persistence in the legume associated with vigorous and stoloniferous grasses was a dominant factor, and unsuccessful experiences by farmers had given grass/legume pastures a bad name.

Each association has a different set of reasons, but the common factor is lack of robustness in the legumes to withstand competition from the companion grass under grazing. However, *A. pintoi* has been found to persist in mixtures with grasses such as *B. humidicola* and *B. dictyoneura* under different grazing managements, ranging from continuous stocking to rotational grazing, and widely differing stocking rates (two to as many as four animals per hectare year-round) (Lascano, 1994). The reasons why it is able to persist have not been determined, but, as with white clover, it seems likely that the prostrate stolons, which largely protect the growing points from consumption, provide at least part of the answer (Clements, 1989).

Physiological differences

With the C_4 pathway of the grasses, the rate of photosynthesis continues to increase; that is, it does not saturate, up to the intensity of full sunlight. In contrast, the C_3 pathway reaches a plateau, or saturates, at about 1300 μmol m^{-2} s^{-1}, or approximately two-thirds of full sunlight. All other things being equal, the grasses therefore have inherently higher growth rates than the legumes. The outcome of this difference is that, unless the legume has some other advantage, it will inevitably be dominated by the grass component (Fisher and Thornton, 1989). Most of the likely factors that could redress the imbalance are usually in favour of the grass component as well. For example, one could hypothesize that the legume could:

- compete more successfully for nutrients
- compete more effectively for light or water

than the grasses.

Competition for nutrients

Grasses with extensive fibrous roots in the soil surface should have a competitive advantage in acquisition of nutrients in the surface soil compared with legumes, most of which do not have this characteristic. Recently, however, Rao and Kerridge (1994) have shown that the tropical legume *A. pintoi* has greater ability to acquire P from low-P soils than *B. dictyoneura*, a grass with which it is commonly grown. This could be another reason why associations of *A. pintoi* with *Brachiaria* spp. have been remarkably stable in the long term under widely differing grazing managements (Lascano, 1994).

Competition for light or water

The thinking behind choosing twining tropical legumes as companion species for vigorous tropical grasses was that they should be able to overtop the grass and so offer it competition for light. In support of this hypothesis, under lenient stocking rates in some places in the humid tropics, the twining legumes *M. atropurpureum* (Walker, 1977) and *Pueraria phaseoloides* can become dominant. There are no published analyses of why this occurs, but it seems plausible that it is an interaction with the light environment. At the light level at which the photosynthesis of C_3 plants saturates, there is little difference between the rates of photosynthesis of either group. In humid environments, where the light levels are reduced by persistent cloud cover, the C_4 grasses may not be able to express the advantage of non-saturating photosynthesis.

Competition for water may act to confer an advantage on either the grass or the legume, depending on their sensitivity to or tolerance of tissue desiccation and their rooting patterns. For example, *M. atropurpureum* is highly desiccation-sensitive in that its tissues die if their water potential falls below the relatively slight level of water stress of −2.2 MPa. Yet it grows well in association with buffel grass (*Cenchrus ciliaris*), which is highly desiccation-tolerant, in a semiarid subtropical environment in southeastern Queensland, Australia, where the rainfall is unreliable and droughts are common. The reason is that the legume is deeper rooted than the grass, so that during drought it is able to maintain its tissues above lethal levels of desiccation by drawing on water to which the grass does not have access (Sheriff and Ludlow, 1985).

Inherent differences between C_3 and C_4 grasses

C_3 grasses are usually of temperate origin while C_4 grasses are mostly tropical or sub-tropical. For this reason it is difficult to separate environmental adaptation from differences in their photosynthetic pathways. Wilson and Hacker (1987) addressed this problem by comparing sympatric grass species from an arid environment in western Queensland, Australia (Table 14.4). The plants were grown in a greenhouse, and the anatomical characteristics that contribute to digestibility were compared. The C_4 species had less mesophyll and more epidermis, bundle sheath, sclerenchyma and vascular tissue. Water stress increased the proportion of mesophyll, but it was concluded that rate of leaf appearance overrode the difference in leaf anatomy. The rate of leaf appearance, and hence growth rate, has an

Table 14.4. Effect of growth rate and water stress on digestibility of leaf and stem tissue of sympatric C_3 and C_4 species (expressed as digestibility %) (Wilson and Hacker, 1987).

Tissue	Treatment	C_4 species Growth rate		C_3 species Growth rate	
		Slow	Fast	Slow	Fast
Leaf	Not stressed	51.4	60.9	63.5	66.5
	Stressed	65.8	69.8	73.6	80.6
Stem	Not stressed	43.2	43.2	50.2	45.5
	Stressed	64.9	48.9	64.7	57.5

important effect on tissue digestibility. Slower-growing species had lower digestibilities of leaf tissue than faster-growing ones.

Weedy characteristics

Many characteristics required of pasture species are similar to those of weeds, and relate to the ability of the component species to persist despite the management applied and in the face of environmental limitations. Species that will prevent invasions by weeds must themselves have many weedy characteristics, aggressiveness or ability to invade open space being one primary requirement. Yet the community at large is increasingly demanding more benign behaviour, which places severe constraints on the introduction and selection process for species for development of pastures to replace native grasslands.

MANAGEMENT METHODOLOGIES

'Envelope' of good management

Spain *et al.* (1985) proposed the concept of an 'envelope' of good management as a tool in evaluation experiments under grazing. Basically, the envelope is determined on a combination of sward yield and legume–grass balance. If the dry-matter yield is either higher or lower than that dictated by the envelope, grazing pressure is either increased or decreased respectively. If the legume contribution is lower or higher than that dictated by the envelope, the rest period in a rotational grazing system is either shortened or lengthened. Kemp (quoted in Michalk and Kemp, 1994) modified this experimental approach for more general use, but using essentially the same management criteria to achieve satisfactory species balance in grazed temperate pastures.

Sward state

The concept of sward state was posed by Hodgson (1985) for temperate pastures, but in the tropics most workers appear to have continued with the stocking-rate concept, whose limitations he exposed. The defined state may be yield of aboveground biomass, although, with dense ryegrass-based pastures, sward height was found to be a satisfactory descriptor because, in these pastures, sward height is strongly correlated with sward yield. The point is that the sward state of pastures under fixed stocking rates varies because growth is variable according to factors such as seasonal conditions and fertilizer application. Consumption asymptotes in grossly under- or overgrazed swards, but it is reasonable to assume that, over fairly broad limits, consumption will depend directly on sward state.

The question is, what gives an adequate measure of sward state in tropical pastures? In sward-forming species, such as *B. humidicola* with *A. pintoi*, sward height may be satisfactory, but it is clearly inadequate for tussocky species such as *A. gayanus*. If the concept of sward state is to be useful with tussocky pastures, suitable descriptors will be needed.

Leaving aside the long-running argument between the proponents of set stocking and those of rotational grazing, the concept of maintaining a fixed range of herbage on offer in an 'envelope of good management' (Spain *et al.*, 1985) is essentially that of maintaining constant sward state. Although Hodgson (1985) also proposed this as an experimental tool to understand the functioning of grazed pastures, it is now being used more widely in the management of grazed temperate pastures, principally in the United Kingdom and New Zealand.

Data from 15-year-old pastures at Carimagua seem to support the concept. Pastures of *B. decumbens*, either alone or with the twining legume *P. phaseoloides*, have been grazed at two beef animals per hectare during the wet season and one animal per hectare during the dry season (Lascano and Estrada, 1989). The pastures were fertilized with (kg ha^{-1}) 44 P, 40 K, 14 magnesium (Mg) and 22 sulphur (S) at establishment and with 10 P, 9 K, 92.5 calcium (Ca), 8 Mg and 11.5 S each second year for the subsequent 9 years. In contrast with many, even most, farm pastures in the neotropics, there is no evidence of degradation. This poses the obvious question, why not? Is it that the site is atypical (the soil is about 37% clay and the site is not well drained), the fertilizer management was appropriate or the grazing management lenient? There are no answers, but the questions are valid. They emphasize that there is very little known about the processes of either sustainable pastures or their degradation in the tropics.

Fire effects

Although sown pastures are rarely burned, except by accident, regular burning is a feature of management of most native savannas. In the neotropics, this occurs on average each 16 months because of the differing vulnerabilities of the component species to fire at different times of the year (see below). Elsewhere, annual burning during the dry season is more common. Undoubtedly, it is a counter-conservation practice and leads to loss of nutrients from the system, particularly N.

The reason native grasslands are commonly burned is to remove the accumulated dead plant material, and to make the emerging new leaf material more accessible to grazing animals. Although there is some evidence to suggest that the newly emergent material after burning in the dry season is no better nutritionally than that which is produced in an unburned sward (Arndt and Norman, 1959), there may be some small increase in the actual amount of forage produced. This is because burning removes most of the transpiring plant tissue so that the plant–water relations of the remaining tissue improve and growth can occur (Fisher, 1978). The effect is transitory, however.

Pasture species have different vulnerabilities to fire at different times of the year. Norman (1965) showed that, at the start of the wet season in the Australian semiarid tropics, all the perennial species were vulnerable either to fire some weeks after the initiation of growth or to heavy grazing during this time. Indeed, it was possible to destroy these native perennials, and their vulnerability was used to introduce the annual legume *S. humilis* (Miller and Perry, 1968). Species of the native savanna communities at Carimagua have differential vulnerability to fire at different times of the year. Fire in either the early wet season or the early dry season favours *Trachypogon vestitus*. Early wet-season burning eliminates *Axonopus purpusii* but burning in mid-wet season allows it to become dominant (G. Rippstein, unpublished data). Clearly, if the same area is burned in consecutive years at the same time, a substantial shift in species composition could be expected. As a consequence, Paladines and Leal (1979) suggested a burning regime at 16-month intervals so that in any 4-year period a complete cycle would be obtained and no one species would be disadvantaged.

Animal management

Jones and Sandland (1974) examined data from a large number of experiments with constant stocking rates throughout the tropics and temperate regions and concluded that there was a linear relationship between live-weight gain and stocking rate over a large range of the latter. But set stocking experiments of necessity imply a range of pasture states, as the growth rate of the pasture changes in response to seasonal conditions, whether caused by temperature or water supply or both. Timing of the application of fertilizer is a further complication, although the latter is so rare in tropical systems that it can be safely ignored.

There has been a long debate over the use of set stocking and rotational grazing. Most of the evidence suggests that there is little advantage to rotational grazing in terms of live-weight gain. However, the question of stability of a mixed pasture under one or other system has received scant attention.

CONCLUSIONS

Our main conclusions are summarized below:

1. Native grasslands have low productivity because they have low growth rates and poor forage quality, which gives low levels of animal production.
2. A number of tropical grasses and legumes have been selected for adaptation to infertile soils and for biotic and abiotic constraints.
3. The sustainability of grass-alone pastures appears to be site- and management-specific. If associative fixation of N could be increased by selection and/or breeding, both quality and productivity of the pasture could be improved, with potential impact on sustainability.
4. Introduced forage legumes may be a viable option for the sustainability of tropical pastures, but maintaining the proper legume balance is a challenge for tropical pasture agronomists, without considering the problems that it presents to farmers and graziers. However, recently selected tropical legumes like *A. pintoi* may offer a simple solution to the problem of legume persistence.
5. Flexible grazing management may contribute to pasture sustainability and production through permitting the maintenance of suitable pasture states.

These conclusions point to a number of clearly defined research needs:

1. There is a need for long-term experiments to develop principles of pasture management of tropical pastures, and to understand the processes that lead to pasture sustainability or its converse, pasture degradation.
2. Producer participation is necessary in the selection of multiple-purpose forages so that legume technology is more readily adopted by both farmers and ranchers.
3. There is a need to develop models of pasture production and sustainability based on a mechanistic understanding of the interactions between soil, plants, the grazing animals and the atmosphere.

REFERENCES

Akin, D.E., Rigsby, L.L., Hanna, W.W. and Gates, R.N. (1991) Structure and digestibility of tissues and brown midrib pearl millet (*Pennisetum glaucum*). *Journal of the Science of Food and Agriculture* 56, 523–528.

Alvarez, A. and Lascano, C.E. (1987) Valor nutritivo de la sabana bien drenada de los Llanos Orientales de Colombia. *Pasturas Tropicales, Boletín* 9(3), 9–17.

Arndt, W. and Norman, M.J.T. (1959) *Characteristics of Native Pasture on Tippera Clay Loam at Katherine, N.T.* Division of Land Research Technical Paper No. 3, CSIRO, Melbourne, Australia, 20 pp.

Boddey, R.M. and Victoria, R.L. (1986) Estimation of biological nitrogen fixation associated with *Brachiaria* and *Paspalum* grasses using ^{15}N labelled organic matter and fertilizer. *Plant and Soil* 90, 265–292.

Boddey, R.M., Rao, I.M. and Thomas, R.J. (1995) Nutrient cycling and environmental impact of *Brachiaria* pastures. In: Miles, J.W., Maass, B.M. and Valle, C.B. do (eds) *The Biology, Agronomy and Improvement of* Brachiaria. Centro Internacional de Agricultura Tropical, Cali, Colombia (in press).

Briske, D.D. and Silvertown, J.W. (1993) Plant demography and grassland community balance: the contribution of population regulation mechanisms. *Proceedings of the 17th International Grasslands Congress, Palmerston North and Rockhampton.* New Zealand Grassland Association, Palmerston North, New Zealand, pp. 291–297.

Burton, G.W., Hart, R.H. and Lowery, R.S. (1967) Improving forage quality in bermudagrass by breeding. *Crop Science* 7, 329–332.

Cárdenas, E.A. and Lascano, C.E. (1988) Utilización de ovinos y bovinos en la evaluación de pasturas asociadas. *Pasturas Tropicales* 10, 2–10.

Caro-Costas, R. and Vicente-Chandler, J. (1972) Effect of heavy rates of fertilization on beef production and carrying capacity over 5 consecutive years of grazing under humid tropical conditions. *Journal of Agriculture of the University of Puerto Rico* 46, 223–227.

Clements, R.J. (1989) Rates of destruction of growing points of pasture legumes by grazing cattle. *Proceedings of the 16th International Grasslands Congress, Nice.* French Grassland Society, Paris, pp. 1027–1028.

Coleman, S.W. (1992) Plant–animal interface. *Journal of Production Agriculture* 5, 7–13.

Crowder, L.V. and Chheda, H.R. (1982) *Tropical Grassland Husbandry.* Longman, Harlow, UK, 562 pp.

Eiten, G. (1972) Cerrado vegetation of Brazil. *Botanical Review* 38, 201–341.

Fisher, M.J. (1970) *Pasture Species for the Tipperary Area, Northern Territory.* Division of Land Research Technical Paper No. 31, CSIRO, Melbourne, Australia, 48 pp.

Fisher, M.J. (1978) The recovery of leaf water potential following burning of two droughted tropical pasture species. *Australian Journal of Experimental Agriculture and Animal Husbandry* 18, 423–425.

Fisher, M.J. and Cruz, P. (1994) Some ecophysiological aspects of *Arachis pintoi*. In: Kerridge, P.C. and Hardy, B. (eds) *Biology and Agronomy of Forage* Arachis. CIAT, Cali, Colombia, pp. 53–70.

Fisher, M.J. and Thomas, D. (1989) Regrowth of the components of two legume/grass associations following grazing on acid soils of the eastern plains of Colombia. *Proceedings of the 16th Grasslands Congress, Nice.* French Grassland Society, Paris, pp. 1035–1036.

Fisher, M.J. and Thornton, P.K. (1989) Growth and competition as factors in the persistence of legumes in pastures. In: Marten, G.C., Matches, A.G., Barnes, R.F., Brougham, R.W., Clements, R.J. and Sheath, G.W. (eds) *Persistence of Forage Legumes.* American Society of Agronomy, Madison, USA, pp. 293–309.

Fisher, M.J., Lascano, C.E., Vera, R.R. and Rippstein, G. (1992) Integrating the native savanna resource with improved pastures. In: *Pastures for the Tropical Lowlands: CIAT's Contribution.* CIAT, Cali, Colombia, pp. 75–99.

Fisher, M.J., Rao, I.M., Ayarza, M.A., Lascano, C.E., Sanz, J.I., Thomas, R.J. and Vera, R.R. (1994) Carbon storage by introduced deep-rooted grasses in the South American savannas. *Nature* 371, 236–238.

Frame, J. and Newbould, P. (1986) Agronomy of white clover. *Advances in Agronomy* 40, 1–88.
Harris, W. (1993) Chairperson's summary paper. Session 11: Plant communities. *Proceedings of the 17th International Grasslands Congress, Palmerston North and Rockhampton*. New Zealand Grassland Association, Palmerston North, New Zealand, pp. 373–375.
Hodgson, J. (1985) The significance of sward characteristics in the management of temperate sown pastures. *Proceedings of the 15th International Grasslands Congress, Kyoto*. Japanese Society of Grassland Science, Nishi-nasuno, Tochigi-ken, Japan, pp. 63–66.
Hodgson, J. (1990) *Grazing Management: Science into Practice*. Longman Handbooks in Agriculture, Longman, Harlow, UK, 203 pp.
Humphreys, L.R. (1991) *Tropical Pasture Utilization*. Cambridge University Press, Cambridge, UK, 206 pp.
Ive, J.R. and Fisher, M.J. (1974) Performance of Townsville stylo (*Stylosanthes humilis*) lines in pure swards and with the annual grass (*Digitaria ciliaris*) under various defoliation treatments at Katherine, N.T. *Australian Journal of Experimental Agriculture and Animal Husbandry* 14, 495–500.
Johnson, R.W. and Tothill, J.C. (1985) Definition and broad geographical outline of savanna lands. In: Tothill, J.C. and Mott, J.J. (eds) *Ecology and Management of the World's Savannas*. CAB, Farnham Royal, UK, pp. 1–13.
Jones, R.J. and Sandland, R.L. (1974) The relation between animal gain and stocking rate: derivation of the relation from the results of grazing trials. *Journal of Agricultutal Science (Cambridge)* 83, 335–342.
Lapointe, S.L. and Miles, J.W. (1992) Germplasm case study: *Brachiaria* species. In: *Pastures for the Tropical Lowlands: CIAT's Contribution*. CIAT, Cali, Colombia, pp. 43–73.
Lascano, C.E. (1994) Nutritive value and animal production of forage *Arachis*. In: Kerridge, P.C. and Hardy, B. (eds) *Biology and Agronomy of Forage* Arachis. CIAT, Cali, Colombia, pp. 109–121.
Lascano, C.E. and Avila, P. (1993) Milk yield of cows with different genetic potential on grass and grass–legume tropical pastures. *Proceedings of the 17th International Grasslands Congress, Palmerston North and Rockhampton*. New Zealand Grassland Association, Palmerston North, New Zealand, pp. 2006–2007.
Lascano, C.E. and Estrada, J. (1989) Long-term productivity of legume-based and pure grass pastures in the eastern plains of Colombia. *Proceedings of the 16th International Grasslands Congress, Nice*. French Grassland Society, Paris, pp. 1179–1180.
Lascano, C.E., Estrada, J. and Avila, P. (1989) Animal production of pastures based on *Centrosema* spp. in the eastern plains of Colombia. *Proceedings of the 16th International Grasslands Congress, Nice*. French Grassland Society, Paris, pp. 1177–1178.
Lascano, C.E., Avila, P., Quintero, C.I. and Toledo, J.M. (1991) Atributos de una pastura de *Brachiaria dictyoneura–Desmodium ovalifolium* y su relación con la producción animal. *Pasturas Tropicales* 13, 10–20.
Leng, R.A., Jessop, N.J. and Kanjanapruthipong, J. (1993) Control of feed intake and the efficiency of utilization of feed by ruminants. In: Farrell, D.J. (ed.) *Recent Advances in Animal Nutrition in Australia 1993*. Department of Biochemistry, Microbiology and Nutrition, University of New England, Armidale, Australia, pp. 70–88.

Long, S.P., Garcia Moya, E., Imbamba, S.K., Kamnalrut, A., Piedade, M.T.F., Scurlock, J.M.O., Shen, Y.K. and Hall, D.O. (1989) Primary productivity of natural grass ecosystems of the tropics: a reappraisal. *Plant and Soil* 115, 155–166.

Long, S.P., Jones, M.B. and Roberts, M.J. (eds) (1992). *Primary Productivity of Grass Ecosystems of the Tropics and Sub-tropics.* Chapman and Hall, London, 255 pp.

Macedo, J. (1994) *Prospectives for the Rational Use of Brazilian Cervades for Food Production.* EMBRAPA/CPAC, Planaltina, Brasil, 19 pp.

McIvor, J.G. (1993) Distribution and abundance of plant species in pastures and rangelands. *Proceedings of the 17th International Grasslands Congress, Palmerston North and Rockhampton.* New Zealand Grassland Association, Palmerston North, New Zealand, pp. 285–289.

Matches, A.G. (1992) Plant response to grazing: a review. *Journal of Production Agriculture* 5, 1–7.

Michalk, D.L. and Kemp, D.R. (1994) Pasture management, sustainability and ecosystems theory: where to from here? In: Michalk, D.L. and Kemp, D.R. (eds) *Pasture Management Technology for the 21st Century.* CSIRO, East Melbourne, pp. 155–169.

Miles, J.W. and Lapointe, S.L. (1992) Regional germplasm evaluation: a portfolio of germplasm options for the major ecosystems of tropical America. In: *Pastures for the Tropical Lowlands: CIAT's Contribution.* CIAT, Cali, Colombia, pp. 9–28.

Miller, C.P. and Stockwell, T.G.H. (1991) Sustaining productive pastures in the tropics 4. Augmenting native pastures with legumes. *Tropical Grasslands* 25, 98–103.

Miller, H.P. and Perry, R.A. (1968) Preliminary studies on the establishment of Townsville lucerne (*Stylosanthes humilis*) in uncleared native pasture at Katherine, N.T. *Australian Journal of Experimental Agriculture and Animal Husbandry* 8, 26–32.

Miranda, C.H.B., Urquiaga, S. and Boddey, R.M. (1990) Selection of ecotypes of *Panicum maximum* for associated biological nitrogen fixation using the ^{15}N isotope dilution technique. *Soil Biology and Biochemistry* 22, 657–663.

Myers, R.J.K. and Robbins, G.B. (1991) Maintaining productive sown grass pastures. *Tropical Grasslands* 25, 104–110

Norman, M.J.T. (1965) Post establishment grazing management of Townsville lucerne on uncleared land at Katherine, N.T. *Journal of the Australian Institute of Agricultural Science* 31, 311–313.

Norman, M.J.T. (1966) *Katherine Research Station 1956–64: A Review of Published Work.* Division of Land Research Technical Paper No. 28, CSIRO, Melbourne, Australia, 84 pp.

Paladines, O. and Leal, J.A. (1979) Pasture management and productivity in the Llanos Orientales of Colombia. In: Sánchez, P.A. and Tergas, L.E. (eds) *Pasture Production in Acid Soils of the Tropics.* CIAT, Cali, Colombia, pp. 311–325.

Parton, W.J., Stewart, J.W.B. and Cole, C.V. (1988) Dynamics of C, N, P and S in grassland soils: a model. *Biogeochemistry* 5, 109–131.

Rao, I.M. and Kerridge, P.C. (1994) Mineral nutrition of forage *Arachis*. In: Kerridge, P.C. and Hardy, B. (eds) *Biology and Agronomy of Forage* Arachis. CIAT, Cali, Colombia, pp. 71–83.

Rao, I.M., Ayarza, M.A., Thomas, R.J., Fisher, M.J., Sanz, J.I., Spain, J.M. and Lascano, C.E. (1992) Soil–plant factors and processes affecting productivity in ley farming. In: *Pastures for the Tropical Lowlands: CIAT's Contribution.* CIAT, Cali, Colombia, pp. 145–175.

Shaw, N.H. and Whiteman, P.C. (1977) Siratro – a success story in breeding a tropical pasture legume. *Tropical Grasslands* 11, 7–14.
Sheriff, D.W. and Ludlow, M.M. (1985) Physiological reactions to an imposed drought by *Macroptilium atropurpureum* and *Cenchrus ciliaris* in a mixed sward. *Australian Journal of Plant Physiology* 11, 23–34.
Slatyer, R.O. (1960) *Agricultural Meteorology of the Katherine Area*. Division of Land Research and Regional Survey Technical Paper No. 13, CSIRO, Melbourne, Australia.
Spain, J., Periera, J.M. and Gualdrón, R. (1985) A flexible grazing management system proposed for the advanced evaluation of associations of tropical grasses and legumes. *Proceedings of the 15th International Grasslands Congress, Kyoto*. Japanese Society of Grassland Science, Nishi-nasuno, Tochigi-ken, Japan, pp. 1153–1155.
Szott, L.T., Fernández, E.C.M. and Sánchez, P.A. (1991) Soil–plant interactions in agroforestry systems. *Forest Ecology and Management* 45, 127–152.
Thomas, R.J. (1992) The role of the legume in the nitrogen cycle of productive and sustainable pastures. *Grass and Forage Science* 47, 133–142.
Thomas, R.J. (1995) Role of legumes in providing N for sustainable tropical pasture systems. *Plant and Soil* 174, 103–118.
Thomas, R.J., Fisher, M.J., Ayarza, M.A. and Sanz, J.I. (1995) The role of forage grasses and legumes in maintaining the productivity of acid soils in Latin America. In: Lal, R. and Stewart, B.A. (eds) *Soil Management: Environmental Basis for Sustainability and Environmental Quality*. CRC Press, Boca Raton, Florida, USA, pp. 61–83.
Thornthwaite, C.W. (1948) An approach to a rational classification of climate. *Geographical Review* 38, 55–94.
Toledo, J.M. (1987) Pasturas en trópico húmedo: perspectiva global. In: *Anais del 1er Simposio do Tropico Umido*. Vol. 5, EMBRAPA/CPATU, Belém, Brazil, pp. 19–36.
Toro, N. (1990) Productividad animal en pasturas de *Brachiaria humidicola* (CIAT 679) solo y en asociación con *Desmodium ovalifolium* (CIAT 13089) bajo un sistema de manejo flexible de pastoreo. Unpublished MS thesis, CATIE, Turrialba, Costa Rica, 111 pp.
Torssell, B.W.R. (1966) Pattern and process in the townsville stylo-annual grass pasture ecosystem. *Journal of Applied Ecology* 10, 463–478.
Torssell, B.W.R., Ive, J.R. and Cunningham, R.B. (1976) Competition and population dynamics in legume–grass swards with *Stylosanthes hamata* (L.) Taub. (sens. lat.) and *Stylosanthes humilis* H.B.K. *Australian Journal of Agricultural Research* 27, 71–83.
Uriquiaga, S., Cruz, K.H.S. and Boddey, R. M. (1992) Contribution of nitrogen fixation to sugar cane: nitrogen-15 and nitrogen-balance estimates. *Soil Science Society of America Journal* 56, 105–114.
Valencia, I.M. (1983) Root competition between *Andropogon gayanus* and *Stylosanthes capitata* in an oxisol in Colombia. Unpublished PhD thesis, University of Florida, Gainesville, 130 pp.
Valle, C.B. do and Miles, J.W. (1994) Melhoramento de gramíneas do gênero *Brachiaria*. In: Peixoto, A.M., Moura, J.C. de and Faria, V.P. de (eds) *Anais do 11o Simpósio sobre Manejo da Pastagem*. Fundação de Estudios Agrários Luis de Queioz, Brasilia, Brasil, pp. 1–23.
Vera, R.R., Thomas, R., Sanint, L. and Sanz, J.I. (1992) Development of sustainable ley-farming systems for the acid-soil savannas of tropical America. *Anais da Academia Brasileira de Ciencias* 64 (Suppl. 1), 105–125.

Walker, B. (1977) Productivity of *Macroptilium atropurpureum* cv. Siratro pastures. *Tropical Grasslands* 11, 79–86.
Williams, M.J. and Hanna, W.W. (1995) Performance and nutritive quality of dwarf and semi-dwarf elephant grass genotypes in the south-eastern USA. *Tropical Grasslands* 29, 122–127.
Williams, J. and Chartres, D.L. (1991) Sustaining productive pastures in the tropics: 1. Managing the soil resource. *Tropical Grasslands* 25, 73–84
Wilson, J.R. and Hacker, J.B. (1987) Comparative digestibility and anatomy of some sympatric C3 and C4 arid zone grasses. *Australian Journal of Agricultural Research* 38, 287–295.
Wilson, J.R. and Minson, D.J. (1980) Prospects for improving the digestibility and intake of tropical grasses. *Tropical Grasslands* 14, 253–259.

Conclusions

Progress in Understanding the Ecology and Management of Grazing Systems

A.W. Illius[1] and J. Hodgson[2]

[1]*Institute of Cell, Animal and Population Biology, Ashworth Laboratories, University of Edinburgh, West Mains Road, Edinburgh EH9 3JT, UK;* [2]*Department of Plant Science, Massey University, Private Bag 11222, Palmerston North, New Zealand.*

INTRODUCTION

Over the last two decades there have been substantial advances in our understanding of the ecology of grazing systems. A major factor contributing to progress has been the developing coordination between the sciences representing interests in the climate, soil, plant and animal components of such systems, and the corresponding focus of collective attention on the critical processes of plant growth and response to herbivory, and dietary choice and herbage consumption by grazing animals. This synergism has been augmented by developing recognition of the complementary nature of evidence from research programmes based on agricultural interests, often essentially manipulative and exploitative in concept, and those reflecting the interests of natural ecologists, more often conservation-oriented and descriptive in nature. Rapid developments in the conceptual framework of mathematical modelling and in computer software have facilitated effective communication between these various interests, and provided an objective basis for establishing and evaluating research programmes.

In contrast to the rapid advances in our understanding of the function of pastoral systems, developments in their management have been relatively slow and often empirical in nature. Sheath and Clark (Chapter 11) provide examples from intensively managed temperate grasslands, but principles have been slower to develop in other climatic zones and in more extensive systems. This is therefore an opportune time to attempt an overview of the level of understanding of the ecology of grazing systems on a world scale, and the extent to which this understanding is being utilized (or may be utilized) in their planning and management. The fourteen review chapters in this book have been chosen with this in mind.

Ecosystem processes have been popular topics for review in recent years, and indeed a number of the authors represented in this book have published authoritative reviews elsewhere (see individual chapters for details). However, as indicated in the Preface, authors were not asked to be comprehensive in reviewing their subject but, rather, to be stimulating (and, if necessary, idiosyncratic) in their appraisal of progress in their chosen field and, if appropriate, their assessment of targets for future attack. The objective in this final chapter is to attempt an overview of all these ideas, and a synthesis of progress and priorities within three broad areas of ecosystem science: the production and dynamics of plant communities; the physiological and behavioural ecology of animals; and the dynamics and heterogeneity of pastoral ecosystems.

PLANTS

There has been something of a dichotomy of approach to research on the ecology and ecophysiology of plant communities, attention focusing either on the population dynamics and demography of plant units (Briske, Chapter 2; Bullock, Chapter 3) or on mass flows of plant tissue or assimilate within herbage canopies (Lemaire and Chapman, Chapter 1). To some extent this reflects regional differences in emphasis on plant characteristics and life histories, the emphasis being on the former set of variables in more rigorous climatic conditions (Archer, Chapter 4; Briske, Chapter 2; Fisher *et al.*, Chapter 14), whereas the latter have been accorded more importance in the context of intensive temperate pastoral systems (Lemaire and Chapman, Chapter 1). However, recent advances in the understanding of some of the principles of meristem dynamics (Matthew, 1992; Hay, 1994) and in size/density compensation (Sackville Hamilton *et al.*, 1995; Matthew *et al.*, 1995) in both gramineous and leguminous plants in temperate systems provide new and fertile ground for further refinement of the theoretical basis for understanding the stability, perenniality and productivity of relatively simple plant communities. These are major developments, but they relate almost exclusively to two temperate plant species, perennial ryegrass (*Lolium perenne* L.) and white clover (*Trifolium repens* L.) which are amongst the most flexible and adaptable to a range of grazing conditions. Adoption of the same procedures to investigate the limits of adaptability in terms of tiller (or node) size/density compensation (Matthew *et al.*, 1995) may well provide an effective means of characterizing the suitability of alternative plant species for agricultural or conservation purposes. In this context, much could be gained from more direct linkage between research projects in different climatic zones offering extended ranges of variation in plant unit (tiller or node) size and morphology.

Bullock (Chapter 3) outlines the effects of climate and biotic variables on plant population density and, using individual tillers as the basis for estimates of population density in the grasses, illustrates both positive and negative effects of grazing on density in different species. The data presented by Matthew *et al.* (1995), confirming earlier evidence from Davies (1988), demonstrates a strong positive relationship between tiller population density and defoliation pressure in perennial ryegrass and, in keeping with the morphological flexibility of the species, there are few examples of circumstances where defoliation pressure is extreme enough to depress tiller density (Bircham and Hodgson, 1983; Grant *et al.*, 1983). The results reviewed by Davies (1988) for perennial ryegrass and white clover, and more recent evidence from Grant (1996a,b) for the native grasses *Nardus stricta* (L.) and *Molinia caerulea* (L.) demonstrate the value of experimental control of sward state in defining the responses of plant populations to defoliation.

Much of the developing understanding of the ecophysiology of pasture plants relates to above-ground parts, and much less is known about the behaviour of root systems in relation to changes in soil and plant management. This is largely a reflection of the difficulty of monitoring below-ground behaviour both objectively and non-invasively. There is, however, considerable scope for improving our understanding of the factors influencing root dynamics and, in particular, the functional links between root and shoot meristem activity (e.g. Atkinson, 1991).

Clearer definition of responses in plant population dynamics and tissue turnover (Lemaire and Chapman, Chapter 1) should help to resolve the current confusion over the issue of plant compensation for grazing damage (Belsky, 1986; Briske and Richards, 1994). Matthew *et al.* (1995) suggest that transient changes in the balance between tiller size and density may enhance plant production beyond the expected size/density compensation limit in perennial ryegrass pastures. However, evidence from studies of tissue dynamics on plants in both intensive and extensive systems suggest that any enhancement in the rate of tissue accumulation following defoliation is most likely to reflect reduction in senescence losses as a consequence of the removal of mature tissue, rather than any enhancement of the rate of growth of new tissue (Hodgson and Grant, 1987).

Briske (Chapter 2) considers grazing impacts on the structure and function of rangeland vegetation in the context of the plant characteristics conferring resistance to grazing, but argues that much of the evidence is empirical in nature. He defines grazing resistance in terms of the balance between avoidance mechanisms and grazing tolerance (regrowth capability). In temperate conditions these attributes may superficially appear to show negative genetic correlation (Hodgson, 1990), but Briske (Chapter 2) presents a more complex picture for mixed rangeland communities. The impact of these concepts is implicit in all of the chapters dealing with vegetation balance (Chapters 1–4 and 9–14), and is considered in the specific context of grass/legume balance in three of them

(Lemaire and Chapman, Chapter 1; Seligman, Chapter 13; Fisher *et al.*, Chapter 14). This balance has proved to be relatively robust in the contrasting conditions of temperate (Lemaire and Chapman, Chapter 1) and mediterranean (Seligman, Chapter 13) conditions, at least for specific species combinations (Parsons *et al.*, 1991). However, the search for a stable balance in both extensive and intensive pastures in the moist tropics has been much less successful (Fisher *et al.*, Chapter 14). This is a potentially vital area for the future of pastoral systems in the tropics, and in this context the development of an effective functional interpretation of vegetation responses to grazing is of major importance (Briske, Chapter 2).

Issues of temporal and spatial heterogeneity are now assuming major importance in theoretical and practical work on grazing systems, and are relevant at all scales of interest from plant-to-plant interactions to landscape variation (Archer, Chapter 4; Bullock, Chapter 3; Tainton *et al.*, Chapter 10). Developments in this area clearly provide the potential for major new insights into the factors influencing time-related changes in plant communities and in the spatial implications of plant–animal interactions (e.g. Schwinning and Parsons, 1996). Archer (Chapter 4) illustrates on the value of alternative historical records in the definition of vegetation change. In the age of Geographical Information System (GIS) technology, this is a timely reminder of the need for accurate long-term records to quantify vegetation change.

ANIMALS

This section seeks to bring together and review the role of diet selection in animal–plant and inter-animal interactions dealt with in the preceding chapters.

Foraging behaviour and diet selection play a pivotal role in grazing systems, not only by linking primary and secondary production, but also because it is the selectivity of herbivores which mediates and localizes their impact on the population ecology of plant species (Grant *et al.*, 1995a; Brown and Stuth, 1993). Yet, despite its importance, the mechanisms underlying diet selection remain obscure. Classical descriptions (e.g. Hunter, 1962) distinguish broad botanical categories according to their use by grazing animals, and allude to the influence of plant phenology and biomass accumulation. Subsequent studies have supported these conclusions, but are largely descriptive and have not succeeded in providing a mechanistic basis for predicting diet choice (e.g. Arnold, 1987; Grant *et al.*, 1985b; Gordon, 1989; Hodgson *et al.*, 1991). Recent experimental approaches have suggested some of the functional and mechanistic bases for diet selection by showing, for example, that diet choice may be influenced in diverse ways by animal state and recent experiences and constrained by the spatial distribution of resources (Newman *et al.*, 1994; Parsons *et al.*, 1994a; Edwards *et al.*, 1994; see Laca and Demment, Chapter 5). Yet the ability to predict diet composition still depends on rather few principles – plants with high concentrations of secondary metabolites are generally avoided (see Launchbaugh, Chapter 6); more is consumed from high biomass swards than from lower biomass swards of the same species and quality (Arnold 1987; Langvatn and Hanley, 1993); there is some trade-off between biomass and digestibility (Illius *et al.*, 1987; Wilmshurst *et al.*, 1995); and there is tendency towards frequency-dependent selection (Illius *et al.*, 1987; Lundberg *et al.*, 1989; Newman *et al.*, 1994; Wallis de Vries and Daleboudt, 1994).

Further understanding of diet selection and herbivore impact requires an analysis of the mechanisms underlying foraging decisions and the cues to which animals respond. For example, how far can diet selection be explained by a strategy that maximizes intake rate? In foraging theory, maximization of the long term average rate of energy intake is generally taken to be equivalent to fitness maximization, and is a currency used to explain the foraging behaviour of a wide range of non-herbivorous animals (Stephens and Krebs, 1986). Although intake rate maximization is assumed to apply in teleonomic models of mammalian herbivore foraging (Belovsky 1978; Owen-Smith and Novellie, 1982; Ungar and Noy-Meir, 1988; Parsons *et al.*, 1994b) there is little direct evidence that herbivores select diets which do actually maximize intake rate (Crawley 1983), even though harvesting difficulty is certainly the special feature of grazing (Ungar, Chapter 7). Amongst the few reports of animals selecting diets which maximized intake rate are those of Black and Kenney (1984) and Demment *et al.* (1993), and in neither case is the evidence particularly striking. More often, herbivore diets are characterized by the diversity of constituents – that is, they contain a mixture of food items with apparent disregard for the intake rate each offers (Westoby, 1974; Crawley, 1983; Illius and Gordon, 1993; Newman *et al.*, 1994). This has led to doubt over the validity of the rate maximization premise for herbivores, and factors such as obtaining a balance of nutrients, dilution of toxins, state dependence of diet selection, sampling and perceptual constraints on recognition of the profitability of food items have been suggested as explanations (Westoby, 1974, 1978; Pulliam, 1975; Belovsky, 1978; Illius *et al.*, 1992; Illius and Gordon, 1993; Parsons *et al.*, 1994a; Newman *et al.*, 1994; Newman *et al.*, 1995). As Stephens and Krebs (1986) point out, none of these factors defy analysis or suggest that herbivore diets fall outwith the paradigm of optimality. Unfortunately, the study of herbivores in complex natural environments, where the scope for selection is likely to be of greatest importance to animal and vegetation alike, must inevitably be limited to a descriptive approach because controlled experimentation is infeasible. It is virtually impossible to infer the functional or mechanistic basis of foraging behaviour, especially when it is highly variable in space and time, from observations which are unsupported by controlled experimentation and detailed measurements. On the other hand, the sharing of experience between research on natural and agricultural ecosystems has shown substantial benefits (Gordon and Lascano, 1994).

Allelochemicals

The occurrence of secondary plant metabolites in woody browse is clearly related to rejection by herbivores, as Launchbaugh shows (Chapter 6), but the very low concentration of these compounds in short-lived gramineous vegetation suggests that they have little influence on selection between grasses (Iason and Waterman, 1988). Other plant taxa show clear cases of deterrence of herbivory by allelochemicals, which act variously as toxins (to which the animal may not be susceptible, other than by having to bear the metabolic costs of detoxification), by reducing energy, protein or mineral digestion or retention, or simply by diluting the nutrient content of plant tissue. Animals facing depletion of food resources during winter or the dry season have often been observed to discard or avoid the current-season's growth of woody browse in preference for more mature growth (Bryant and Kuropat, 1980; Palo *et al.*, 1983; Provenza and Malachek, 1984; Reichardt *et al.*, 1984). Current-season's growth is less lignified than older twigs, but while having a higher digestibility and N content, it also has higher allelochemical concentration, which acts as a strong deterrent to browsing (Bryant, 1981). Illius and Jessop (1995) modelled the interaction between allelochemical concentration and nutrient content, and suggested that the reduction in energy yield as a result of the dilution effect and the metabolic costs of detoxification is not substantial in comparison with the potential effects of lignification on daily intake. Instead, it seems more likely that allelochemical deterrence acts via amino acid and nitrogen loss, arising as costs of controlling acid–base balance. The interaction between nutrient intake rate and detoxification capacity was hypothesized to result from the limited ability of animals to detoxify allelochemicals when under nutritional stress.

The main problem in achieving a balanced understanding of the importance of secondary plant metabolites for diet selection is in knowing how the animal integrates or balances the various positive and negative attributes of the vegetation. How much of the variation in diet composition, after controlling for variation in allelochemical content, is due to variation in digestibility and nutrient content? How do animals respond to increases in the rate at which they can find and ingest a plant species which imposes a mild toxic cost? What is the response of animals which take mixed grass-browse diets (and this seems to include all classes of livestock at some time or place) to the seasonal changes in the digestibility/allelochemical trade-off of grasses and browse plants? Do animals integrate all positive and negative effects of dietary constituents to maximize long-term rate of energy and nutrient intake, subject to avoiding being poisoned? Clearly, to put the biochemical aspects of diet selection in context requires consideration of both the deterrent (allelochemical-related) and attractive (nutrient-related) plant attributes, and of the possible interactions between the two. Allelochemicals are just one of a number of classes of plant attribute, along with the physical, structural and chemical properties, which influence intake rate (Burlison *et al.*, 1991; Laca *et al.*,

1992; Illius *et al.*, 1996; Ungar, Chapter 7) and digestion and metabolism (Dove, Chapter 8).

Nutrient balance and diet selection

Less has been said about the positive goals which animals seek to achieve when selecting a diet than about the avoidance of negative effects (but see reviews by Provenza, 1995; Illius and Jessop, 1996). Evidence that animals can select diets to meet their potential rates of nutrient utilization comes from experiments on the effect of animal state on diet selection under choice feeding. These show clearly that pigs can avoid nutrient imbalances and can select a diet which matches their changing requirements (Kyriazakis *et al.*, 1990). It might be thought that ruminant animals face a more complex problem in selecting a diet, since digestion and metabolism in the rumen is interposed between the nutrient ratio of foods eaten and the nutrient ratio absorbed – a point made forcibly by Dove (Chapter 8). However, sheep can select a diet whose protein content varies consistently with degree of maturity and with time (Cropper, 1987; Hou, 1991), and are prepared to perform work to maintain the composition of their diet (Hou *et al.*, 1991). Further evidence of the association between nutrient requirements and diet selection in sheep was provided by Kyriazakis and Oldham (1993), who offered lambs one of six isocaloric diets with metabolizable protein:energy ratios ranging from 4 to 14 g MJ^{-1} either alone or as a choice with the highest-protein feed. Comparison of protein deposition in the sheep offered single diets with that in sheep selecting between two feeds showed that, given a choice, sheep selected diets which maximized their growth rates by avoiding diets too high or low in protein relative to energy.

The ability of animals to avoid nutrient imbalances may involve the same process as the avoidance of toxins, namely, via sensations of malaise induced by post-ingestional consequences (Provenza, 1995). Animals appear to use these sensations and other aspects of 'post-ingestional feedback' to avoid nutrient imbalances and toxins and to modify nutrient intake if they are given the opportunity to select from an adequate range of dietary ingredients (Burritt and Provenza, 1992; Provenza *et al.*, 1994; Launchbaugh, Chapter 6). For ruminants, the response to absorbed energy and nutrients is indeed mediated by rumen processes, and the synchrony of supply of metabolites to rumen microbes can exert a considerable effect on nutrient supply (Dove and Milne, 1994). In this sense, as Dove (Chapter 8) makes clear, the distinction between energy and protein consumed is arbitrary, given the intervention of microbial metabolism and N recycling. This can cause an increase in fermentable energy intake to result in increased microbial protein supply to the small intestine.

Dietary divergence between species

Sympatric animal species show differences in diet composition, and hence a degree of niche separation (e.g. Gwynne and Bell, 1968; Jarman and Sinclair, 1979; Murray and Brown, 1993) and there is some evidence that ungulate species diversity is related to habitat heterogeneity (Turpie and Crowe, 1994). Species differences in foraging strategy are presumed to arise from differences in the efficiency or manner in which they obtain their metabolic and habitat requirements from a given vegetation type. For example, goats eat more gorse than sheep do when stocked together (Clark et al., 1982), and this can partly be explained by differences in their digestive physiology. The superiority of goats in consuming and digesting gorse and other low-quality forages is attributed to greater particulate outflow rate from the rumen, and to higher rumen NH_3 from better N recycling supporting higher fibre digestion rates (Howe et al., 1988; Domingue et al., 1991). Murray and Illius (Chapter 9) discuss some of the ways in which animals are adapted to a particular feeding niche where their foraging efficiency is, hypothetically, maximized (see also Gordon and Illius, 1995). Whether or not dietary specialization implies some degree of past or present competition for resources, increasing specialization will reduce niche overlap and thus may reduce competition with other species. It is not clear how far the evolution of foraging efficiency has been affected by interspecific competition, nor is there particularly good evidence, because of the difficulty of achieving effective experimental control, for feeding competition being an important factor determining herbivore community structure. Predation, disease and parasitism may be more important determinants of animal abundance (see Sinclair, 1985).

The existence of intraspecific competition is clearly seen in the response of individual animal performance to increased stocking rate (Fig. 12.2), recalling the adage that a sheep's worst enemy is another sheep. But even here, the degree of competiton, and hence the slope of the response, will be a function of the intraspecific variance in diet composition. Given additional variance in herbivore diets, due to the presence of other animal species, the slope of the response would be expected to be shallower still. The slope can thus be regarded as a coefficient of competition. In one of the few attempts to quantify such effects, Connolly (1987) and Nolan and Conolly (1989) analysed a series of mixed grazing experiments with sheep and cattle, using response equations with coefficients both for the effect of species density on itself and for the effect of companion species. Their results showed that the performance of each species was more affected by the presence of its own than by the other species, when compared at equivalent metabolic mass. Mixed stocking improved output over equivalent stocking with single species, suggesting a degree of niche separation. It appeared that sheep might be more sensitive to the presence of cattle than vice versa. Further evidence is required of the response of one species, in a given environment, to increasing numbers of other species; of the ability of a species to diversify its diet under feeding competition; and of the seasonal incidence of dietary overlap and therefore of potential competition. It may then be possible to move further towards matching animal species mixtures to vegetation structure.

Spatial scale

While it might be expected that the degree of heterogeneity affects the scope for complementary use of vegetation by different animal species, much of the evidence for niche separation comes from apparently quite homogeneous grassland systems. Such a bland assertion ignores the crucial question of scale. The problem of what is the appropriate scale for measuring and studying ecological processes has long been recognized (e.g. Greig-Smith, 1952). The importance of spatial scale for herbivore foraging is now well estabished, and is having an impact on modelling (Senft *et al*., 1987; Coughenour, 1991; Laca and Demment, 1991; Laca and Ortega, 1995). 'Heterogeneity' is itself scale-dependent; that is, what may be perceived at one scale as heterogeneous may be homogeneous at other levels – the difference between the beetle's and the buffalo's view of the prairie. This requires that space be defined from the herbivore's point of view (Morse *et al*., 1985; Pickett *et al*., 1989; Kotliar and Weins, 1990).

Variation exists at each of a range of spatial scales – from the local (variation between successive bites) through the scale of whole plant communities and up to landscape and regional scales (see Tainton *et al*., Chapter 10). This suggests the existence of an optimal spatial and temporal scale for patch assessment and decision-making that is characterized by the animal and vegetation properties which produce variance in intake rate. In other words, it is worth more to the animal to make decisions at some spatial scales than at others, and such decisions require sampling of the vegetation for long enough to provide sufficiently reliable information (Gordon and Illius, 1992). Just as body size limits the animal's ability to gather and process food, it constrains the recognition of heterogeneity in vegetation and this in turn must constrain the animal's ability to exploit such information in foraging (Illius and Gordon, 1993). For example, sheep selected patches of ryegrass and clover that were of intermediate clover content over those with either low or high clover, but did not exploit small-scale heterogeneity in the surface clover content of each patch (Illius *et al*., 1992).

Herbivore impacts are also scale sensitive, being dependent on the scale of the analysis, spatial distribution of plants, and distribution of defoliation (Brown and Allen, 1989; Brown and Stuth, 1993; Laca and Ortega, 1995). Moreover, since much of the difference between animal species in foraging behaviour and mechanisms of diet selection can be attributed to variation in body size, it can be predicted that species will differ in the spatial scale at which heterogeneity in the vegetation can be exploited, according to their body size and incisor dentition. Therefore species differing in size will have different effects on habitat heterogeneity, especially at the boundaries between vegetation types, where the colonizing plant species is present at low biomass, and likely only to be accessible to the smaller animal species.

Two recent examples of the importance of spatial scale are the studies of Scoones (1993, 1995), on the use by cattle of heterogeneous habitats, and the study by Ash and Stafford Smith (1995), both of which suggest that the decline in animal performance during drought or with increasing stocking rate is sensitive to the spatial scale of the area in which animals are permitted to forage. Scoones (1993, 1995) argued that livestock survival was critically dependent on access to localised key resources during the dry season. Ash and Stafford Smith (1995), comparing studies from a number of sites, found that the decline in cattle gains with increasing stocking rate was steeper in 5–15 ha paddocks than in 15–50 ha paddocks, while in 400–1200 ha paddocks there was no significant decline. The likely explanation is that there is greater habitat heterogeneity at larger spatial scales, which serves to buffer variation in herbivore competition for resources and impact. Walker *et al.* (1987), comparing ungulate mortality from a two-year drought in four wildlife conservation areas in southern Africa, concluded that mortality was greater where the establishment of watering points led to reduced spatial heterogeneity of grazing impacts and abolition of reserve stands of lightly grazed grassland. Although this effect was confounded by the higher animal biomass in watered sites, it was argued that spatial patterning of resources acted as a buffer during drought.

Implications for future work

Despite recent progress, there is still a rather poor understanding of the functional and mechanistic bases of foraging behaviour and diet selection in domestic and wild herbivores. This is well illustrated by the fact that species differences in feeding niche can be described, but, at best, can be explained only tentatively. There is a correspondingly poor ability to predict, in any detail, animal responses to vegetation. Investigations of diet choice and foraging behaviour must therefore seek explanations, and not be content with description. The biochemical and physiological background to diet selection has been clarified sufficiently to justify much closer integration of these fields with behavioural studies. The animal's perception of its environment, at many spatial scales, constrains its response to heterogeneity, and remains a considerable challenge for researchers.

SYSTEMS

Pastoral systems show tremendous diversity, from the cool-temperate systems with dependable rainfall and intensive physical and management inputs, to the arid rangelands with capricious rainfall and low productivity. There is commensurate diversity in the management approaches to such systems, as the contrast provided by the treatments of Sheath and Clark (Chapter 11) and Stafford Smith (Chapter 12) illustrates. The former, writing of the pastoral systems of New Zealand, demonstrate what can be achieved by scientific input to systems which are predictable and productive: the development and prescription of sophisticated and reliable techniques for managing systems for desired ends. These systems are controllable. In arid and semiarid rangeland systems, which have quite different dynamics, the inadequacy of the paradigms of equilibrium and succession have led to the development of the state-and-transition approach (Tainton *et al.*, Chapter 10; Stafford Smith, Chapter 12). As a management technique, this addresses as threats and opportunities the series of stochastically driven changes of state which characterize the dynamics of such systems. The ability of managers to exert control is subject to a number of constraints, such as the unpredictability of events, the adverse economics of extensive systems and their enormous physical scale. An important difference between systems at each end of the continuum between intensive and extensive systems lies in the tempo of management and human perception. What is easily handled in terms of spatial scale, temporal frequency of intervention and investment cost per unit area in intensive systems may not be feasible in rangelands. Informed management depends, ultimately, on the considerable body of technical knowledge about the components and process of the system. Ungar (Chapter 7) sounds a note of caution that the application of knowledge about some small-scale processes, such as grazing mechanics, is likely to be restricted to controllable grazing systems. This is perhaps an extreme judgement, but if understanding of the micro-scale mechanisms of grazing is to feed through to agricultural improvement, intensive systems will be the first and rangelands the last to benefit.

One of the obvious contrasts between intensive and extensive grazing systems is in their dynamical response to, and likelihood of experiencing, perturbations. Much of this is accounted for by differences in rainfall, the primary determinant of plant growth, and by the generally negative relationship between the mean and variance of rainfall. With increasing aridity comes increasing stochastic domination of the conditions for plant growth, and hence of ecosystem dynamics. High variability implies the occurrence of rare, extreme events, such as a drought year followed by one of unusually high rainfall. These can drive episodic changes in plant species composition, and are therefore thought to underlie the discontinuous shifts in state documented for semi-arid rangelands (Griffin and Friedel, 1985; Westoby *et al.*, 1989; Walker, 1993; O'Connor and Roux, 1995; Weigand *et al.*, 1995; Tainton *et al.*, Chapter 10; Stafford Smith, Chapter 12). For example, fire or exceptional rainfall may be prerequisites of germination. In eastern Australian shrub grassland, unusually (i.e. 2–5 times per century) good rainfall in two or more years allows the establishment of many shrub seedlings and a substantial fuel of ephemerals and perennial grasses. Removal of the fuel by heavy grazing prevents fire, and this allows shrubs to grow, establish a seed bank, suppress grass growth and become dominant within 10–20 years (this and many other examples in Westoby *et al.*, 1989).

Events and behaviour such as this are described in terms of grazing ecosystems experiencing transitions between alternative stable states. In contrast, stable climatic conditions, with dependable rainfall, are often thought of as conferring greater stability on ecosystems, in the sense of their experiencing minimal perturbation, and their responses to disturbance being only gradual. That is not to say that the responses cannot eventually be large: upland Britain, for example, has seen considerable change in vegetation and soils in the last few hundred years, resulting largely from increased sheep and deer numbers and from atmospheric pollution (Birks, 1988). Such systems, when maintained under constant herbivore pressure, are unlikely to show changes induced primarily by climatic extremes, but do show successional changes between dominance of *Calluna* heath, *Agrostis–Festuca* grassland or *Betula* or *Pinus* woodland over periods of 2–15 years in response to altered grazing pressure (Miles, 1988). Similar dynamics would be expected of most lowland temperate grazing systems (Bullock, Chapter 3; Sheath and Clark, Chapter 11) and, perhaps, mesic systems on volcanic soils in the humid tropics and sub-tropics. Note that the timescale over which the agent of change (or event) is effective may be only slightly longer than in more obvious candidates for a state and transition model, and here the contrast with the gradualist successional model becomes blurred. Thus, it could be said that cool, slow-moving systems with low environmental variability exhibit some continuity in the changes in vegetation state with changing intensity of herbivory, while hot, rapidly-moving systems in capricious environments may flip between a range of possible states in an event-driven manner. Ironically, one of the properties of the hot end of the spectrum is said to be 'resilience'. This is usually meant in the sense that such systems are more resistant to degradation than was first thought, especially by the colonial administration (see Abel and Blaikie, 1989; Behnke and Scoones, 1993; Tapson, 1993), despite exhibiting some of the worst examples of apparent degradation. This rather calls into question what purpose is being served in the current debate amongst ecologists by the classical concepts and terminology from theoretical studies of ecosystem stability. The following is an attempt to review the origin and meaning of the concepts of stability, non-equilibrium and resilience, and to ask what part they play in our current view of grazing ecosystem function.

Stability

There seem to be two ways of establishing an ecosystem's stability and resilience. The most widely practised is to look out of the window periodically and see if it is still there after all these years, and, if so, conclude that it must be persistent, is probably stable and therefore has resilience. Quantitative techniques applied to field data have advanced considerably in recent years, but have still some way to go (see Gaston and McArdle, 1994), and Archer (Chapter 4) provides a stimulating assessment of alternative procedures for investigating historical changes in vegetation characteristics. The second way is to start from a theoretical viewpoint, derive the conditions for stability and related concepts, and then apply these in the field. An uneasy tension exists between the two, and it is far from certain that the concepts survive the transition from theory to application.

Theoretical investigations of the dynamics of Lotka–Voltera models of predator–prey systems form the basis of our understanding of the conditions for the stability of ecosystems (Rosenzweig and McArthur, 1963). For a given model of the predator–prey (or, more generally, consumer–resource) interaction, there generally exists a parameter space which yields some type of stable solution, in the form of a dynamic equilibrium. Crawley (1983) gives a helpful account of how, in a simple model linking plant growth and consumption to plant biomass, and animal population growth to intake, the stability of the system is increased by increasing plant intrinsic growth rate or by reducing asymptotic plant biomass (thereby increasing density-dependent regulation of growth) or by decreasing herbivore growth efficiency. There exist various combinations of parameters which have stable equilibria, stable limit cycles, or violent oscillations which lead to herbivore or plant extinction. Stability, as a theoretical concept, is thus intimately tied to the deterministic dynamic equilibrium, where the mutually dependent consumer and resource are in balance. Pimm (1984) provides a clear set of definitions of ecosystem variables. 'A system is deemed stable if and only if the variables all return to the initial equilibrium following their being perturbed from it. A system is locally stable if this return is known to apply only certainly for small perturbations and globally stable if the system returns from all possible perturbations.'

The dynamic equilibrium of a model system is obviously dependent on the structure of the model and is altered by, for instance, choosing a different type of herbivore functional response or introducing plant reserves which allow plant recovery from total depletion. Take, for example, Noy-Meir's (1975) treatment of plant–herbivore dynamics, which provides a celebrated example of the way modelled interactions can predict the existence of more than one stable state, and which is cited as explanation of why rangeland systems may occupy more than one state (Westoby et al., 1989). In using this type of model to ask how many steady states there are, we must be careful to make the distinction, often overlooked, between alternative models with either constant or variable herbivore numbers. For fixed stocking rates, as in Noy-Meir's (1975) original model, or where herbivore population dynamics are not linked to vegetation state, the quantity of vegetation at which the herbivore population is static (the herbivore isocline, $dH/dt = 0$) is perpendicular to the herbivore axis in H-V space, and can intersect the vegetation isocline ($dV/dt = 0$) either once or three times (Fig. 15.1), producing alternative stable states. The existence of two alternative stable states also depends on the vegetation isocline being convoluted (as in Fig. 15.1), as can result from various assumptions about density-dependent plant growth, plant reserves and the exact shape of the herbivore functional response (Noy-Meir, 1975). In an interesting recent development, Reitkerk and van de Koppel (1996)

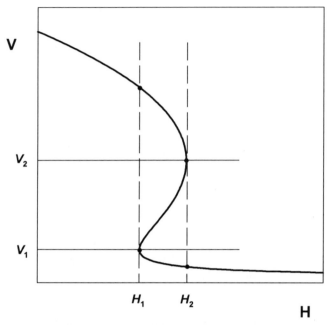

Fig. 15.1. Vegetation isocline (curve) intersected by herbivore isoclines (straight lines) when vegetation (V) is plotted against herbivore numbers (H). In the first case (dashed lines), where herbivore numbers are fixed, or at least not linked to V, isoclines at less than H_1 or greater than H_2 intersect the vegetation isocline only once, either at high or low vegetation. Herbivore isoclines at numbers intermediate between H_1 and H_2 have two alternative stable states and one unstable state. In the second case (solid lines), where herbivore numbers vary with V, then the herbivore isocline depends on the particular quantity of vegetation needed to maintain numbers; quantities less than V_1 or greater than V_2 produce stable dynamic equilibria (see Rosenzweig and McArthur, 1963), while intermediate quantities produce stable limit cycles.

argue that a humped resource isocline may underlie vegetation growth, as a result of feedbacks between vegetation and rainfall infiltration rate, and between vegetation and nutrient loss via erosion. Now consider the alternative model of the herbivore population that allows their numbers to vary, by linking fecundity to vegetation biomass. In this case, the herbivore isocline is perpendicular to the vegetation axis (and equal to the quantity of vegetation that allows the herbivores to maintain their numbers; see Noy-Meir, 1982). Therefore, the herbivore and vegetation isoclines can only intersect once, at what must be a stable equilibrium or stable limit cycle, according to the conditions established by Rosenzweig and McArthur (1963).

The unsurprising conclusion of this comparison is that the choice of model determines the stability conditions revealed. Wang and Gutierez (1980) argue forcefully that stability analyses reflect model assumptions rather than testable field biology, having shown the arbitrariness of stability characteristics of models with varying degrees of realism – age structured populations and seasonality, for instance. It would not seem unfair to conclude that the correspondence between the stability conditions of simple models and the set of meta-stable 'states' observed in real systems is largely metaphorical. This is not to say that simple models are wrong, for they are successful in their own terms at describing the behaviour of coupled plant–animal interactions in an imaginary homogeneous environment. However, one must be cautious about extrapolating beyond these conditions. Indeed, the problem lies with the very simplicity of such models, which do not incorporate quite obvious influences on real ecosystems, such as spatial and temporal variation, which affect their stability (see DeAngelis and Waterhouse, 1987; Crawley, 1983, p. 213). The question here concerns what modifications are required to model the essential properties of real systems, both natural and managed.

Transitions between states in rangeland systems do not involve merely quantitative changes in vegetation biomass, but involve qualitative changes in plant community structure and in the potential for future changes of the same order. For modelling, the crucial difference between classical multiple equilibria and the catalogue of alternative states in rangeland systems is that the former can be expressed in terms of a continuous vegetation axis (e.g. plant biomass) whereas the latter need to be represented either on a catalogue of new axes, corresponding to each new state, or some fairly creative thinking has to be done to represent them on a single new axis that can then be related to herbivore response. The work of Hardy and Mentis (1986; see Tainton *et al.*, Chapter 10), relating plant community composition (the vegetation axis) to its effects on animal production, is a guide to how this might be attempted.

Whatever links there might have been between theoretical conceptions of stability and that observed in nature were severely questioned by Connell and Sousa (1983). In a review of 49 sets of long-term census data, they sought to assess whether, in populations followed for at least one complete turnover of all individuals, there was a subset that exists in an equilibrium state or states. The results showed a continuum of temporal variability in dynamics, only a few cases of stable limit cycles, and no evidence of multiple stable states in unexploited natural populations or communities. They found 'no evidence to show that following a disturbance any community has adjusted back to an original species configuration which then resisted change beyond one complete turnover' (Connell and Sousa, 1983, p. 806). The study could be criticised for ignoring, or not having access to, suitable data. For example, Schoener (1985) responded by showing that some lizard populations were unusually constant over time, by comparison with other taxa considered by Connell and Sousa. Silvertown (1987) pointed out that data from the Park Grass Experiment at Rothamsted, UK, had been overlooked.

These provide good *prima facie* evidence of stability in plants grouped as grasses, legumes and miscellaneous, although there is considerable variability at finer scales of taxonomic resolution. Stability, like so much else, is a scale-sensitive phenomenon. By excluding cases where populations were kept at different, possibly stable states by human intervention (e.g. Noy-Meir, 1975), and demarcating 'appropriate scales of space and time', Connell and Sousa's study could also be criticized for being too stringent in its criteria. These exclude the timescales relevant to rangeland management, which recognizes changes of state (Laycock, 1991) whether or not these correspond to stable equilibria.

Non-equilibrium

The foregoing considered some of the deterministic and biotic elements in system stability. A second class of disruptive influence on ecosystems, and one which is thought to raise questions about the usefulness of equilibrium concepts themselves, is that due to stochastic or abiotic effects on systems which would be stable under constant conditions. In seasonal systems which are continuously grazed, herbivore numbers are usually limited by their ability to survive periods of weight loss during the season of plant dormancy, both by using plant residues from the growing season and body reserves established then (e.g. Milner and Gwynne, 1974; Sinclair, 1975; Ellis and Swift, 1988). Density-independent population processes are relatively important (e.g. Langvatn *et al.*, 1996; Albon *et al.*, 1996), often associated with climatic extremes such as poor winter weather or lateness of rains in arid systems. Animal populations are therefore small, relative to that supportable during the growing season, with the result that there is weak or no density-dependent regulation of either animal or vegetation population processes occurring then. Vegetation biomass shows wide positive and negative excursions from the annual mean, and only a small proportion of it is consumed by mammalian herbivores, the remainder being lost through invertebrate herbivory and decomposition. In such systems, the coupling of animal and plant processes appears to be weak, compared with climatic and density-independent determinants of animal and plant abundance. The weakness of linkages between systems processes, due both to extreme seasonality and to stochastic environmental factors, has given rise to the view that such systems do not obey classical equilibrium dynamics, as defined by coupled consumer–resource dynamics, but exhibit 'non-equilibrial' dynamics (Chesson and Case, 1986; DeAngelis and Waterhouse, 1987; Ellis and Swift, 1988; Behnke *et al.*, 1993). Weins (1984) suggested that all ecological systems fall somewhere on a continuum from equilibrial to non-equilibrial, and characterized the latter as showing weak biotic coupling, independence of species, abiotic limitation rather than resource limitation, density independence and large stochastic effects.

Non-equilibrial systems could be defined as those lacking or having only weak inter-dependence of consumer and resource dynamics, although the term is generally used to describe systems which are merely not at equilibrium. However, there is a sense in which all natural systems are non-equilibrial, insofar as environmental fluctuations and stochastic variation in parameter values, such as birth or death rates, are constantly redefining the dynamic equilibrium – sometimes at a rate faster than the system can respond (according to its resilience – see below). Then, the system will be in permanent orbit around a moving attractor, never coming to rest at an equilibrium which, nevertheless, does exist (see Fig. 15.2). This is the soft case of non-equilibrium, where density-dependent effects on herbivore fecundity and mortality are present, but their severity is defined by current seasonal conditions. They will be more severe than average about half the time, in years when plant growth is poor and the animal and plant processes become tightly coupled, and relaxed or lacking in the remaining better-than-average years, when coupling of the processes becomes looser. For example, Caughley and Gunn (1993) showed that short-term unpredictable fluctuations in rainfall are the cause of long-term aperiodic fluctuations in kangaroo numbers. Note that their model, in which vegetation and animal dynamics are linked, would, under constant rainfall, have a stable point. In this case, as in that illustrated in Fig. 15.2, equilibrial systems can appear to be 'non-equilbrial' if sufficiently perturbed. The Turkana rangeland system in East Africa is considered to be an example of a non-equilibrium system, due to its extreme seasonality, with animals having minimal impacts on plants, and with animal mortality being determined largely by the length of the dry season (Ellis and Swift, 1988). Coppock (1993), comparing the Borana and Turkana systems, suggests that drought frequency and degree of aridity mark the border between quasi-equilibrial and non-equilibrial systems. Likewise, Ellis (1994) argues that environments experiencing rainfall with greater than 30% interannual coefficient of variation are so dominated by variability that they can be distinguished from equilibrial systems. This defines non-equilibrium in rather arbitrary, quantitative terms, according to whether stochastic influences are considered to be dominant.

The hard case of non-equilibrium is where, under conditions of extreme seasonality, predation or disease, there is really no relationship between the vegetation and the rate of change of herbivores, nor vice versa. These are, simply, non-interactive systems (Caughley and Lawton, 1981) for which no equilibrium state exists strictly as a result of plant–animal interactions. What actually does govern the dynamics must lie in relations with other agents. In terms of animal numbers, humans supported and economic output (if not land occupied), most rangeland systems must, by definition, fall well short of this extreme and are quasi-equilibrial, or, to put it more simply, seldom at equilibrium. The management implications of variability are examined in Scoones (1994).

A qualitative definition of non-equilibrium addresses the disparity between population-regulating processes at small and large spatial scales. At small spatial scales, even small perturbations can cause local extinctions. The population could then be regarded as comprising a collection of transient patches, whose dynamics can best be described by gap models in which colonization and extinction are

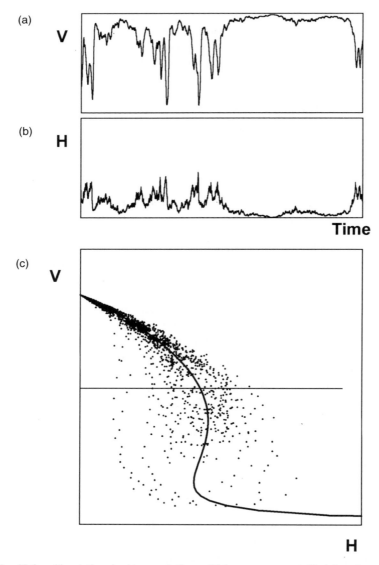

Fig. 15.2. Vegetation–herbivore relations which appear non-equilibrial, under conditions where herbivore mortality is stochastic. The model, relating herbivore intake and population growth to vegetation biomass, and with density-independent mortality (Crawley, 1983, eqn 4.7), was modified by setting the rate of per capita mortality to a positive random variable with cv = 50%. (a) Changes in vegetation biomass and (b) herbivore numbers over 2000 timesteps. (c) Departures of the state variables (V and H; •) from the stable equilibrium which, under a constant mortality rate, would lie at the intersection of the vegetation isocline (curve) and the herbivore isocline (horizontal line). Note that the herbivore numbers tend to be lower than the equilibrium, and vegetation higher.

probabilistic. It can be shown that the stable equilibrium state is a property that can emerge asymptotically from extrapolation to sufficiently large spatial scales (DeAngelis and Waterhouse, 1987). A non-equilibrial community is one where non-deterministic population fluctuations on a small spatial scale are an essential part of community phenomena at a larger scale (Chesson and Case, 1986). Alternatively, population fluctuations at the larger scale may be the result of processes at the same scale. Grazing systems may possess some of the features of both, for while plants and the impact of herbivory may be highly localized, the mobility of herbivores allows them to avoid the consequences of purely local plant extinctions and be more affected by community-scale plant population dynamics.

Resilience

Much is made of the 'resilience' of rangeland systems to disturbance, but here again, the analogy with the original concept may be misleading. Resilience was first defined by Holling (1973) as 'a measure of the ability of these systems to absorb changes of state variables, driving variables, and parameters, and still persist'. Without a means of quantifying resilience, the concept remains merely a useful metaphor which 'characterizes the magnitude of the population perturbations the system will tolerate before collapsing into some qualitatively different dynamical regime' (May, 1981). Beddington *et al.* (1976) used the rates of return to equilibrium after a perturbation as a measure of resilience, thus defining it in terms of the steepness, rather than the size, of the basin of attraction. The concept of return time is, of course, meaningless for unstable equilibria. A useful corollary of this definition is that highly resilient systems will show less 'non-equilibrial' properties, because they track more rapidly the environmentally induced vagueness in the equilibrium point. Pimm (1984) defines resilience as 'the characteristic return time, being the time taken for a perturbation to return to $1/e$ ($\approx 37\%$) of initial value'. As such, it is amenable to field determination. DeAngelis (1992) applies this definition to a wide range of consumer–resource models. Most recently, Ives (1995) has developed a method of estimating resilience in stochastic systems as the variability in population densities relative to environmentally driven variability in population growth rates.

The quantitative definition of resilience is somewhat at odds with continued use in its original, qualitative sense (e.g. Walker and Noy-Meir, 1982; Holling, 1986). This definition incorporates, and thus confuses, notions of stability, the size of the domain of attraction, the resistance of a system to perturbation, and its variability. For example, Abel and Blaikie (1989, p. 114) interpreted resilience as the ability to recover from disturbance, and to be a property of unstable systems. Of semiarid rangeland, they concluded (p.120): 'because it is unstable, rangeland is also intrinsically "resilient" compared to more "stable" ecosystems. Degradation occurs when rangeland is perturbed beyond its ability to recover. Resilience varies with land type.' Faced with this confusion over definitions, one is inclined to redirect Caughley's (1979) famous question about carrying capacity and ask 'what is this thing called resilience?'. In its loose definition, expressing the ecologist's intuitions about semiarid rangeland, resilience seems to come down to the greater susceptibility of herbivores to starvation than of plants to herbivory or soil to erosion. This differential mortality of the components of the system in response to a stress, together with the relatively slow rate of recovery of the herbivore population, has the effect of allowing some sort of plant community to regenerate or establish from seed banks at low herbivore pressure. Systems with plants less tolerant of severe herbivory and so having a narrower differential between animal and plant mortality would be expected to show more comprehensive collapses during drought.

Implications

The implications of all this are as follows. Theoretical approaches reveal the range of possible ecosystem behaviour, but field ecologists perhaps take too lightly the difficulties of translating these insights to the real world. Theoretically-derived concepts of stability, equilibrium and resilience may not be useful and may indeed be quite misleading in the search for understanding of real systems, especially those under management. The conditions for stability of simple models are highly sensitive to model assumptions, and it is arguable that the multiple stable states predictable by them have little bearing on the significant qualitative changes that characterize shifts between vegetation communities in many rangeland ecosystems. Instead of using such concepts as metaphors for system behaviour, and then seeking to match these to the real world, perhaps endeavour should focus on a mechanistic understanding of the underlying nature of grazing systems, in all their specificity. For example, O'Connor (1991) adopted Connell and Sousa's (1983) proposal 'to study the broader class of mechanisms which ensure population persistence regardless of whether equilibria can be identified' to elucidate the conditions for local extinction in perennial grassland. This book provides much evidence of similar endeavours. Stafford Smith's (Chapter 12) call for a functional classification of rangeland types is an example of how progress is likely to be made, by seeking the particular from which to generalize. Likewise, the point of elucidating the concept of resilience is to enlighten the investigation of the biological basis or nature of resilience in grazing systems. In other words, what properties of the system determine how fast it recovers from perturbation? As our mechanistic understanding of the systems components and processes improves, there is obviously increased scope for learning from mechanistic modelling of systems responses to environmental and management inputs. It is this approach, allied to field experimentation, which is likely to produce the fastest progress in the future. Lastly, spatial scale makes all the difference to the perception of ecological concepts, and its importance is often overlooked. There is no spatial uniformity in stability, persistence or resistance. This is important because degradation is a process which starts locally and expands. Stafford Smith (Chapter 12) points out that state-and-transition models are deficient in assuming a landscape with a uniform management impact, and blind to the greater susceptibility of spatially diverse landscapes to localized perturbations, which may become foci of degradation. The challenge here is to relate the spatial scale of diversity to the appropriate scale and tempo of management.

CONCLUSIONS

Detached evaluation of all the factual information presented in this book is difficult. However, despite the risk of over-simplification, we draw the following general conclusions:

- Grazing systems ecology has advanced through improved mechanistic understanding of the key processes at plant, animal and systems levels. It is assuming its rightful place as an objective, quantitative science, and is proving to be a particularly fruitful area for the development of conceptual understanding through mathematical modelling.
- There is much to be gained from linking the research approaches developed in natural and managed systems. Of the main constituent processes in grazing systems, the most rapid progress over recent years has probably been made in understanding the behaviour and nutrition of grazing animals. This is a direct reflection of the close linkage of research results from natural and agricultural ecosystems and the development of an effective series of workshops dealing with common interests. There is scope for similar initiatives in other areas of ecosystems work.
- The investigation of systems heterogeneity – the scales, processes and impacts of spatial and temporal variability on system characteristics, stability and productivity – is clearly a major growth area. It offers new insights into ecosystem behaviour, and makes new demands on data collection and analysis, and on modelling.
- A better understanding is needed of the factors offering scope for the control of state in grazing systems particularly those in more extreme climates.
- Whatever the progress made in understanding ecosystem behaviour, there is limited time for the development of objective, simple and effective management strategies to protect the world's depleting pastoral resources in the face of increasing population pressure and resource exploitation, and the potential impacts of climate change. We reiterate the serious concerns expressed by plenary speakers at the 17th International Grassland Congress (Hadley, 1993; Mooney, 1993) and elsewhere, for the urgency of the need cannot be denied.

ACKNOWLEDGEMENTS

Helpful discussions with Andrew Ash, Roy Behnke, Julian Derry, John Fryxell, Iain Gordon, Emilio Laca, Martyn Murray, Beate Nurnberger, Max Rietkerk and Ian Scoones.

REFERENCES

Abel, N.O.J. and Blaikie, P.M. (1989) Land degradation, stocking rates and conservation policies in the communal rangelands of Botswana and Zimbabwe. *Land Degrdation and Rehabilitation.* 1, 102–123.

Albon, S.D., Coulson, T. and Clutton-Brock, T.H. (1996) Demographic constraints in red deer: can the past predict the future? *Journal of Animal Ecology* (in press).

Arnold, G.E. (1987) Influence of the biomass, botanical composition and sward height of annual pastures on foraging behaviour by sheep. *Journal of Applied Ecology* 24, 759–772.

Ash, A. and Stafford Smith, M. (1995) How recommendations derived from point-based forage production and stocking rate models can lead to resource degradation in Australia's rangelands. *Proceedings of the World Conference on Natural Resource Modeling*, Pietermaritzburg, South Africa.

Atkinson, D. (1991) *Plant Root Growth, An Ecological Perspective.* Special Publication No. 10, British Ecological Society. Blackwell Scientific Publications, Oxford.

Beddington, J.R., Free, C.A. and Lawton, J.H. (1976) Concepts of stability and resilience in predator-prey models. *Journal of Animal Ecology* 45, 791–816.

Behnke, R.H. and Scoones, I. (1993) Rethinking range ecology: implications for rangeland management in Africa. In: Behnke, R.H., Scoones, I. and Kerven, C. (eds) *Range Ecology at Disequilibrium.* Overseas Development Institute, London, pp. 1–30.

Behnke, R.H., Scoones, I. and Kerven, C. (1993) *Range Ecology at Disequilibrium.* Overseas Development Institute, London.

Belovsky, G.E.(1978) Diet optimization in a generalist herbivore: the moose. *Theoretical Population Biology* 14, 105–134.

Belsky, A.J. (1986) Does herbivory benefit plants? A review of the evidence. *The American Naturalist* 127, 870–892.

Bircham, J.S and Hodgson, J. (1983) The influence of sward condition on rates of herbage growth and senescence in mixed swards under continuous stocking management. *Grass and Forage Science* 38, 323–331.

Birks, H.J.B. (1988) Long-term ecological change in the British uplands. In: Thompson, D.B.A. and Usher, M.B. (eds) *Ecological Change in the Uplands.* Blackwell Scientific Publications, Oxford. pp. 37–56.

Black, J.L. and Kenny, P.A. (1984) Factors affecting diet selection by sheep. II Height and density of pasture. *Australian Journal of Agricultural Research* 35, 565–578.

Briske, D.D. and Richards, J.H. (1994) Physiological responses of individual plants to grazing: current status and ecological significance. In: Vavra, M., Laycock, W. and Pieper, R. (eds) *Ecological Implications of Livestock Herbivory in the West.* Society for Range Management, Denver, Colorado, pp. 147–176.

Brown, B.J. and Allen, T.F.H. (1989) The importance of scale in evaluating herbivory impacts. *Oikos* 54, 189–194.
Brown, J.R. and Stuth, J.W. (1993) How herbivory affects grazing tolerant and sensitive grasses in a central Texas grassland: integrating plant response across hierarchical levels. *Oikos* 67, 291–298.
Bryant, J.P. (1981) Phytochemical deterrence of snowshoe hare browsing by adventitious shoots of four Alaskan trees. *Science* 213, 889–890.
Bryant, J.P. and Kuropat, P.J. (1980) Selection of winter forage by subarctic browsing vertebrates: The role of plant chemistry. *Annual Review of Systems Ecology* 11, 261–285.
Burlison, A.J. Hodgson, J. and Illius, A.W. (1991) Sward canopy structure and the bite dimensions and bite weight of grazing sheep. *Grass and Forage Science* 46, 29–38.
Burritt, E.A. and Provenza, F.D. (1992) Lambs form preferences for non-nutritive flavors paired with glucose. *Journal of Animal Science* 70, 1133–1136.
Caughley, G. (1979) What is this thing called carrying capacity? In: Boyce, M.S. and Haydn-Wing, L.D. (eds) *North American Elk: Ecology, Behaviour and Management.* University of Wyoming, Laramie, pp. 2–8.
Caughley, G. and Lawton, J.H. (1981) Plant-herbivore systems. In: May, R.M. (ed.) *Theoretical Ecology.* Blackwell Scientific Publications, Oxford, pp. 132–166.
Caughley, G. and Gunn, A. (1993) Dynamics of large herbivores in deserts: kangaroos and caribou. *Oikos* 67, 47–55.
Chesson, P.L. and Case, T.J. (1986) Nonequilibrium community theories: variability, history, and coexistence. In: Diamond, J. and Case, T.J. (eds) *Community Ecology.* Harper & Row, New York, pp. 229–239.
Clark, D.A., Lambert, M.G., Rolston, M.P. and Dymock, N. (1982) Diet selection by goats and sheep on hill country. *Proceedings of the New Zealand Society of Animal Production* 42, 155–157
Connell, J.H. and Sousa, W.P. (1983) On the evidence needed to judge ecological stability or persistence. *American Naturalist* 121, 729–824.
Connolly, J. (1987) On the use of response models in mixture experiments. *Oecologia* 72, 95–103.
Coppock, D.L. (1993) Vegetation and pastoral dynamics in the southern Ethiopian rangelands: implications for theory and management. In: Behnke, R.H., Scoones, I. and Kerven, C. (eds) *Range Ecology at Disequilibrium.* Overseas Development Institute, London, pp. 42–61.
Coughenour, M.B. (1991) Spatial components of plant–herbivore interactions in pastoral, ranching and native ungulate ecosystems. *Journal of Range Management* 44, 530–542.
Crawley, M.J.(1983) *Herbivory.* Blackwell, Oxford.
Cropper, M.R. (1987) Growth and development of sheep in relation to feeding strategy. PhD thesis, University of Edinburgh.
Davies, A. (1988) The regrowth of grass swards. In: Jones, M.B. and Lazenby, A. (eds) *The Grass Crop.* Chapman and Hall, London, pp. 85–127.
DeAngelis, D.L. (1992) *Dynamics of Nutrient Cycling and Food Webs.* Chapman and Hall, London.
DeAngelis, D.L. and Waterhouse, J.C. (1987) Equilibrium and non-equilibrium concepts in ecological models. *Ecological Monographs* 57, 1–21.

Demment, M.W., Distel, R.A., Griggs, T.C., Laca, E.A. and Deo, G.P. (1993) Selective behaviour of cattle grazing ryegrass swards with horizontal heterogeneity in patch height and bulk density. *Proceedings of the XVII International Grassland Congress* 712–714.

Domingue, M.F., Dellow, D.W. and Barry, T.N. (1991) Voluntary intake and rumen digestion of low quality roughage by goats and sheep. *Journal of Agricultural Science, Cambridge* 117, 111–120.

Dove, H. and Milne, J.A. (1994) Digesta flow and microbial protein production in ewes grazing perennial ryegrass. *Australian Journal of Agriculture Research* 45, 1229–1245.

Edwards, G.R., Newman, J.A., Parsons, J.A. and Krebs, J.R. (1994) Effects of the scale and spatial distribution of the food resource and animal state on diet selection: an example with sheep. *Journal of Animal Ecology* 63, 816–826.

Ellis, J.E. (1994) Climate variability and complex ecosystem dynamics: implications for pastoral development. In: Scoones, I. (ed.) *Living with Uncertainty* Intermediate Technology Publications, London, pp. 37–46.

Ellis, J.E. and Swift D.M. (1988) Stability of African pastoral ecosystems: Alternate paradigms and implications for development. *Journal of Range Management* 41, 450–459.

Ellis, J.E.,Coughenour, M.B. and Swift, D.M. (1993) Climate variability, ecosystem stability and the implications for range and livestock development. In: Behnke, R.H., Scoones, I. and Kerven, C. (eds) *Range Ecology at Disequilibrium*. Overseas Development Institute, London, pp. 1–30.

Gaston K.J. and McArdle, B.H. (1994) The temporal variability of animal abundances: measures, methods and patterns. *Philosophical Transactions of the Royal Society of London* B 345, 335–358.

Gordon, I.J. (1989) Vegetation community selection by ungulates on the Isle of Rhum. II. Vegetation community selection. *Journal of Applied Ecology* 26, 53–64.

Gordon, I.J. and Illius, A.W. (1992) Foraging strategies of sheep and goats: from monoculture to mozaic. In: Speedy, A.W. (ed.) *Progress in Sheep and Goat Research*. CAB International, Wallingford, pp.153–178.

Gordon, I.J. and Illius, A.W. (1995) The nutritional ecology of ruminants: a reinterpretation. *Journal of Animal Ecology* 64, 18–28.

Gordon, I.J. and Lascano, C. (1994) Foraging strategies of ruminant livestock on intensively managed grasslands: potential and constraints. *Proceedings of the XVII International Grassland Congress*, pp. 681–689.

Grant, S.A., Barthram, G.T., Torvell, L., King, J. and Smith, H.K. (1983) Sward management, lamina turnover and tiller population density in continuously stocked *Lolium perenne*-dominated swards. *Grass and Forage Science* 38, 333–344.

Grant, S.A., Bolton, G.R. and Torvell, L.J. (1985a) The response of blanket bog vegetation to controlled grazing by hill sheep. *Journal of Applied Ecology* 22, 739–751.

Grant, S.A., Torvell, L., Sim, E.M., Small, J.L. and Armstrong, R.H. (1996a) Controlled grazing studies on *Nardus* grassland. 1. Effects of between-tussock sward height and species of grazer on *Nardus* utilisation and floristic composition. *Journal of Applied Ecology* (in press).

Grant, S.A., Torvell, L., Common, T.G., Sim, G.L. and Small, J.L. (1996b) Controlled grazing studies on *Molinia* grassland. I. Effects of different seasonal patterns and levels of defoliation on *Molinia* growth and responses of swards to controlled grazing by cattle. *Journal of Applied Ecology* (in press).

Grant, S.A., Suckling, D.E., Smith, H.K., Torvel, L.J., Forbes, T.D.A. and Hodgson, J. (1985b) Comparative studies of diet selection by sheep and cattle: the hill grasslands. *Journal of Ecology,* 73, 987–1004.

Greig-Smith, P. (1952) The use of random and contiguous quadrats in the study of the structure of plant communities. *Annals of Botany* 16, 293–316.

Griffin, G.F. and Friedel, M.H. (1985) Discontinuous change in central Australia: some implications of major ecological events for land management. *Journal of Arid Environments* 9, 63–80.

Gwynne M.D. and Bell R.H.V. (1968) Selection of vegetation components by grazing ungulates in the Serengeti National Park. *Nature* 220, 390–393.

Hadley, M. (1993) Grasslands for sustainable ecosystems. *Proceedings of the XVII International Grassland Congress, New Zealand and Queensland,* SIR Publishing, Wellington, New Zealand, pp. 21–27.

Hardy, M.B. and Mentis, M.T. (1986) Grazing dynamics in sour grassveld. *South African Journal of Science* 82, 566–572.

Hay, M.J.M. (1994) Autecology of white clover (*Trifolium repens* L.) with special reference to the effect of stolon burial on branch formation. PhD thesis, Massey University, Palmerston North, New Zealand.

Hodgson, J. (1990) Plants for grazing systems. *Proceedings of the New Zealand Society of Animal Production* 50, 24–29.

Hodgson, J. and Grant, S.A. (1987) Plant responses to defoliation. In: Rose, M. (ed.) Herbivore Nutrition Research. Research papers presented at the Second International Symposium on the Nutrition of Herbivores, Brisbane, July 1987, pp. 1–2.

Hodgson, J., Forbes, T.D.A., Armstrong, R.H., Beattie, M.M. and Hunter, E.A. (1991) Comparative studies of the ingestive behaviour and herbage intake of sheep and cattle grazing indigenous hill plant communities. *Journal of Applied Ecology,* 28, 205–227.

Holling, C.S. (1973) Resilience and stability of ecological systems. *Annual Review of Ecology and Systematics* 4, 1–23.

Holling, C.S. (1986) The resilience of terrestrial ecosystems: local surprise and global change. In: Clark, W.C. and Munn, R.E. (eds) *Sustainable Development of the Biosphere.* Cambridge University Press, Cambridge.

Hou, X.Z. (1991) Diet selection in sheep. PhD thesis, University of Edinburgh.

Hou, X.Z., Lawrence, A.B., Illius, A.W., Anderson, D. and Oldham, J.D. (1991) Operant studies on feed selection in sheep. *Proceedings of the Nutrition Society* 50, 95A.

Howe, J.C., Barry, T.N. and Popay, A.I. (1988) Voluntary intake and digestion of gorse (*Ulex europaeus*) by goats and sheep. *Journal of Agricultural Science, Cambridge* 111, 107–114.

Hunter, R.F. (1962) Hill sheep and their pasture. A study of sheep-grazing in south east Scotland. *Journal of Ecology* 50, 651–680.

Iason, G.R. and Waterman, P.G. (1988) Avoidance of plant phenolics by juvenile and reproducing female mountain hares in summer. *Functional Ecology* 2, 433–440.

Illius, A.W. and Gordon, I.J. (1993) Diet selection in mammalian herbivores – constraints and tactics. In: Hughes, R.N. (ed.) *Diet Selection* Blackwell Scientific Publications, Oxford.

Illius, A.W. and Jessop, N. S. (1995) Modelling metabolic costs of allelochemical ingestion by foraging herbivores. *Journal of Chemical Ecology* 21, 693–719.

Illius, A.W. and Jessop, N.S. (1996) Metabolic constraints on voluntary intake in ruminants. *Journal of Animal Science* (in press).

Illius, A.W., Wood-Gush, D.G.M. and Eddison, J.C. (1987) A study of the foraging behaviour of cattle grazing patchy swards. *Biology of Behaviour* 12, 33–44.

Illius, A.W., Clark, D.A. and Hodgson, J. (1992) Discrimination and patch choice by sheep grazing grass-clover swards. *Journal of Animal Ecology* 61, 183–194.

Illius, A.W., Gordon, I.J., Milne, J.A. and Wright, W. (1996) Costs and benefits of foraging on grasses varying in canopy structure and resistance to defoliation. *Functional Ecology* 9, 894–903.

Ives, A.R. (1995) Measuring resilience in stochastic systems. *Ecological Monographs* 65, 217–233.

Jarman, P.J. and Sinclair, A.R.E. (1979) Feeding strategies and the pattern of resource partitioning in ungulates. In: Sinclair, A.R.E. and Norton-Griffiths, M. (eds) *Serengeti: Dynamics of an Ecosystem.* University of Chicago Press, Chicago, pp. 130–163.

Kotliar, N.B. and Weins, J.A. (1990) Multiple spatial scales of patchiness and patch structure: a heirarchical framework for the study of heterogeneity. *Oikos* 59, 253–260.

Kyriazakis, I. and Oldham, J.D. (1993) Diet selection in sheep: the ability of growing lambs to select a diet that meets the crude protein nitrogen $\times$ 6.25 requirements. *British Journal of Nutrition* 69, 617–629.

Kyriazakis, I., Emmans, G.C. and Whittemore, C.T. (1990) Diet selection in pigs: choices made by growing pigs given foods of different protein concentrations. *Animal Production* 51, 189–199.

Laca, E.A. and Demment, M.W. (1991) Herbivory: the dilemma of foraging in a spatially heterogeneous food environment. In: Palo, R.T. and Robbins, C.T. (eds) *Plant Defenses Against Mammalian Herbivory*. CRC Press, Boca Raton, Florida, pp. 29–44.

Laca, E.A., Ungar, E.D., Seligman, N. and Demment, M.W. (1992) Effect of sward height and bulk density on bite dimensions of cattle grazing homogeneous swards. *Grass and Forage Science* 47, 91–102.

Laca, E.A. and Ortega, I.M. (1995) Integrating foraging mechanisms across spatial and temporal scales. *Proceedings of the Fifth International Rangeland Congress*, Salt Lake City, Utah, 23–28 July, 1995.

Langvatn, R. and Hanley, T.A. (1993) Feeding-patch choice by red deer in relation to foraging efficiency. An experiment. *Oecologia* 95, 164–170.

Langvatn, R., Albon, S.D., Clutton-Brock, T.H. and Burkey, T. (1996) Climate, plant phenology and variation in age at first reproduction in a temperate ungulate. *Journal of Animal Ecology* (in press).

Laycock, W.A. (1991) Stable states and thresholds of range condition on North American rangelands: a viewpoint. *Journal of Range Management* 44, 427–433.

Lundberg, P., Åström, M. and Danell, K. (1989) An experimental test of frequency-dependent food selection: winter browsing in moose. *Holarctic Ecology* 13, 177–182.

Matthew, C. (1992) A study of seasonal root and tiller dynamics in swards of perennial ryegrass (*Lolium perenne* L.). PhD thesis, Massey University, Palmerston North, New Zealand

Matthew, C., Lemaire, G., Sackville Hamilton, N.R. and Hernandez-Garay, A. (1995) A modified self-thinning equation to describe size/density relationships for defoliated swards. *Annals of Botany* 76, 579–588.

May, R.M. (1981) Models for two interacting populations. In: May, R.M. (ed.) *Theoretical Ecology*. Blackwell Scientific Publications, Oxford, pp. 78–104.

Miles, J. (1988) Vegetation and soil change in the upland. In: Thompson, D.B.A. and Usher, M.B. (eds) *Ecological Change in the Uplands*. Blackwell Scientific Publications, Oxford, pp. 57–70.

Milner, C. and Gwynne, D. (1974) In: *Island Survivors* Jewell, P.A., Milner, C. and Morton Boyd, J. (ed.) The Soay sheep and their food supply. Athlone Press, London, pp. 160–194.

Mooney, H.A. (1993) Human impact on terrestrial ecosystems – what we know and what we are doing about it. *Proceedings of the XVII International Grassland Congress, New Zealand and Queensland*, SIR Publishing, Wellington, New Zealand, pp. 11–14.

Morse, D.R., Lawton, J.H., Dodson, M.M. and Williamson, M.H. (1985) Fractal dimension of vegetation and the distribution of arthropod body lengths. *Nature* 314, 731–733.

Murray, M.G. and Brown, D. (1993) Niche separation of grazing ungulates in the Serengeti: an experimental test. *Journal of Animal Ecology* 62, 380–389.

Newman, J.A., Penning, P.D., Parsons, A.J. Harvey, A. and Orr, R.J. (1994) Fasting affects intake behaviour and diet preference of grazing sheep. *Animal Behaviour* 47, 185–193.

Newman, J.A., Parsons, A.J., Thornley, J.H.M., Penning, P.D. and Krebs, J.R. (1995) Optimal diet selection by a generalist grazing herbivore. *Functional Ecology* 9, 255–268.

Nolan, T. and Connolly, J. (1989) Mixed versus mono grazing of steers and sheep. *Animal Production* 48, 519–533

Noy-Meir, I. (1975) Stability of grazing systems: an application of predator–prey graphs. *Journal of Ecology* 63, 459–481.

Noy-Meir, I. (1982) Stability of plant–herbivore models and possible applications to savanna. In: Huntly, B.J. and Walker, B.H. (eds) *Ecology of Tropical Savannas* pp. 591–609.

O'Connor, T.G. (1991) Local extinction in perennial grasslands: a life-history approach. *American Naturalist* 137, 753–773.

O'Connor, T.G. and Roux, P.W. (1995) Vegetation changes (1947–71) in a semi-arid, grassy dwarf shrubland in the Karoo, South Africa: influence of rainfall variability and grazing by sheep. *Journal of Applied Ecology* 32, 612–626.

Owen-Smith, N. and Novellie, P. (1982) What should a clever ungulate eat? *American Naturalist* 119, 151–178.

Palo, R.T., Pehrson, A. and Knutsson, P.G. (1983) Can birch phenolics be of importance in the defence against browsing vertebrates? *Finnish Game Research* 41, 75–80.

Parsons, A.J., Harvey, A. and Johnson, I.R. (1991) Plant–animal interactions in a continuously grazed mixture. 2. The role of differences in the physiology of plant growth and of selective grazing on the performance and stability of species in a mixture. *Journal of Applied Ecology* 28, 635–658.

Parsons, A.J., Newman, J.A., Penning, P.D., Harvey, A. and Orr, R.J. (1994a) Diet preference of sheep: effects of recent diet, physiological state and species abundance. *Journal of Animal Ecology* 63, 465–478.

Parsons, A.J., Thornley, J.H.M., Newman, J.E. and Penning, P.D. (1994b) A mechanistic model of some physical determinants of intake rate and diet selection in a two-species temperate grassland sward. *Functional Ecology* 8, 187–204.

Pickett, S.T., Kolasa, J., Armesto, J.J. and Collins, S.L. (1989) The ecological concept of disturbance and its expression at various heirarchical levels. *Oikos* 54, 189–194.

Pimm, S.L. (1984) The complexity and stability of ecosystems. *Nature* 307, 321–326.

Provenza, F.D. (1995) Postingestional feedback as an elementary determinant of food preference and intake in ruminants. *Journal of Range Management* 48, 2–17.

Provenza, F.D. and Malachek, J.C. (1984) Diet selection by domestic goats in relation to blackbrush twig chemistry. *Journal of Applied Ecology* 21, 831–841.

Provenza, F.D., Scott Ortega-Reyes, C.B., Lynch, J.J. and Burritt, E.A. (1994) Antiemetic drugs attenuate food aversions in sheep. *Journal of Animal Science* 72, 1989–1994.

Pulliam, H.R. (1975) Diet optimization with nutrient constraints. *American Naturalist* 109, 765–768.

Reichardt, P.B, Bryant, J.P., Clausen, T.P. and Weiland, G.D. (1984) Defense of winter-dormant Alaska paper birch against snowshoe hares. *Oecologia* 65, 58–69.

Rietkerk, M. and van de Koppel, J. (1996). Alternative stable states and threshold effects in semi-arid grazing systems. *Journal of Ecology* (in press).

Rosenzweig, M.L. and McArthur, R.H. (1963) Graphical representation and stability condition of predator–prey interactions. *American Naturalist* 97, 209–223.

Sackville Hamilton, N.R., Matthew, C. and Lemaire, G. (1995) In defence of the −3/2 boundary rule: a re-evaluation of self-thinning concepts and status. *Annals of Botany* 76, 569–578.

Schoener, T.W. (1985) Are lizard population sizes unusually constant through time? *American Naturalist* 126, 633–641.

Schwinning, S. and Parsons, A.J. (1996) Analysis of the coexistence mechanisms for grasses and legumes in grazing systems. *Journal of Ecology* (in press).

Scoones, I. (1993) Coping with drought: responses of herders and livestock in contrasting savanna environments in southern Zimbabwe. *Human Ecology* 20, 31–52

Scoones, I. (1994) *Living with Uncertainty*. Intermediate Technology Publications, London.

Scoones, I. (1995) Exploiting heterogeneity: habitat use by cattle in dryland Zimbabwe. *Journal of Arid Environments* 29, 221–237.

Senft, R.L., Coughenour, M.B., Bailey, D.W., Sala, O.E. and Swift, D.M. (1987) Large herbivore foraging and ecological heirarchies. *BioScience* 37, 789–799.

Silvertown, J. (1987) Ecological stability: a test case. *American Naturalist* 130, 807–810

Sinclair, A.R.E. (1975) The resource limitation of trophic levels in tropical grassland ecosystems. *Journal of Animal Ecology* 44, 497–520.

Sinclair, A.R.E. (1985) Does interspecific competition or predation shape the African ungulate community? *Journal of Animal Ecology* 54, 899–918.

Stephens, D.W. and Krebs, J.R. (1986) *Foraging Theory*. Princeton University Press, New Jersey.

Tapson, D. (1993) Biological sustainability in pastoral systems: the KwaZulu case. In: Behnke, R.H., Scoones, I. and Kerven, C. (eds) *Range Ecology at Disequilibrium*. ODI, London. pp. 118–135.

Turpie, J.K. and Crowe, T.M. (1994). Patterns of distribution, diversity and endemism of larger African mammals. *South African Journal of Zoology* 29, 19–32.

Ungar, E.D. and Noy-Meir, I. (1988) Herbage intake in relation to availability and sward structure: grazing processes and optimal foraging. *Journal of Applied Ecology* 25, 1045–1062.

Walker, B.H. (1993) Rangeland ecology: understanding and managing change. *Ambio* 22, 80–87.
Walker, B.H. and Noy-Meir, I. (1982) Aspects of the stability and resilience of savanna ecosystems. In: Huntly, B.J. and Walker, B.H. (eds), *Ecology of Tropical Savannas*, Springer-Verlag, Berlin, pp. 556–590.
Walker, B.H., Emslie, R.H., Owen-Smith, R.N. and Scholes, R.J. (1987) To cull or not to cull: lessons from a southern African drought. *Journal of Applied Ecology* 24, 381–401.
Wallis de Vries, M.F. and Daleboudt, C. (1994) Foraging strategy of cattle in patchy grassland. *Oecologia* 100, 98–106.
Wang, Y.H. and Gutierez, A.P. (1980) An assessment of the use of stability analyses in population ecology. *Journal of Animal Ecology* 49, 435–452.
Weigand, T. Milton, S.J. and Wissell, C. (1995) A simulation model for a shrub ecosystem in the semiarid karoo, South Africa. *Ecology* 76, 2205–2221.
Westoby, M. (1974) An analysis of diet selection by large generalist herbivores. *American Naturalist* 108, 290–304.
Westoby, M. (1978) What are the biological bases of varied diets? *American Naturalist* 112, 627–631.
Westoby, M., Walker, B.H. and Noy-Meir, I. (1989) Opportunistic management for rangelands not at equilibrium. *Journal of Range Management* 42, 266–274.
Wiens, J.A. (1984) On understanding a non-equilibrium world: myth and reality in community patterns and processes. In: Strong, D.R., Simberloff, D., Abele, L.G. and Thistle, A.B. (eds) *Ecological Communities: Conceptual Issues and the Evidence*. Princetown University Press, New Jersey, pp. 439–457.
Wilmshurst, J.F., Fryxell, J.M. and Hudson, R.J. (1995) Forage quality and patch choice by wapiti (*Cervus elaphus*). *Behavioral Ecology* 6, 209–217.

Index

Allelochemicals (*see* Plant secondary
    compounds) 148, 159–177,
    434
Animal production
    growth 231–235, 335, 397, 408,
    420
    lactation 233, 264, 303–323, 409
    maintenance 231, 254, 259, 285,
    305–323
    mortality 335, 445–446
    reproduction 254
    wool production 311, 338, 368

Biodiversity (*see* Heterogeneity; Mixed
    Grazing)
Biomass 38–53, 70–93, 116, 121,
    397, 408, 418, 433–443
    accumulation 432

Carbon (*see also* Photosynthesis;
    Respiration) 3–10, 18
    accumulation 115–118
    cycle 3
    flux 3–10

    stable isotopes 113–115
Competition in plant communities
    40–58, 69–94, 257
    coexistence 88–93, 104–106
    competitive ability 40–58, 80–86,
    228–291
    competitive exclusion 86–88, 103
    competitive strategy 40–58,
    78–80
    density dependence 75–77
    dynamics 89–93, 103–126
    equilibrium 89, 111
    intensity 80–83
    inter-plant 77, 361, 415
    inter-specific 78–80, 87–89
    relative competitive ability
    80–86, 331
    for resources
        light 5, 18, 81–84, 415
        limiting resource 6, 80–86
        nutrients 81, 91, 407, 415
        water 81, 91, 415
    stress and disturbance 81–84
Competition–stress–disturbance
    models 81–86, 281
Conservation 247, 302, 308, 314, 320

Consumption of herbage (*see* Herbage intake)

Defoliation
frequency   21–28, 288
intensity   21–28, 40–57, 288, 308
intermittent   21–28
pattern   204–207
severity   23, 281, 289
Dendroecology   118
Density dependence   70–77, 198, 444

Ecosystems
arid   101, 276, 282, 289, 325, 438
mediterranean grasslands   359–391, 432
mesic grasslands   46, 54, 325, 331, 348
natural grasslands   37–58, 108, 115–123, 288–292
rangelands   101–126, 274, 294, 325–357
savanna   108, 115, 120, 247, 278, 294, 325, 365, 278, 394–420
semi-arid   54, 101, 106, 276, 282–291, 325, 368, 379, 394, 439
shrublands   116, 249, 283, 291, 330, 347, 359–384
temperate grasslands   3–29, 69–94, 279, 302, 406, 410, 438
tropical grasslands - well watered   393–420, 440
woodland   101–126, 250, 381, 396, 440
boundaries   116–126, 249
Ecosystem biogeochemistry   124
Ecosystem dynamics   101–126, 290, 326, 384, 411, 429–449
degradation   113, 248, 283, 293, 341, 369, 375, 440, 449
desertification   115
equilibrium   102–111, 282, 294, 330, 444
stability   105, 282–292, 343, 440–448
succession   107–126, 249, 266, 279, 330, 382, 439
Ecosystem state
complexity   275–292, 335, 338, 359
diversity   247, 259, 275, 278, 282, 293, 326, 381
hierarchy   301, 320
Environment   (*see* Heterogeneity)
Equilibrium/disequilibrium theory   282, 330–332, 440–447

Fertiliser   283, 292, 312, 347, 364, 376, 396
Fire   57, 81, 109, 117, 124, 254–258, 266, 288, 292, 327, 331, 359–380, 393–420
Flowering   52
Forage (*see* Herbage)
Foraging (*see also* Herbage intake; Selective grazing)   137–154
foraging strategy   137–138, 140–144, 149, 153, 251, 432
chewing   140–146, 151, 190, 193–199
digesting   147, 219–240
energy budget   137, 146, 152
feeding station   206
fitness   145–152
foraging ability   140–147, 196, 251
grazing   137, 153, 409, 418
handling   190
harvesting   433
migration   248, 265–266, 267
opportunity cost   147
predator avoidance   145–149, 153
prehending   374
resting   148
rumination   151, 250
searching   144, 153, 185, 208
selecting   144, 147, 159, 168, 250, 257, 265, 335
swallowing   190

Foraging contd
  foraging strategy models 137–154
  ingestive behaviour 185–211
    bite area 193, 198, 202, 208, 251, 257
    bite depth 198, 203, 255
    bite dimensions 189, 199, 207
    bite mass 147, 151, 193, 255
    bite rate 149, 204, 437
    biting effort 204
    grazing time 139, 146–153, 189
    head movement 194
    intake rate 134, 145–154, 185, 193, 202–208
    jaw movement 194, 203
    teeth 196, 201
    tongue 196
  Optimal Foraging Theory 139, 144–154, 206, 334, 433

Grasslands/rangelands
  conservation 110, 247
  degradation 102, 247, 283, 341
  management 106, 248
Grazing
  avoidance 38–48, 50–53, 86, 163–169
  deterrence 40–45, 161
  patchy 26, 139, 146, 206, 280, 308
  preference (*see also* selective grazing) 162, 222, 236, 280, 307
  pressure 262, 281, 340, 375, 383, 407
  resistance 38–58, 431, 440
  selection 40, 50, 56, 84, 250, 257, 265, 335, 409, 432–438
  severity 176
  tolerance 38–48, 53–55, 84
Grazing management (*see also* Systems management)
  continuous stocking 7, 21–26, 41, 53, 306, 375
  deferment 309, 375
  defoliation interval 21–29, 209
  efficiency of utilization 24–29, 248, 303, 376
  frequency 21–26
  herbage allowance 21
  intensive 21–29, 254, 280, 285, 294, 303, 325, 337, 261
  rotational stocking 7, 207, 263, 306–311, 315, 373, 415–419
  seasonality 29, 56
  set stocking 22–29, 208, 418
  severity 12, 21–29, 176, 286
  stocking rate 56, 292, 303–323, 332–336, 340–343, 373, 394–420, 436
  strip grazing 21, 209, 319

Herbage (*see also* Leaf/leaves)
  accumulation 9, 24–29
    net 9, 308
  growth 3–29, 39, 42, 308–321
  growth rate 26–29, 53
  mass 306–318
  production 26–29
  senescence 308–315
  utilization 24–28, 308, 340, 369, 376, 396, 411
Herbage intake (*see also* Foraging; Herbivore nutrition) 185–211, 222, 227–236, 251–258, 301–321, 334, 433, 437, 441
  effect of animal variables 57, 147–150
    acclimation and learning 150, 153–169
    body mass 250–264, 437
    dietary experience 150
    energy reserves 150, 265
    incisor breadth 251–260
    metabolic rate 251, 265
    mouth shape 258, 265, 279
    physiological state 149, 174, 265
  effect of plant and plant community variables
    availability 147, 334
    bulk density 193, 199–207, 254

Herbage intake *contd*
  height   193, 199–210, 252, 255, 259
  nutritive value   285, 393
  quality   255, 279, 306–312, 319, 373
  spatial variability   47, 259, 277, 325, 334, 364, 437
  structure   197, 249, 279
Herbicide   367, 377
Herbivore   248, 276–294, 334, 432–449
  browser   278, 294
  grazer   278
  species   57, 278
Herbivore nutrition   219–240
  digestion   219–240, 251, 260, 434
    microbial fermentation   221–240
  energy budget   255
  feeding value   222, 236, 238
  food comminution   219
    comminutive energy   226, 237
    compression forces   196
    fibre   223, 226, 235
    particle size   224
    shear strength   193, 202
    tensile properties   203, 281
  gut volume   150, 224, 251
  nutrients
    energy   220, 224, 236, 400, 433
    lipids   221, 227
    minerals   227, 259, 263–267, 393, 434
    non-protein nitrogen   227, 379
    protein   219–240, 393–397, 434
    volatile fatty acids   235
  nutritive value   227–237
    cell contents   227
    cell wall content   223–240, 397
    digestibility   225, 249–255, 277, 308, 379, 393–420
    metabolizable energy   221, 228, 263, 310
  nutritional ecology   219, 251
  rumen metabolism   229, 233, 238, 309, 401, 435
  rumination   222–24
  stem:leaf ratio   223, 252–259, 403
  voluntary feed consumption
    forage consumption constraint   224–246
    grasses and legumes   224, 230, 231
    nutritional balance   227
    seasonal effects   223, 229, 254
Herbivory   429, 434
Heterogeneity
  climatic   103–110, 302, 312, 328, 394, 430
    rainfall   249, 263, 275–282, 289–293, 395–420, 338–350
    season   326, 445
    temperature   277, 326
  edaphic   124
  foraging   92
  historical   326, 347, 363, 430, 440
  parameters
    autocorrelation   348
    kurtosis   348
    mean   326, 348
    scale   278, 281, 334, 341, 348, 437
    variance   326, 348
  spatial   47, 87, 91, 124, 139, 206, 259, 277–280, 289, 325–350, 364, 432, 438
  temporal   91, 276, 280, 325–350, 432

Ingestive behaviour (*see* Foraging; Grazing)

Leaf/leaves
  leaf area index (LAI)   4–29
  leaf dynamics   4–16
    age structure   13–17
    appearance   13–17, 416
    appearance interval   13–17
    death   9, 29
    defoliation   11–15
    elongation   10, 53
    emergence   13–17
    growth   15–18

Leaf/leaves *contd*
    life span   10, 13, 24–29
    production   24–29
    senescence   9–15, 28, 399, 419
   leaf sheath   18
   leaf size   10–18, 24
Legume/legumes   405–420, 430
   shrubby   413
   stoloniferous   415
   twining   414
Light
   intensity   6, 230, 413
   interception   5–8
   photosynthetically active radiation (PAR)   8
   phytochrome mediation   20
   red/far red ratio   17–20
   solar radiation   5
Lottery models   92

Mixed species grazing   57, 247–267, 277–293, 307, 372, 408, 420, 431, 436
   coexistence   88–94, 247, 259, 278
   competitive exclusion   86
   facilitation   248
   feeding niche   247, 266, 278, 436
   grazing succession   263
   interspecific competition   77, 87, 247–251, 257–260
   resource partitioning   80–86, 248

Nitrogen (*see also* Soil)   10–12, 400, 406
   accumulation   10
   allocation   7
   concentration in plant tissue   10
   cycling   10, 235, 401, 404
   deficiency   7, 264
   fertilizer use   16, 229, 308, 316, 367, 377
   fixation   12, 229, 367–382, 405, 420
   fluxes   4, 10
   losses   12
   metabolically active   173
   recycling   10, 29, 220
   storage   10
   uptake   10

Patches   (*see also* Heterogeneity)   85, 90, 293, 341, 348, 437, 445
Photography   120–126
Photosynthesis (*see also* Light)   6–8, 401, 415–416
   gross   4–9, 311
   net assimilation   7
   radiation use efficiency   6
Plant/plants
   accessibility   40
   adaptation   40–58, 160
   community   430
   composition (*see also* Herbivore nutrition)
      biogenic opals   118
      nitrogen   221–240
      structural carbohydrates   221–240
   defences
      biochemical   40–52, 159–178
      mechanical   40–52
   growth   3–29, 37–58, 73–77
      compensatory growth   55
   meristems   53
   morphogenesis   13, 53
   palatability   40, 53, 159, 285, 409
   persistence   19, 57, 415, 420
   phenology   13–21, 40, 52
   phenotypic plasticity   13–21, 50
   population density   18, 75, 444
   population dynamics   13, 69–77, 441
      birth   69–75
      death   69–75
      net reproductive rate   70
   production models   329–332
   reproduction   69–77
      seeds   54, 73, 230
      seedlings   54, 73, 80, 85
      vegetative reproduction   73, 77
   secondary compounds   51, 148, 159–168

Plant/plants contd
  alkaloids 49, 162, 166, 172
  cyanogenic glucosides 49, 52, 162, 166, 175
  phenols 49, 52
  phytotoxins 159–168
  tannins 49, 164, 171, 175, 238, 309, 379
  survival 37–58, 160
Plant–animal interface (*see also* Foraging; Grazing) 14, 58, 185, 240, 277, 432, 443
Plant communities (*see also* Swards)
  biomass 70, 82, 93, 116, 121, 278, 376
  composition 80–84, 276–279, 283, 313, 328, 347, 370–376
  compositional change 41–58, 69, 110–1117, 328, 340, 345
  degradation 283, 330, 333, 345, 359, 396–420
  disturbance 447
  diversity 275
  dynamics 69–94, 102–107, 359, 430
  equilibrium 88, 106, 111
  establishment 328
  gaps 85, 90
  grass/legume balance 376, 431
  grazing lawns 255, 279
  mosaics 124
  persistence 150, 366, 379, 449
  productivity 328, 345, 370
  resilience 340, 360, 370, 374, 411, 440, 447
  responses to grazing (*see also* Herbage) 37–58, 324
  succession 101–126
Plant growth/regrowth 3–29, 37–58
  functional response 38–47
  internode elongation 16
  patterns of 302, 380
  reproductive 273
  seasonal 249, 302, 311, 361, 371, 418
  vegetative 373
Plant species
  annual 361–384, 406

browse 152
early successional 38–52
functional groups 38, 105
late-successional dominants 38–58
mid grasses 50
perennial 288, 361, 371, 374, 377, 394–419, 430, 439, 448
ruderal 41–45, 81, 365
shallow rooted ephemerals 292, 439
short grasses 50, 54, 279
stress tolerators 38–52, 81
subordinate 57
tall grasses 50, 54, 279, 407
tropical 230, 407
woody 41, 101–126, 250, 259–362, 374, 383
Productivity
  primary 397, 432
  secondary 432
Protein 220, 229, 309, 379

Resource/resources
  allocation 80–86
  availability 80–86, 259, 263, 275, 280, 301–323, 432
  use 80–86, 248
Respiration (in plants) 4–29
Root/roots 431
  root–shoot interactions 431

Seed/seeds 401, 413
  bank 54, 73, 733, 365, 439, 448
  dispersal 90, 307, 373, 380
  production 57, 73, 77, 291, 373
  viability 54, 77
Seedling/seedlings
  establishment 54, 77, 85, 328, 374, 380
  growth 77, 80
  recruitment 54
  survival 74
Selective grazing (*see also* Foraging)
  adaptive mechanisms
    avoidance 167

Selective grazing *contd*
   detoxification 167, 170–175, 434
   tolerance 174
  dietary variety 160, 166, 175, 433
  excreta 221, 280, 404, 411
   nutrient balance 175, 435
  palatability 159, 163, 281, 285
   flavour 162–170, 178
   odour 162, 170
   specific hunger 168, 176
  partial preference 149, 433
  preference/avoidance mechanisms 159–178
   acquired aversion 166–169
   conditioned flavour aversion 161–165
   dietary experience 168, 174
   digestive effects 164, 166
   gastrointestinal feedback 165, 169
   gastrointestinal malaise 162
   instinctive avoidance 167
   nutrient dilution 165
   social models 168
Self-thinning law 18, 77
Senescence 9–15, 24–29, 397
Size/density compensation 16–22, 75, 78, 430
Socio-economics 326, 338, 367
  advisory services 311, 344
  economic viability 284, 294, 306, 313, 320, 327, 335–346, 382–439, 445
  government 293, 339, 342
  land tenure 367
  nomadism 342, 396
  subsistence 336, 342, 367, 383, 405–411
Soil
  aluminium saturation 397
  chemical properties 278
  disturbance 391, 309, 360
  erosion 330, 345, 31, 375, 442, 448
  hydrology 328, 377, 442
  nutrient cycling 397, 419
  nutrient status 376

  nitrogen 228, 397–399
  phosphorus 397–399
  pH 278
  treading damage 307, 319, 414
  water relations 255, 309, 375
  water runoff 309
Sown pastures 282, 293, 407
  establishment 328, 374, 380
  management (*see* Grazing management)
State and Transition models 106–111, 330, 439, 449
Stem (*see* Plant/plants)
  culm 48, 53
Stochastic Dynamic Programming 146–151
Stochastic events 439–447
Supplements 293, 302, 310, 317, 347, 365, 370, 375, 380, 383, 441
Sward/swards (*see also* Plant communities)
  dead matter 15–29
  density 254
  growth 13–29
  state 418, 431
  structure 13–29, 249, 263, 279
  surface height 23, 252–259, 418
  vegetative 251
Systems management (*see also* Grazing management) 294, 301–321, 331, 336, 393
  business plan 302
  complexity 277, 369
  efficiency 248, 303, 309
  fencing 267, 283, 292, 310, 312, 325, 335, 376
  flexibility 303
  goals 343, 369
   mixed land use 247–272
   soil conservation 345
   sustainability 309, 344, 362, 382, 405–420
  management envelope 417
  monitoring
   condition scoring 311, 318
   feed budgeting 310–321
   pasture cover 310–321
   sward height 311

Systems management *contd*
  product   303
  seasonality   302–323
  stock policy   303
  water supply   310, 344, 356, 376, 419

Tiller/tillers (*see also* Leaf/leaves; Plant/plants)
  apex   17
  appearance   13–17, 69–77
    rate   14–17
  density dependence   75–79
  leaf number   13–17
  mortality   16, 69–77
  population   69–73
  density   13–17, 20–24, 69–77, 431
  dynamics   16–20, 69–77
  site filling   16
  size   17, 20–24, 430

Tissue flux   4–29
  accumulation   4–10, 431
  decomposition   4–10
  growth   13–29, 419
  senescence   4–10, 24–29, 419, 431
  synthesis   13–29
  turnover   276

Vegetation (*see* Plant communities)
Voluntary forage consumption (*see* Herbage intake)